Icons of Evolution:
Pacific Island tree-snails of the family Partulidae

Justin Gerlach

Phelsuma Press, Cambridge, U.K.

http://islandbiodiversity.com/partulapages.htm

ISBN: 978-0-99322-033-3

This publication was made possible by the support of Peter Kistler, Geoffrey Macaulay, Aroldo de Rienzo, Roger Thoma and Anna L. Wood.

Contents

1. Introduction

As a result of the dramatic extinction of a third of the species, tree snails of the family Partulidae are regarded as a flagship for Pacific Island invertebrate conservation (Cowie & Cook 2001; Cowie *et al.* 2002; Lydeard *et al.* 2004). Prior to the family's decline in the 1980s partulids had been regarded as highly significant model organisms for evolutionary research, starting with the work of Henry Crampton. In 1916 Crampton published the first volume of what was planned to be a series on *Partula* snails. The monograph on the species of Tahiti was highly influential in the development of evolutionary theory throughout the 20th century. In particular, Crampton's conclusion that *Partula* features were 'indifferent characters', not subject to natural selection, was cited in support of the concept of evolutionary drift in some of the most important works in biology (Dobzhansky 1931; Huxley 1940, 1942; Mayr 1942, 1947; Cain & Sheppard 1950; Ford 1964). Most significantly, Crampton's quantification of morphological variation in different populations was instrumental in developing the idea of population (and later genetic) based species, rather than the 'typological' species concepts (Mayr *et al.* 1953).

Although Crampton's monographs were of great historical importance, they were only a small part of what he had intended. This first great study of evolution in the field was intended to cover all of the species of the Society islands and elsewhere. The Tahiti (1916) volume was followed by one on the species of the Marianas islands (1925) and one on Moorea (1932), but the series remained unfinished at the time of his death in 1956.

In studying *Partula* Crampton aimed to "determine the rôle played by environment as a whole in the differentiation of species and varieties existing under entirely natural conditions" (Crampton 1916). "Directed by an interest in the general subject of molluscan distribution, and specifically by the brilliant and suggestive research of Mayer upon the Partulæ of Tahiti, as well as by Mayer's personal advice, the writer undertook the close and detailed study in the field of that genus of terrestrial pulmonates. It may be stated at this juncture that no region could possibly be more favourable than Polynesia for the study of the problem in question; the writer cordially indorses [sic] the opinion of Mayer that the volcanic and other islands of the South Seas offer a field that in almost every respects attains the ideal of the biologist."

Crampton's work inspired the geneticists Bryan Clarke and Jim Murray to follow up the study: "The polymorphic land snails of the genus *Partula* provide unusual opportunities for the study of evolution in natural populations. At the species level, the diversity of forms permits the investigation of the control of gene frequencies in populations. Above the species level, the occurrence of a number of similar species invites the study of speciation and reproductive isolation in the group" (Murray & Clarke 1968). They found Moorea in particular to be "both a museum and a laboratory of speciation" (Murray & Clarke 1980).

Fig. 1.1. Henry Crampton in 1927 (http://embryo.asu.edu/handle/10776/2917) and the last wild *Partula navigatoria* on Raiatea in 1992

Fig. 1.2. One of the plates from Crampton's 1932 Moorean monograph

In general snails make good subjects for evolutionary study as they may be abundant and conspicuous, are relatively easy to identify and are generally sedentary, and so "genetic changes are highly localized and selective agents are possible to identify" (Ożgo 2014). This has led to notable studies of evolution of polymorphism in the European banded snails *Cepaea nemoralis* and *C. hortensis* but, as Clarke & Murray (1969) recognised, the Partulidae had even more potential: "*Partula* offers several advantages for investigating micro-evolutionary change. It is highly variable, both within and between populations. Its mobility is low, so that genetic differences occur between populations only short distances apart (20 m or less). It is ovoviparous, and easy to rear in the laboratory, so that the differences can be investigated experimentally. Finally it shows, at least in some places, a very curious pattern of speciation." Their study of Moorean species "revealed in exceptional detail the pattern of interactions between incipient species, and have presented some fascinating paradoxes" (Murray & Clarke 1980).

Of all the evolutionary questions that can be addressed through studying partulids, speciation has received most attention, but interpretations have differed. In evaluating his data from Tahiti Crampton (1916) concluded that "The originative influence of the "environment" seems to be little or nothing. Isolation proves to be a "condition" and not a "factor" in differentiation of forms belonging to this genus." His interpretation that selection was a minor influence on *Partula* meant that subsequently (1928) he had to clarify his position: "To forestall any possible misapprehension later, let me state at once that I hold myself to be a true believer in orthodox natural selection. Darwin clearly separates this from congenital variation and takes the latter as given, so to speak. Others of the neo-Darwinian school attempted to expand natural selection so as to make it originative in effect as well as discriminatory, notably those who propound the theory of germinal selection." Later studies have sought to interpret the patterns in Moorean partulids in the light of selection and drift (e.g. Goodacre 2001). The Moorean species were chosen as a focus because of the presence of numerous discrete colour patterns (as in *Cepaea*) and chirally variable species. These characteristics were also present to a slightly lesser degree on Tahiti and some comparative studies were undertaken there, but species of the other islands have been considered only in phylogenetic terms. The lack of a clear taxonomy for most species has so far prevented a full evolutionary analysis. The work presented here completes and updates Crampton's study by resolving the taxonomy of the family as a whole.

Geography of the Partulidae

By the time Crampton started working on Partulidae their general distribution was well known, with species recorded across the south Pacific, from 18 degrees north (Pagan island, Marianas) to 28 degrees south (Rapa, Austral islands). The islands occupied by the family are outlined here. They can be divided into three groups: Melanesia, Micronesia and Polynesia. These are cultural divisions rather than biogeographical ones but are the most convenient groupings for the Pacific.

Fig. 1.3. Biogeography of partulids: species numbers in different island groups across the Pacific Ocean. The enormous diversity of species in the Society islands is highlighted

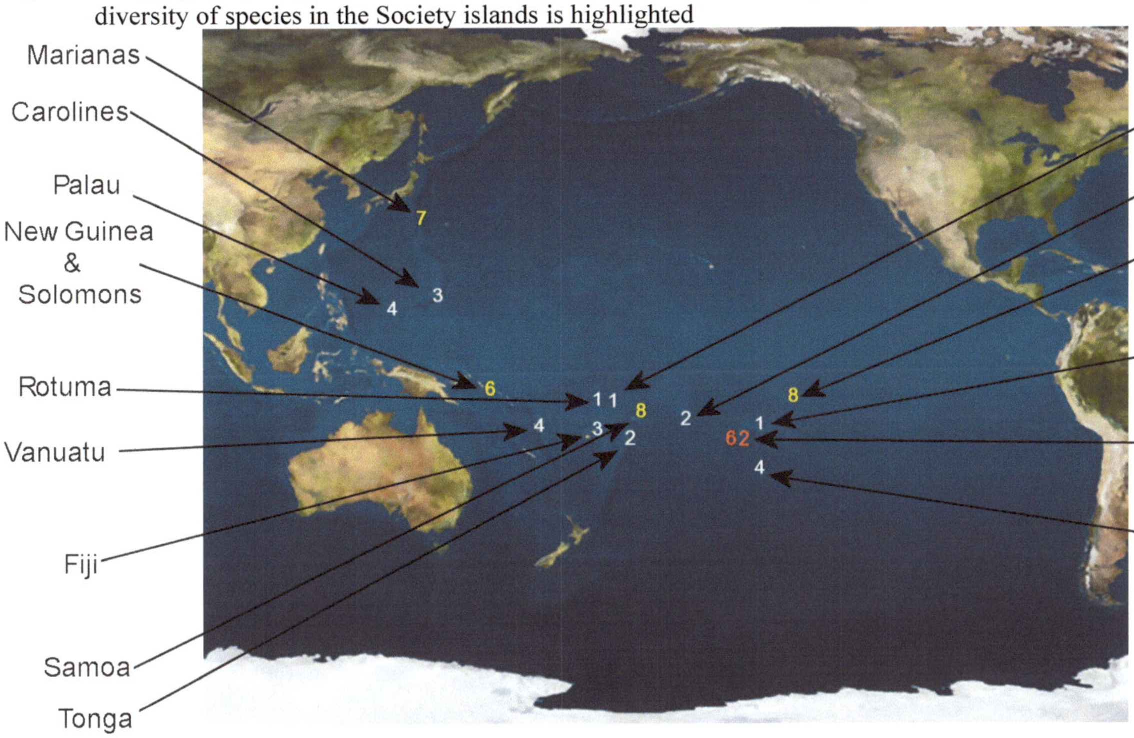

Most islands occupied by partulids are volcanic, formed as the result of the subduction of the Pacific Plate under the Philippine Plate (Marianas) and under the Indo-Australian Plate (Tonga, Fiji, Samoa). Some islands in the west are continental, Palau being the eastern edge of the Philippine continental plate. Other western islands are remnants of volcanoes on ancient continental platforms, most notably the Solomon islands on the 122 million year old Ontong Java plateau. The most recent of all the islands is Rarotonga in the Cooks, formed 2.3 million years ago, followed by the Marquesas (less than 3.8 million years old). All islands occupied by partulids are high islands (e.g. Tahiti 2,241 m and Guadalcanal 2,331 m) with the exception of the smaller islands of the Solomons. The smallest and lowest island with partulids is Ontong Java, with a land area of just 12 km^2 and a maximum altitude of 13 m.

Almost all of the islands are occupied by humans; the age of human settlement varies widely, from prehistoric occupation 30,000 years ago in the Solomon islands and the Marianas, 5,000 years ago in the Samoan islands, to the Cook and Austral islands which have been occupied since about 500-800 AD. Some islands were substantially altered by major human populations in the past; the Marquesas are thought to have been occupied by around 100,000 people in the 16th century, but now supports fewer than 9,000.

The outlier is Salibabu (=Salibaboe) in the Indonesian Talaud islands. This is the most westerly record of the family and although the habitat of the island is suitable the record is based on a single specimen, here considered to be a misidentification (see p. 118).

Fig. 1.4. Partulid islands: a – Bismarcks (New Britain); b – Vanuatu (Efate); c – Fiji islands; d – Marianas (Guam); e – Palau (rock islands); f – American Samoa (Ofu); g – Tonga ('Eua); h – Cooks (Rarotonga); i – Australs (Raivavae); j – Marquesas (Nuku Hiva); k – Societies (Maupiti); l – Societies (Mt. Aorai, Tahiti) (photos: a – P. Currie; b – Daderot; c – J.P. McVey; d – P. Craig; e – D. Wright; f – M. Gleining; g – Vaia98; h – S. Berardi; j – R. Jouan)

2. Historical summary

The first record of a partulid was Thomas Martyn's figure of '*Limax faba*' in 1784 (Fig. 2.1). This was part of the material collected in the Society islands during the three Pacific voyages under Captain James Cook. Although the exact details are not recorded it is probable that the shell was collected by Daniel Solander on the first expedition in 1769 (Gerlach 2014). Other specimens from Solander were bought by Lorenz Spengler and kept in his collection in Denmark, and later incorporated into the Danish Royal Natural History Museum, subsequently the Zoology Museum of the University of Copenhagen. This material was figured and described by Johann Hieronymus Chemnitz in 1795, including *Partula faba* and *P. otaheitana* (Gerlach 2014). Spengler also had *P. affinis* specimens which were not figured by Chemnitz (Fig. 2.2). Hendrik Hendriksen Beck later (1837) catalogued the collection.

Beck (1837) made a distinction between the Chemnitz and Martyn figures, noting that Martyn's was dark brown with a pale medial band (actually a reddish band) and Chemnitz's reddish with a blackish border before the suture. The figures

Fig. 2.1 The first *Partula*: Martyn (1784) *Partula faba*, Chemnitz (1795) *P. faba* and *P. otaheitana* (edited for position)

Fig. 2.2 Spengler specimens: top left – *Partula faba* ZMIC-GAS-431 (shell illustrated in Chemnitz 1795); top centre – *P. faba* ZMUC-GAS-433 (Mörch label); top right – *P. faba* ZMUC-GAS-432; bottom left – *P. affinis* ZMUC-GAS-435 (Spengler label); bottom right – *P. otaheitana* ZMUC-GAS-434 not known to be Spengler's (Beck label)

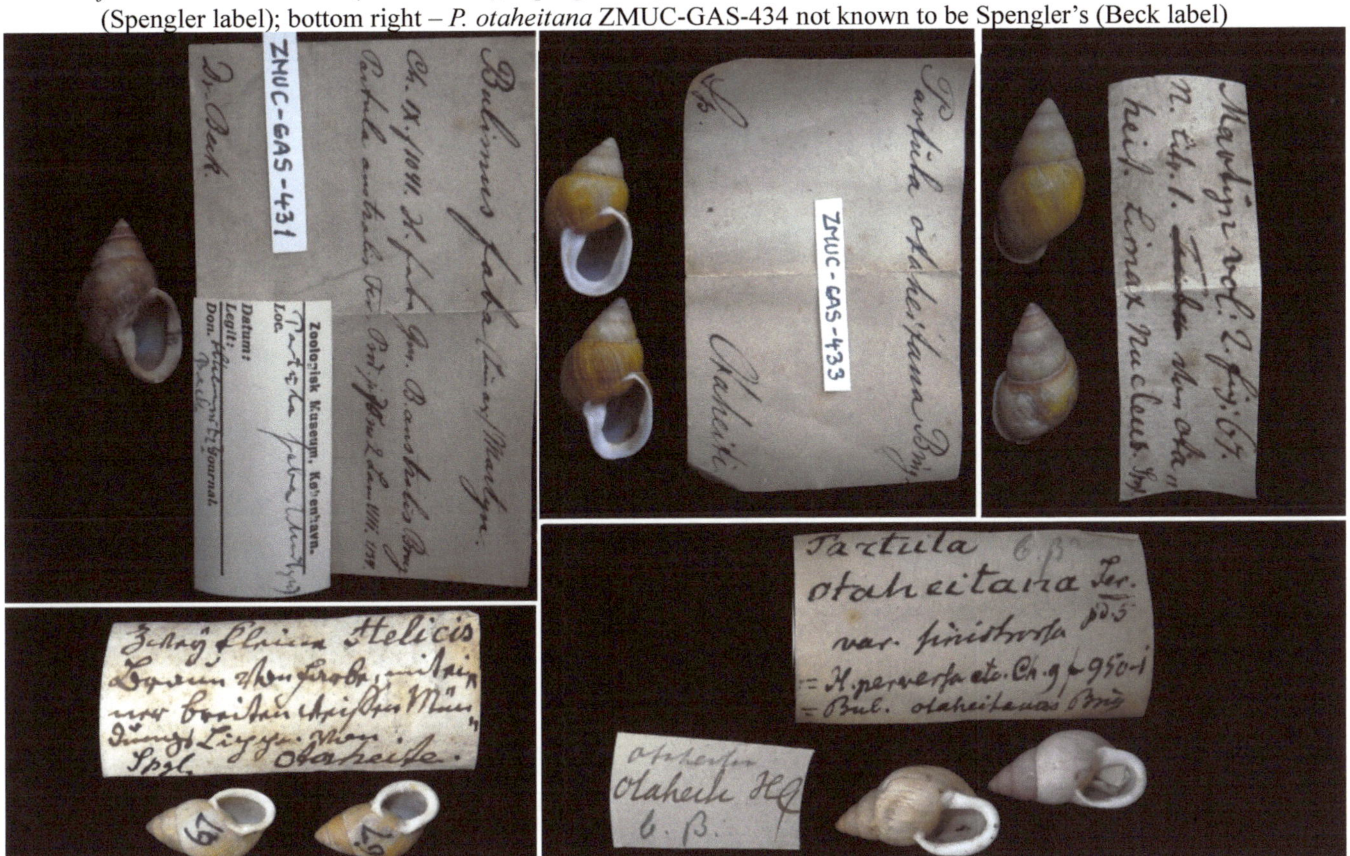

differ in the lack of the medial band in Chemnitz's figure and a wider subsutural band. The general colour difference is not apparent in the figures, suggesting that Beck had at least one of the specimens, leading him to consider that Chemnitz's specimen was redder than Martyn's figure. As Martyn and Chemnitz were declared unavailable works by ICZN (1957) *faba* became available by Gmelin's (1791) listing of Martyn's name (as *Helix faba*), but the Chemnitz names were disregarded.

Other specimens from Cook's expeditions were in the Duchess of Portland's collection (Lightfoot 1786) but have not been located. Other notable early works from this period are those by Bruguière and Férussac. Bruguière (1792) gives the first detailed description of *Partula faba* (as *australis*) and *P. otaheitana.* Although Férussac (1819) gave no detailed descriptions of the species he did provide the first description of partulid anatomy from the one pickled specimen that had been sent to him. A wider range of species were collected by the French expeditions of the Uranie in 1818-20 (Quoy & Gaimard 1824), the Coquille in 1822-5 (Lesson 1831) and the Astrolabe in 1826-9 (Quoy & Gaimard 1832), the British expedition on the Blossom in 1825-8 (Gray 1839) and the US Exploring Expedition of 1838-42 (Gould 1848, 1852, 1862). Smaller collections were made by Graeffe (Mousson 1865, 1899, 1871).

The largest mid-19th century collections were amassed by Hugh Cuming who collected widely throughout the Pacific during three expeditions, starting in 1828. His earliest material was described by William John Broderip (1832) who gives the first ecological data, describing *Partula rosea* from Huahine as "Found by Mr. Cuming on the *Te* plant." Most of Cuming's material was described by Lovell Augustus Reeve (1842, 1843, 1849-50) and especially by Ludwig Pfeiffer (30 publications between 1841 and 1881). 53 species were described from Cuming material, of which 23 are valid names (22% of currently accepted species) and of the 60 nominal species collected by Cuming, 31 are valid species (with a further 4 valid subspecies). Unfortunately, many of Cuming's localities are clearly incorrect and this can be identified as the start of a period of great confusion in partulid taxonomy.

At this time shells were usually mounted by gluing them directly on to pieces of wood or card. The mounts were labelled and these mounted specimens either displayed on tables or kept in drawers. In all cases they were exposed to the air and in the major urban museums accumulated a layer of soot during the 19th and 20th centuries. Many are discoloured as a result. The glue has become brittle with time and many shells have become detached from the mounts. This may mean that the shells are now disassociated from whatever limited data were originally available. The most distinctive mounts were probably those from the MacAndrew collection, much of which is in the University Museum of Zoology, Cambridge. In these a

Fig. 2.3 Early figures of *Partula* species: Lesson 1831 (6-7 – *Partula otaheitana*, 8-9 – *P. lineata*, 11 – *P. grisea*); Quoy & Gaimard 1832 (12-17 – *P. vanikorensis*, 18-22 – *P. gibba*); Gray 1839 (4 – *P. faba*); Gould 1852 (88 – *Samoana conica*, 89 – *Eua zebrina*, 91- *P. otaheitana*, 92 – *P. varia*, 93 – *P. vanikorensis*); Mousson 1865 (*P. obesa*)

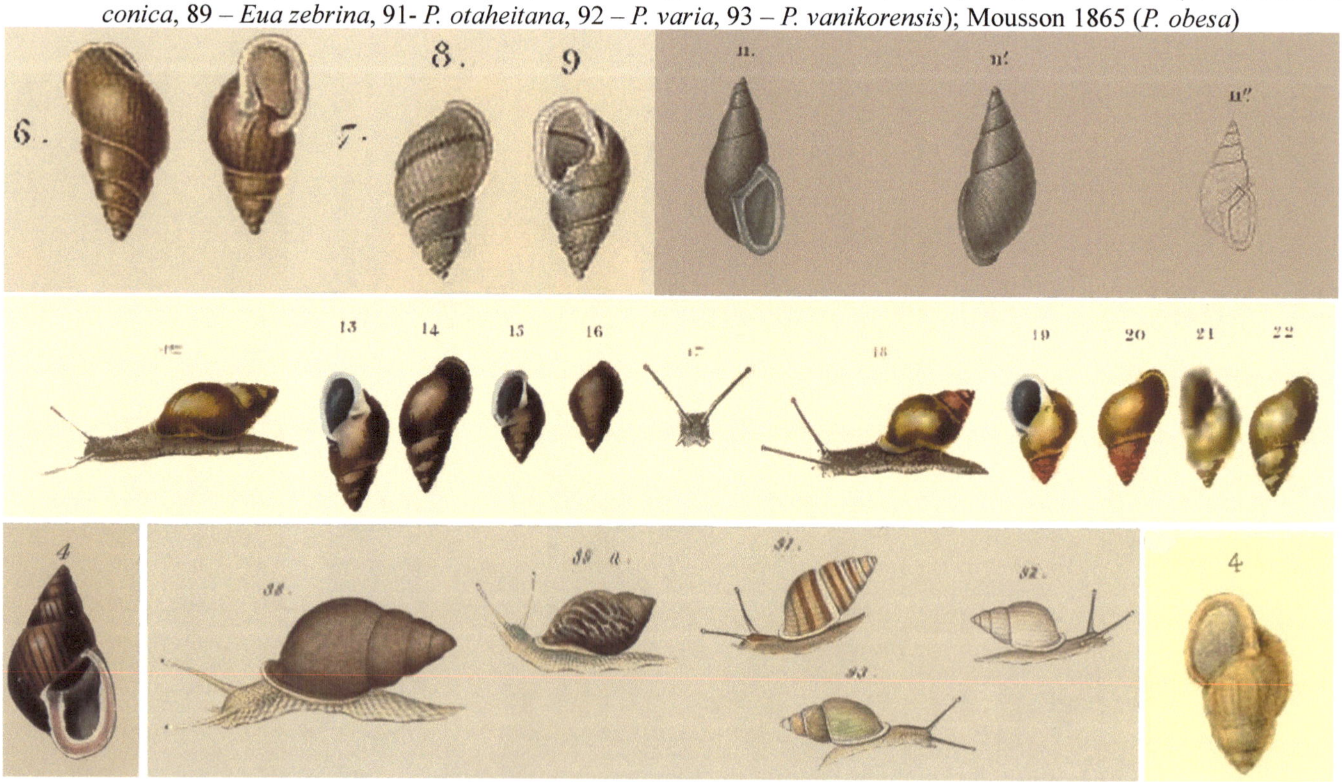

blue card was used and has fortunately remained clean (Fig. 2.6). These shells are well preserved but were obtained from William Harper Pease without any locality data.

Pease was the most prolific namer of partulids, with 48 names to his credit (of which 18 appear to be valid species and 6 valid subspecies). His material was largely provided by Andrew Garrett who undertook several collecting trips to different archipelagos, especially to the Society islands (in 1860-3). He later settled there, living on Huahine from

Fig. 2.4 Reeve's figures from 1842 (rows 1-2) and 1849-50 (rows 3-5): rows 1-2: 1-2 *Partula hyalina*, 3-4 – *P. gibba*, 5-6 – *P. varia*, 7-8 – *P. auriculata*, 9-10 – *P. rosea*, 11-12 *Samoana inflata*, 13-14 *P. faba*, 15-16 – *P. otaheitana*; rows 3-5 – 1a-c – *P. rosea*, 2 – *P. guamensis*, 3a-b – *S. inflata*, 4 – *P. guamensis*, 5a-b – *P. faba*, 6a-b – *P. radiolata*, 7-13a-b – *P. otaheitana*, 14 – *P. hyalina*, 15a-b – *P. gibba*, 16 – *S. ganymedes*, 17a-c – *P. varia*

Fig. 2.5 Pfeiffer's figures (1-50 – 1857, 51-56 – 1859): 1-3 *Partula faba*, 4-5, 31-32, 45-46, 53-54 *P. otaheitana*, 6-7 *P. dentifera*, 8-9, 39-40 *P. varia*, 10-11 *Eua zebrina*, 12-13 *P. nodosa*, 14-15 *Samoana inflata*, 16-17 *P. diminuta*, 18-19 *P. hebe*, 20-24 *P. gibba*, 25-26 *P. auriculata*, 27-28 *P. radiolata*, 29-30 *P. hyalina*, 33-38 *P. rosea*, 41-42 *P. taeniata*, 43-44 *S. gonochila*, 47-48 *S. decussatula*, 49-50 *P. micans*, 51-52 *P. pacifica*, 55-56 *P. suturalis*

Fig. 2.6 19th century museum mounts: left – MacAndrew collection (UMZC); right – more conventional board (NHMUK)

1870, and as a result of his close connection to these islands he paid particular attention to the distribution of Society islands snails. He recorded (letter of 1875, quoted in Smith 1903) "When I explored the islands for *Partula*, I kept a daily journal and gave each species, when discovered, a provisional name, and kept the species from each valley separate. After I sold the collection to Mr. Pease I put up nine sets for him to send to London. He adopted nearly all my provisional names. Carpenter, Adams and Cuming compared the shells with those in the Cumingian collection, and the former published the result of their determinations… I reserved a similar set for myself of all except five or six species of which there were no duplicates left after Pease made his selection for his cabinet. Ten years after my first explorations I went over the same localities a second time, so I certainly am pretty well posted on the Partulas of this group." This description records Garrett's care, but also what was to become a great source of confusion: the use of 'provisional' names. Here, Garrett was identifying species in the field, sending them to Pease with a provisional name. Pease either used the same name or changed it to one he thought more accurate, meaning that multiple names apply to the same species.

Despite this nomenclatural problem Garrett left remarkably good data on distribution and variation of the partulids of Society islands. As the first reliable recorder of Pacific snail localities Garrett (1884) had been frustrated by the confusion of Cuming's species: "it really seems as if he intentionally gave the wrong habitats of some of the shells collected by himself, or marked them 'hab.?' in order to mislead the monographers and make as many species as possible. He did not appear to care so much about shells in a scientific point of view as to possess the largest collection in the world. Nothing vexed him so much as having any one doubt the validity of any of his new species… He was also in the habit of changing the author's types whenever he obtained what he supposed to be better specimens. No doubt in most cases he was right in replacing the same species. Still, they were not the author's types."

Garrett (quoted in Smith 1903) was also frustrated by Pease who "was very careless about localities, as any one can see by studying his papers. He was also careless about his duplicates, which he kept in cigar-boxes; and, as he once wrote me, his 'little daughter amused herself in arranging (?) his duplicates.' After his death, those who packed his collection to be sent to Boston must have made sad confusion by the admixture of species."

In addition to sending material to Pease, Garrett was also employed as a collector for the Museum of Comparative Zoology, Harvard, and the Godeffroy Museum. Many of the specimens he sent to the museums bore label ('manuscript') names attributed to Pease. Paetel's (1873, 1883) and Schmeltz's (1865, 1869, 1874) Godeffroy collecton catalogues created confusion by publishing these provision names, adding to the already extensive list of nomina nuda created by Carpenter from Pease's specimens. In total Carpenter, Schmeltz and Paetel created 62% of the nomina nuda.

Garrett had also preserved the bodies of some specimens, some of which were used in Binney's (1876) first comparative anatomical study. By the time of Garrett's death in 1887 major collections of partulids had been established in several museums: Zoologisches Museum Hamburg (Garrett's Godeffroy Museum material); Natural History Museum, London (Cuming, Garrett, Pease); Museum of Comparative Zoology, Harvard (Garrett, Pease's own collection); Bernice Puahi Bishop Museum (Garrett's own collection); Carnegie Museum (Garrett, Pease); Academy of Natural Sciences, Philadelphia (Garrett); American Museum of Natural History (Garrett). The material in the Carnegie Museum formed the basis of the first attempts at a comprehensive revision of *Partula* by William Dell Harman (1880, 1881a, b, 1882).

As Smith (1902) summarised: "When Dr. Hartman began his special study of Partula, about 1877, the genus was in a condition of almost hopeless confusion. Cuming's species, described by Pfeiffer and others, were assigned to islands often thousands of miles from their true habitats, and in some cases it really seemed as if the errors were intentional. Pease was generally correct in his deterimination [sic] of species and varieties, but very careless about localities; many of his MS. names were supposed to indicate valid species, though he had never intended them to be more than provisional or the marks of varieties. Carpenter, who examined a series of the Pease duplicates, referred them to other species, and almost always erroneously. Scores of varieties had been described as species; many species were regarded as varieties; hybrids and decorticated shells were described as species; known species were erroneously recorded from different islands, and described as new by naturalists who regarded the habitat as a sufficient proof of validity. Even reference to type specimens was not always conclusive; in more than one instance the original type had been replaced by a shell of another species."

Hartman compared the specimens he had access to (mainly Pease and Garrett material) with named specimens in other museums. He also received a few new species from collectors such as John Brazier and Edgar Leopold Layard. Hartman (1881b) introduced numerous subgenera but most were preoccupied and so unavailable. He subsequently changed his mind and as Dall (1882) recorded "We are authorized, on the part of the author (who as much as any one else regrets the circumstance), to state that he desires to withdraw, as far as lies in his power, the names applied to the sections of Partula, which he is now of the opinion are perhaps hardly important enough, from a systematic point of view, to deserve naming".

In 1899 Alfred Goldsborough Mayer (originally Mayor) collected *Partula* on Tahiti in the first study of geographical variation of the genus. Along with C.S. Kempff and H.E. Moore he collected 735 snails in six valleys on Tahiti aiming to determine the distinctiveness of the island's species: "A great deal has been written concerning the classification of the species of *Partula* inhabiting the Islands of the Tropical Pacific. Unfortunately, however, the various species have been distinguished only by inspection of the color, form, etc., of the adult shells; and no attempt has been made to dissect the young out from the fullgrown snails and thus determine, by direct evidence, whether or no the so-called "species" intergrade, and if so to what extent" (Mayer 1902). By collecting every snail he could find and dissecting out the embryos and scoring their colour pattern Mayer hoped to determine how the colour patterns (then the primary taxonomic feature) were inherited. He found that the young often differed from their parents in patterns or in chirality.

Mayer's 1902 paper inspired Henry Edward Crampton to use snails to investigate the application of Mendelian inheritance to speciation. Initially Crampton had thought to work on the Hawaiian Achatinellidae that had been highlighted by John T. Gulick (1872), but Gulick and Mayer persuaded him to expand the *Partula* study. Accordingly he started investigating the species of Tahiti and Moorea in 1907. He wrote (Crampton 1916) that "Three primary subdivisions of the investigation were made at the outset. The first was a study of the problem of isolation and environmental influences as "conditions" or "factors" of biological differentiation. These have been variously estimated in the writings of naturalists from Darwin, Wagner, Murray, Wallace, Gulick, and Romanes to Allen, Jordan, Ortmann, and others, who have dealt with various groups of animals, and who have accorded to the "environment" almost all degrees of efficiency from omnipotence to impotence. In pursuance of this purpose it was planned to collect the snails as extensively as possible from all the valleys of the islands, thus to extend the observations of Mayer in Tahiti, and to make a complete survey of the whole group for comparison with the earlier observations of Garrett, who of all systematists concerned with the fauna of Polynesia has given the fullest and most circumstantial account of the distribution in his time of Partulæ and of other terrestrial mollusca. Secondly it was desirable to carry further the work begun by Mayer upon the inheritance of shell and color characteristics, which are so important for the determination of the continuity or discontinuity of related forms. The material collected for the first purpose fortunately serves the second as well, for the snails are viviparous and many of the adult characters under investigation may be determined with complete or substantial accuracy for the young carried in the brood-pouch for weeks after development is begun, contrary to the opinions of Hartmann [sic] and others who have not been familiar with freshly caught material. In the third place it was planned to bring back living specimens to see if they would serve for laboratory experiments upon the course of inheritance in pure and hybrid strains…"

In 1908 Crampton collected on Huahine, Raiatea, Tahaa and Bora Bora. He visited almost every valley and collected as many snails as he could find at each site. These he measured, assigned to a colour pattern, recorded chirality and usually preserved the body separately. From these preserved bodies he extracted the embryos and noted the chirality and colour. Comparing his collections to those of Garrett in Hawaii, Crampton realised that he needed to sample much more comprehensively than he had been doing. In 1909 he returned to the islands, collecting on Moorea and Raiatea. His first

Fig. 2.7 Mayer's (1902) figure (edited for position): 1-2 – *Partula hyalina*, 3-8 and 11-13 *P. otaheitana*, 9-10 *P. diminuta*, 14-15 *P. nodosa*

publication on partulids came in 1916 with the Tahiti volume. This covers only eight species, but runs to 400 pages, with comprehensive tables of measurements and detailed figures.

A visit to the Marianas islands in 1923 resulted in the second volume (published 1925). He returned to Moorea in 1923 and 1924, before finally publishing in 1932. Following the completion of the Moorean volume Crampton turned his attention to Raiatea where he collected in 1908, 1909, 1935 and 1947. He published some papers with Cooke (Cooke & Crampton 1930; Crampton & Cooke 1953) and started writing up the Raiatean monograph, but never completed it. He last worked on the monograph in November 1953 and died in 1956.

Most of the delay in Crampton's publication was due to his obsessive collecting and data recording. Further delay was added by distractions: he did publish some new species descriptions on the partulids of the Austral islands (Crampton & Cooke 1953) and also made some smaller scale collections on some other Pacific islands. By the time of his death he had recorded measurements of 113,348 individual adult shells collected by himself and during the course of his research he collected around 400,000 specimens (Crampton 1928).

At the same time as Crampton was collecting *Partula* two notable names in mollusc taxonomy were working on wider projects. At the Academy of Natural Sciences, Philadelphia Henry August Pilsbry was investigating a wide range of taxa for his contributions to the 'Manual of Conchology' (1909). He created the family Partulidae and divided them into different genera. Pilsbry had also collected new partulid species himself, on the Pinchot South Seas Expedtion to the Marquesas in 1929, although he never described them. Some of Cramptons' work was influenced by discussions with Pilsbry and, more notably, with C. Montague Cooke who was the malacologist at the Bernice Puahi Bishop Museum in Hawaii.

One of Cooke's most significant contributions to malacology was organising the Mangareva Expedition in 1934. This was particularly important for the study of partulids, collecting new species in the Australs and Tahiti. The labels of many important specimens bear the names of collectors on this expedition: Alexander R. Anderson, Donald Anderson, C. Montague Cooke, Kenneth P. Emory, Clifford F. Gessler, Yoshio Kondo, Harold St. John and Elwood C. Zimmerman. Cooke was a firm advocate of the value of anatomy in molluscan taxonomy and was intending to revise all the Pacific land snail families based on anatomical studies. He gave the task of revising some families to Yoshio Kondo. Work on the Partulidae obviously had to involve Crampton and Cooke persuaded him that a study of the anatomy of the Partulidae was worthwhile, although Crampton was sceptical of the value of anatomical characters. Kondo started the project in 1947 and completed it as a PhD study at Harvard in 1954-5. Kondo continued adding to the information on partulid anatomy with occasional annotations to his thesis from 1964 until 1986. By the 1970s he intended to publish an emended version, but it was never completed.

Kondo's frequent revisions arose from his intention to expand on Cooke's grand project, undertaking an anatomically based revision all Pacific island snails. In order to realise this he collaborated with a number of researchers collecting in different island groups. He collected on many islands, most notably the Carolines and Marquesas. The latter, along with John Peake, expanded on the 1930 collectons made in the Pacific Entomological Survey by Guillaume LeBronnec, a long -term resident of the islands. Significant additions to partulid material were made by John Bayard Burch. He collected on many island groups and made particularly important collections on Tahiti, Moorea, Raiatea and Bora Bora with John B. Burch Jr., Evin Oshima and C.M. Patterson. Kondo's dissection of some of this material led to new species descriptions (Kondo 1973, 1980) and an unpublished revision of the Tahitian species (Kondo & Burch MS). Kondo's death in 1990 left many projects unfinished.

Crampton's work had included two main threads: taxonomy and Mendelian inheritance. While Kondo kept the taxonomic study going fitfully, the genetic aspect was taken up by Bryan Clarke and Jim Murray in 1962. Crampton had inspired them to investigate inheritance of colour patterns in Moorean species and in the 1960s they produced a series of papers identifying gene interactions using the now classic methods of laboratory crosses that they had developed earlier for the European banded snails of the genus *Cepaea* (Murray 1993). As the field of genetics developed, the techniques available expanded and Clarke and Murray were joined by Michael Johnson who expanded the molecular aspect of the study. By 1991 they were using gene sequencing with a further generation of PhD students and postdoctoral researchers, including in 1996 Sara Goodacre and Chris Wade (Goodacre 2001, 2002; Wade & Mordan 2000; Wade *et al.* 2001).

The Society islands were the main focus of partulid field research until the 1970s when the carnivorous snail *Euglandina rosea* was introduced to the islands in a catastrophic attempt at biological control. Despite objections from Burch, Clarke, Murray, Johnson and others, *Euglandina* were released on Tahiti (1974) and Moorea (1977). By 1992 Society island partulids were almost all extinct and research now focuses largely on conservation, coordinated by Paul Pearce-Kelly and Dave Clarke (Partula Propagation Group 1991; Pearce-Kelly & Clarke 1992; Pearce-Kelly *et al.* 1994; Gerlach 1994; Coote 2007, 2011, 2013, 2014). The captive populations in genetics research populations became the core of the conservation breeding programme. Additional species were added to this through expeditons to the Society islands in 1991 (Raiatea and Huahine – Operation Partula Expedition), 1992 (Raiatea – J. Gerlach), 1994 (Huahine – Pacific Islands Land Snail Group [PILSG]) and 1995 (Tahiti and Marquesas – PILSG). Since 2003 Trevor Coote has monitored relict wild populations. Additional populations have been found by various naturalists in the islands (including Eric Loève, Jean-François Butaud, Fred Jacq, Jean-Yves and Meyer).

Other island groups have also been surveyed relatively reently: Cooks (by Fred Brook); Fiji (Gilianne Brodie and others); Wallis and Futuna (Ira Richling, Jean-Yves Meyer and others); Samoa and American Samoa (Fred Brook, Robert Cowie, Robert Cook and others); Tonga (Bryan and Anne Clarke and Chris Wade); New Guinea and the Caroline islands (Kurt Auffenberg, Michael Hadfield, Brenden Holland, Fred Kraus, Peltin Pelep, Rebecca Rundell, John Slapcinsky, John Starmer, Fred G. Thompson and others). The Mariaians have been particularly well studied with recent surveys by Michael Hadfield, David Sischo and others, and monitoring of populations by Scott Bauman, G. Curt Fiedler, David Hopper, Alex Kerr and Barry Smith. Multiple islands were surveyed by the Museum Nationale d'Historie Naturelle, with Philippe Bouchet, Olivier Gargominy and Benoît Fontaine leading snail research on the Australs, Marquesas and Vanuatu.

Burch's 1970 collections from Tahiti and Moorea led to phylogenetic research at the University of Michigan led by Diarmaid Ó Foighil and Taehwan Lee (Lee *et al.* 2007b, 2008, 2009, 2014). This led to additional research by Cindy Bick, focussing on the Solomon islands.

This monograph of the Partulidae is based on examination of around 200,000 specimens dating from the collectons of Daniel Solander through to the new material and studies of the past few years. The information associated with this material varies greatly; collection data have been kept in the original form (for example in the use of feet or metres) except where interpretation has been required.

Fig. 2.8 Crampton's *Partula* collections in the Academy of Natural Sciences, Philadelphia: left – shells; centre – shells and a jar of spirit preserved bodies; right – Crampton's spirit preserved samples

3. Taxonomy, nomenclature and hybrids

Taxonomic and nomenclatural issues

The taxonomy of the genus *Partula* has been a source of confusion since the early 18th century collections of Hugh Cuming (described by Broderip [1832], Pfeiffer [1854a-e, 1855a, 1857b-e, 1860, 1862] and Reeve [1842, 1843, 1849-50]) were largely wrongly localised. Similarly, Pease (1866-72) gave relatively accurate identifications but questionable localities; in some cases Garrett (who collected the specimens) was able to correct these errors, but many species remain confused. Further confusion was added by Carpenter (1865), Paetel (1873, 1883) and Schmeltz (1865, 1869, 1874) whose catalogues of species listed many varieties, hybrids and anomalous shells as species. The names they used were those provided by Pease on the labels, but many had not actually been described. This confuses the nomenclature, although as none of these species were described Carpenter, Paetel and Schmeltz's names remained nomina nuda (ICZN [1999] article 12.1) except when subsequently described properly by Pease. Another nomenclatural issue affecting the 19th century taxa is the date of publication; several publications, often proceedings of scientific meetings, were dated but may not have been published until one or more years later. This affected many names published by Pease, where the date of the species should be the date of publication, and not that printed on the publication itself. In some cases this means that the name was actually published first by a different author, even though they cite Pease as the authority.

Fig. 3.1 A shell list written by Pease and sent to a collector, showing his casual use of manuscript names (BPBM)

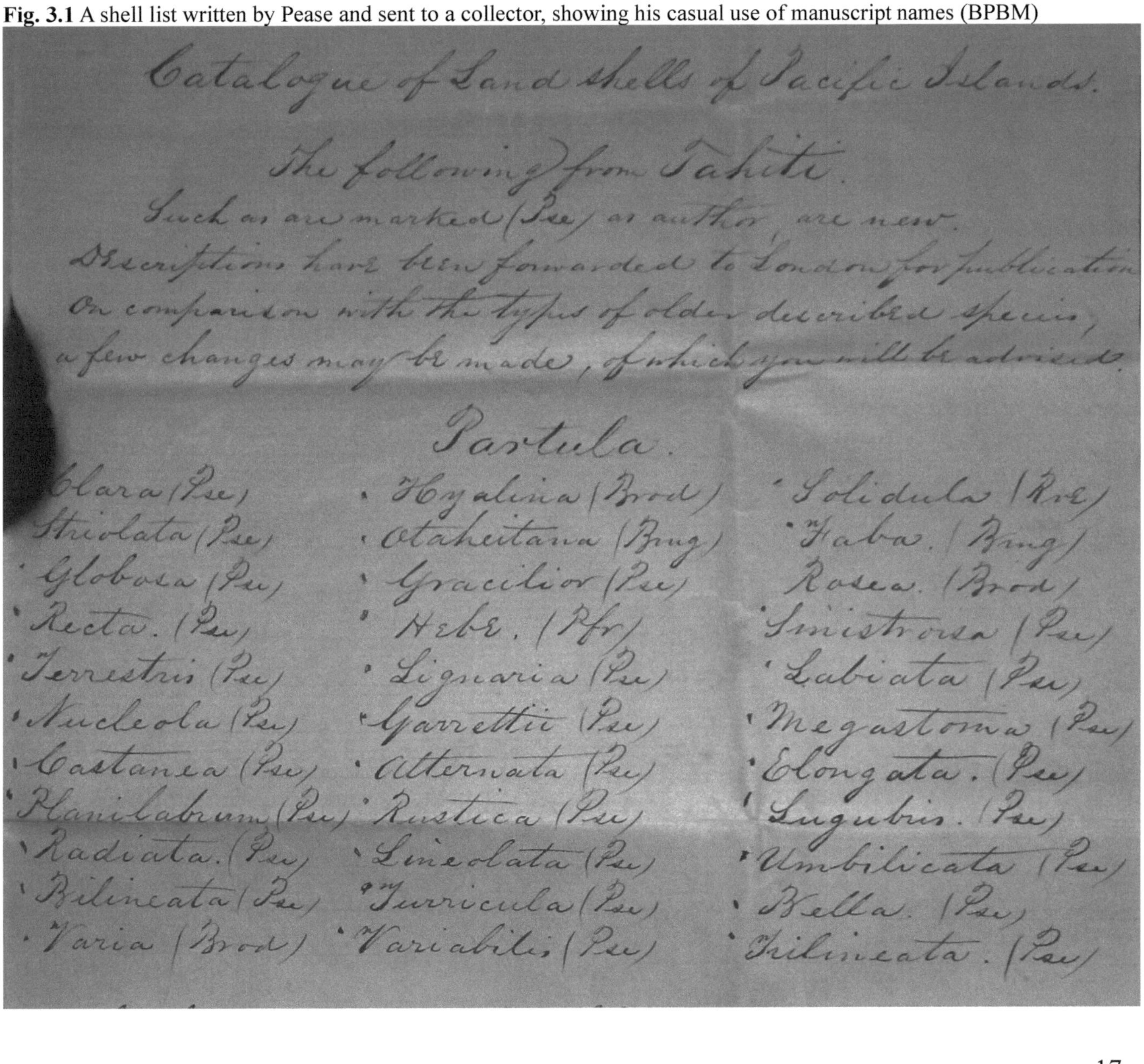

Catalogue of Land shells of Pacific Islands.
The following from Tahiti.
Such as are marked (Pse) as author are new.
Descriptions have been forwarded to London for publication
On comparison with the types of older described species,
a few changes may be made, of which you will be advised.

Partula.

Clara (Pse) · Hyalina (Brod) · Solidula (Rve)
Striolata (Pse) · Otaheitana (Brug) · Faba. (Brug)
Globosa (Pse) · Gracilior (Pse) · Rosea. (Brod)
Recta. (Pse) · Hebe. (Pfr) · Sinistrorsa (Pse)
Terrestris (Pse) · Lignaria (Pse) · Labiata (Pse)
Nucleola (Pse) · Garrettii (Pse) · Megastoma (Pse)
Castanea (Pse) · Alternata (Pse) · Elongata. (Pse)
Planilabrum (Pse) · Rustica (Pse) · Lugubris. (Pse)
Radiata. (Pse) · Lineolata (Pse) · Umbilicata (Pse)
Bilineata (Pse) · Turricula (Pse) · Bella. (Pse)
Varia (Brod) · Variabilis (Pse) · Trilineata. (Pse)

Some of Pease's names were used by Binney (1876) in his description of the anatomy of several species before Pease had published them. Binney is not the authority for these names as he explicitly stated that his publication was not a taxonomic work (ICZN [1999] article 8.3).

Hartman attempted to clarify the situation starting from around 1877 (Smith 1902), sometimes adding confusion of his own by labelling specimens "type" indicating that the specimen agreed with the original description rather than being a true type. Sometimes he amended this to "typical" but others were left with confusing labelling. The same applies to shells in the Pease and Garrett collections, even though some are the genuine types. According to Smith (1902), in some cases type specimens were swapped for other specimens, confusing all identification.

Species concepts

An essential background to defining the species within a taxonomic group is an understanding of the concept of 'species' being applied. This is a highly contentious area of biology with a remarkable number of definitions; Mayden (1999) listed 22, although many of these can be combined.

Traditionally species were not recognised based around a specific definition, this Morphological, Linnean or Classical Species Concept was effectively formalised in the Phenetic Species Concept (Michener 1970) where the species can be defined quantitatively. These concepts have been largely abandoned in favour of more biologically meaningful approaches, of which the most familiar is the Biological Species Concept (Wright 1940), defined by Mayr (1942) as "*groups of actually or potentially interbreeding natural populations, which are reproductively isolated from other such groups*". This concept works well generally but does encounter problems when reproductive isolation is not absolute (i.e. hybrids form between conventionally accepted 'species') and where differentiation occurs even with interbreeding. It also only applies to sexually reproducing organisms, all asexual individuals being effective species under this concept. It does not explicitly deal with evolutionary lineages, implies group selection and may confuse isolating mechanisms with isolating effects (Mayden & Wood 1995). Leaving aside asexual species, which can be treated separately by an Agamous Species Concept, no selection (group or otherwise) is included in the biological species concept and conflating isolating mechanisms with effects is not a major issue as the definition merely seeks to describe the pattern and not provide its underlying explanation. The issue of hybridisation is more problematic and prompted the development of other species concepts emphasising reproductive aspects: the Recognition Species Concept (Paterson 1985) and the Reproductive Competition Species Concept (Ghiselin 1974). A purely evolutionary approach is taken by the Evolutionary Species Concept (Simpson 1951) which accommodates asexual species, hybrids and ancestral species. Unlike the Biological Species Concept it sees reproductive isolation as the result of speciation, rather than the cause. This can be related to Evolutionary Significant Units (Waples 1991). The Unified Species Concept (de Quieroz 2005) may be a reformulation of the Evolutionary Species Concept (Naomi 2011), seeing all evolutionary lineages as species and considers a species to be a group of separately evolving metapopulation lineages. The different concepts were reviewed by Mayden (1997).

The lack of a clear solution to the 'species problem' has led some to propose alternatives to 'species'; Grant (1981) proposes "biospecies" for interbreeding species and "ecospecies" for ecological species; Ereshefsky (1992) proposes "phylospecies" for phylogenetic species and Pleigel & Rouse (2000) modify this to "Least-Inclusive Taxonomic Units". The most extreme view is the call for species to be abandoned altogether (Hendry *et al.* 2000; Mishler 2003; Fisher 2006).

Despite the great controversy that remains around species concepts none of this is new. As Charles Darwin wrote to Joseph Hooker in 1856: "It is really laughable to see what different ideas are prominent in various naturalists' minds, when they speak of "species"; in some, resemblance is everything and descent of little weight — in some, resemblance seems to go for nothing, and Creation the reigning idea — in some, sterility an unfailing test, with others it is not worth a farthing. It all comes, I believe, from trying to define the indefinable." (in Darwin 1887). What Darwin saw as the indefinable nature of species arises from the fact that what we understand to be species do not exist in nature; they are human constructs imposed on evolving lineages. As such, no species concept can accurately define all species. If the artificial nature of the species is recognised then the only requirement for a species concept is that it is useful. As evolution dominates all aspects of biology today species have to have some evolutionary meaning, thus the classical morphological and phenetic concepts have no place while many other concepts fail the utility test. Some variant of the Biological Species Concept (along the lines of the Evolutionary version) remains the most practical, and a relaxed biological species concept has been advocated (Orr 2001; Noor 2002; Coyne & Orr 2004). This is the approach followed here; a species unites all individuals thought to be potentially interbreeding and genetically distinct from other populations. This genetic distinction may be demonstrated through molecular data or by analysis of morphological characters thought to result from underlying genetic factors. The flexibility of this approach means that every 'species' needs to be justified as no arbitrary criterion is being used. In taking this approach I am explicitly recognising that that the distinctions between a variety, subspecies and species are completely artificial. Even if they are delimited genetically, the thresholds used would remain arbitrary. As Darwin (1856, in 1887) observed: "the only distinction between species and well-marked varieties is, that the latter are known, or believed, to be connected at the present day by intermediate gradations, whereas species were formerly thus connected."

For partuids where complete data (shell, anatomical and genetic lineage) are available I have required species to be recognisably distinct in all three types of data. Subspecies form a distinct genetic lineage and are visibly recognisable on shell characters but are anatomically indistinguishable, or differ less than similar species. Varieties are not recognised as they do not form distinct genetic or anatomical groups. In a small number of cases distinct species have been accepted despite the lack of clear shell and anatomy differences because a comprehensive molecular phylogeny is available that identifies at least two well structured clades. The interpretation of such genetic data suffers from a major limitation in the available sampling, with only one nuclear and one mitochondrial gene used in the most comprehensive study (Lee *et al.* 2014). In reconstructing phylogeny and delimiting taxa a distinction needs to be made between gene trees and species trees. Single gene phylogenies are only that, a single gene tree which is unlikely to reflect phylogeny accurately in all areas. A multi-gene tree may equate to a species tree, but this is not available for partulids.

Evaluating partulid species

The majority of partulid taxa were described in the 19th century using conchological characters alone. These relied on general shape, degree of closure of the umbilicus, colour and the form of the mouth opening (see p. 28). Apertural dentition is a major distinguishing character for the more distinctive of these species. In most species the lip is thick and expanded; there is also a notable internal thickening, often forming a shelf-like internal labrum. This ridge may be characteristically angled and its exact form distinguishes some populations. In the palatal region the lip may be expanded into an obtuse bulge or a prominent denticle. Above this the lip may be noticeably thinner, resulting in a sinus and a general ear-like (auriform or obauriform) shape to the aperture. The columella may be reflected over the umbilicus, resulting in different populations having characteristically open, obscured or closed umbilical regions; there seems to be little individual variation in this character. In some populations the columella is distinctively twisted or folded. Rarely a denticle is present on the columella (Pilsbry's 'pillar tooth'). The parietal region is usually covered by a callus extending from the labial insertion to the columella. In some species this is thick and opaque, in others it is a thin, shining and transluscent. It may carry a slight nodular thickening which may be developed into a distinct pearl-like tooth or a prominent denticle. The exact development and position of this denticle is variable, but valley populations can usually be characterised. Shell sculpture shows little variation: most species have irregular growth lines and spiral striae on the spire. In most species the spiral striae are indistinct on the body whorl. Colouration is highly variable in most species, although valley colour types can usually be described.

Through the 20th century snail descriptions relied increasingly on anatomy as an important component in species definition. The reproductive anatomy (principally penis form and position of the spermatheca) has been seen as the most informative anatomical region, reflecting reproductive isolating mechanisms. Features of the mantle, including the pericardium and the kidney have also been used, but mainly in higher level taxonomic classification, and these are rarely of use at species level. In some taxa the radula has also been useful in recognising distinct evolutionary units (species or subspecies). However, radula variation has rarely been addressed; in some cases it has been assumed to be a constant character and in others an unusably variable character. For Partulidae Binney (1876) described the radula of a number of species but since then it has not been investigated systematically. Radula variation was investigated as part of the present study, with the number and form of the teeth in a row determined for one specimen of most species and a wide range of individuals from different populations in some taxa. As has previously been noted (Binney 1875; Pilsbry 1900, 1909) three distinct tooth forms can be found in partulids in general: a central tooth, 6-13 laterals and a variable number of marginals. The number of laterals within a population was found not to be highly variable, but nor was it completely consistent. Marginal numbers tended to vary widely. Thus tooth counts are of limited taxonomic value in Partulidae, although notable differences were found between some genera, most notably in *Palaopartula* where no distinction could be made between laterals and marginals. The radula was not used as a defining character in the present study.

The reproductive anatomy was first described for partulids by Férussac (1821). The systematic utility of the genitalia was advocated by Cooke who encouraged Kondo to undertake a comprehensive investigation of the partulids. Kondo's 1955 PhD thesis on partulid anatomy is the basis for all subsequent work. Kondo found the form of the penis and its internal ornamentation to correspond to distinct populations. Not all populations could be distinguished but Kondo made no attempt to revise the taxonomy in 1955, accepting Crampton's conchologically based identifications. Subsequently Kondo did start to reassess the taxonomy based on dissections (1968) but publication of this work never progressed beyond a preliminary stage. Kondo's demonstration that partulid populations can usually be distinguished anatomically is convincing and was confirmed by the present study, dissecting a wide range of Crampton's samples. All species where bodies have been preserved were dissected and where possible several individuals were examined to account for abnormalities and to evaluate individual variation. Kondo (1955) had dissected 25 *P. faba* for this reason, to allay Crampton's concerns that they would be too variable to be meaningful. Kondo found little variation that could not be explained simply by ontogeny. This is not always the case; Kondo himself noted some degree of variation in *P. citrina* and a few species where two or three anatomical types could be recognised (e.g. *P. gibba*). These differences were limited though, expect in *P. otaheitana* where several radically different forms could be identified. This was later shown to be partly due to misidentification of some populations (Kondo & Burch 1983), the remaining anatomical diversity in this taxon is here suggested to be the result of hybridisation of different populations (see p. 116).

Molecular data have been available for the phylogenetic study of partulids since 1970, initially based on proteins (Schwable & Murray 1970; Johnson *et al.* 1986a, b), then nuclear and mitochondrial DNA (Murray *et al.* 1991; Goodacre 2001, 2002; Goodacre & Wade 2001a, b; Johnson *et al.* 2000; Lee *et al.* 2007, 2008, 2009, 2014). Of these, the most extensive studies have been mitochondrial and these have indicated widespread hybridisation in Tahitian and Moorean species. It is not known whether this was also the case on other islands as suitably preserved samples are relatively scarce elsewhere.

Hybrids

Extensive hybridisation in the Society islands *Partula* species has been suspected for over 100 years based on shell morphology (Garrett 1884; Smith 1902; Pilsbry 1909). Dissections and molecular phylogenies suggest that there has been extensive hybridisation in *Partula*, at least on Tahiti and Moorea (Schwable & Murray 1970; Murray & Clarke 1968, 1969, 1980; Lee *et al.* 2007b, 2009). Some hybridisation is also apparent in Huahine species (p. 161, 164). Laboratory crosses (Murray & Clarke 1980) have confirmed that hybridisation is possible, and provide informative data on hybrid morphology. The morphology of some of the experimental hybrid crosses are shown in Fig.3.2 and Table 3.1.

In the Moorean *Partula* not all species could be hybridised experimentally but all species could be united in a chain of hybridisation links (Fig. 3.3). Murray & Clarke (1980) recognised two major groups of Moorean *Partula* within which inter-fertility was almost complete: the *suturalis* group (*P. suturalis, P. aurantia, P. tohiveana/olympia, P. mooreana*) and the *taeniata* group (*P. taeniata, P. exigua*). *P. mirabilis* formed a link between the two, being able to hybridise freely with *P. taeniata* and *P. aurantia.* Hybridisation could be extensive; based on dissections and protein electrophoresis Murray & Clarke (1968) wrote that "we cannot recognise any pure populations of *P. aurantia*." They also found probable hybrids with *P. taeniata* dominating most *P. 'exigua'* populations. Other species were more equivocal in the field; Clarke & Murray (1969) suggested that *P. mooreana* and *P. suturalis* hybridised above Haapiti. On southern slopes of Mount Tohivea *P. mooreana* and *P. tohiveana* hybridised to "a limited degree".

Fig. 3.2 Hybrids: a-c – *Partula mirabilis* × *taeniata* (ZSL 102); d – *P. suturalis* × *aurantia* (Clarke & Murray, NHMUK); e-f – *P. rosea* × *varia* (ZSL 1461); g-k – Murray & Clarke (1980) experimental hybrids (ANSP specimens, Clarke & Murray numbering): g – S99 (*olympia* × *otaheitana amabilis*); h – S22 (*olympia* × *aurantia*); i – S12 (*mooreana* × *olympia*); j – S65 (*mirabilis* × *mooreana*); k – S107 (*taeniata striolata* × *diminuta*)

Specimens of *P. mirabilis* × *taeniata* are easily divisible into two conchological categories: *mirabilis* types (e.g. ANSP CSJ 001) with modified *P. taeniata* or *P. mirabilis* anatomy, and *taeniata* types (e.g. ZSL 102) with *P. mirabilis* anatomy. These forms are also found in the wild, such as the population of *P. mirabilis/taeniata propinqua* (see p. 152).

Application of species definitions to Partulidae

The complexity of the hybridisation on Moorea highlights the issues in any strict application of species concepts. As there are no completely isolated species on Moorea, all could be combined into a single 'species'; however, combining the highly divergent, reproductively isolated *P. taeniata* and *P. tohiveana* in a single species would be nonsensical. Accordingly, here taxa are recognised as species if they are conchologically, anatomically and genetically distinguishable. Genetic distinctiveness has been relaxed as a criterion when the distribution of genotypes can be reconciled with hybridisation between distinct species, or with incipient, but almost complete, speciation. Where a taxon (species or lineage) occurs on more than one island it will inevitably drift to some degree of genetic distinctiveness without necessarily leading to meaningful speciation. Allopatric taxa have only been taken to be distinct species where phenotypic distinctions are apparent or where there is evidence of high levels of genetic divergence. For taxa where only conchological characters are available their taxonomic distinctiveness has to be taken as provisional. Further difficulties are created by species known only from a single specimen, these have only been recognised as species where they lie outside of the range of variation of the most similar taxon.

Another vexed issue in partulid taxonomy is the status of subspecies. Although a great many taxa were described as varieties only a small number of names have been applied at a subspecies level, almost exclusively on Tahiti and Moorea. These were assigned (mostly by Crampton) subjectively on the basis of being conchologically distinct allopatric

Table 3.1 Experimental crosses of *Partula* species (Johnson *et al.* 1977; Murray & Clarke 1980), taxa as named by Murray & Clarke (1980). Samples with an asterisk are illustrated in Fig. 3.2. Anatomical types are based on Kondo (1955) and dissections present in subsequent sections. Excludes crosses between subspecies.

Species crossed		Number of crosses	hybridisation	Sample	maternal	paternal	anatomy
aurantia	*mirabilis*	8	38% matings (n=8)	S63B	*mirabilis*	*aurantia*	normal *mirabilis*
aurantia	*olympia*	6	50% matings (n=6)	S22 *	*olympia*	*aurantia*	normal *olympia*
aurantia	*suturalis*	3	free				
exigua	*taeniata*	3	67% probable (n=3)				
filosa	*taeniata*	3	some	S107 *	*filosa*	*taeniata*	long *filosa*
mirabilis	*mooreana*	5	20% possible (n=5)	S65 *	*mirabilis*	*mooreana*	moderate *mooreana* or abnormal
mirabilis	*olympia*	13	15% matings (n=13)				
mirabilis	*propinqua*	4	free, ¾ non-reciprocal				
mirabilis	*suturalis*	7	33% possible (n=3)				
mooreana	*olympia*	13	8% (n=13)	S12 *	*mooreana*	*olympia*	short *mooreana*
mooreana	*otaheitana*	2	none				
mooreana	*suturalis*	5	20% possible (n=5)				abnormal
mooreana	*tohiveana*	7	none	S74			short *tohiveana* (self-fertilised?)
olympia	*otaheitana*	2	50% possible (n=2)	S99 *	*olympia*	*otaheitana*	short *olympia*
olympia	*suturalis*	13	15% possible (n=13)	S36	*suturalis*	*olympia*	*suturalis* (self-fertilised?)
olympia	*tohiveana*	9	free	S5	*olympia*	*tohiveana*	
otahietana	*suturalis*	3	none				
propinqua	*taeniata*	5	free, $^2/_5$ non-reciprocal				
suturalis	*tohiveana*	11	none				

Fig. 3.3 Hybridisation patterns. Left: chain of hybridisation on Moorea, size of line indicates relative significance of hybridisation, grey lines are possible hybrids (based on data in Table 3.1). Right: Anatomy of Huahine hybrids

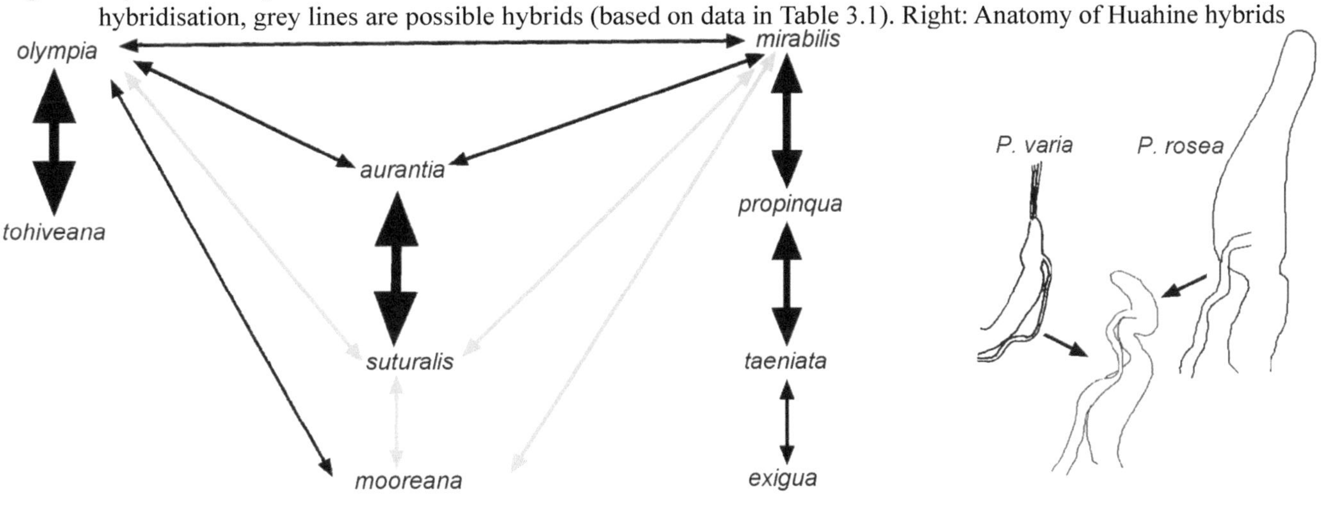

forms with strong affinity to other populations. In the case of *Partula otaheitana* on Tahiti some of Crampton's forms appear to be valid, being anatomically distinctive although conchologically the differences are minor. Others are not supportable either on phenotype or genotype. For example, most of Crampton's varieties of *P. clara* seem to have little standing. Existing subspecies have been retained only where the populations are distinctive both conchologically and anatomically. Some of the allopatric Raiatean 'species' have been reduced to subspecific rank where conchological features enable reliable identification but anatomical differences are minor or non-existent.

Genera of Partulidae

Pilsbry (1900) created the family Partulidae specifically for *Partula*. *Partula* had previously been placed in the Dicerae Menke, 1828, an invalid group established as a division of the "Heliceae" for *Vertigo* and *Partula*. A superfamily was created for Partulidae, the Partuloidea, by Baker (1963) and Partulidae were placed in the monotypic infraorder Partuloinei Schileyko & Starobogatov, 1989 (in Golikov & Starobogatov 1989). These taxa are in the larger grouping of Orthurethra; this was rejected by Nordsieck (1986) but supported by Wade *et al.* (2001).

The diagnosis of Partulidae remains that of Pilsbry (1900): "Orthurethra in which the kidney is triangular, equal to the pericardium in length. Genitalia of the haplogonous type, the penis without an appendix. Shell Bulimoid, with the outer lip expanded or reflexed in known forms; the columellar lip reflexed, curving into the basal margin, simple or nodulous; parietal wall often dentate, but the aperture never has entering lamellae.

"Jaw very thin, formed of many narrow flat plaits, converging downward towards the middle. Radula broad; the central tooth is not much smaller than the laterals, and has a long middle and small side cusps; lateral teeth with the mesocone long, ectocone well developed, but no entocone; marginal teeth oblique with three cusps directed outward, the inner one largest."

Pilsbry (1909) placed all partulids in a single genus with several subgenera (*Carolinella Leptopartula, Marianella, Marquesana, Melanesica, Palaopartula, Partula, Samoana* and *Thakombaua*). Pilsbry & Cooke (1934) described *Eua* and raised *Samoana* to a genus. This was adopted by Kondo (1955, 1968) but the old useage was followed by Vaught (1989) who listed one genus with multiple subgenera. Conversely, Schileyko (1999) included four genera: *Eua* (with two subgenera: *Nesanassa* and *Thakombaua*), *Samoana* (with two subgenera: *Samoana* s.s. and *Marquesana*), *Partula* (with five subgenera: *Partula* s.s., *Melanesica, Carolinella, Marianella* and *Leptopartula*) and *Rennellia*. Here four genera are recognised: *Eua, Partula, Palaopartula* and *Samoana*. The molecular phylogeny (Lee *et al.* 2014) gives little support to any subgenus.

Questionable species

Historically there has been some confusion between *Partula* and *Placostylus*. For example, *Placostylus remotus* Hedley, 1898 was considered a *Partula* by Schileyko (1999) although it is more usually placed in *Scilistylus* (as the type of the genus) and has little resemblance to a partulid.

Several species have long been regarded as unidentifiable. Examination of all material (specimens, description and/or figures) and evaluation of the range of variation in all species has allowed all but one of these to be identified. The remaining species is *Partula bulimoides* Lesson, 1831. This is known only from the original description of a shell collected on the Coquille expedition. It has been assumed to be a synonym of *P. faba* (following Pfeiffer 1881), but Raiatea was not visited by the Coquille; Lesson cited it from New Guinea. Hartman (1881) stated without explanation 'I have no hesitation in pronouncing *upolensis, canalis, semi-lineata,* and *conica* varieties of this species.' He later (1885: 222) stated that "Although Lesson's brief diagnosis of *bulimoides* approximates nearer to some varieties of *conica* than to any known species, his measurements of the shell and the locality induce me to restore the species." Pilsbry (1909: 305) was unable to place it with any known species. The description is ambiguous and it may not be a *Partula*.

A small number of fossil Partula species have been described based exclusively on general shell shape. These are not true partulids: *americana* Heilprin, 1887 [= *Tocobaga*], *vicentina* Oppenheim, 1890 [= *Procersatus*], *dautzenbergi* Cossmann, 1906 [= *Procerastus*] and are excluded from the present publication. Similarly supposed partulids now placed in other families are also excluded (*australis* Bowdich, 1822; *bataviate* Grateloup, 1840; *delatouri* Hartman, 1886; *densilineata* Reeve, 1850; *dumartroyi* Souleyet, 1842; *flavescens* King, 1831; *glaber* Hartman, 1885; *labrella* Grateloup, 1840; *layardi* Brazier, 1890; *maclearyi* Brazier, 1876; *maximiliana* Poitiez & Michaud, 1838; *peasei* (Cox, 1871; *pudica* Férussac, 1821; *rohri* Pfeiffer, 1846; *salomonis* Pfeiffer, 1852; *unidentata* Sowerby, 1852; *virgulata* Mighels, 1845 [list corrected after Richardson 1990]). Numerous publications and internet databases erroneously list *Patula* (and sometimes *Partulina*) species as *Partula*.

4. Anatomy, shell characters and methods

Anatomically partulids are typical gastropods (Fig. 4.1), with notable features found in the excretory, nervous, digestive and reproductive systems as well as the shell. Characters useful for higher-level taxonomy include the kidney and ureter, used by Pilsbry (1900) to define the major pulmonate categories used today. Partulidae fall into the Orthurethra (Kondo 1948) on the basis of their short urethra and elongate kidney. Solem (1962) erroneously thought partulids had an unusually shortened triangular kidney for a orthurethran, this is similar to Pilsbry's (1909) description of the kidney being "about equal to the pericardium"; Pilsbry & Cooke (1934) described it as "dilated and triangular posteriorly, but become narrow and bandlike in the anterior moiety. There is no ureter, the kidney opening directly into the lung by a lateral pore at the apex of the long anterior extension and remote from the hindgut." Thus it is a long orthurethran kidney, but easily misidentified due to the presence of a conspicuous triangular section.

The nervous system was described by Tillier (1990) who noted that the cerebro-pedal commissures are not particularly long, with the right being shorter than the left; the pleural ganglia are situated close to the pedal ganglia; the visceral ganglion is medial, touching the left parietal ganglion and fused with the right; the left parietal ganglion is sometimes in contact with the left pleural ganglion. These features are largely primitive for orthurethrans. Tillier (1990) also recorded the presence of a digestive crop in *Eua* and *Partula* (but not *Samoana*). In the case of *Eua* this was accompanied by a 'fleshy diverticulum'.

Radula

Radular characteristics have been used extensively in mollusc systematics and the form and number of radular teeth is a standard part of modern taxonomic descriptions. In terrestrial pulmonates there tends to be little structural variation in teeth within genera but the numbers of teeth per row is a useful characteristic. Most species descriptions indicate that there is little or no variation in the number of teeth, although it is known that tooth number does increase ontogenetically in

Fig. 4.1 General anatomy of a partulid: digestive system in green, nervous system in blue, reproductive system in brown, circulatory system in red, major muscles in black, with cerebral ganglia (right) in dorsal (top and centre) and posterior (bottom) views

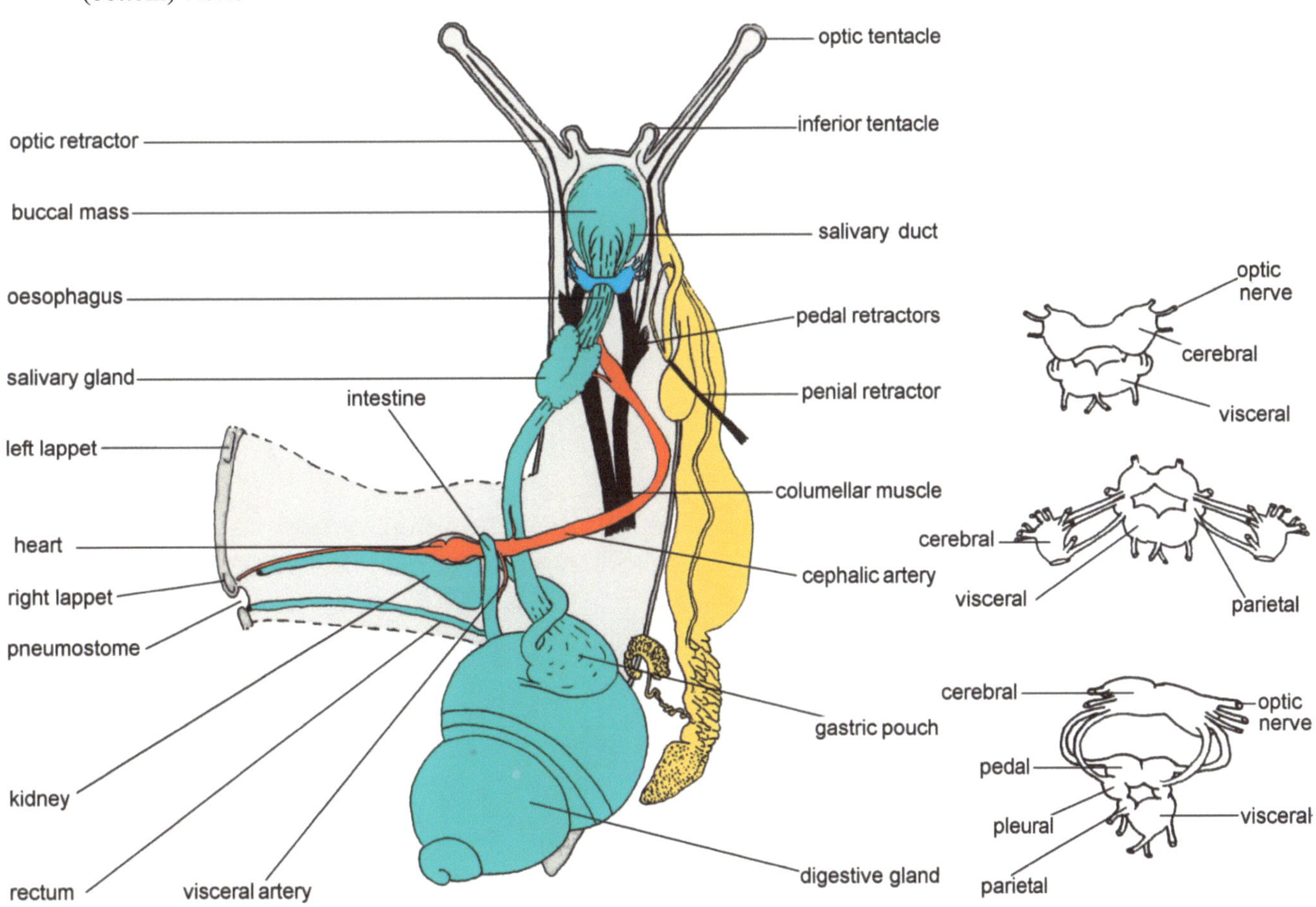

opistobranch and nudibranch molluscs (e.g. Sterki 1894; Bertsch 1976). Variation is known from adults of some taxa, e.g. Endodontidae (Solem 1982), and particularly in arboreal *Drymaeus* species (Solem 1955) and Achatinellidae (Cooke & Kondo 1960). For the majority of species where variation has been reported individuals vary by 1-5 teeth per half row), but in some Achatinellidae variation can be as high as 35 (*Achatinella concavospira*: Cooke & Kondo 1960).

The only published comparative analyses of partulid radulae are those of Binney (1875), Semper (1894) and Pilsbry (1909) covering 15 taxa. The radula of single species were figured by Heyneman (1867), Rensch (1935) and Kondo (1955) and a single specimen was examined by Crampton (MS) (Fig. 4.2). As observed by earlier workers the central tooth has a broad, slightly pointed mesocone and indistinct ectocones, laterals have a broad, outwardly angled mesocone and a smaller pointed ectocone. Marginals have a pointed mesocone and an ectocone divided into two denticles. Towards the outer edges the marginals become sinuous. There have been no comprehensive studies of radular variation in the family. In the present study 921 individuals were examined of 74 species (excluding laboratory hybrids). Variation was high in most species (varying by up to 62 teeth per half row); lateral teeth varied little in number and could be characterised geographically (Table 4.1).

Of the partulid genera *Palaopartula* is the only one with a distinctive radula with a high laterals count (13-14) but

Fig. 4.2 Historical figures of partulid radulae. a – Heyneman 1867 (*Partula lirata*); b – Binney 1875 (*P. amanda*); c – Semper 1894 (*P. vanikorensis*); d – Rensch 1935 (*P. regularis*); e – Pilsbry 1909 (*P. varia*); f – Pilsbry 1909 (*P. rosea*); g – Pilsbry & Cooke 1934 (*Eua globosa*); h – Kondo 1955 (*P. faba*); i – Crampton MS (*P. faba subangulata*).

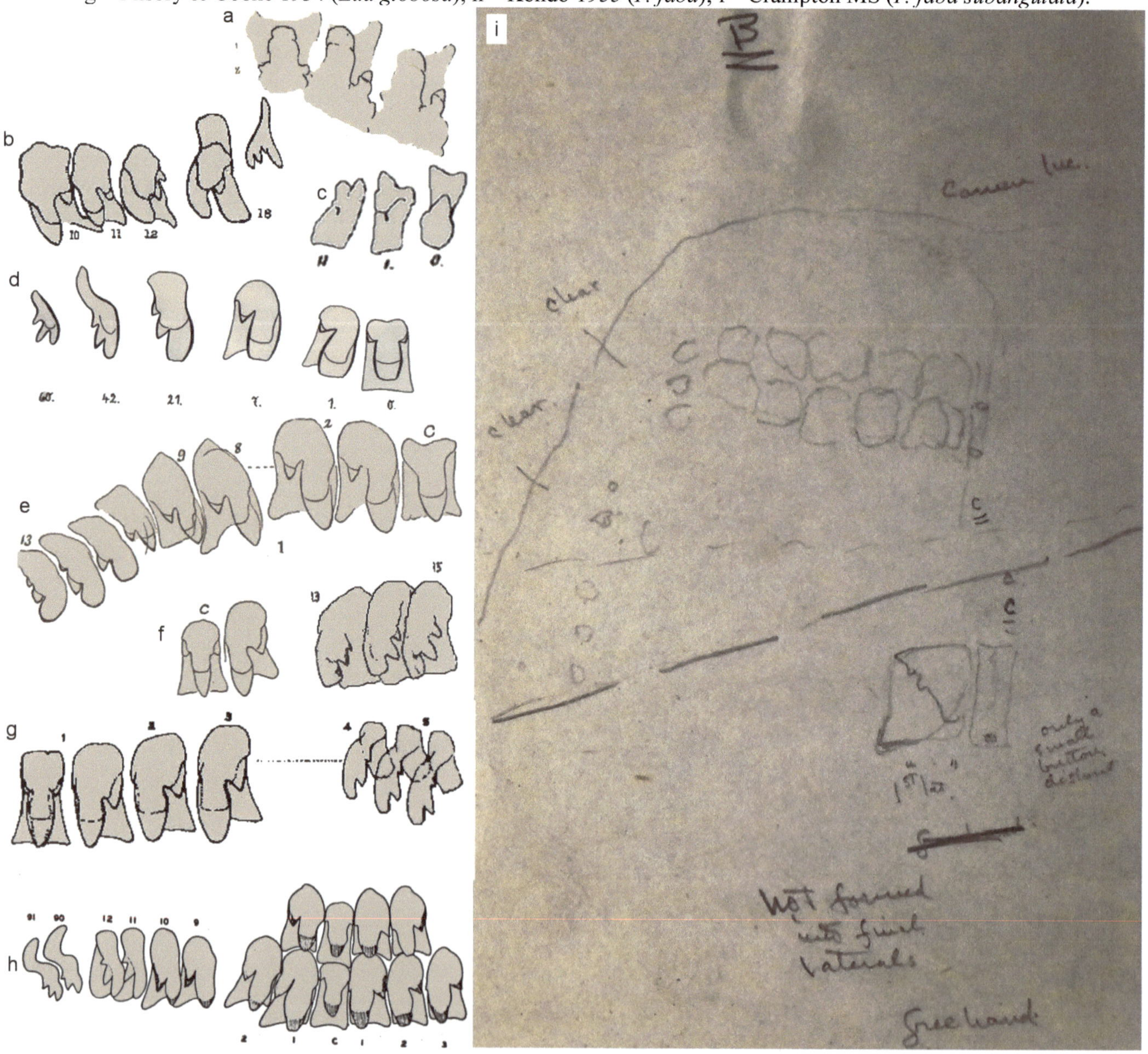

Fig. 4.3 Radula types: *Partula* half row; *Eua zebrina*; *Palaopartula thetis*; *Samoana attenuata*; *Partula hyalina*; *Sphendone insolita* (after Slapcinsky & Kraus 2016) showing lateral teeth, (central tooth marked 'C', last lateral numbered), central and laterals, typical marginals and final marginals. Scale bar for *Eua* 50 μm.

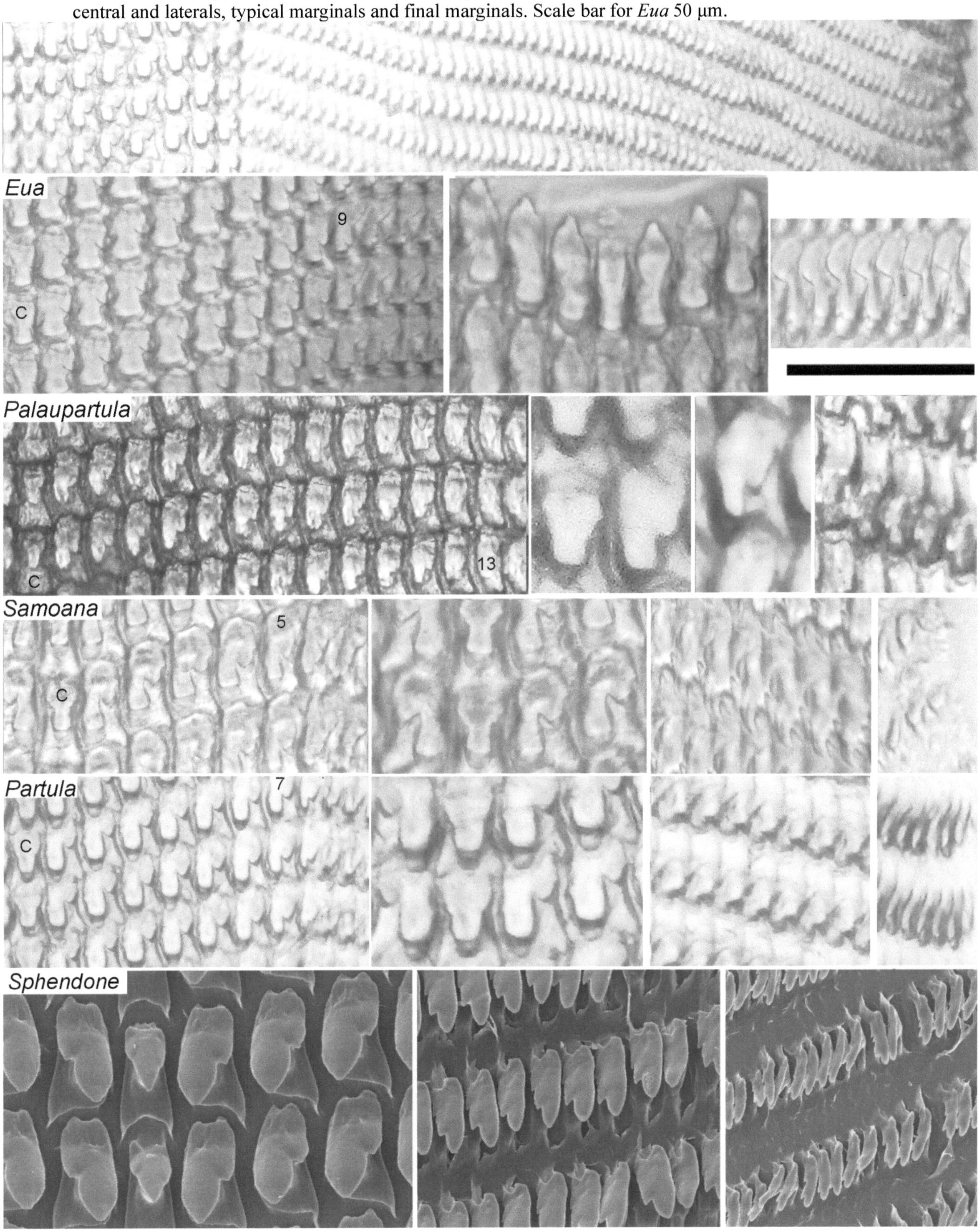

with very little distinction between laterals and marginals. The other genera have clear distinctions between laterals and marginals and have 5-12 laterals. *Samoana* has the lowest numbers of marginals and show relatively little specific variation. In *Partula* low numbers of laterals and marginals are found in the west of the range: 5-10 laterals and 53-96 marginals in New Guinea and the Solomons. In the east tooth counts were generally higher and variation was usually moderate (6-12 laterals and 56-131 marginals, standard deviations of marginal tooth counts varied from 12.15 in *P. coxi* to 2.32 in *P. pacifica*). Tahitian *Partula* were more variable, especially in the widespread *P. affinis* (marginals s.d. 14.96 overall, most variable in Faaroa – 20.51), *P. clara* (10.73, greatest variation in Tereehia – 14.46) and *P. otaheitana* (9.79, most varied on Mt. Marau where different subspecies overlap – 11.49). *P. hyalina* was only moderately variable (7.60) and species with restricted ranges showed much less variation: *P. producta* (5.91), *P. diminuta* (4.93) and *P. nodosa* (4.21).

Moorean species show highest levels of variation in *P. tohiveana* (9.90; *tohiveana* – 10.61, *olympia* 4.32, Fareaito – 15.52), followed by *P. mirabilis* (9.45; Maramu – 12.58) and *P. exigua* (9.21; Paraoro – 12.28) and *P. aurantia* (8.27). The two most widespread species had only moderate levels of variation: *P. suturalis* (7.02, varying from 5.90 in Faatoafe to 9.90 in Maharepa, with no geographical pattern), *P. taeniata* (7.32, varying from 6.65 in Atimaha to 10.29 in Fareaito; most extreme in south-central populations). *P. mooreana* was the least variable (6.34).

On Huahine moderate variation was recorded in *P. rosea* (6.70) and *P. arguta* (7.89), and higher levels in *P. varia* (10.25, highest in Faahiti [13.28] and Tepua [13.04], lowest in Atita [0.60]). Raiatean species had moderate variation (7.00-8.99) except for high variation in *P. navigatoria* (13.08; Tepua 14.48), *P. hebe* (9.81; highest in Farepaiti – 12.34), *P. crassilabris* (10.97), *P. garrettii* (10.92). Unusually low variation was found in species with restricted ranges (except *P. crassilabris*): *P. lugubris* (5.30), *P. leptochila* (4.51, except for a possible hybrid population at Temehani – 20.82) and *P. remota* (2.87). A similar range effect was found in *P. garrettii* as, although the species as a whole had high variaiton, this resulted from three partially separate distributions, each with relatively little variation (true *garrettii* – 6.23, *thalia* 6.16 and *rustica* 2.88). The widespread *P. faba* is moderately variable (8.47, Averarahi and Faaharoto 15.05). Of the Tahaa species the closely related *P. bilineata* and *P. umbilicata* are well separated: *P. bilineata* 86-93 marginals (s.d. 8.02) and *P. umbilicata* with 72-86 (6.73). *P. planilaburm* is more varied (10.91) and *P. faba* has very high levels of variation (22.82), with lowest values Hamene (7.55), Hurepiti (9.68) and Patio (9.29). The single species on Bora Bora, *P. lutea*, varied by 12.08.

A genetic component to tooth numbers is indicated by the loss of any variation in inbred captive populations (n=85). The species data are summarised in Appendix II.

Reproductive anatomy

The reproductive anatomy of *Partula* was briefly described by Férussac (1821). The first figure was that of Semper (1865). Ontogenetic change and individual variation in partulid reproductive anatomy was first evaluated by Kondo (1955)

Table 4.1 Geographical variation in lateral teeth in partulids. Red – main counts, yellow – uncommon counts

Genus	Island group	5	6	7	8	9	10	11	12	13	14
Eua	Samoa		yellow	yellow	yellow	yellow	red				
	Tonga		red	yellow							
Samoana	Samoa	yellow	yellow	red	yellow						
	Tonga		red								
	Marianas		red								
	Societies	yellow	red								
	Marquesas	yellow	red								
	Australs	yellow	red	yellow							
Palaopartula	Palau									red	red
Sphendone	Palau						red				
Partula	New Guinea			yellow	yellow		yellow				
	Solomons	yellow	red	yellow	yellow	yellow					
	Vanuatu			red	yellow	yellow					
	Fiji				yellow	yellow					
	Marianas		yellow	red	yellow	yellow			yellow		
	Carolines					red					
	Wallis & Futuna		yellow	yellow	yellow						
	Cooks				yellow	red					
	Tahiti		yellow	red	red	yellow	yellow	yellow			
	Moorea			red	red	yellow	yellow	yellow			
	Huahine		yellow	yellow	red	red	yellow	yellow	yellow		
	Raiatea			yellow	yellow	red	red	yellow	yellow		
	Tahaa		yellow	yellow	yellow	red	red	yellow	yellow		
	Bora Bora				yellow	red	yellow				
	Australs		yellow		yellow		yellow				

who dissected 25 *Partula faba*; the only notable variation was the under-developed nature of three immature specimens. For most other species he dissected at least five individuals. A more extensive study of Tahitian species (Kondo & Burch MS) covered 712 specimens of 13 species. The most recent study provided a phylogenetic context (Schileyko 1979).

As with the majority of gastropod taxa there is little variation in the female parts; spermatheca position varies to some degree between species. A spermatophore was described (Kondo & Burch MS) as "consisting of head, body and tail (confirming H.B. Baker's description (pers. comm.) to Kondo)... The spermatophore is composed of a horny frontal head (A) within which is visible a narrow, elongate twisted core (B). The midbody (C) is a white lumpy substance without any horny covering and appearing like coarse bread dough, ending with a tapering tail (D)." None were located in the present study.

The male structures are of more taxonomic and phylogenetic significance. There has been some confusion over the homology of different structures, with Pilsbry (1909), Pilsbry & Cooke (1934) and Kondo (1955) describing parts of the system as a penis and a variably developed epiphallus. Later Kondo (MS) modified this to an epiphallus and 'subphallus', although the latter term has no real meaning. In contrast Schileyko (1999) considered partulid male genitalia to comprise a penis and flagellum, due to the blind-ending nature of the 'epiphallus' in many species. Anatomically the gastropod penis should comprise a muscular basal part terminating at the insertion of the penial retractor muscle. The vas deferens enters the penis here, and the 'epiphallus' is an expansion of the vas deferens at this point, the function of which is the formation of the spermatheca. There may be a flagellum or caecum (depending on whether this is a slender, flagellate-shaped structure or a pocket) coming off from the 'epiphallus'. The 'epiphallus' narrows abruptly to form the vas deferens proper. Although the vas deferens is slightly inflated in some partulids it does not form a distinct epiphallar structure, thus Tillier & Mordan's (1995) contention that there is no epiphallus in the family is correct. As with *Draparnaudia* (Tillier & Mordan 1995) the vas deferens often enters the penis medially, resembling an accessory structure. This led Schileyko (1999) to identify the 'epiphallus' as a flagellum, but it is not flagellar-shaped and is muscular, along with the rest of the penis. Furthermore, the penial retractor muscle attaches at the end of this structure, confirming that it is part of the penis. Internally the distal structure is almost always separated from the main part of the penis by a constriction and has a distinct internal ornamentation. Slapcinsky & Kraus (2016) referred to this structure as an 'apical chamber'. It is here termed the penial caecum, to avoid confusion with a true epiphallus or a functional flagellum.

The internal ornamentation of the penis consists of a series of folds of tissue which enable it to be everted during mating. In the retracted state the folds form ridges. Typically there are one or two main ridges and several smaller ones, these may be linked by cross-ridges in some species. The terminal lobe of the penis is usually smoother internally, and the ridges are often broken into rows of papillae. Kondo (1955) described *Partula* penial anatomy as "having a variety of shapes from large-clavate, subcylindrical, vermiform, to a short protuberance... Interior of penis with several vertically arranged (rarely laterally) small pilasters which interconnect with each other latero-diagonally by means of fleshy spurs."

Kondo (1955) divided penial anatomy into several distinct anatomical forms: *Eua*, *Samoana* and seven main *Partula* forms (Fig. 4.4): *faba*-type (considered primitive by Kondo; robust, several modified types: *citrina* – caecum reduced, *aurantia* – main part reduced, *arguta* – narrowed, *bilineata* – caecum narrowed, *robusta* – reduced, caecum short, club-shaped, *gibba* – reduced, caecum broad); *remota*-type (baseball-bat shaped); *hebe*-type (papillae much reduced, widespread but poorly defined); reduced *hebe*-type (as *hebe*-type but extremely reduced); *tohiveana* (or vermiform) type (long, vermiform, a shortened form was categorised as '*turneri*-subtype'); *leucothoe* type (distinctively marlinspike-shaped). Kondo (1955)

Fig.4.4 *Partula faba* reproductive and Kondo's *Partula* penis types (after Kondo 1955)

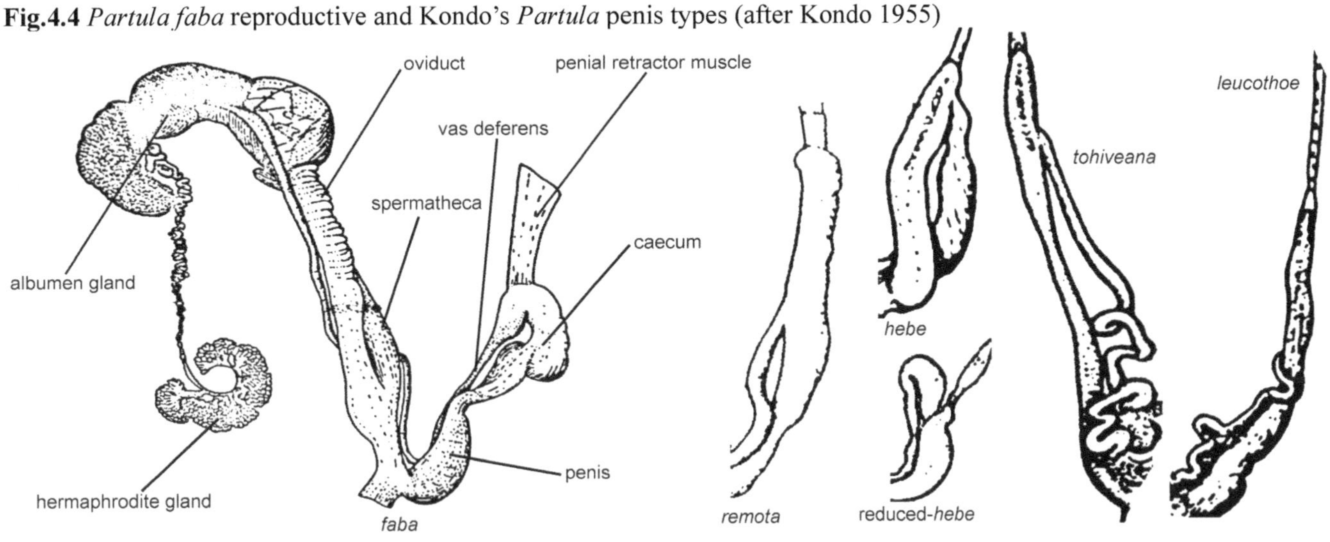

attempted to interpret the anatomical forms in a phylogenetic context, however, more recent data suggests that anatomy is useful for taxonomic purposes but is too labile for wider phylogenetic analysis.

Partulidae are ovoviviparous and embryos are initially covered by shells, except in *Partula hyalina*, *P. clara*, some *P. taeniata*, and *Eua globosa*. In all other species the egg shell dissolves before birth. Kondo (Mannotated thesis) noted that in some samples (Crampton and some others) the formalin was acidic and had dissolved the egg shells, so calcified and uncalcified eggs were not necessarily correctly distinguished in older material. The embryo is born at a shell size of about 2½ whorls. At this stage it is some 30% of the adult length.

Crampton and Kondo noted differences in development in some species, ascribing individual forms to developmental stages: embryonic (after dissolution of egg shell, before birth), aneanic (new-born), metaneanic (juvenile), paraneanic (subadult – full sized but without lip development), adult. These correspond to the familiar neonate, juvenile, subadult and adult.

Shell

The majority of partulid taxa were described on shell features alone, often with a heavy reliance on colour pattern. Colour patterns are extremely variable, particularly in the Society Islands. Crampton (1916, 1925, 1932) defined 37 colour and pattern varieties, to which he gave Latin names (Fig. 4.5). Confusingly these often look like species or varietal names and Cockerell (1925) thought that they could not be excluded from taxonomic nomenclature, although fortunately this view was subsequently ignored. Murray & Clarke (1966) and Johnson *et al.* (1993) considered that some of Crampton's colour forms were indistinguishable (Table 4.2). Crampton himself seems to have had some difficulty in assigning shells to the forms as his categories in publications and manuscripts are often contradictory. A standardised system used here is shown in Fig 4.5. The colour is largely restricted to the periostracum and the pattern is modified in some species by loss of this layer. This 'decortication' is often apparent apically, but may occur in streaks over the entire shell, sometimes very extensively.

Partulid shells are typically conical but vary widely, from ovate to elongately conical (Fig. 4.6a-d) and from extremely thin and fragile to very thick. Almost all partulids have a protoconch of 2½ whorls, the first of which is smooth, the remainder ornamented with fine spiral punctuate striae. *Partula lirata* is an exception, with 3 apical whorls. Sculpture takes the form of incised spiral striae, which are most pronounced towards the apex (Fig. 4.6e-i). In most species they become indistinct on the body whorl. At the junction between each whorl the suture may be impressed, forming a shallow groove, or adpressed, (closely overlapping). It may also be incised (indented where the growth lines intersect with it). These features vary by genus or species. The umbilicus may closed by the columella, largely obscured by it (rimate), narrowly perforate or broadly open. The lip may be narrowly or broadly expanded, reflected and thickened internally. There is often a shining callus on the parietal region between the lip insertion and the columella. There may be a nodule or denticle on the parietal region (Fig. 4.6s), a broad palatal denticle or tooth on the inner margin of the lip. This may compress the aperture, making it auriculate (Fig. 4.6q). In some species the columella is arcuate or nodular, and straight or dilated. It may also be grooved (impressed) at its insertion with the body whorl, or with an internal fold (plicate or subplicate, if the fold is indistinct).

Table 4.2 Colour patterns: Crampton's colours and banding types, with colour explanations from Murray & Clark (1966) and the numerical categories of Johnson *et al.* (1993). Banding used here marked in bold.

Colours	Explanation	Banding		Explanation
inornata	unbanded, any colour	cingulata	vera	one median narrow dark band (= narrow zonata)
pallida/pallidor	white (W) or pale		alternata	cingulata with indistinct bands
yellow	yellow (Y1-2)	unilineata		one submedian dark band, most intense on spire (= cingulata)
pink/rosea	pink (P1-3)	**cestata**	vera	one median wide dark band, plus sutural and basal bands
fusca	pale brown (N1)		helix	cestata with medial band absent from body whorl
carnea	pale red-brown (N1)	**frenata**	vera	two narrow dark bands
strigata	fusca to fulva, dark striations		alternata	frenata with indistinct bands
fulva	yellow-brown (N2)		basalis	frenata with faint bands
castanea	deep chestnut brown (N3)	trilineata		three dark bands, subsutural, median and basal (= frenata var.)
phaea	dark brown (N3)	concrescens		trilineata with subsutural and medial bands fused
phaea-purpurea	dark purple-brown (N3-4)	**lyra**		frenata, with basal band and dark diffuse area above upper band
purpurascens	purple	quadriga		lyra but upper diffuse area forms a distinct band
atra	very dark, almost black (N4)	assymetrica		lyra with two lower bands confluent
		zonata	vera	brown with two pale bands
Spire colour			alternata	zonata with indistinct bands
apex	pallida to fusca, tinted apex		subzonata	brown with one pale medial band (= bisecta with sutural band)
mitella	fusca to fulva, spire pink (P2)	**bisecta**	vera	dark with light medial band, may also have narrow sutural band
mitella-rubra	fusca to fulva, spire orange (O2)		alternata	bisecta vera with indistinct bands
bicolor	pale with dark spire		lata	bisecta with broad medial band
		nodosa		uniform dark brown with white sutural area

Abbreviations and conventions

In the following accounts all shell figures are accompanied by a vertical scale in mm. Museum acronyms are:

AM – Australian Museum, Sydney, Australia
AMNH – American Museum of Natural History, New York, USA
ANSP – Academy of Natural Sciences, Philadelphia, USA
BPBM – Bernice P. Bishop Museum, Honolulu, USA
CAS – California Academy of Sciences, San Francisco, USA
CMNH – Carnegie Museum of Natural History, Pittsburgh, USA
DMNH – Delaware Museum of Natural History, Wilmington, USA
FMNH – Field Museum of Natural History, Chicago, USA
HNHM – Hungarian Natural History Museum, Budapest, Hungary
IRSNB – Institut Royale des Sciences Naturelles de Belgique, Brussels, Belgium
JG – author's collection
LACM – Natural History Museum of Los Angeles County, Los Angeles, USA
MCZ – Museum of Comparative Zoology, Harvard, USA
MNHN – Museum Nationale d'Histoire Naturelle, Paris, France
NHMUK – Natural History Museum, London, UK
NMB – Naturhistorisches Museum Basel, Basel, Switzerland
NMW – National Museum of Wales, Cardiff, UK
QM – Queensland Museum, Brisbane, Australia
RMNH – Naturalis Biodiversity Centre, Leiden, Netherlands
SMF – Senckenberg Museum, Frankfurt, Germany
UF & FLMNH – Florida Museum of Natural History, University of Florida, Gainesville, USA
UGI – University of Guam, Guam, USA
UMMZ – University of Michigan Museum of Zoology, Ann Arbor, USA
UMZC – University Museum of Zoology, Cambridge, UK
USNM – National Museum of Natural History, Smithsonian Institution (formerly NMNH), Washington, USA
UZH – Zoological Museum, Universität Zürich, Zurich, Switzerland
YPM – Yale University Peabody Museum of Natural History, New Haven, USA
ZMB – Zoologisches Museum Berlin, Germany
ZMH – Zoological Museum Hamburg, Germany
ZMUC – Zoological Museum, University of Copenhagen, Denmark
ZSL – Zoological Society of London, London, UK
ZSM – Zoologische Staatssammlung Munchen, Munich, Germany

Fig. 4.5 Colour patterns from Crampton (1916, 1932), excluding intermediate shades mitella-rubra and phaea-purpuraea. Standard patterns used in the present work are shown with representative photographs on the right

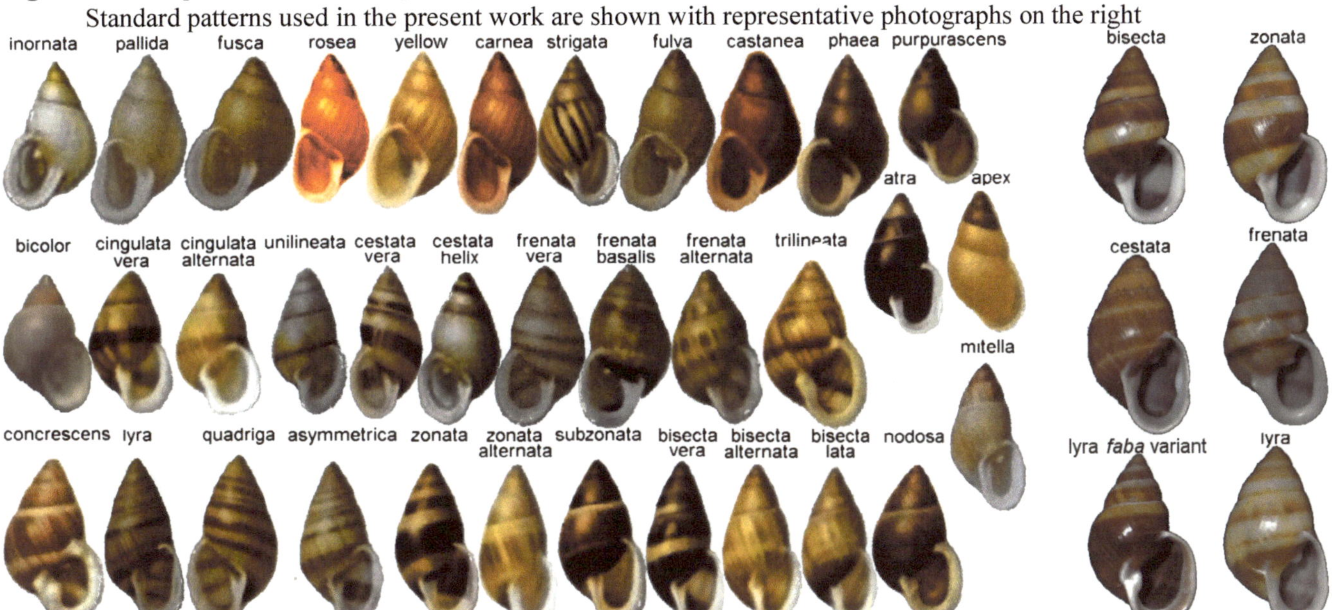

Fig. 4.6 Conchological features. a-d – dextral shell forms: a – globose-ovate; b – bulimoid; c – ovately conical; d – elongately conical; e – embryonic whorls; f-i – ornamentation: f – smooth; g – strong wavy spiral striae; h – granular; i – lirate; j-l – suture: j – not impressed; k – impressed; l – adpressed; m-p – lip: m – not expanded; n – slightly expanded; o – broadly expanded; p – carinate; q-u – apertural features: q – palatal (labial) denticle, impressed columella and partietal callus translucent; r – straight, dilated columella and opaque parietal callus; s – angled columella with parietal denticle; t – nodular columella; u – columella plicate; v – carinate lip, with ridge marked in red. w-z – umbilicus: w – open umbilicus; x – perforate umbilicus; y – rimate umbilicus; z – closed umbilicus

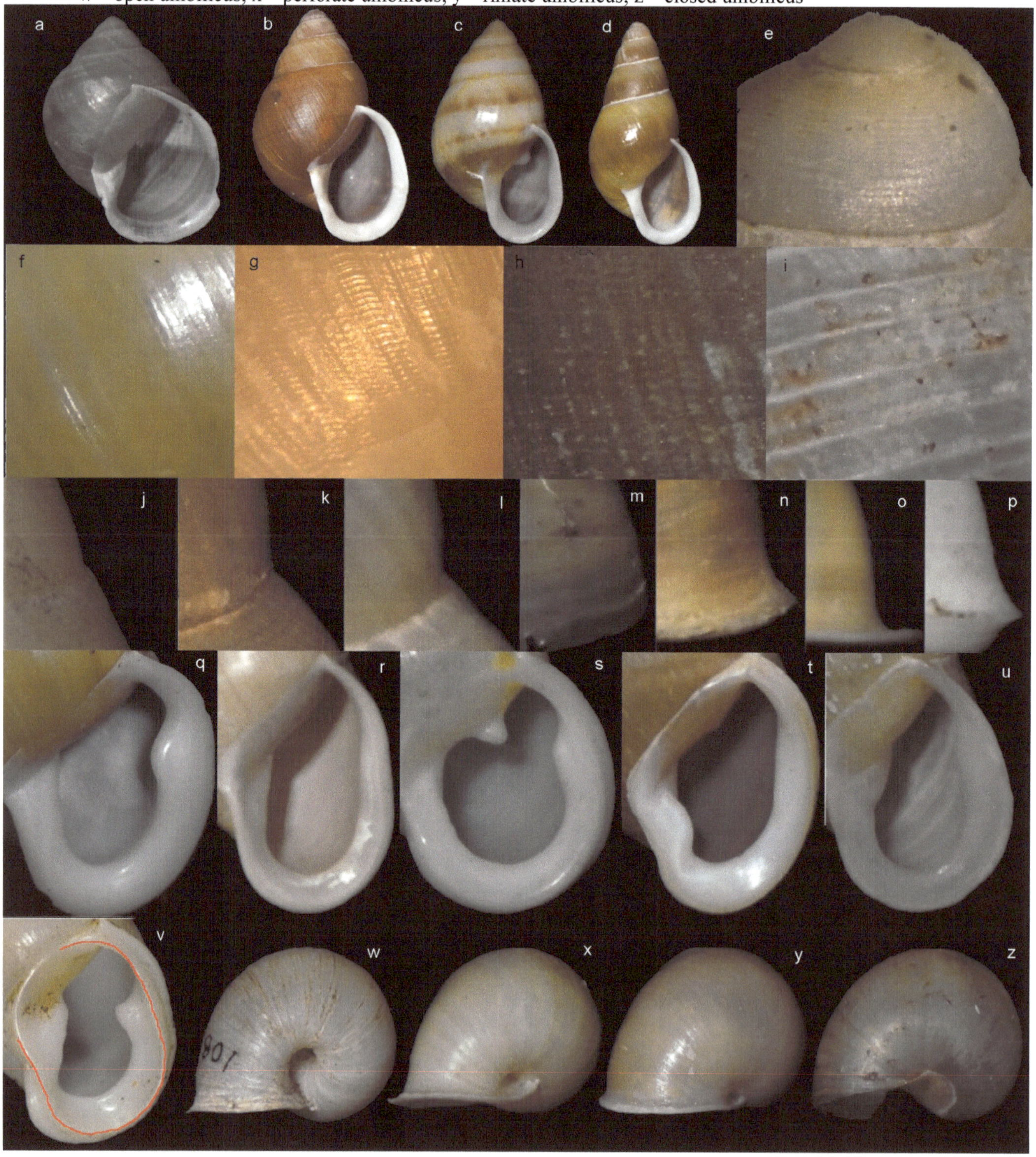

5. ***Eua*** Pilsbry & Cooke, 1934

Eua Pilsbry & Cooke, 1934: 4
Eua Pilsbry & Cooke, 1934: 18 (subgenus)
Nesanassa Pilsbry & Cooke, 1934: 18 (subgenus)
Eua Pilsbry & Cooke, Kondo 1968: 75, 1973: 19. Schileyko 1999: 263. Richardson 1990: 3
Eua (*Eua* s. str.) Schileyko 1999: 263
Eua (*Nesanassa* Pilsbry & Cooke, 1934) Schileyko 1999: 264

Dextral. Shell globular to oblong-conic. 4½ whorls. Umbilicus open, narrow to wide. Sculptured with spiral striae. Penis short, stout, containing a prominent pilaster which folds over a smaller pilaster, both connected by a lateral transverse ridge. Vas deferens terminal, entering penis alongside or within, retractor muscle. Penial caecum minute. Egg capsules vary in calcification, being fully calcified except in *E. globosa* and in some *E. montana*.
Radula: number of teeth variable, laterals 7-16, usually at least 10.
Type species: *Eua globosa* Pilsbry & Cooke, 1934
Distribution: restricted to Tonga and Samoa. 4 species recognised, with an additional undescribed species. Each of the larger Samoan islands supports two species. Kondo (MS) thought that all Samoan species could be synonymised, but this was based purely on external appearance.

Fig. 5.1 Range of all *Eua* species

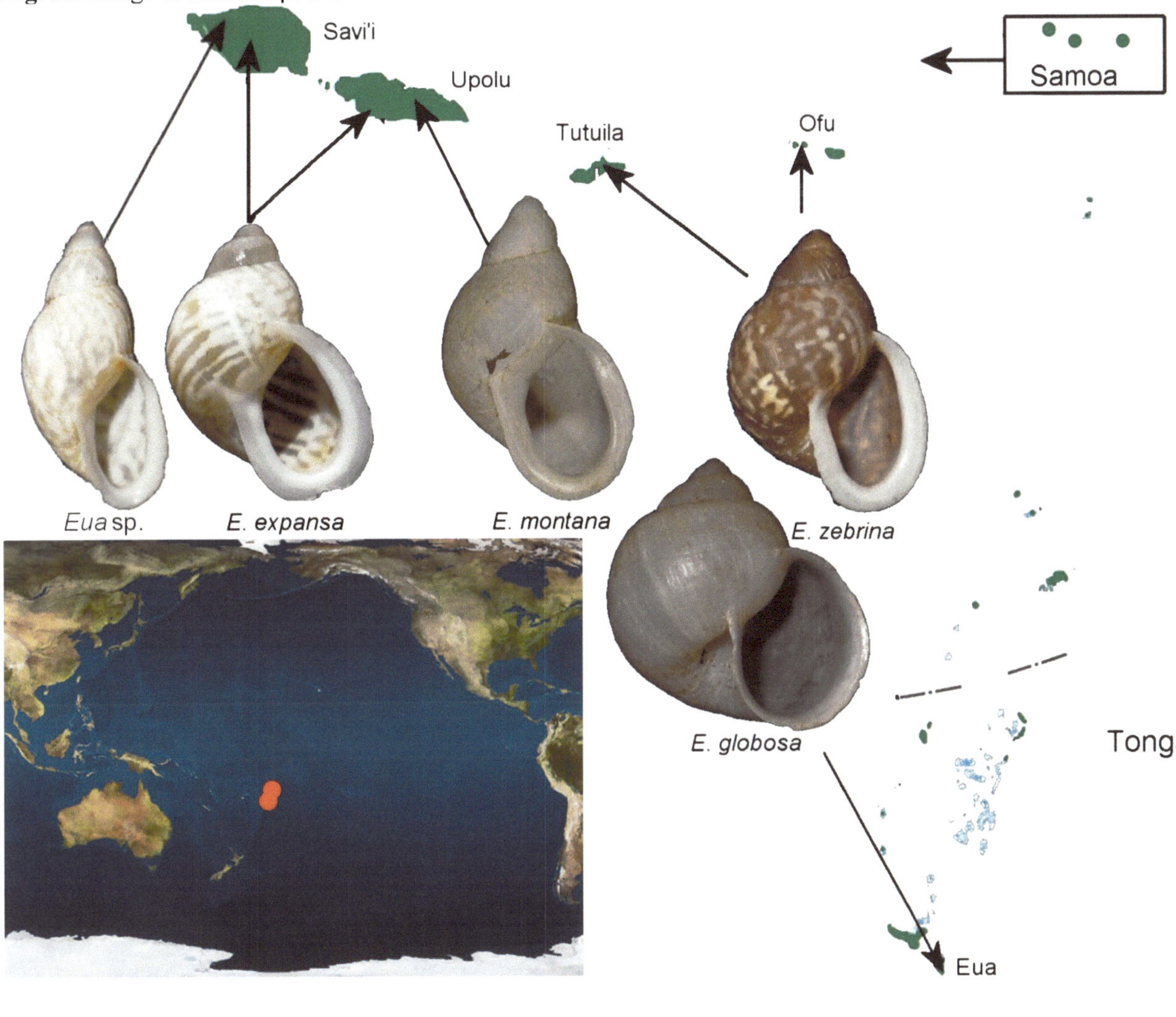

Tonga islands

A single *Eua* species has been recorded from one island where it is sympatric with one *Samoana* species.

Eua globosa Pilsbry & Cooke, 1934

Eua globosa Pilsbry & Cooke, 1934: 4

Eua globosa Pilsbry & Cooke, Zilch 1959: fig. 457. Kondo 1955: 64, 1968: 75. Richardson 1990: 3

Dextral. Globose-conic, broadly umbilicate. Thin, surface dull, with fine spiral striae on spire, fading on body whorl. Some

Table 5.1 *Eua* shell measurements

	N	L			W			L aperture			W/L		
		range	mean	sd	range	mean	sd	range	mean	sd	range	mean	sd
E. globosa	25	20.1-23.7	21.22	0.558	17.1-20.0	18.98	0.962	13.0-15.1	13.31	0.346	80.6-91.7	89.57	2.637
E. expansa	27	16.8-19.0	17.37	0.675	12.7-21.5	14.79	0.598	11.0-13.5	11.46	0.411	72.7-92.3	75.39	2.566
E. sp.	2	19.6-19.8	19.70	-	12.0-12.3	12.15	-	12.1-12.2	12.05	-	61.0-62.2	61.60	-
E. zebrina	7077	14.7-23.8	18.90	1.171	9.3-14.9	12.10	0.698	9.1-14.9	11.78	0.856	54.5-76.5	63.98	2.728
E. montana	17	16.5-20.4	18.06	0.975	11.5-13.2	12.27	0.559	7.2-11.9	9.02	1.572	59.3-78.2	69.43	4.276

Fig. 5.2 *Eua* reproductive anatomy, inset: interior of penis, showing penial pilaster and opening to penial caecum; and interior of caecum (after Kondo 1955 except *E. montana* – Kondo MS, *E.* sp. – BPBM 108175). Scale bar 5 mm

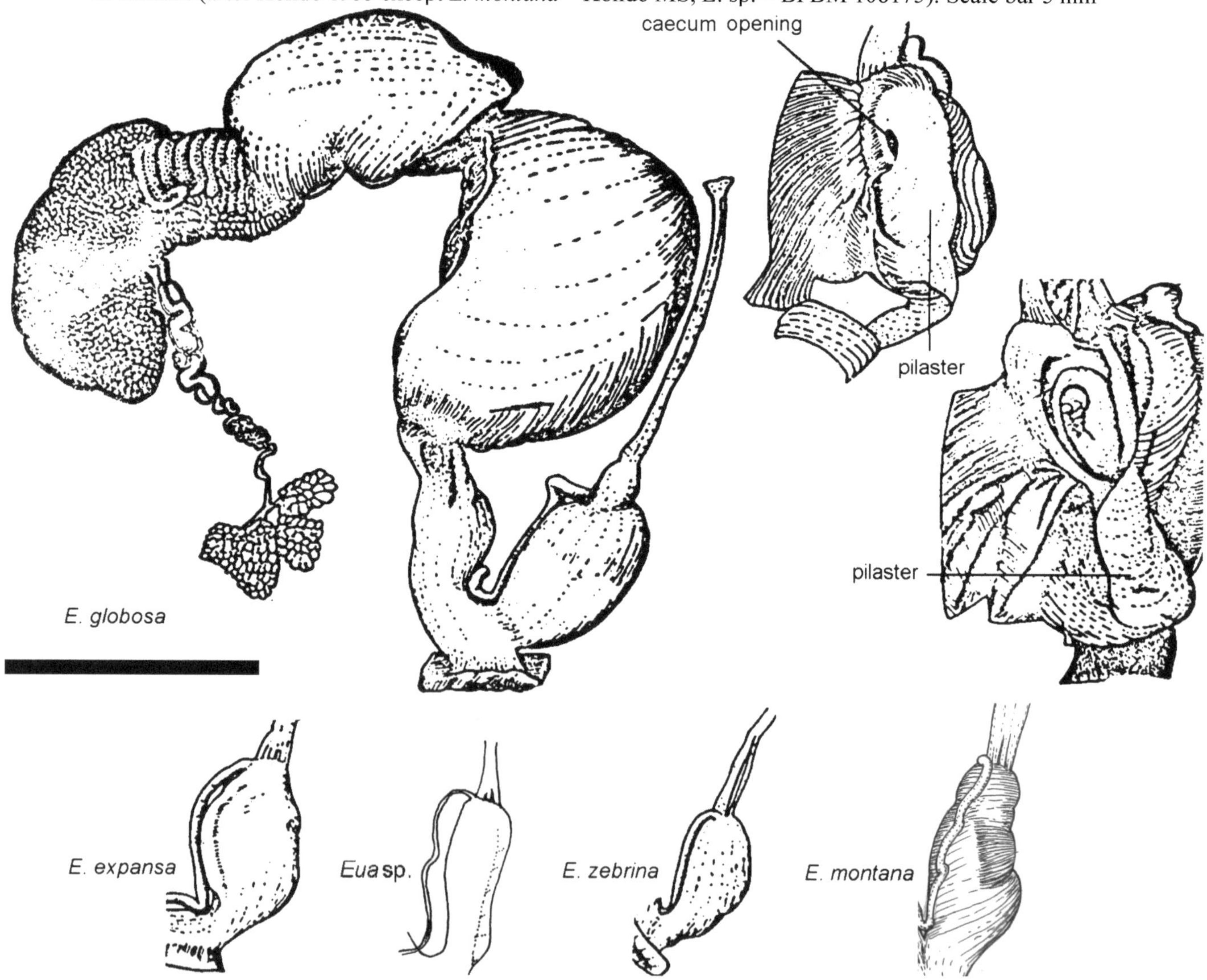

striae may remain visible on upper half of body whorl only. 4½ very convex whorls. Apex obtuse. Body whorl broad, globose. Aperture broad, ovate, edentulous. Lip moderately expanded, only slightly thickened within. Columella vertical, margin projecting, inner edge receding. Parietal callus thin, shining.
Uniformly creamy white to pale buff.
Body grey, mantle mottled with black (Pilsbry & Cooke 1934).
Anatomy: Penis $^{1}/_{5}$ oviduct length, short but broad; subglobose. Retractor expanded to contain vas deferens at apex of penis. Interior of penis highly complex with a large pilaster adnate to penial sheath on one side, with a series of vertical ridges to the right side. Main pilaster with a basal groove between the left face and penis wall, with an apical cavity. Penial caecum rudimentary, opening elliptical, near top of pilaster, ornamented with concentric ridges around a central prominence. Eggs uncalcified.
Type specimens: Holotype and paratype ANSP 162130; paratypes ANSP 295203-4 and BPBM 115426.
Distribution: 'Eua island. Recorded at 100 m at Bugai.

Samoa islands

Four *Eua* species have been recorded on 7 islands in Samoa and American Samoa.

Eua expansa (Pease, 1872)

Partula zebrina Gould, Mousson 1865: 173 (misidentification), 1869: 339 (misidentification). Garrett 1887: 133. Pilsbry 1909: 270
Partula extensa Pease, 1871b: 473 (nomen nudum)
Partula expansa Pease, 1872: 26
Partula expansa Pease, Pfeiffer 1877: 203. Hartman 1881a: 182, 1881b: 13. Garrett 1887: 133. Paetel 1889: 266. Smith 1903: 438. Pilsbry 1909: 269. Thiele 1928: 124. Kondo 1968: 75. Richardson 1990: 3
Partula extensa Pease, Pfeiffer 1877: 204 (misspelling)
Eua (*Nessaassa*) *expansa* (Pease), Pilsbry & Cooke 1934: 19. Cowie 1998: 54

Dextral. Obliquely conically ovate, rather thin. Dull. Broadly and deeply umbilicate. Spire very short, conic; 4-4½ convex whorls. Spiral striae restricted to upper part of spire. Body whorl inflated, base very convex. Aperture elongately ovate, oblique. Lip broadly expanded, flat, thickened within.
Off-white with grey and white spirally diagonal bands and spots.
Anatomy: As *E. globosa*, penis slightly longer (¼ oviduct length), base narrow, pilaster smaller, ridges more developed and fusing with pilaster. Caecum larger, opening into penis with an elongate aperture, partially covered by a membrane. Interior

Fig. 5.3 *Eua globosa* type ANSP 295203 (photo – F. Borrero)

Fig. 5.4 *Eua expansa*: a – holotype ANSP 59841; b – specimen NHMUK 92.6.9.2-3; c – specimen NHMUK Savai'i 1995

with complex ridges, some spiral. Eggs calcified.
Type specimens: Holotype ANSP 59841 (erroneously listed as 59453 by Baker [1963]).
Distribution: Samoa – Savai'i and 'Upolu islands. Widely recorded above 50 m on both islands. Arboreal on shrubs and on ground. Originally (but erroneously) cited from Tutuila, which is occupied by *E. zebrina.*

***Eua* sp.**

Eua sp. 1 Brook 2012: 149

Dextral. Elongately conical. Umbilicus open, narrow. Thin but strong. 4½ convex whorls. Protoconch flattened. Spiral striae on spire, fading on body whorl. Suture slightly impressed. Aperture elongately ovate, vertical. Lip expanded, flattened, inwardly thickened. Columella straight, receding.
Off-white with grey and white spirally diagonal bands and spots.
Body of only preserved adult specimen yellow, with contrasting brown mantle.
Anatomy: As *E. zebrina*; penis triangular in section. Pilaster well developed, with reduced accessory ridges.
Distribution: Samoa – Savai'i island. Recorded from Matavanu at 450 m (1 live adult, 1 shell: E. Christensen and E. Hume 1934 – BPBM 108175-6) and above 950 m (2 adult shells, 2 live juveniles: Brook 2012).

Eua zebrina (Gould, 1848)

Partula zebrina Gould, 1848: 196
Partulus actor Albers, 1850: 187
Partula Recluziana Petit de la Saussaye, 1850: 170
Partula zebrina Gould, Gould 1852: 82, 1862: 33. Pfeiffer 1853a: 450, 1859: 513, 1877: 206. Adams 1855: 145. Mousson 1869: 339. Paetel & Schaufuss 1869: 83. Schmeltz 1869: 71. Pease 1871b: 474. Paetel 1873: 105, 1885: 220, 1889: 268. Schmeltz 1874: 91. Hartman 1881a: 189, 1881b: 12. Pfeiffer & Clessin 1881: 303. Garrett 1887: 133. Smith 1903: 482. Pilsbry 1909: 268
Partula actor Albers, Pfeiffer 1853a: 450, 1857: 266. Adams 1855: 145. Pfeiffer 1856a: 162, 1856b: 266, 1859: 513, 1868: 160, 1877: 206. Hartman 1881a: 179, 1881b: 12, 1885: 220, 1892: 98. Pfeiffer & Clessin 1881: 303. Garrett 1887: 133. Paetel 1889: 266. Smith 1903: 459. Pilsbry 1909: 268
Partula Recluziana Petit, Pfeiffer 1853a: 452, 1856: 162. Adams 1855: 145. Mousson 1859: 339. Schmeltz 1869: 71, 1874: 91. Pease 1871b: 474. Paetel 1873: 104, 1883: 153. Semper 1874: 158. Kobelt 1880: 18
Partula Tryoni Hartman, 1885: 204
Partula tryoni Hartman, Paetel 1889: 204. Smith 1903: 460. Pilsbry 1909: 268. Clench 1941: 21
Partula zebrina var. *recluziana* Petit, Pilsbry 1909: 269
Eua (*Nessanassa*) *zebrina* (Gould), Pilsbry & Cooke 1934: 18. Cowie 1998: 54
Eua zebrina (Gould), Zilch 1959: 458.Kondo 1955: 70, 1968: 75. Richardson 1990: 3
Eua (*Nessanassa*) *zebrina recluziana* Petit de la Saussaye, Cowie 1998: 54

Dextral. Oblong-conic. Umbilicus open, rimate. Moderately thin, only very slightly shiny. Apex obtuse. 4½-4¾ convex whorls with close, fine spiral striae on all whorls. Second whorl disproportionately large. Body whorl inflated. Aperture subvertical. Lip flatly reflexed, thickened within with a strong rib ascending the columellar margin, often forming a vertical ridge near the dilated upper end.
Corneous to brown with off-white spots, flecks or stripes, or off-white with corneous markings, rarely uniform. Lip white.

Fig. 5.5 *Eua* sp. (BPBM 108175)

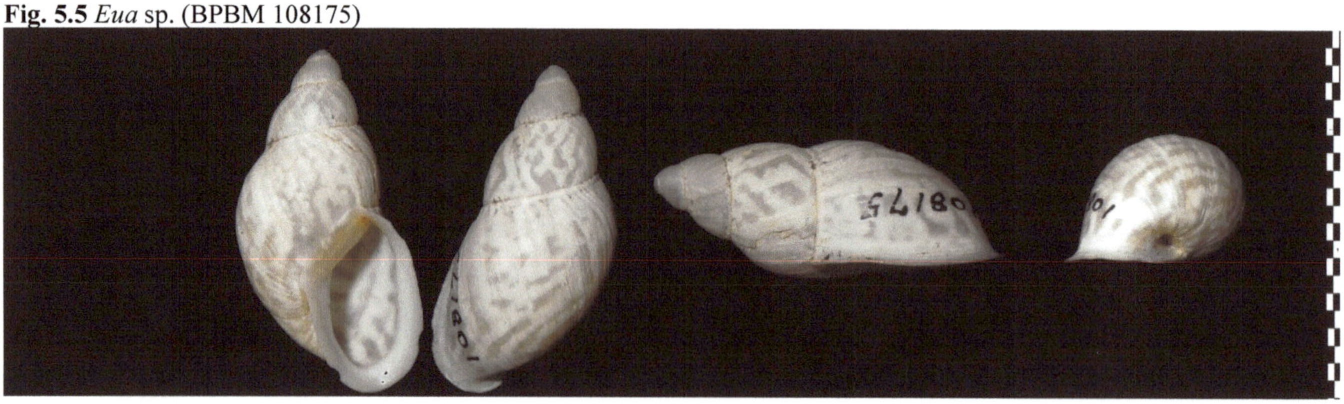

Anatomy: As *E. expansa* except penial retractor muscle split into two unequal branches. An elongate form from Tutuila (BPBM 84467) was identical to the typical form except in having a longitudinal depression on one side of the penis, corresponding to the edge of the pilaster, this is probably individual variation. Eggs calcified.
Type specimens:
Partula zebrina Gould, 1848 – Syntypes: 'holotype' USNM 5491; 'paratype' MCZ 156376, 169428.
Partulus actor Albers, 1850 – Type not located; NHMUK Cuming specimens used in Pfeiffer's (1853a) redescription.
Partula Recluziana Petit de la Saussaye, 1850 – Type not located; more ovate than typical *zebrina.*
Partula tryoni Hartman, 1885 – Holotype CMNH 62.4261.
Distribution: American Samoa – Tutuila, Nu'usetoga (islet off Tutuila) and Ofu (Cowie & Cook 1999). Erroneously listed from 'Upolu by Schmletz (1869). On vegetation and on ground.

Fig. 5.6 *Eua zebrina*: a – *zebrina* syntype USNM 5491; b – *tryoni* holotype CMNH 62.4261; c – *recluziana* type (from Petit 1850); d – two '*recluziana*' UMZC 75324; e – *actor* from Albers collection (from Pfeiffer 1857); f – '*actor*' NHMUK; g – colour variation in a single population (ANSP 161267) (photo – B. Murphy; c – C. Sturm)

Eua montana (Cooke & Crampton, 1930)

Partula montana Cooke & Crampton, 1930: 7

Eua (*Nessanassa*) *montana* (Cooke & Crampton), Pilsbry & Cooke 1934: 19. Cowie 1998: 54

Eua montana (Cooke & Crampton), Kondo 1955: 71, 1968: 75. Richardson 1990: 3

Dextral. Ovately conical. Umbilicus open. Thin. $4^1/_3$ very convex whorls. Protoconch strongly flattened. Spiral striae on spire, fading on body whorl but still visible near base. Suture slightly impressed. Spire narrowly convex. Aperture elliptical, slightly oblique. Lip thin, flattened, narrowed at outer and inner insertions, inwardly thickened. Columella straight.

Whitish with traces of pale brown epidermis. Dull whitish with tawny epidermis but extensively decorticated.

Anatomy: As *E. expansa*, penis shape may be elongate (Kondo MS) or more globose (FMNH 152833). The internal ridges may be reduced (Kondo MS) or well developed (BPBM 1967-144). Egg capsule calcified (Cooke & Crampton 1930) or uncalcified (FMNH 152833).

Type specimens: Holotype ANSP 292306.

Distribution: Samoa – 'Upolu. Always a localised, high-altitude species, recorded from the Afiamalu region, on the mountains around Tiapapala Pass.

Note: A junior primary homonym of *montana* Möllendorff, 1900, an unnecessary replacement name for *guamensis* Pfeiffer, 1846.

Fig. 5.7 *Eua montana* a – holotype ANSP 292306; b – specimen FMNH 152833 (photo: a – F. Borrero)

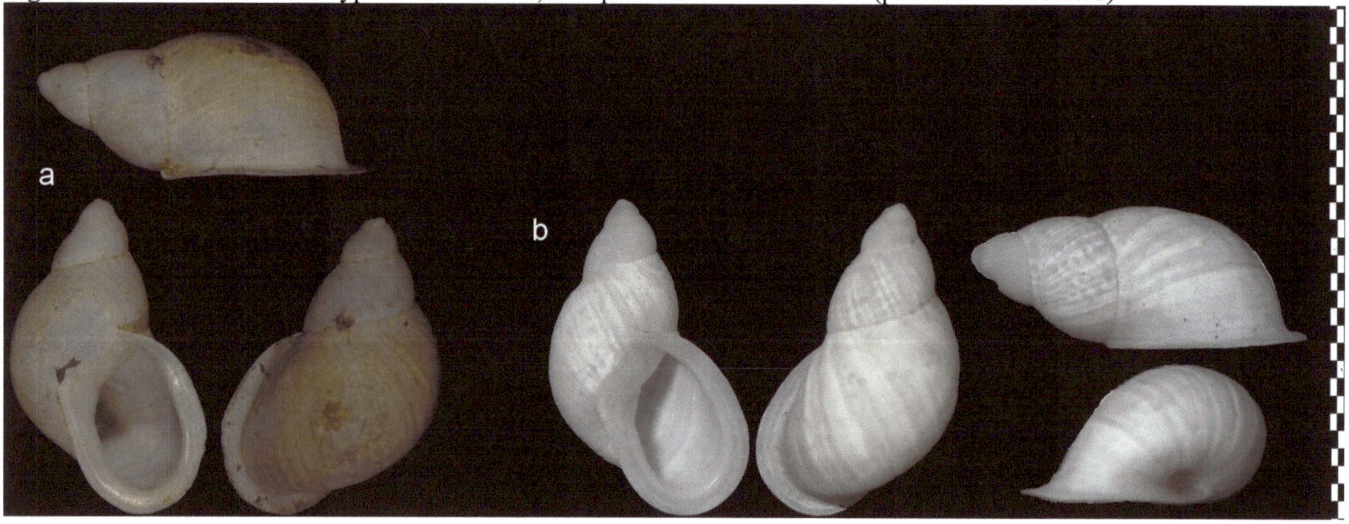

Fig. 5.8 *Eua zebrina* live animals showing colour vaation (photo: R. Cowie)

6. ***Samoana*** Pilsbry, 1909

Æega Hartman, 1881b: 11 (*Partula* subgenus) [preoccupied]
Evadne Hartman, 1881b: 12 (*Partula* subgenus) [preoccupied]
Latia Hartman, 1881b: 12 (*Partula* subgenus) [preoccupied]
Samoana Pilsbry, 1909: 165 (*Partula* subgenus)
Marquesana Pilsbry, 1909: 165 (*Partula* subgenus)
Samoana Pilsbry, Schileyko 1999: 266. Richardson 1990: 68
Samoana (*Samoana*) Pilsbry, Schileyko 1999: 266
Samoana (*Marquesana*) Pilsbry, Schileyko 1999: 267

Dextral or sinistral. Umbilicus open, often rimate. Sculpture of spiral striae, fading on body whorl. Lip often broadly expanded, simple. Penis short, moderately stout, containing a large pilaster folding over a smaller pilaster and overhanging a cavity leading into the caecum. Penial caecum usually well developed but slender, containing papillae and ridges. Penial retractor attaches to caecum, usually with a vestigial retractor running on to the apex of the main part of the penis. Vas deferens enters penial caecum subapically. Eggs fully calcified.
Radula: few laterals; 5-8, usually 6.
Type species: *Partula canalis* Mousson, 1865 (junior synonym of *Samoana conica* (Gould, 1848))
Distribution: Marianas, Fiji, Tonga, Samoa, Australs, Marquesas and Societies. Taxa clearly distinct except in the Marquesas. All *Samoana* are arboreal.

Table 6.1 *Samoana* shell measurements

	N	L			W			L aperture			W/L		
		range	Mean	sd	range	mean	sd	range	mean	sd	range	mean	sd
fragilis dextral	48	13.0-19.3	15.14	0.761	8.5-14.2	10.27	0.606	7.3-11.6	8.74	0.505	63.5-73.6	67.80	2.397
fragilis sinistral	1	-	12.00	-	-	9.80	-	-	7.90	-	-	82.00	-
alabastrina	20	18.9-28.8	21.81	0.751	11.5-13.9	12.87	0.587	10.3-13.5	12.39	0.507	54.5-63.5	59.01	2.140
cramptoni	9	18.5-19.0	18.68	0.262	10.5-11.8	11.28	0.371	8.6-9.5	9.08	0.113	57.9-63.0	60.17	2.108
conica	891	19.8-29.0	23.07	1.192	12.3-18.0	14.59	0.693	11.9-15.9	13.85	0.690	48.3-71.5	63.27	2.501
abbreviata	350	16.8-22.9	19.63	0.967	11.3-16.7	13.68	0.927	10.5-14.5	12.53	0.796	63.5-78.9	69.64	2.375
stevensoniana	29	21.0-24.7	22.94	0.646	11.1-14.9	13.54	0.614	9.6-13.9	11.93	1.004	53.4-67.1	59.27	2.053
thurstoni	4	17.8-21.7	19.00	1.824	11.9-13.1	12.95	1.066	8.8-11.7	10.30	1.453	62.7-79.2	68.50	7.454
dryas	83	13.3-16.6	14.95	0.258	8.2-10.7	9.61	0.201	8.2-10.3	9.11	0.193	56.9-79.5	65.86	0.455
margaritae	245	19.0-25.9	22.32	0.276	10.3-14.5	12.61	0.210	9.7-14.1	8.22	0.266	51.5-61.5	56.28	0.266
gonochila	369	15.5-21.0	17.84	1.068	8.4-13.5	10.92	0.696	7.4-13.0	10.38	0.807	50.0-68.8	58.32	2.241
bellula	225	12.7-23.6	15.34	1.470	6.6-13.5	9.24	0.818	6.2-13.4	8.68	0.810	50.0-65.5	60.28	2.222
decussatula	66	10.4-17.5	14.58	0.833	6.2-12.6	9.86	0.645	8.2-11.4	9.90	0.570	54.6-77.6	69.66	4.063
pilsbryi	93	17.8-20.0	18.79	0.617	10.8-12.2	11.47	0.435	10.4-12.0	11.21	0.451	57.1-63.4	60.70	2.182
medana	192	10.0-15.0	12.40	0.902	5.7-10.0	7.99	0.464	5.9-9.4	7.81	0.710	66.7-72.7	69.20	2.723
inflata	117	17.9-23.4	21.05	0.762	10.6-17.4	15.15	1.001	9.8-14.2	12.91	0.659	67.4-79.5	71.96	2.893
ganymedes	275	17.8-24.6	21.38	0.978	8.8-15.0	13.38	0.947	7.9-14.0	11.91	0.782	44.3-71.0	62.74	3.065
minor	70	10.5-15.0	13.43	0.672	7.0-11.5	10.63	0.592	7.0-10.0	9.52	0.491	66.7-82.1	79.24	1.960
attenuata	338	13.4-19.0	15.61	0.556	6.3-9.2	8.25	0.328	7.3-9.8	8.73	0.333	43.0-59.4	52.757	1.717
annectens	58	13.1-14.4	13.92	0.390	7.9-8.2	8.08	0.288	8.0-8.4	8.25	0.245	56.3-60.3	57.91	1.300
burchi	19	16.0-19.1	17.59	0.756	7.1-10.6	9.05	0.734	9.1-10.6	9.52	0.384	41.8-56.1	51.50	4.104
diaphana	39	12.4-19.1	13.71	0.752	6.8-9.7	7.74	0.505	6.5-8.3	7.29	0.731	41.8-65.6	57.791	4.278

Fig. 6.1 Reproductive anatomy of *Samoana* (after Kondo 1955) based on *S. cramptoni*, with inset details of the penial caecum and the pilasters within the penis. Scale bar 5 mm

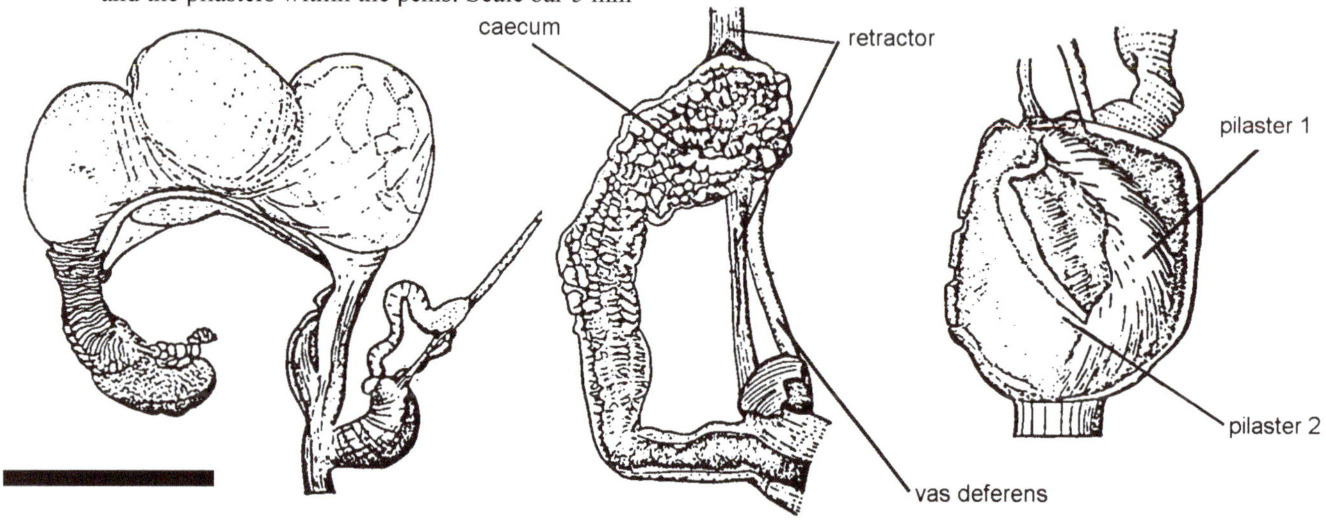

Fig. 6.2 *Samoana* range

Marianas islands

Samoana fragilis (Férussac, 1821)

Partula fragilis Férussac, 1821: 66

Partula fragilis Férussac, Quoy & Gaimard 1824: 486. Pfeiffer 1842: 114, 1848: 74, 1853a: 452, 1859: 510, 1868: 159, 1877: 199. Adams 1855: 145. Pease 1871b: 473. Paetel 1889: 267. Smith 1903: 478. Pilsbry 1909: 318. Crampton 1925: 30

Partula rubens 'Müller', Anton 1839: 40 (as syn. of *fragilis*, nomen nudum)

Patula quadrasi Möllendorf, 1894: 14

Partula quadrasi Möllendorf, Pilsbry 1909: 318. Crampton 1925: 30

Samoana fragilis (Férussac), Kondo 1955: 103, 1968: 75, 1970: 86. Bauman 1996: 21. Richardson 1990: 73

Dextral. Ovate conic. Thin and fragile, translucent. Umbilicus partially covered. 4 convex whorls separated by adpressed, marginated suture. Spiral striae present on all whorls. Body whorl slightly angled. Aperture oblique, oval. Lip thin, barely reflected. Columella dilated above, recurred, forming a distinct angle with parietal wall.

Shell colourless to faintly corneous, lip off-white.

Body white with narrow darker maculations.

Anatomy: Penis ½ oviduct length, elongate. Vas deferens terminal. Caecum small, with thin sinuous ridges in 6 vertical rows, faint lateral ribs near base. Penis with principal pilaster curved, secondary pilaster reduced to a thickening of the penial sheath.

Type specimens:

Partula fragilis Férussac, 1821 – MNHN, not located

Patula quadrasi Mollendorff, 1894 – Syntypes SMF 165564, BPBM-MO 86198 (labelled paratypes of Möllendorff, Quadras, and Boettger species), 87465 (labelled 'paratype material of Möllendorf, Wagner, etc., from Pacific Is. from Senckenbergische Museum, Frankfurt, Germany. Through Dr. F. Haas.').

Distribution: Guam and Rota, first found on Rota in 1959 despite collections from 1925-6 and 1952.

Fig. 6.3 *Samoana fragilis* penial anatomy (after Kondo 1955) and range

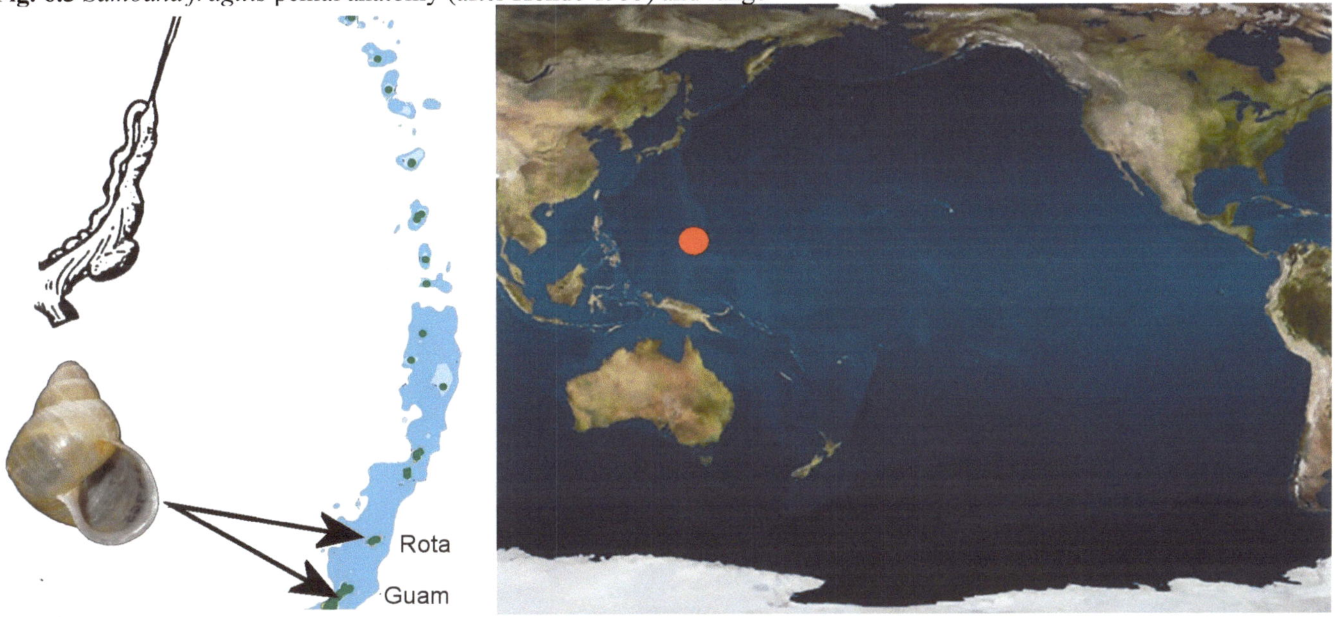

Fig. 6.4 *Samoana fragilis*: a – *quadrasi* syntype (BPBM 86198); b – sinistral specimen (BPBM 75352); c – Guam (FMNH 24965); d – Rota (UF 481329)

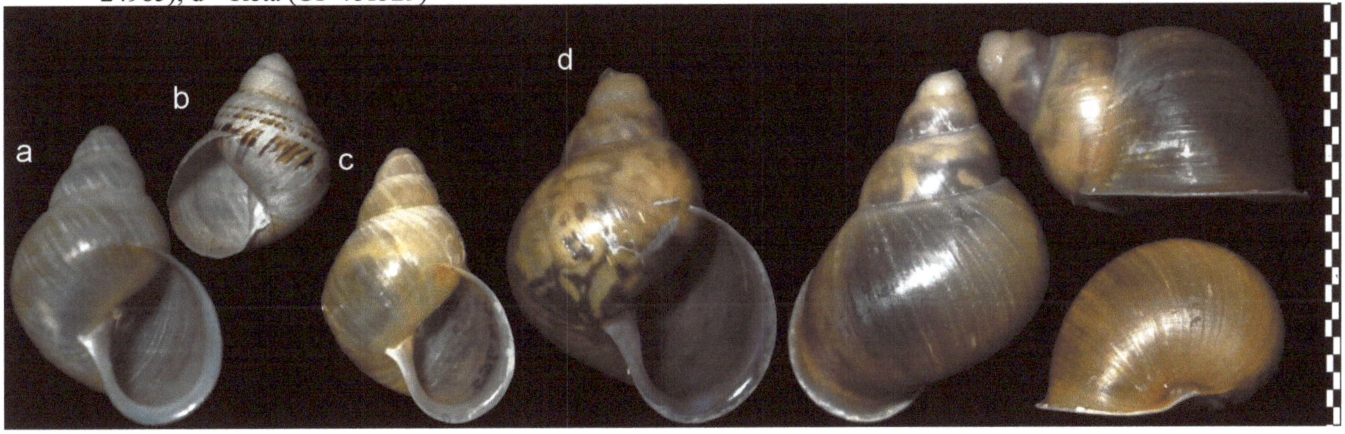

Fig. 6.5 Crampton's (1925) *Samoana fragilis* figures (edited for position)

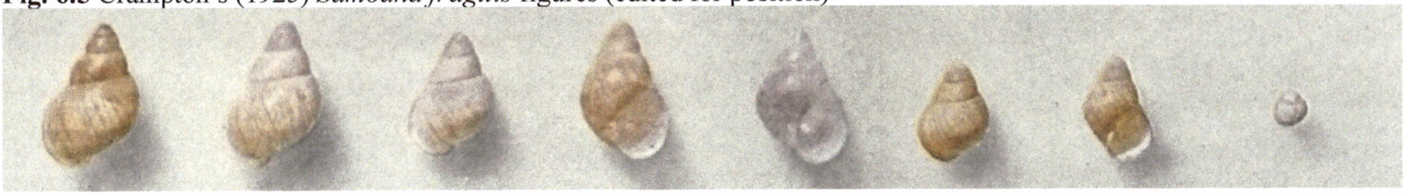

Fig. 6.6 Range of Samoan, Fijian and Tongan *Samoana* and details of penial anatomy (anatomy after Kondo 1955 except *S. stevensoniana* – FMNH 152740). Fiji islands: 1 – Moala, 2 – Totoya, 3 – Matuku. Scale bar for anatomy 5 mm

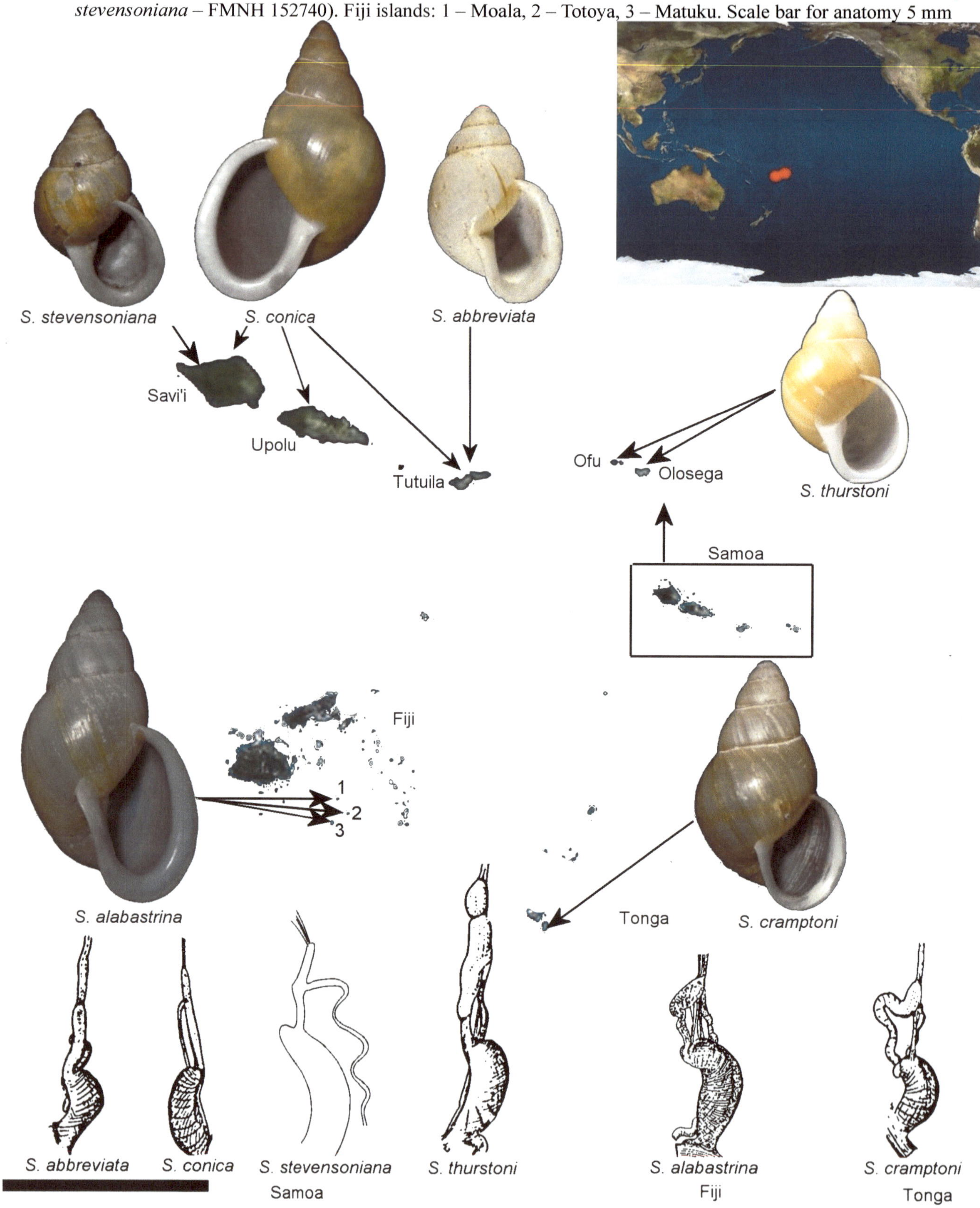

Fiji islands

A single *Samoana* species is found on two Fijian islands, it is not sympatric with any other partulid.

Samoana alabastrina (Pfeiffer, 1857)

Partula alabastrina Pfeiffer, 1857d: 390

Partula alabastrina Pfeiffer, Pfeiffer 1859a: 509, 1868: 157, 1877: 197. Paetel 1873: 104, 1883: 152, 1889: 266. Kobelt 1880: 18. Hartman 1881a: 179, 1881b: 14, 1885: 223, 1892: 99. Pfeiffer & Clessin 1881: 301. Smith 1903: 468. Pilsbry 1909: 294

Partula eburnea Hartman, 1886: 33

Partula eburnea Hartman, Hartman 1892: 99. Smith 1903: 467. Pilsbry 1909: 281. Solem 1959: 69

Partula nematoraphe Pilsbry, 1909: 279

Partula nematoraphe Pilsbry, Cooke & Crampton 1930: 5. Solem 1959: 74

Samoana alabastrina (Pfeiffer) Pilsbry & Cooke 1934: 17. Clench 1941: 20. Kondo 1955: 86, 1968: 75. Richardson 1990: 68

Dextral. Oblong-conic. Thin, but strong, slightly shiny. Umbilicus open, compressed. 5½ convex whorls, apex obtuse. Surface with faint spiral striae. Body whorl slightly longer than spire, tapering to base, slightly compressed, right margin broadly expanded. Aperture slightly oblique, oblong. Columella straightly receding, margin very broad. Lip margins united by a very thin callus.

White to yellowish-buff, with a yellowish green periostracum. Spire sometimes rose tinted. Lip white.

Anatomy: Penis ¼ oviduct length. Caecum short, narrow below, expanded above. Vas deferens apical. Vestigial retractor unusually well developed, divided into two or three parts. Caecum internally ornamented with semi-lunate tubercles on 9-10 ridges. Secondary penial pilaster greatly reduced.

Type specimens:

Partula alabastrina Pfeiffer, 1857 – Syntypes NHMUK un-numbered. The largest is significantly larger than Pfeiffer's measurements (28 mm long, compared to 23 mm) but they are Cuming collection specimens and there is no reason to think they are not the types. It is possible that Pfeiffer's measurement is a typographical error. There are also two shells not in the type collection labelled Matuku, Fiji, August 1855 which may also have been used by Pfeiffer.

Partula eburnea Hartman, 1886 – Syntypes AM un-numbered; NHMUK 96.11.30-1-2.

Partula nematoraphe Pilsbry, 1909 – Syntypes CMNH 62.4293.

Distribution: Moala islands: Moala, Matuku and Totoya. Type specimens (Cuming collection) supposedly from 'Salomon's Islands' (Pfeiffer 1857d).

Fig. 6.7 *Samoana alabastrina* type specimens: a – *eburnea* NHMUK; b – *nematoraphe* CMNH 624293; c – *alabastrina* NHMUK (photo: b – C. Sturm)

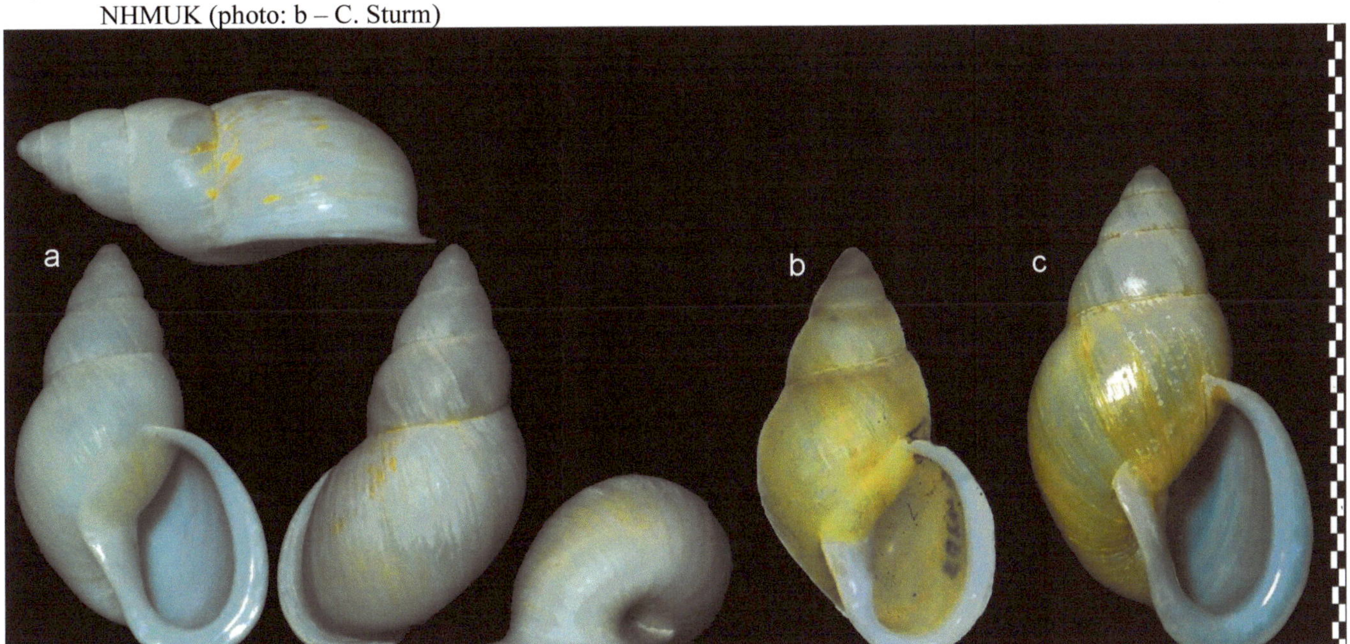

Tonga islands

One *Samoana* species is found on 'Eua island where it is sympatric with *Eua globosa.*

Samoana cramptoni Pilsbry & Cooke, 1934

Samoana cramptoni Pilsbry & Cooke, 1934: 8

Samoana cramptoni Pilsbry & Cooke, Kondo 1955: 75, 1968: 75. Richardson 1990: 72

Dextral. Oblong-conic, narrowly umbilicate and rimate. Shell thin, translucent, glossy with fine spiral striae on upper part of spire, fading on body whorl where only a few visible around umbilicus. 5 convex whorls. Apex obtuse. Aperture nearly vertical, shortly oval. Lip narrowly expanded, dilated at columella which is slightly convex in oblique view.

Greenish yellow, a faint chestnut band sometimes present on body whorl. Aperture slightly bluish. Lip faintly flesh-tinted.

Anatomy: Penis ½ oviduct length, short, stout, bent inward. Caecum tubular, longer than penis, large and wrinkled, vas deferens entering caecum subapically. Penis with one large internal pilaster partially obscuring a smaller pilaster, both connected by small muscular spurs, joined at base and apex. At apex opens into caecum by a small orifice. Caecum ornamented with small sinuous ridges, thickening apically and gradually becoming papillate.

Type specimens: Syntypes ANSP 162332 (and A1763B), 295205; BPBM 162332.

Distribution: 'Eua island, at Fuai (200 m).

Samoa islands

Four species have been recorded on six islands in Samoa and American Samoa.

Samoana conica (Gould, 1848)

Partula conica Gould, 1848: 196

Partula conica Gould, Gould 1852: 81, 1862: 33. Pfeiffer 1853a: 445, 1859a: 507, 1859b: 120, 1868: 155, 1877: 194. Adams 1855: 145. Gould 1862: 33. Mousson 1865: 171, 1869: 338. Paetel & Schaufuss 1869: 83. Schmeltz 1869: 71, 1874: 91. Pease 1871b: 473. Paetel 1873: 104, 18893: 152, 1889: 266. Pfeiffer & Clessin 1881: 300. Smith 1903: 470. Pilsbry 1909: 265

Partula conica Gould (*upolensis*) Schmeltz, 1865: 25 (nomen nudum), 1874: 91. Paetel 1873: 104, 1883: 153, 1889: 268

Partula canalis Mousson, 1865: 172

Partula canalis Mousson, Pfeiffer 1868: 155. Mousson 1869: 337. Paetel & Schaufuss 1869: 83. Schmeltz 1869: 71, 1874: 91. Pease 1871b: 473. Paetel 1873: 104, 1883: 152, 1889: 266. Pfeiffer 1877: 194. Pfeiffer & Clessin 1881: 300. Garrett 1887: 134. Smith 1903: 470. Pilsbry 1909: 263. Thiele 1931: 659. Zilch 1959: 456

Partula canalis var. *semilineata* Mousson, 1869: 338

Partula canalis var. *semilineata* Mousson, Schmeltz 1869: 17, 1874: 91. Pease 1871b: 473, Semper 1874: 158. Paetel 1883: 153. Smith 1903: 470. Pilsbry 1909: 264

Partula bulimoides Lesson, Hartman 1881a: 180 (non Lesson, nomen nudum). Garrett 1887: 134. Pilsbry 1909: 264

Evadne bulimoides Lesson, Hartman 1881b: 12 (non Lesson; nomen nudum)

Partula canalis var. *biconica* Pilsbry, 1909: 264

Fig. 6.8 *Samoana cramptoni* holotype ANSP 162332

Fig. 6.9 *Samoana conica*: a – *semilineata* syntype UMZ 518634; b – *canalis* syntype UMZ 518630; c – *biconica* holotype ANSP 59844; d – *canalis* ANSP 59838 (Schmelz specimen); e – *conica* syntypes USNM 5490 and 5490a; f – dark variety, Tutuila (BPBM 1961-98); g – abnormal aperture, 'Upolu (BPBM 1967-128) (photo: e – B. Murphy)

Samoana conica (Gould), Pilsbry & Cooke 1934: 16. Kondo 1955: 84, 1968: 75. Richardson 1990:70. Cowie 1998: 55

Samoana canalis (Mousson) Pilsbry & Cooke 1934: 17. Kondo 1955: 83, 1968: 75. Solem 1973: 168. Cowie 1998: 55

Samoana canalis biconica Pilsbry, Cowie 1998: 55

Samoana canalis semilineata Mousson, Cowie 1998: 55

Sinistral. Elongately conical. Umbilicus open, narrow. Thin, slightly shiny. Surface with fine spiral striae, smooth on apex and penultimate whorl, recurring as wavy spirals on body whorl, strongest towards base. Spire conic; apex subacute. 5½-5¾ nearly straight-sided whorls, body whorl large, almost equal to spire, most convex at umbilicus (except var. *biconica*, widest below periphery). Aperture ovate-oblong, subvertical, expanded below. Lip flatly reflexed, moderately expanded, margins subparallel. Columella somewhat folded and impressed above (guttered). Parietal callus very thin, transparent.

Yellow to greenish yellow suffused chestnut, occasionally brown. Yellow shells are sometimes streaked, spire darker, often worn white. Suture margined with a white line; lip white, sometimes flesh tinted.

Var. *semilineata* is relatively small.

Anatomy: Penis $^1/_5$–$^1/_3$ oviduct length. Vestigial retractor conspicuous, connecting caecum below vas deferens, near midpoint, to apex of penis. Penial pilasters typical. Caecum long; base narrow, smooth or with semi-lunate ridges on basal third. Basal third broad, diaphanous, ornamented with low, lateral ridges, separated from middle third by large, oblong papillae. Medial third with vertical tubercular ridges. Apical third with sparse papillae.

Type specimens:

Partula conica Gould, 1848 – Syntypes ANSP 2687; SMF 157957; USNM 5490 (erroneously labelled 'holotype').

Partula canalis Mousson, 1865 – Syntypes SMF 157957; UZH 518630, 518633.

Partula canalis var. *semilineata* Mousson, 1869 – Syntypes UZH 518634-6.

Partula canalis var. *biconica* Pilsbry, 1909 – Holotype ANSP 59844a.

Distribution: Samoa and American Samoa – Savi'i, 'Upolu and Tutuila. *canalis* recorded from Savi'i and 'Upolu , *conica* and *semilineata* from Tutuila. Gould (1848) listed "Raraka" atoll of the Tuamotus, which Pilsbry (1909) noted must have been a mistake. Recorded from sea level to 350 m. On Tutuila recorded from the Tiatauala, Toa, Faiga, Vatia and Alava ridges (Cowie & Cook 1999, 2001).

Samoana abbreviata (Mousson, 1869)

Partula abbreviata Pease 1864: 675 (nomen nudum)

Partula abbreviata Mousson, 1869: 339

Partula abbreviata Mousson, Schmeltz 1869: 71, 1874: 91. Pease 1871b: 473. Paetel 1873: 104, 1883: 152, 1889: 266. Pfeiffer 1877: 200. Gloyne 1878: 338. Graeffe 1879: 85. Hartman 1881a: 179, 1881b: 13, 1885: 221, 1892: 98. Garrett 1887: 134. Smith 1903: 470. Pilsbry 1909: 267

Samoana abbreviata (Mousson) Pilsbry & Cooke 1934: 16. Kondo 1955: 80, 1968: 75. Richardson 1990: 68. Cowie 1998: 54

Dextral. Ovate. Thin but strong. Umbilicus open, rimate. Spire convexly-conic, obtuse. Suture slightly impressed. 5-5½ slightly convex whorls ornamented with dense minute way spiral striae. Body whorl $^2/_3$ of shell height, inflated, rounded, swollen at base. Aperture subvertical. Lip acute, widely and flatly reflexed. Parietal callus very thin, indistinct. Columella

Fig. 6.10 *Samoana abbreviata*: a – syntype MNHN 2000.27657; b- dark variety BPBM 1967-98 (photo: a – M. Caballer)

somewhat folded deep within, margin widely reflexed, slightly impressed at insertion.
Greenish yellow to olive brown, spire and suture dectoricated white. Lip white
Anatomy: Penis ¼ oviduct length, caecum short, semi-cylindrical. As *S. thurstoni* except lower half of caecum with 6-7 vertical rows of minute ridges.
Type specimens: Syntypes MNHN-IM-2000-27657; SMF 164994; UZH 518640.
Distribution: American Samoa – Tutuila. Recorded above 250 m on Leano Peak, Alava/Maugaloa, Faiga, Vatia, Toa, Maugaloa, Pagatatua, Levaga and Polauta ridges (Cowie & Cook 1999, 2001).

Samoana stevensoniana (Pilsbry, 1909)

Partula cepolensis Mousson, Paetel & Schaufuss 1869: 83 (nomen nudum)
?Partula upolensis 'Mousson', Garrett 1887: 134 (as syn. of *conica*, nomen nudum)
Partula conica Gould, Garrett 1887: 134 (misidentification)
Partula stevensoniana Pilsbry, 1909: 266
Samoana stevensoniana (Pilsbry) Pilsbry & Cooke 1934: 16. Kondo 1955: 82, 1968: 75. Richardson 1990: 75. Cowie 1998: 55

Dextral. Ovately conic. Thin. Umbilicus open. Spire short, conical. 5¼ convex whorls, body whorl evenly convex except near aperture where it is a little expanded towards base. Protoconch deeply and densely spirally punctuate-striate, except first half-whorl. Spiral striae continuing on spire and upper half of body whorl, wavy spirals visible on basal half of body whorl. Aperture wide, only very slightly oblique. Lip expanded, reflexed, a little impressed at columellar insertion.
Periostracum greenish-yellow, often partially decorticated. Spire pale corneous or brown. Suture edged with a white line. Inside of aperture flesh-coloured. Lip white.
Anatomy: Penis, with one large pilaster, caecum long, moderately slender. Vestigial retractor may be absent. Kondo (1955) described the penis as much reduced but his specimen was in poor condition.
Type specimens: Lectotype ANSP 77306a ('holotype' according to Baker 1963) and paralectotypes ANSP 77306.
Distribution: Samoa – Savai'i and 'Upolu . Found in forests at mid- to high-altitudes.
Note: *Partula cepolensis* was considered a synonym of '*Partula bulimoides'* by Richardson 1990.

Fig. 6.11 *Samoana stevensoniana* (a – type ANSP 77306; b – BPBM 108306; c – FMNH 152740) and (d) *S. thurstoni* (type BPBM 83121)

Samoana thurstoni (Cooke & Crampton, 1930)

Partula thurstoni Cooke & Crampton, 1930: 6

Samoana thurstoni (Cooke & Crampton), Pilsbry & Cooke 1934: 16. Kondo 1955: 78, 1968: 75. Richardson 1990: 75. Cowie 1998: 55

Dextral. Ovately conical. Umbilicus open. Thin, shiny. 5¼ convex whorls, ornamented with faint growth striae and indistinct spiral striae. Spire conic. Body whorl inflated. Aperture ovate, lip only slightly expanded with a very weak internal thickening towards columellar insertion. Columella flaring above. Parietal callus very thin.

Dark olive brown, paler on protoconch. Suture margined white, outer rim of lip white.

Anatomy: As *S. cramptoni*, differing in smaller secondary pilaster, caecum constricted near apex to form a distinct apical lobe. Caecum ornamentation shows an abrupt change from ridges to papillae. Vestigial retractor almost indistinguishable from nerve fibres. Vas deferens medial.

Type specimens: Holotype BPBM 10853; paratypes BPBM 83119, 83121, 189706.

Distribution: American Samoa – Ofu and Olosega, from above 100 m (Cowie & Cook 1999, 2001; Cowie *et al.* 2002).

Austral islands

Four *Samoana* species have been described from the Austral islands, these are reduced to two here.

Samoana dryas (Crampton & Cooke, 1953)

Partula dryas Crampton & Cooke, 1953: 141

Partula oreas Crampton & Cooke, 1953: 143

Partula hamadryas Crampton & Cooke, 1953: 146

Samoana dryas (Crampton & Cooke), Kondo 1955: 87, 1968: 75. Richardson 1990: 72

Samoana oreas (Crampton & Cooke), Kondo 1968: 75. Richardson 1990: 75

Samoana hamadryas (Crampton & Cooke), Kondo 1968: 75. Richardson 1990: 74

Dextral. Ovate conic, thin, fragile and slightly translucent. Umbilicus narrowly perforate, rimate. 4½-5 moderately convex whorls. Body whorl narrowed towards base. Suture of body whorl narrowly impressed. Decussating spiral lines on spire, fading on body whorl and may be visible only on base. Aperture ovate, narrow. Lip thin, slightly expanded, narrowed at outer insertion where it curves inwards. Columella slenderly triangular, strongly incrassate with a distinct channel. Parietal callus very thin, shiny, with or without an outer border, ornamented with fine spiral striae.

White to dull brown, darker on spire, or bicolored: upper half of each whorl white, lower half hyaline. A narrow white line is present above suture (not distinguishable in white shells).

Body pale yellow.

Fig. 6.12 Anatomy of Australs *Samoana*: a – Kondo 1955; b – MNHN Rv32; c – MNHN Rr32a; d – *oreas* (Kondo 1955, BPBM 135981); e – *hamadryas* (BPBM 144843, MNHN Rv62-3); f – *dryas* (BPBM 135957.1-2, MNHN Rv50)

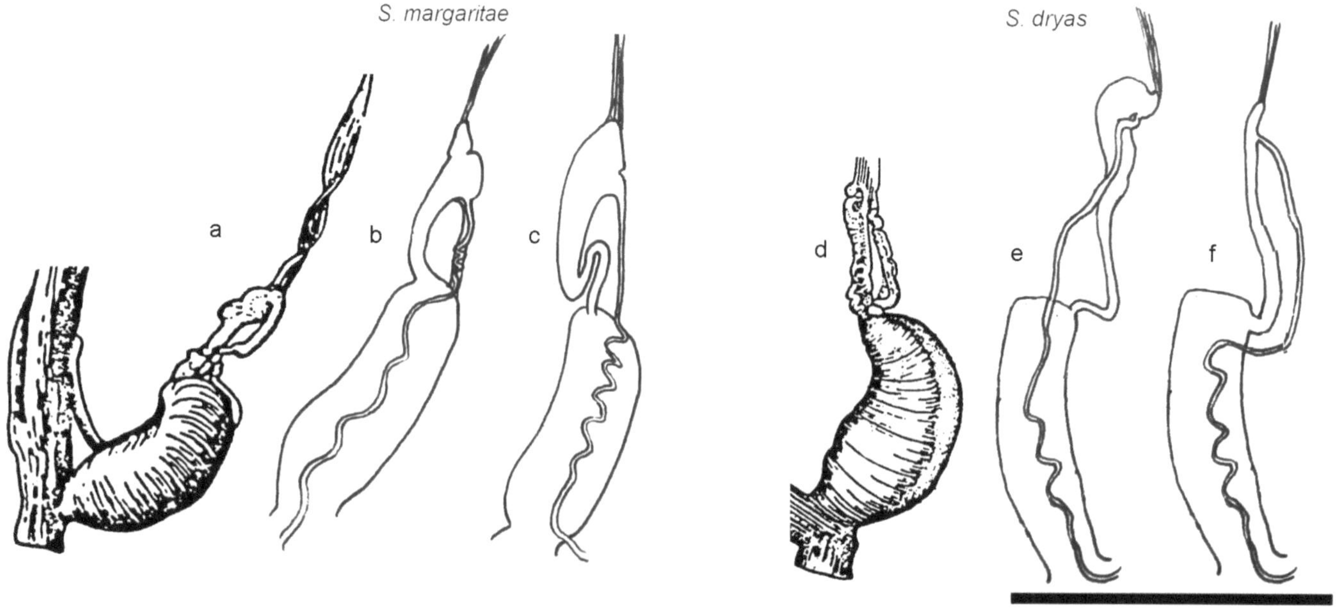

Fig. 6.13 Distribution of Australs *Samoana*. Raivavae *S. dryas*: red – original 1934 records, green – *hamadryas* morphotype 2002, blue – *oreas* morphotype 2002, yellow – unassigned morphotype 2002. Rapa *S. margaritae*: blue – original 1934 records, red – 2002 records.

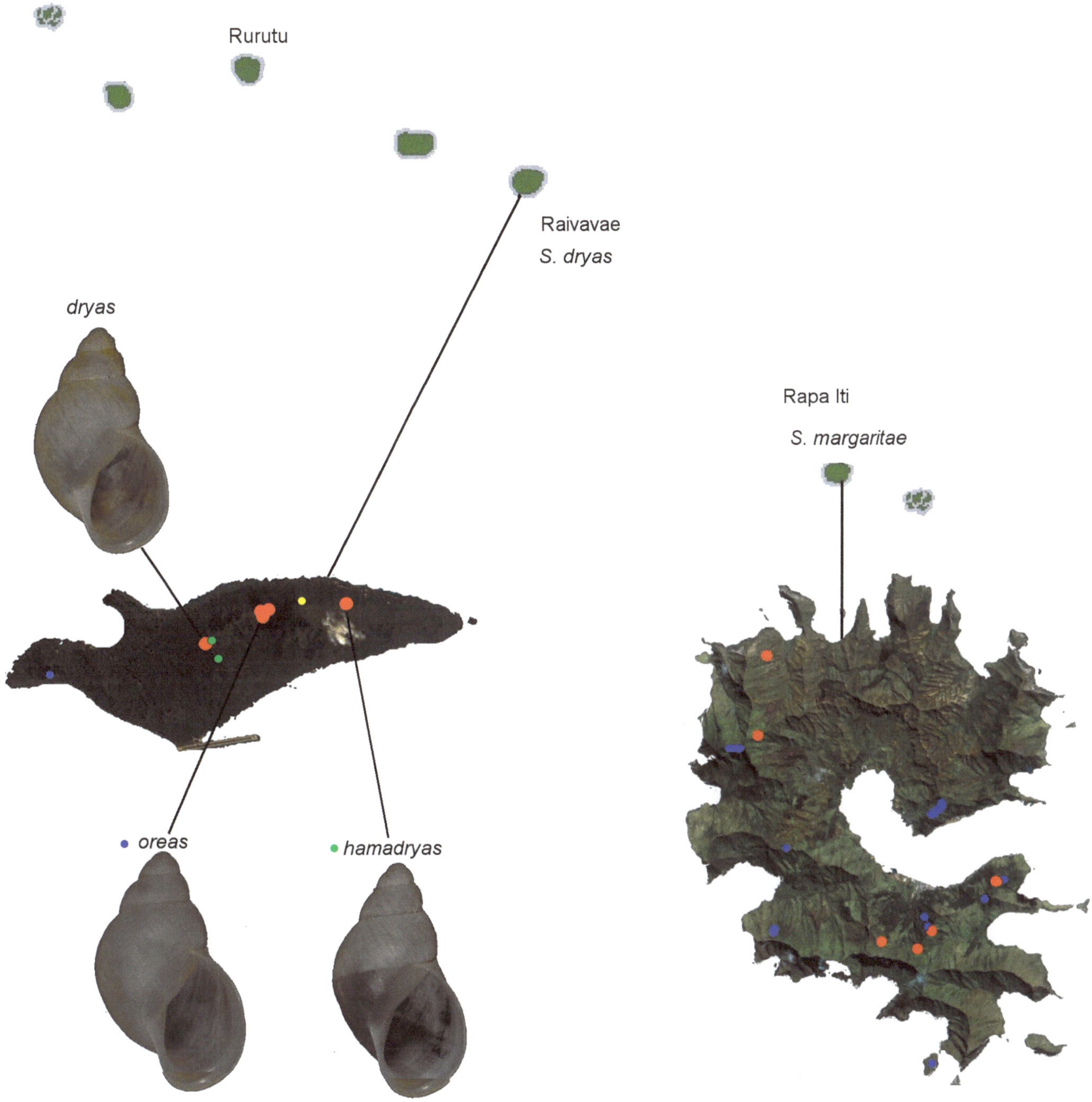

Anatomy: Penis ¼ oviduct length, highly muscular, distal end distinctly angled, junction of penis and caecum constricted more than in any other group of partulids. No vestigial retractor. Caecum long, slender and terminally inflated, internally ornamented with irregular ridges. Vas deferens subterminal. No significant variation between the nominal species.

Type specimens:

Partula dryas Crampton & Cooke, 1953 – Syntypes ANSP 294788; BPBM 9980.

Samoana hamadryas Crampton & Cooke, 1953 – Syntypes ANSP 294790; BPBM 9978.

Samoana oreas Crampton & Cooke, 1953 – Syntypes ANSP 294789; BPBM 9979.

Distribution: Raivavae – isolated populations have been located across the island. Distinct species were described from Mt.

Muani (*dryas*), the saddle between Mounts Muanui and Hiro (*oreas*), and south of Anatonu (*hamadryas*). Also recorded from Mt. Matotea, Mt. Raraterepa and Mt. Maunanui. Arboreal.

Note: The species described by Crampton & Cooke (1953) are local shell varieties rather than true species and cannot be distinguished anatomically. Cooke (in litt. to Crampton, BPBM archives) was not convinced that the 'species' were distinct, but deferred to Crampton's opinion. Crampton & Cook's bicoloured *hamadryas* from south of Anatonu was more rimate than the other populations, with 5 whorls rather than 4½-4¾, and a body whorl that was not impressed at the suture. The white *oreas* population from the saddle region had a broader columella and a more pronounced parietal callus border, although this latter feature is not constant in the original specimens. The brown *dryas* population from Mt. Muani lacked the border to the parietal callus. Other material shows that the typical *oreas* forms from Mt. Matotea (MNHN Rv50) are anatomically slightly different from the *hamadryas*–like forms from Mt. Raraterepa (MNHN Rv62) and Mt. Maunanui (MNHN Rv63). These *hamadryas* are strongly striate and, unlike Crampton & Cook's specimens, are distinctly impressed at the suture.

Samoana margaritae (Crampton & Cooke, 1953)

Partula margaritae Crampton & Cooke, 1953: 136

Samoana margaritae (Crampton & Cooke), Kondo 1955: 89, 1968: 75. Richardson 1990: 75

Dextral. Elongately conic. Solid. Umbilicus open. 5-5½ moderately convex whorls. Suture slightly impressed, indistinctly margined. Surface dull, with distinct growth lines, spiral striae dense on spire, fading on body whorl except around umbilicus. Aperture elongate, oblique. Lip expanded, evenly thickened, narrowing towards insertion, smooth and polished. Columella straight. Parietal callus thin, outer border raised.

Light cinnamon brown, to chestnut or rarely purplish brown. Rarely bicoloured, with lower half of whorls dark. All adult shells lack epidermis. Lip light to deep purple brown, pale along extreme margin.

Fig. 6.14 *Samoana dryas* specimens: a – *dryas* holotype (ANSP 294788); b – *hamadryas* holotype (ANSP 294790); c – *oreas* holotype (ANSP 294789); d – Mt. Matotea (MNHN Rv 50.1); e – Mt. Raraterepa (MNHN Rv 62.1); f – Mt. Maunanui (MNHN Rv 63.1) (photos: a-c – F. Borrero)

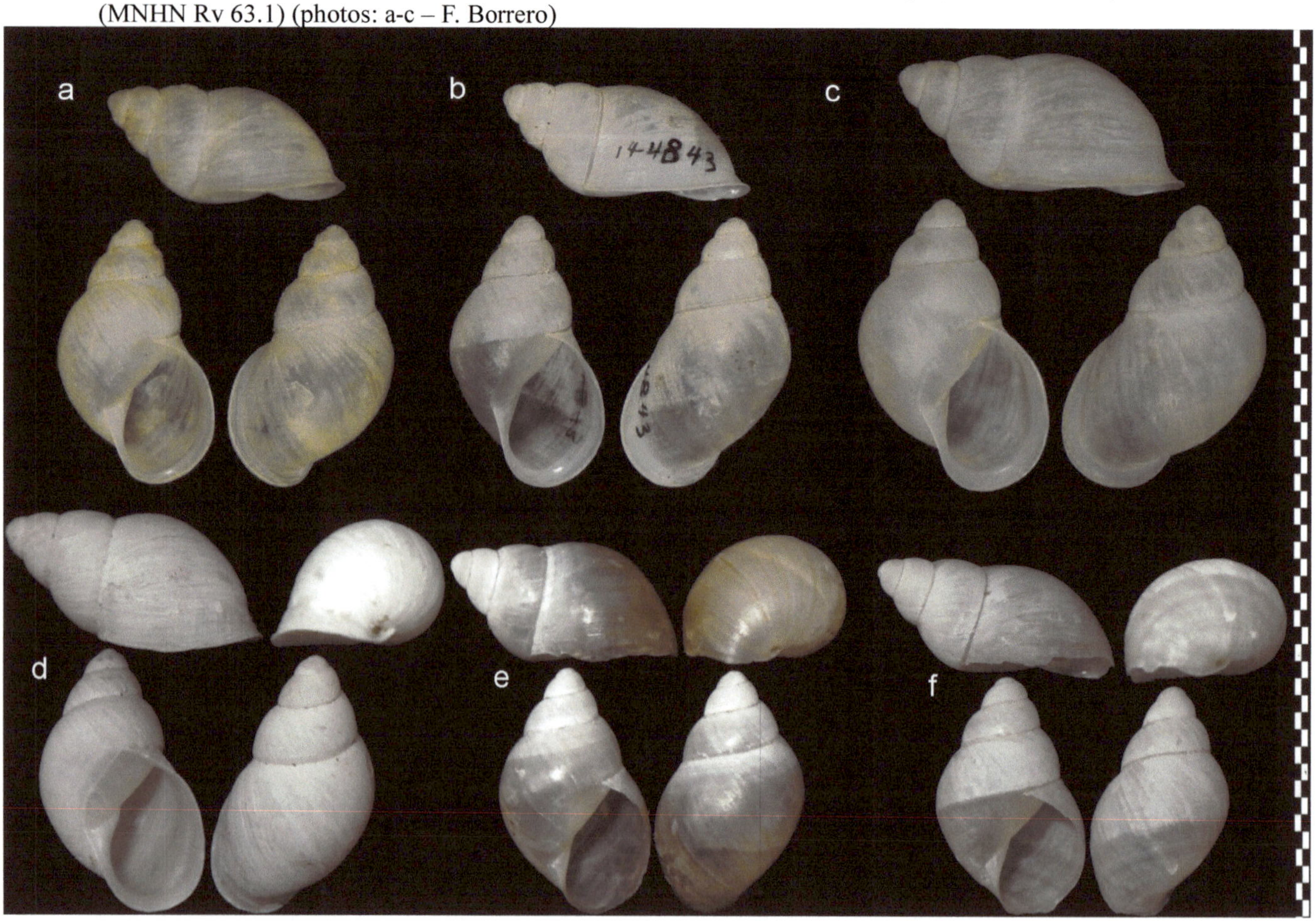

Foot greyish fawn, mantle largely dull purplish brown.
Anatomy: Penis ¼ oviduct length, as long as penis, inflated at apex, ornamentation reduced to sparse tuberculae in upper and lower parts and low ridges in medial part. Vas deferens terminal, sometimes coiled within a membrane running. Secondary penial pilaster reduced to a slight thickening of penial sheath.
Type specimens: Syntypes ANSP 294786, 294787, 456636; BPBM 138209, 136711 (=9981-2).
Distribution: Rapa – Mostly recorded from higher areas, Mt. Pukumape 165 m, Peupeu 200 m, Mount Tanga 200-250 m, Mt. Maugaoa 265-310 m and Mt Namuere 425 m (from BPBM and MNHN collections). Also recorded from Karaporahi islet (MNHN).
Arboreal, 95% of specimens (n>300) collected on the Mangareva Expedition exclusively on *Celtis* sp., but found on a variety of plants by Fontaine & Gargominy (2002) An unusually large subfossil specimen was located in Tupua bay. A specimen labelled Rurutu (MNHN Rr32a) may be mislabelled.

Marquesas islands

Kondo started a revision of the Marquesas *Samoana* and published one new species in 1989 (Kondo & Burch 1989) but the full revision never progressed beyond a draft in 1979. The draft manuscript was itself based on unfinished work by Cooke from the 1940s, who had intended to describe four new species based on the collections made by G. LeBronnec of the Bishop Museum's Pacific Entomological Survey (1929-32). Although Kondo appeared unaware of it, Cooke's plans were also based on two unpublished species set aside as potential new species by Pilsbry from his collection on the Pinchot Expedition, in 1928.

Kondo (MS) thought that the Marquesas species had speciated through "The fractionation of the original founder stock into seven kinds of *Samoana* may hypothetically be synchronized with the fragmentation of the (postulated) original larger single 160 × 150 mile island into ten smaller islands, six of them now inhabited by *Samoana*. Fragmentation (by postulated subsidence) would result in cut-off of gene flow, producing isolation one of the prime causes of changes leading to speciation." Kondo (MS) also considered that "*Samoana magdalinae* is a biological species. *Samoana bellula, strigata,*

Fig. 6.15 *Samoana margaritae*: a – type (BPBM 138209); b – supposed Rurutu specimen (MNHN Rr32a); c – 'Rapa' (MNHN Rp32.1)

Fig. 6.16 *Samoana* of the Marquesas islands with details of ranges on each island (below)

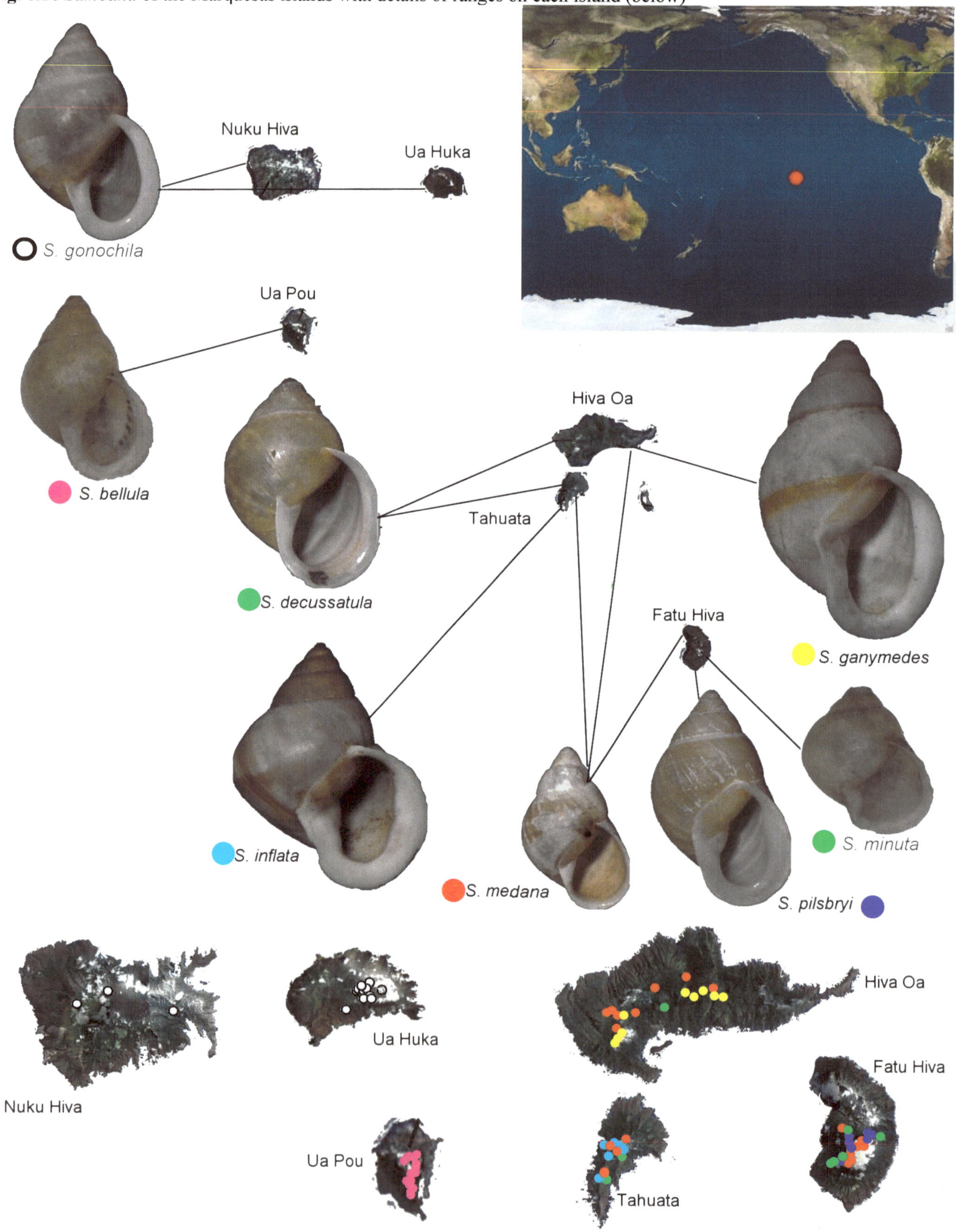

ganymedes, inflata, decussatula, and *medana* are super geographical races ranked here as species." In the present revision seven distinct species are recognised.

Samoana gonochila (Pfeiffer, 1847) **comb. nov.**

Bulimus gonochilus Pfeiffer, 1847: 82
Bulimus gonochilus Pfeiffer, Pfeiffer 1848: 69
Partula gonocheila Reeve 1851: pl4 f19
Partula gonochila Pfeiffer, Pfeiffer 1853a: 448, 1856: 162, 1857f: 272, 1859a: 512, 1868: 160, 1877: 203. Adams 1855: 145. Schmeltz 1869: 72, 1874: 92. Pease 1871b: 473. Paetel 1873: 104, 1883: 153, 1889: 267. Gloyne 1878: 338. Pilsbry 1909: 272
Partula gonocheila Pfeiffer, Hartman 1881b: 12, 1885: 220, 1892: 98. Pfeiffer & Clessin 1881: 302. Smith 1903: 437. Richardson 1990: 30
Partula repanda Pfeiffer, 1855a: 98
Partula repanda Pfeiffer, Pfeiffer 1856: 162, 1859a: 512, 1868: 160, 1877: 203. Kobelt 1880: 19. Hartman 1881a: 186, 1881b: 12, 1885: 220, 1892: 98. Pfeiffer & Clessin 1881: 302. Paetel 1883: 153, 1889: 268. Smith 1903: 459. Solem 1959: 74
Partula recta Pease, 1864 (nomen nudum)
Partula recta Pease, 1868b: 155
Partula strigata Pease, 1868b: 155
Partula strigata Pease, Pease 1871b: 473. Pfeiffer 1877: 202. Hartman 1881a: 187 (as syn. of *P. varia*), 1881b: 14, 1885: 220. Garrett 1884: 80. Paetel 1889: 268. Smith 1903: 459. Pilsbry 1909: 171
Partula recta Pease, Pease 1871b: 473. Pfeiffer 1877: 202. Graeffe 1879: 85. Hartman 1881a: 186 (as syn. of *repanda*), 1881b: 12, 1885: 220. Pfeiffer & Clessin 1881: 302. Garrett 1887b: 25. Smith 1903: 458. Kilias 1969: 248
Partula strigata obesior Pilsbry, 1909: 172
Partula strigata recta Pease, Pilsbry 1909: 172
Samoana strigata (Pease), Kondo 1968: 75. Richardson 1990: 75. Johnson *et al.* 2000: 580

Fig. 6.17 Anatomy of Marquesas *Samoana.* a – Tahuata (Kondo 1955, BPBM 997355); b-f – Hiva Oa (b – ZSL 694a & 833, BPBM 98547, 98599; c – ZSL 833; d – 694b; e – Kondo 1955; f – subadult BPBM 1969-46); g-i – Ua Huka (g – BPBM 10363; h & i – BPBM 105048); j & k – Hiva Oa (j – Kondo 1955; k – Kondo & Burch 1989); l – Ua Pou (Kondo 1955 mislabelled '*strigata*', and BPBM 1970-36); m – Ua Pou (ZSL 661); n – Fatu Hiva (Kondo 1955, BPBM 100103); o – Fatu Hiva (BPBM 11874) & Nuku Hiva (BPBM 96132); p – Tahauta (Kondo & Burch 1989, BPBM 99823) and Hiva Oa (BPBM 1969-41), Fatu Hiva (BPBM 1969-4); q & r – Hiva Oa (BPBM 11876); s – Tahauta (BPBM 99794); t-w – Fatu Hiva (t-v – BPBM 11875; w – BPBM 100068)

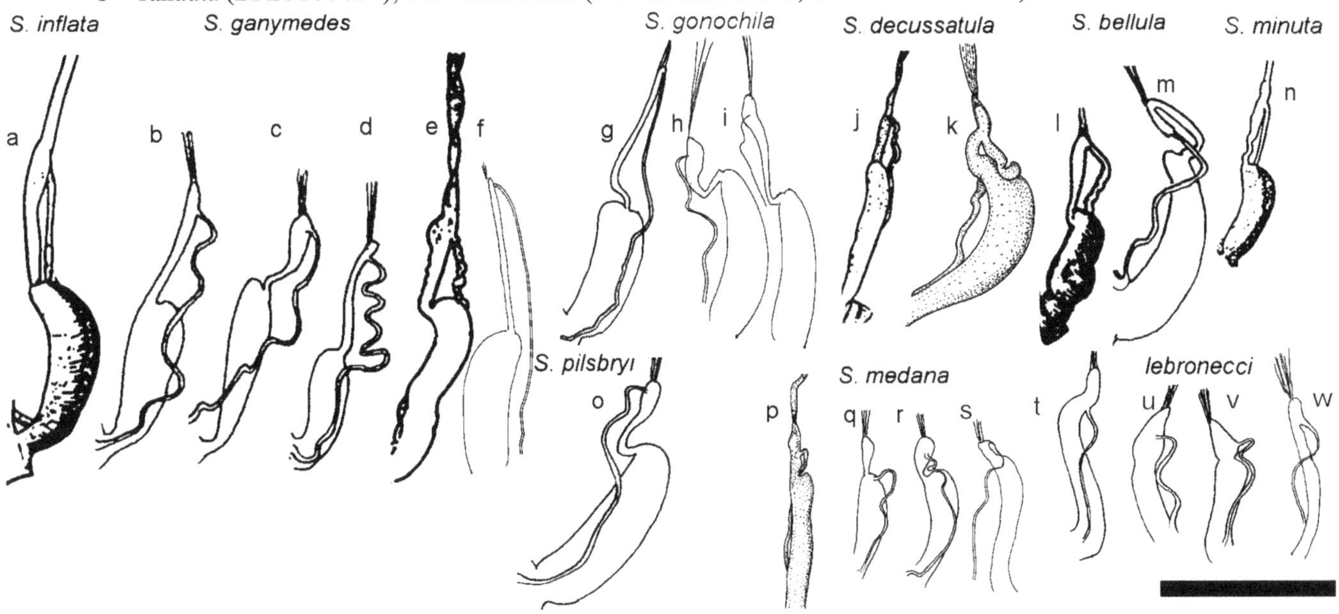

Dextral. Ovately conical, varying from globose (*obesior*) to elongate (*recta*). Moderately thin to thick. Umbilicus open. 4½ convex whorls, body whorl evenly rounded. Surface ornamented with fine wavy spiral striae, most conspicuous on body whorl. Aperture ovately rectangular, obliquely truncate above. Columella indistinctly plicate above. Lip flatly reflexed, thickened within except near the upper margin.

Off-white to dull olive-brown, body whorl pale with a few darker oblique streaks, spire darker. Lip flesh-tinted.

Body mottled, visible through the shell in juveniles.

Type specimens:

Bulimus gonochilus Pfeiffer, 1847 – Syntypes NHMUK un-numbered (Cuming specimens, Marquesas).

Partula repanda Pfeiffer, 1855 – Syntypes NHMUK un-numbered (Cuming specimens supposedly from New Hebrides, but labelled '? Cook group' and, noted by Solem 1959 as more likely from the Societies).

Partula strigata Pease, 1868 – Lectotype 59785; paralectotype 463796. Lectotype designation by Baker (1963).

Partula recta Pease, 1868 – Lectotype ANSP 59789; paralectotypes ANSP 463799; MCZ 25338, 25343; ZMB Moll-74326. Lectotype designation by Baker (1963).

Partula strigata obesior Pilsbry, 1909 – Syntypes ANSP 59788.

Distribution: Ua Huka (*strigata*), Nuku Hiva (*strigata, obesior* and *recta*). It was listed from Ua Pou by Pilsbry (1909) in error, deriving from Pease's type material collected by "a native missionary" on 'Woapo' (Garrett 1884; Kondo MS). This error was further repeated by Kondo (1955) with a misidentification of a specimen of *S. bellula*. Arboreal; described as abundant on trees (Garrett 1887b; Kondo MS).

Note: Garrett (1887b) observed that this species varies greatly in size, shape and colour. On Ua Huka most populations are ovate but some are dominated by more conical individuals. The ovate ones may resemble *S. decussatula* and some have been labelled "decussatula stimulants" (BPBM 1970-69 – Vaikivi: with 5/6 resembling *decussatula*; and BPBM 1970-74, Mt Taua, with 3/5 resembling *decussatula*). 43 of the conical specimens from Vaikivi and Tukituki were set aside by Cooke as a new species. Kondo (MS): "This is a minor variant of *strigata*, uniformly corneous and visually different from its ovate compatriots in its small size and lesser convexity but do not qualify as a separate species from *strigata*, not even a geographical race." Both forms have been found more recently; the globose form from the south: Hane valley (115-150 m) and on the trail to the Hitakait cliffs, Hokatu, and the more typical conical *strigata* from the ridge at 250-400 m on the trail to Matukiaoha (specimens collected by T. Coote).

The Nuku Hiva shells are smaller and thinner. Nuku Hiva "appear to be unfavorable for *strigata*. Each colony constitutes scarce individuals. Shells are small and ovate but tendency towards conical is evident." (Kondo MS). The ovate shells are the *obesior* form, the conical ones *recta*. The patchy nature of their distribution on this island probably led to local

Fig. 6.18 *Samoana gonochila*: a – *gonochila* type (NHMUK); b – *strigata* type (ANSP 59785); c – *obesior* type (ANSP 59788); d – *recta* type (ANSP 59789); e – *repanda* type (NHMUK); f – *strigata* specimen (NHMUK); g-h – Nuku Hiva (g – BPBM 96132; h – BPBM 96128); i-j – Ua Huka (i – BPBM 104828; j – BPBM 104853)

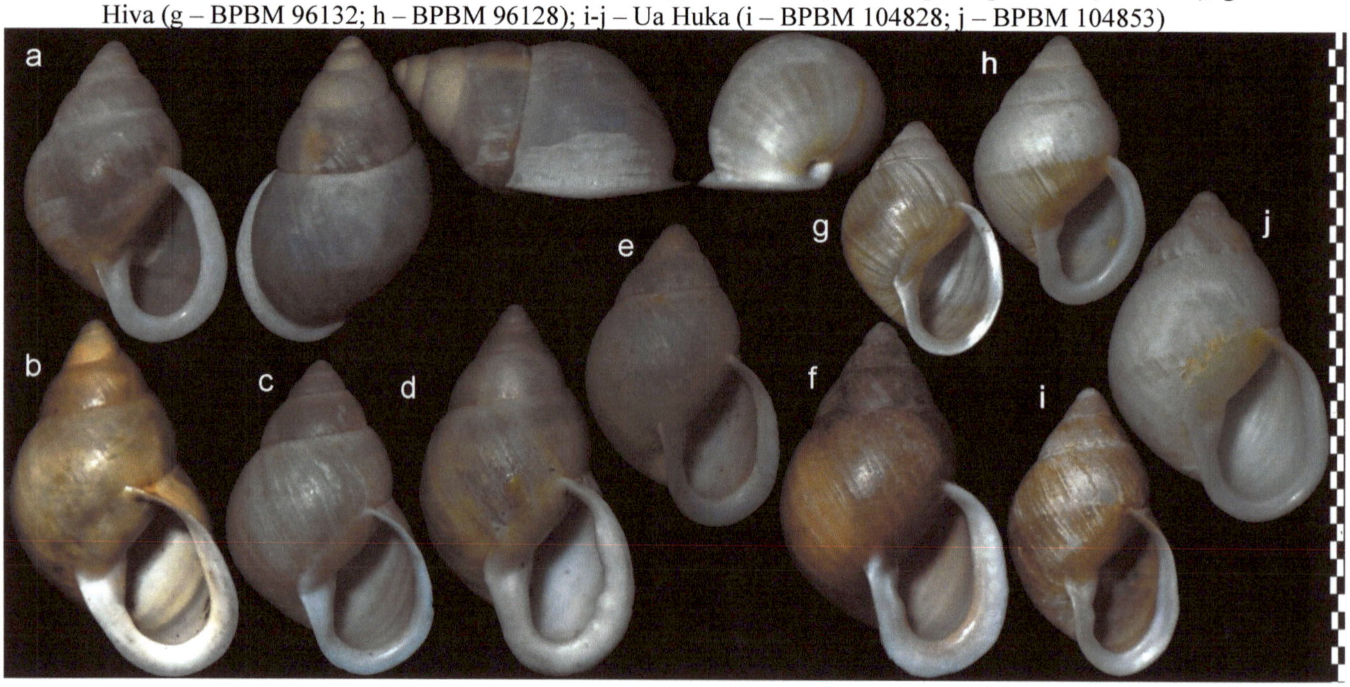

variation in shell shape, resulting in confusion over identifications and the species boundaries. Of LeBronnec specimens 19 adults and 26 juveniles were labelled *S. decussatula*, from Tapuaooa (31600 ft, n=17) and Ooma 3000 ft, n=2) from 3000-3100 ft but are identifiable as *S. gonochila* (four of which are like *S. medana*, including BPBM 96133, these were "set aside by Cooke as a possible impoverished new variety of *strigata*, even a possible new species" Kondo MS). All '*decussatula*' (e.g. BPBM 96132), *recta* or *obesior* on Nuku Hiva were anatomically identical.

P. gonochila was labelled as being from the Marquesas, although Hartman (in Smith 1903) thought it from Samoa.

***Samoana pilsbryi* sp. nov.**

Dextral. Conical, with a slight broadening in middle. Umbilicus open, narrow. 4½ almost straight-sided whorls, body whorl moderately inflated at periphery. Surface with fine spiral striae on spire. Aperture narrowly ovate. Lip widely and flatly expanded, noticeably thickened within.

Light brown with some streaking, sometimes a faint narrow brown peripheral band, usually with a white subsutural band caused by decortications. A few bicolorous: light brown with wide supraperipheral white band but this is due to decortication, not pigmentation. A very uniform species.

Mantle in juveniles is blackish with white maculae, becoming white with black threads and finally fully white in adults (based on BPBM 96130, 96132, 96082, 96128).

Anatomy: As *S. gonochila.*

Distribution: Fatu Hiva, recorded from 650-3100 ft, not occurring in the lowest areas occupied by *S. minuta* (Kondo MS). Arboreal, present on *Meterosideros*. Forms small colonies (Kondo MS, data from 1969), colonies cover 50×50 yards (containing 26 adults, 34 juv) to 10×10 yards (5 individuals).

Type specimens: Holotype ANSP 251988, collected by H.A. Pilsbry, 1923.

Nomenclature: *pilsbryi* after the first collector Henry A. Pilsbry. This was the name that Cooke (MS) intended to use for the species, and is validated here.

Notes: First collected by H.A. Pilsbry on the Pinchot South Seas Expedition in 1923, who set aside his specimens as a new species but never described them. 93 collected by LeBronnec, these specimens selected by Cooke as a new species, '*Samoana pilsbryi*', but never described. 132 collected by Kondo in 1969.

Kondo's confusion with the Marquesas partulids is exemplified by BPBM 11874 which he labelled "Samoana pilsbryi Ms. Cooke = small ganymedes, should name it ganymedes pilsbryi" in 1962, but relabelled in 1972 "No! it is Samoana strigata."

Similar to *S. gonochila* but more slender, with less convex whorls and a narrower aperture.

Samoana decussatula (Pfeiffer, 1850)

Bulimus decussatulus Pfeiffer, 1850: 131

Partula decussatula (Pfeiffer), Reeve 1850: pl4 f23. Pfeiffer 1853b: 453, 1856: 162, 1857f: 274, 1859: 515, 1868: 161, 1877: 208. Pease 1864: 675. Paetel 1873: 104, 1883: 152, 1889: 266. Kobelt 1880: 18. Hartman 1881a: 181, 1881b: 303, 1885: 217, 1893: 98. Garrett 1887b: 24. Smith 1903: 461. Pilsbry 1909: 174

Samoana decussatula (Pfeiffer) Kondo 1955: 96, 1968: 75. Richardson 1990:72

Dextral. Ovate-conic. Umbilicus narrow. Thin, fragile, translucent, slightly shiny. Very minutely decussated with growth-lines and spiral striae. Spire short, conic, obtuse. Suture slightly impressed. 4½-5 convex whorls, body whorl rounded. Columella slightly folded at base, receding. Aperture angularly oval. Lip thin, narrowly expanded, the margins converging, right margin with a bell-like expansion. Columellar margin reflexed over umbilicus.

Fig. 6.19 *Samoana pilsbryi*: a – holotype (ANSP 251988); b – three juvenile specimens from Pilsbry's series; c – BPBM 11874 ex 99965; d – BPBM 99966

Diaphanous fulvous-whitish to corneous, spire tinted brown. Pale decorticated subsutural band and narrow faint brown peripheral band on body whorl. Very rarely (BPBM 99877) the peripheral band may be distinct.
Body light brown, mantle (visible through the shell) light brown with white mottling and large blackish maculations.
Anatomy: Penis ½ oviduct length, long, distally expanded. Vas deferens submedian. Caecum short, papillose, papillae fusing into ridges at base. Principal penial pilaster broad apically, a short spur points towards the short secondary pilaster.
Distribution: Hiva Oa (Vaei and Tahauku valley, 250 ft, recorded recently at 900 m on Mt. Temetiu) and Tahuata. Nuku Hiva specimens are misidentifications of *S. gonochila*. Arboreal; described as abundant on leaves (Garrett 1887b). On both islands *S. decussatula* was found at low altitudes, with the similar *S. medana* at higher altitudes, while the very different *S. ganymedes* and *S. infl*ata were not altitude restricted.
Type specimens: Syntypes NHMUK un-numbered (Cuming specimens).
Notes: Tahuata specimens are thinner than those on Hiva Oa and often resemble *S. medana* but are anatomically typical *S. decussatula* (BPBM 99879 and 1969-48).

Samoana inflata (Reeve, 1842)

Partula inflata Reeve, 1842: 83

Partula inflata Reeve, Reeve 1843: 197, 1850: pl1 f3. Catlow & Reeve 1845: 163. Pfeiffer 1853a: 452, 1855b: 267. 1856: 162, 1859: 514. Albers 1850: 186. Paetel & Schaufuss 1869: 83. Paetel 1873: 104. Hartman 1881a: 183, 1881b: 12, 1885: 219, 1892: 98. Pfeiffer & Clessin 1881: 303. Garrett 1887b: 25. Smith 1903: 479. Pilsbry 1909: 169

Bulimus thersites Pfeiffer, 1842: 123

Bulimus thersites Pfeiffer, Pfeiffer 1848: 75. Férussac & Deshayes 1851: 123

Samoana inflata (Reeve), Kondo 1955: 98, 1968: 75. Richardson 1990: 74. Johnson *et al.* 2000: 580

Samoana ganymedes (Pfeiffer), Johnson *et al.* 2000: 580 (misidentification, partim)

Dextral. Globosely conic to ovate. Moderately thin to thick, dull. Umbilicus open. 5½ convex whorls. Spire with dense wavy spiral striae, fading on body whorl. Suture moderately impressed. Body whorl swollen at periphery, last ¼ whorl compressed, giving a distinctively humped appearance (this is well developed and accompanied by a widely flared lip in 73% of Garrett's sample BPBM 94671; LeBronnec's specimens are mainly flared, with only slight humping). Specimens lacking the humped profile resemble *S. ganymedes*. Aperture ovate; lip broadly expanded, flatly reflexed, thickened within. Columella broad, with a deep apical fold.
Anatomy: Penis ½ oviduct length, large. Caecum long, narrow. Vestigial retractor muscle much reduced. Caecum with only sparse papillae around vas deferens as vertical or semi-lunate short ridges. Vas deferens subapical, entering caecum as a conical projection. Penis with a well developed secondary pilaster. Primary pilaster with three strong diagonal spurs.

Fig. 6.20 *Samoana decussatula*: a – type (NHMUK un-numbered); b-c – Garrett specimens from Hiva Oa (b – ANSP 59911; c – BPBM 4833); d – Tahuata (BPBM 99806); e – Hiva Oa (BPBM 1969-48)

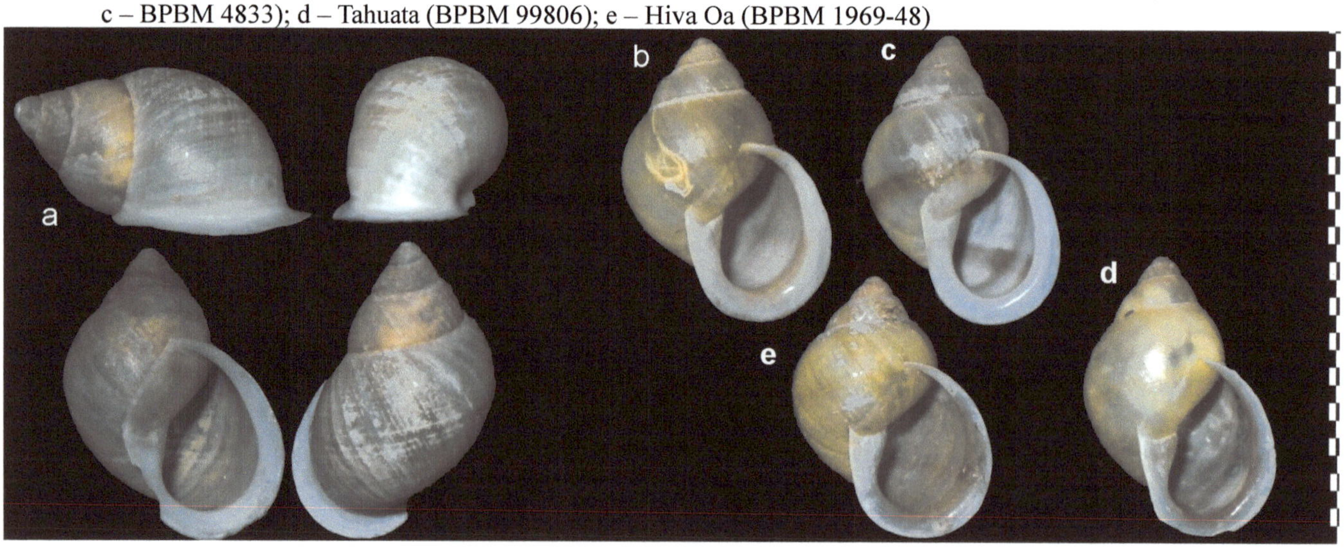

Fig. 6.21 (opposite) *Samoana inflata* and *S. ganymedes* (a-h – *S. inflata*; i-y – *S. ganymedes*). a – *inflata* type NHMUK 1996129; b – specimen NHMUK; c-f – UMZC 75307; g-h – ANSP 59781; i-l – ZSL 833; m-p – *ganymedes* type NHMUK; r-s – *fasciata* types MCZ 155152; t-w – ANSP 294796; x – ZSL 534; y – BPBM 98599

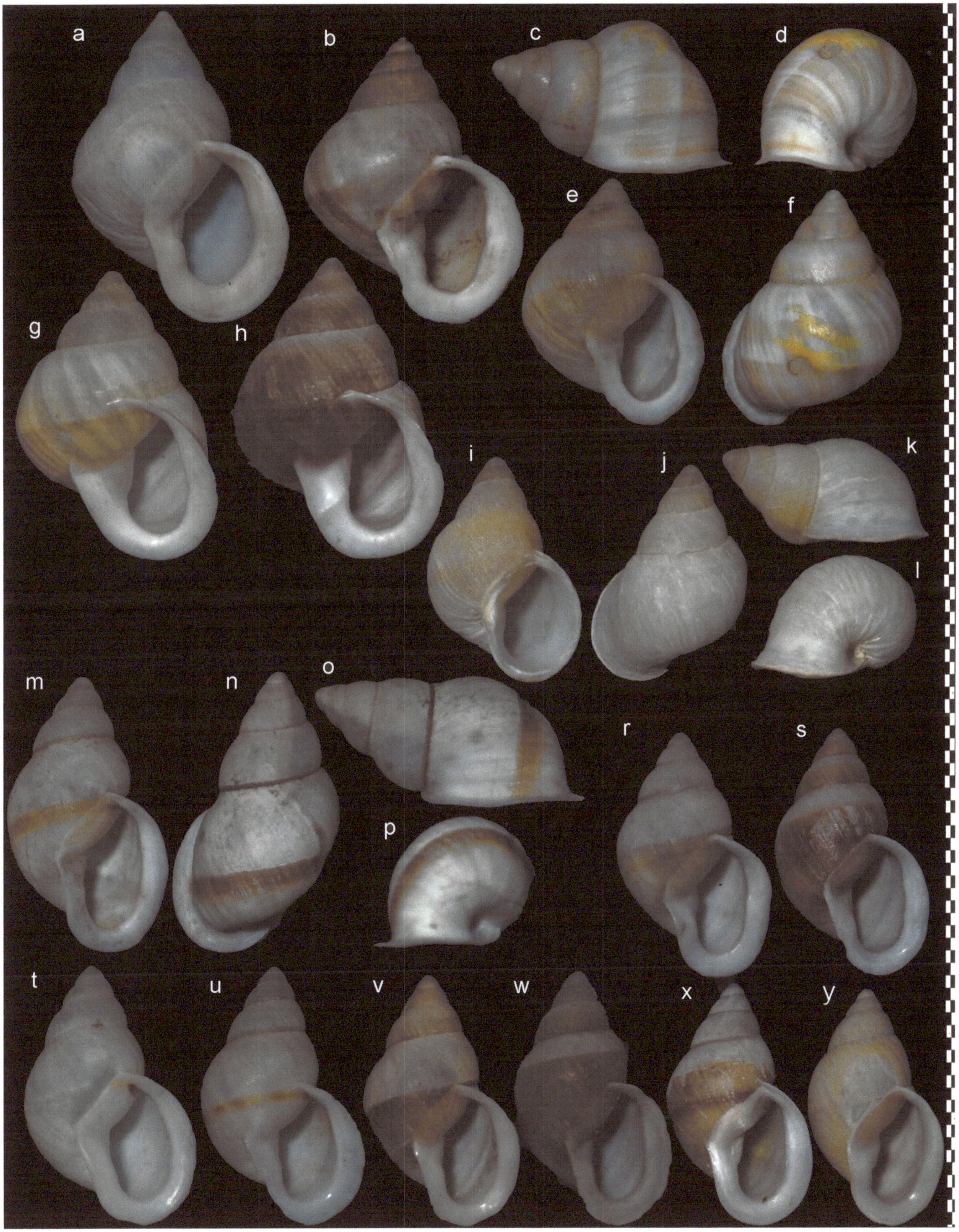
a
b
c
d
e
f
g
h
i
j
k
l
m
n
o
p
r
s
t
u
v
w
x
y

Type specimens:

Partula inflata Reeve, 1842 – Syntypes MCZ 297850; NHMUK (un-numbered), based on Cuming specimens supposedly from Hiva Oa. The locality is presumably an error for the nearby island of Tahuata.

Bulimus thersites Pfeiffer, 1842 – Not located, supposedly from 'Insulae Sandwich' ex Petit de la Saussaie collection (Pfeiffer 1842).

Distribution: Tahuata, recorded from 400-500 m. Arboreal; common on trees at around 500 m (Garrett 1887b). Recorded from Hiva Oa but these are *S. ganymedes*.

Note: The *S. ganymedes*-like forms are most conspicuous at Hanatetena (BPBM 99887) where they form 14% of the sample (n=78; Kondo MS).

Bulimus thersites and *Partula inflata* were first regarded as conspecific by Pfeiffer (1848) and as he had experience of the types of both names this is presumably correct. This synonymy has been followed by all subsequent authors, but never discussed. Férussac & Deshayes (1851) considered Pfeiffer's name to be the senior, but all other authors have adopted Reeve's name.

Samoana ganymedes (Pfeiffer, 1846)

Bulimus ganymedes Pfeiffer, 1846: 39

Bulimus ganymedes Pfeiffer, Pfeiffer 1848: 72

Partula ganymedes (Pfeiffer), Reeve 1850: pl3 f16. Pfeiffer 1853a: 447, 1856: 162, 1858a: 511, 1868: 160, 1877: 202. Pease 1864: 675, 1866: 202, 1871b: 473. Hartman 1881a: 182, 1881b: 11, 1885: 219, 1892: 98. Pfeiffer & Clessin 1881: 302. Paetel 1883: 153, 1889: 267. Garrett 1887b: 26. Smith 1903: 457. Pilsbry 1909: 170. Thiele 1931: 658. Zilch 1959: 459

Partula fasciata 'Pease' Carpenter in Pease 1865: 675 (nomen nudum, in syn. of *ganymedes*)

Partula fasciata Pease, 1866: 202

Partula fasciata Pease, Pease 1871b: 473

Partula gonocheila (non Pfeiffer) Hartman 1881a: 12, 1885: 219 (all nomen nudum)

Samoana ganymedes (Pfeiffer), Kondo 1955: 100, 1968: 75. Richardson 1990: 73. Johnson *et al.* 2000: 580

Samoana inflata (Reeve), Johnson *et al.* 2000: 580 (misidentification, partim)

Dextral. Oblong-conic. Thickness variable. Umbilicus open. Spire conic, rather acute. Ornamented with fine wavy spiral striae, surface often spirally decussate. 5½ moderately convex whorls. Body whorl indistinctly angular in the middle. Columella straight. Aperture oblong, obliquely truncated above. Lip simple, thin, broadly expanded.

Periostracum greenish-yellow, often decorticated. A moderately wide chestnut band is usually present (80% of Garrett's duplicates in BPBM from one location; n=128) often present, this may be basal or peripheral. Form *fasciata* has an olive periostracum, usually decorticated to just a subperipheral band and brown spire.

Anatomy: Penis ⅓ oviduct length, expanded terminally. Vas deferens subapical. Vestigial retractor greatly reduced and indistinct. Caecum small, with small ridges apically, a strong vertical ridge and vas deferens entering caecum as a conical projection, minute tubercles on lower half, becoming semi-lunate near base. Penial pilasters curved.

Type specimens:

Bulimus ganymedes Pfeiffer, 1846 – Syntypes CMNH 62.4263; NHMUK un-numbered.

Partula fasciata Pease, 1865 – Syntypes MCZ 155152. Johnson (1972) listed MCZ 25113 and 25324 as syntypes, both are ex Pease collection but it is not clear that they are types.

Distribution: Hiva Oa (Mauna-o-Fefe, Mt Kahana). Arboreal; abundant on trees (Garrett 1887b). Found mainly at lower altitudes. Recorded on Tahuata but these are *S. inflata*.

Note: Kondo (annotated thesis) notes BPBM specimens include intermediates with *ganymedes/inflata/strigata* and a *ganymedes* var. from Hiva Oa, including "*strigata*-like" specimens (e.g. BPBM 98685) and "*inflata*-like" (e.g. BPBM 98707). 79% of specimens were typical (n=78; Kondo MS).

Samoana bellula (Hartman, 1885)

Partula bellula Hartman, 1885: 203

Partula bellula Hartman, Garrett 1887b: 27. Pilsbry 1909: 173

Partula crassiuscula 'Garrett', Smith 1903: 477 (nomen nudum)

Samoana bellula Hartman, Kondo 1955: 95, 1968: 75. Richardson 1990: 70. Johnson *et al.* 2000: 580

Samoana strigata Pease, Kondo 1955: 94 (misidentification)

Dextral. Ovate-conic, rarely conic. Umbilicus narrow. Thin, translucent, shiny with fine spiral striae, wavy on base. 4½-5⅓ moderately convex whorls, body whorl inflated. Aperture oblique, ovate. Lip reflected. Callus thin but distinctly bordered, not extending to upper termination of the outer lip, a barely indicated convexity on palatal labial callus. Columellar lip dilated.

Periostracum very pale olive or greenish-yellow cuticle with inconspicuous darker streaks, very rarely with a darker peripheral band. One individual (BPBM 54317) is darker than all others. Body whorl corneous, spire finely reddish brown, apex pale flesh-tinted. Suture white. Lip grey, callus white edged.

Body with distinct, large dark maculae on the mantle, some individuals are uniformly grey.

Anatomy: Penis $^{1}/_{3}$ oviduct length, caecum long and narrow. Vas deferens apical or subapical. Caecum with reduced ornamentation of lateral ribs, delicate ridges distally, penis with secondary pilaster well developed or reduced to a low ridge.

Type specimens: Holotype CMNH 62.4264.

Distribution: Ua Pou, above 330 m (recorded from Hakahetau, Vaihakaatiki, Mt. Poumaka, Teavakaakiti and Mt. Poutetainui). Garrett (1887b) recorded only two specimens, one collected on a bush at around 800 m. Kondo (MS) recorded the species from "all inhabitable altitudes from 1000 to 3300 ft… The habitable forests are few and sparse. It is therefore remarkable that *bellula* is prolific and collected in relatively large numbers by LeBronnec and Peake."

Note: Some specimens are flared and resemble *S. decussatula* (e.g. BPBM 108377).

Samoana medana Kondo & Burch, 1989

Samoana medana Kondo & Burch, 1989: 131

Dextral. Ovately conical to conical (forms 'diaphana', 'kavaana' and 'alloides'). Fragile or moderately fragile, thin and translucent. Umbilicus narrowly perforate, rimate. Surface dull, ornamented with fine spiral striae and fine wavy growth lines, forming a decussate pattern. 4½ slightly convex whorls. Suture only slightly impressed. Body whorl rotund. Aperture large, ovate. Lip expanded, thin, inner side with a slight ridge, outer side forming a broad flange. Columella simple, forming a narrow triangularly curved flap near the umbilicus. Parietal callus very thin but distinct.

Light to dark brown, light individuals may have a dark peripheral band (form 'cincta'). Lip white.

Body grey-brown with white markings on mantle.

Anatomy: Penis long, with constriction at junction with caecum. Vas deferens subterminal. Caecum with 5-6 vertical zigzag ridges, one strong. Penis with a single vertical pilaster, right side swollen and conspicuous. A variant ('lebronneci') has a reduced caecum, with a less pronounced constriction between penis and caecum.

Type specimens: Holotype BPBM 11879 ex 94655; paratypes BPBM 94655-94658 (11 specimens).

Distribution: Hiva Oa, Tahuata, Fatu Hiva. Widespread on Hiva Oa (Vaiepoepo on shrubs at 710-732 m, Kapaafaa, Tepuna, Teipunui, Maunaofefe, Temetiu, Kaava, Ootua at 610-1,158 m). Recorded on *Calophyllum* ('ata'), *Freycinetia, Weimannia* and *Meterosideros.*

Note: Cooke divided the Tahuata samples into two forms; typical *medana* and 'diaphana' from Amatea and Vaitupaehi, with 'kaavana' (from Kaava ridge and Mt. Temetiu) and 'alloides' from Hiva Oa (both seem to be minor variants of 'diaphana') and 'lebronecci' from Fatu Hiva. The main shell forms, *medana*, 'diaphana' and 'lebronneci', may form a continuum from 'diaphana' to *medana* to 'lebronneci' and to *S. decussatula.* Anatomically a distinction may be made between *medana* and 'lebronneci', but these may integrade. 'diaphana' and 'cincta' are anatomically identical to *medana.* There is a distinct anatomical form in *medana* on Tahuata (possibly an individual oddity Fig. 6.17s). It appears that 'lebronneci' is the Fatu Hiva form and could be regarded as taxonomically distinct but approaches *medana* in anatomy and overlaps in shell form.

Typical *medana*, 'kaavana' and 'cincta' have been found on all three islands occupied by the species. 'lebronneci' is known

Fig. 6.22 *Samoana bellua*: a – type (CMNH 624264); b-e – individual variation (b – ZSL 661; c – BPBM 108377; d – BPBM 54317; e – BPBM 108549; f – BPBM 1970-36) (photo: a – C. Sturm)

from Fatu Hiva, 'diaphana' from Hiva Oa and Tahuata and 'alloides' only from Hiva Oa. This may arise from an ancestral form colonising all three islands and diverging allopatrically (medana on Tahuata, 'lebonneci' on Hiva Oa and 'diaphana' on Fatu Hiva), but subsequent movement between islands has mixed the incipient species. This movement may have been human mediated, these small snails being well adapted to accidental transport on plants.

Fig. 6.23 *Samoana medana*: a-l – Hiva Oa; m-o – Fatu Hiva, p-q – Tahuata. a – holotype (BPBM 11879); b – 'cincta' Mt. Ootua (BPBM 98584); c – typical form (BPBM 9585); d – 'alloides' Tepuu (BPBM 11877); e – 'alloides' Teipunui (BPBM 94885); f – 'kaavana' (BPBM 11876); g – 'diaphana' (ANSP 251990); h-j – juvenile 'diaphana' (ANSP 251990); k – 'diaphana' (BPBM 115662); l – 'diaphana' Mt. Temetiu (BPBM 98627); m – 'lebronneci' Oomoa (BPBM 11875); n – typical Mt Teamet (BPBM 1967-7); o – 'cincta' Tevaione (BPBM 100522); p – typical (BPBM 99879); q – 'diaphana' Vaitupaahea (BPBM 99823)

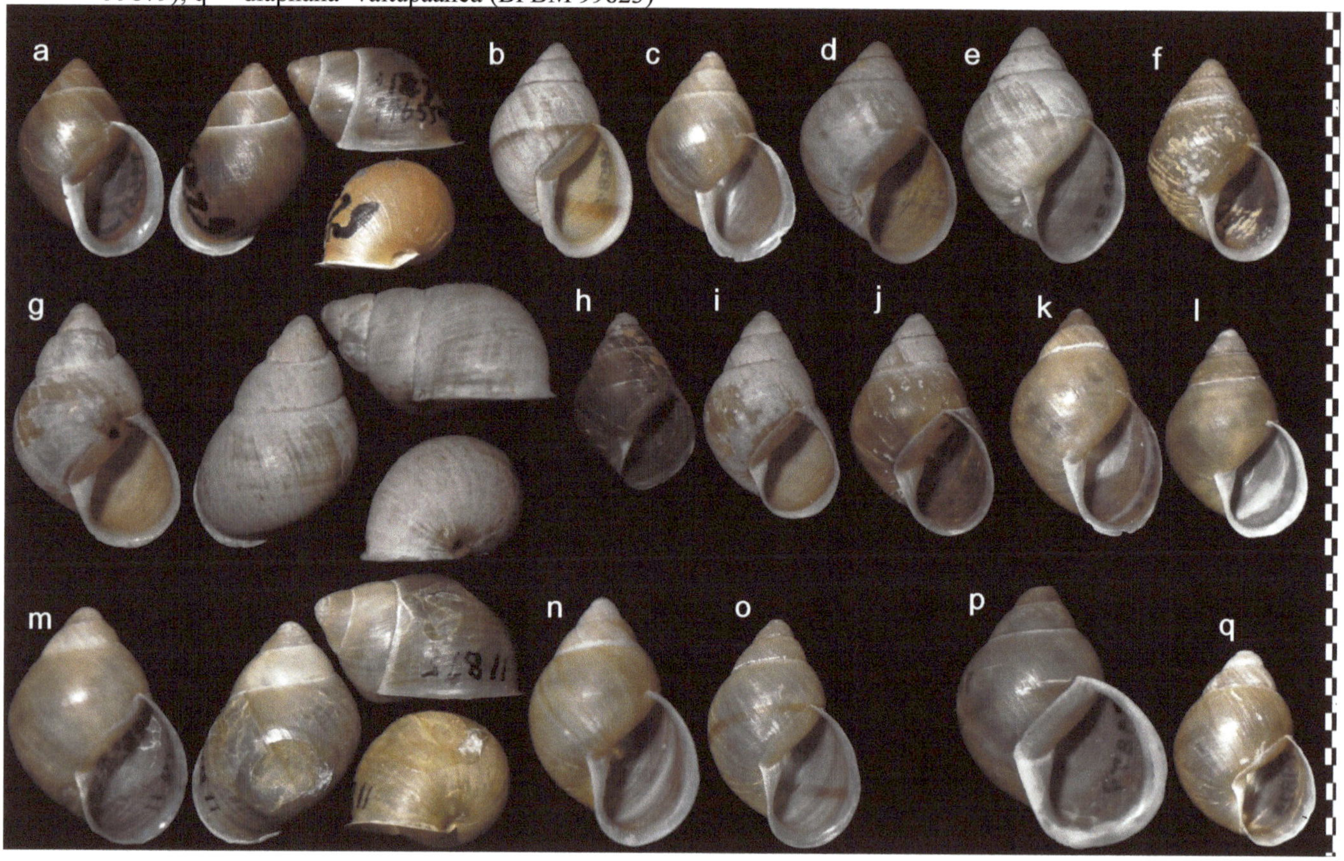

Fig. 6.24 Distribution of *Samoana medana* variants. Right map shows Kondo's field locality map allowing precise collection points to be identified (background removed for clarity)

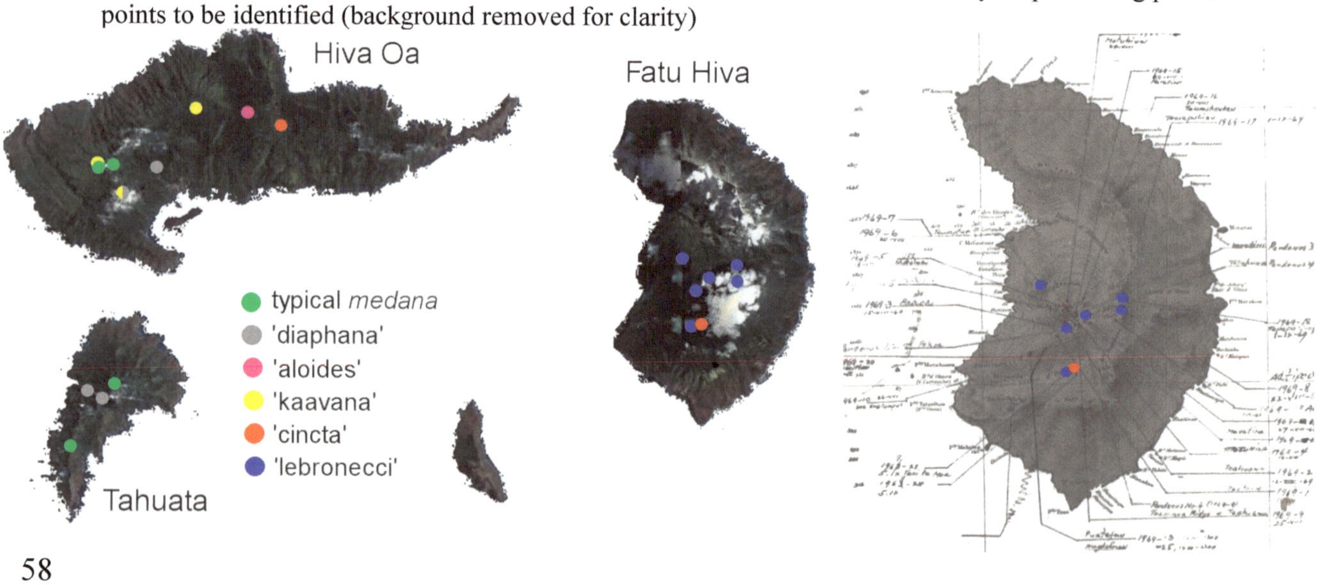

Samoana minuta (Pfeiffer, 1857) **comb. nov.**

Partula minuta Pfeiffer, 1857a: 244

Partula minuta Pfeiffer, Pfeiffer 1857c: 384, 1859a: 514, 1868: 161, 1877: 208. Kobelt 1880: 16. Hartman 1881a: 184, 1881b: 11, 1885: 18. Pfeiffer & Clessin 1881: 303. Paetel 1889: 267. Smith 1903: 457. Pilsbry 1909: 301. Rensch 1937: 582. Richardson 1990: 41

Partula Magdalinae Hartman, 1885: 203

Partula magdalinae Hartman, Garrett 1887b: 27. Hartman 1892: 98. Smith 1903: 460. Pilsbry 1909: 175

Samoana magdalinae (Hartman), Kondo 1955: 102, 1968: 75. Johnson *et al*. 2000: 580

Samoana magdelinae (Hartman), Richardson 1990: 74 (misspelling)

Dextral. Globose. Thin, fragile and translucent. Umbilicus narrow. Short, conic spire with sculpture of minute, close spiral striae, weak above the periphery. 4-4½ moderately convex whorls, body whorl very inflated, prominent in the peripheral region. Aperture oblique, ovate, usually appearing angular due to the upper part of the columella twisting over the umbilicus and inwards, leaving an angular notch at the base of the aperture. Lip thin, reflexed with a slight internal thickening.

Thin pale yellow periostracum over white shell with corneous streaks, faint subperipheral corneous band. Periostracum usually deciduous above the periphery of the body whorl. Lip grey to white.

Anatomy: Penis ½ oviduct length, long. Caecum moderate, vas deferens subapical. Vestigial retractor reduced and adnate to vas deferens. Caecum with only small ridges on upper portion, basal third with large ridges. Penis with well developed pilasters.

Type specimens:

Partula minuta Pfeiffer, 1857 – Holotype NHMUK un-numbered.

Partula magdalinae Hartman, 1885 – Syntypes CMNH 62.4263.

Distribution: Fatu Hiva, in all areas. Originally highly abundant and widespread but restricted to Teavapuhia plateau by 1995. The type was supposedly part of a collection of Admiralty island species collected by Cuming, but including species from the Society Islands. The origin of the *P. minuta* holotype is therefore uncertain. There are two supposed '*minuta*' specimens in Hamburg (ex Museum Klagenfurt) also from the Admiralty islands, but these are misidentified. Arboreal; recorded on shrubs (Garrett 1887b) and the leaves of *Hibiscus tiliaceus, Freycinetia, Piper latifolium, Musa fehi, M. cavendishii, Morinda citrifolia, Inocarpus edulis*, coffee, ferns and *Cyclophyllum barbatum* ('toauta') trees (Kondo MS). Noted to be exceptionally sticky, like *S. annectens*, and this was thought to relate to Fatu Hiva being the wettest island in the Marquesas (Kondo MS).

Fig. 6.25 *Samoana minuta*: a – *magdalinae* type (CMNH 624263); b – *minuta* type (NHMUK un-numbered); c-d – specimens (c – NHMUK; d – BPBM 100103) (photo: a – C. Sturm)

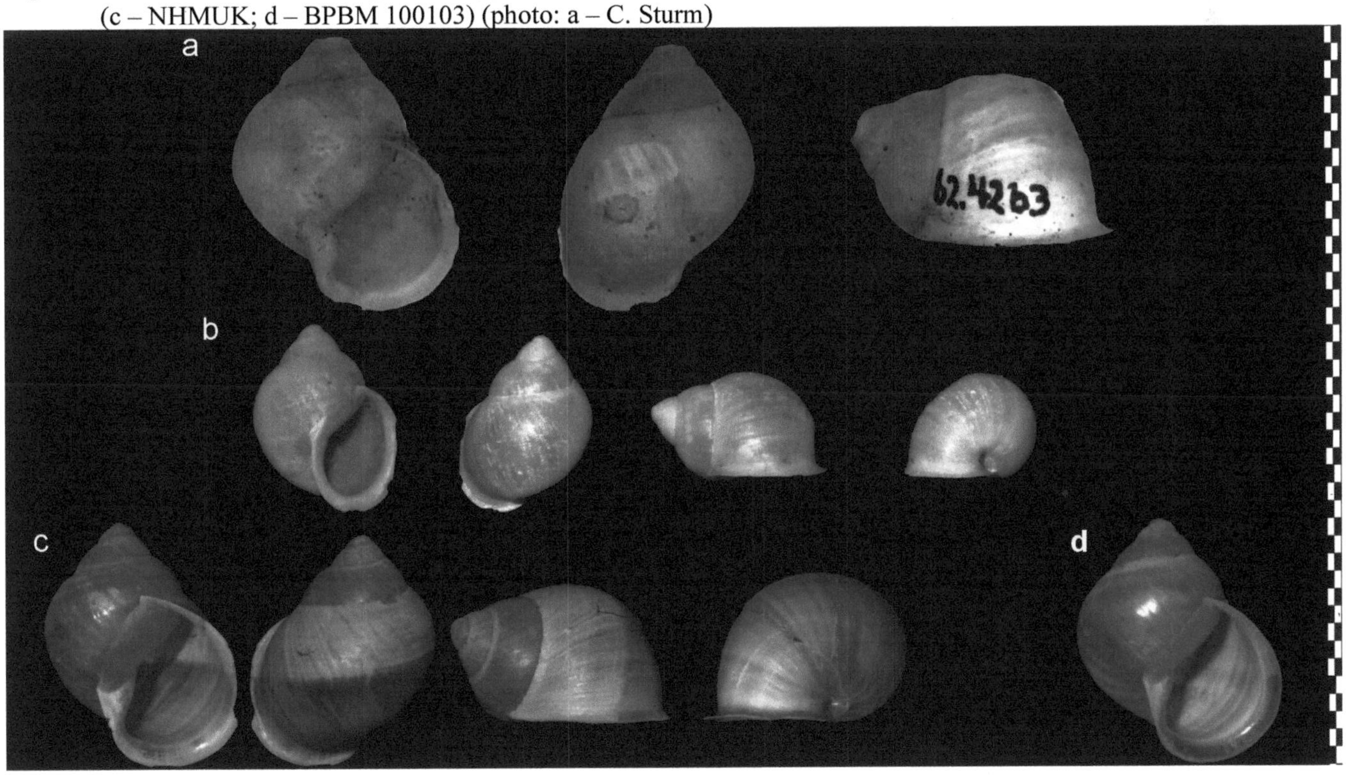

Society islands

All four Society Island *Samoana* are anatomically similar but are clearly distinct genetically (Lee *et al.* 2009, 2014) and conchologically. Lee *et al.* (2009) determined that on mitochondrial DNA evidence *S. diaphana* and *S. burchi* are sister species, with *S. attenuata* the basal Society Island *Samoana*. *S. annectens* was not included in their study.

Fig. 6.26 Society Islands occupied by *Samoana*, with detail of recorded populations on each island

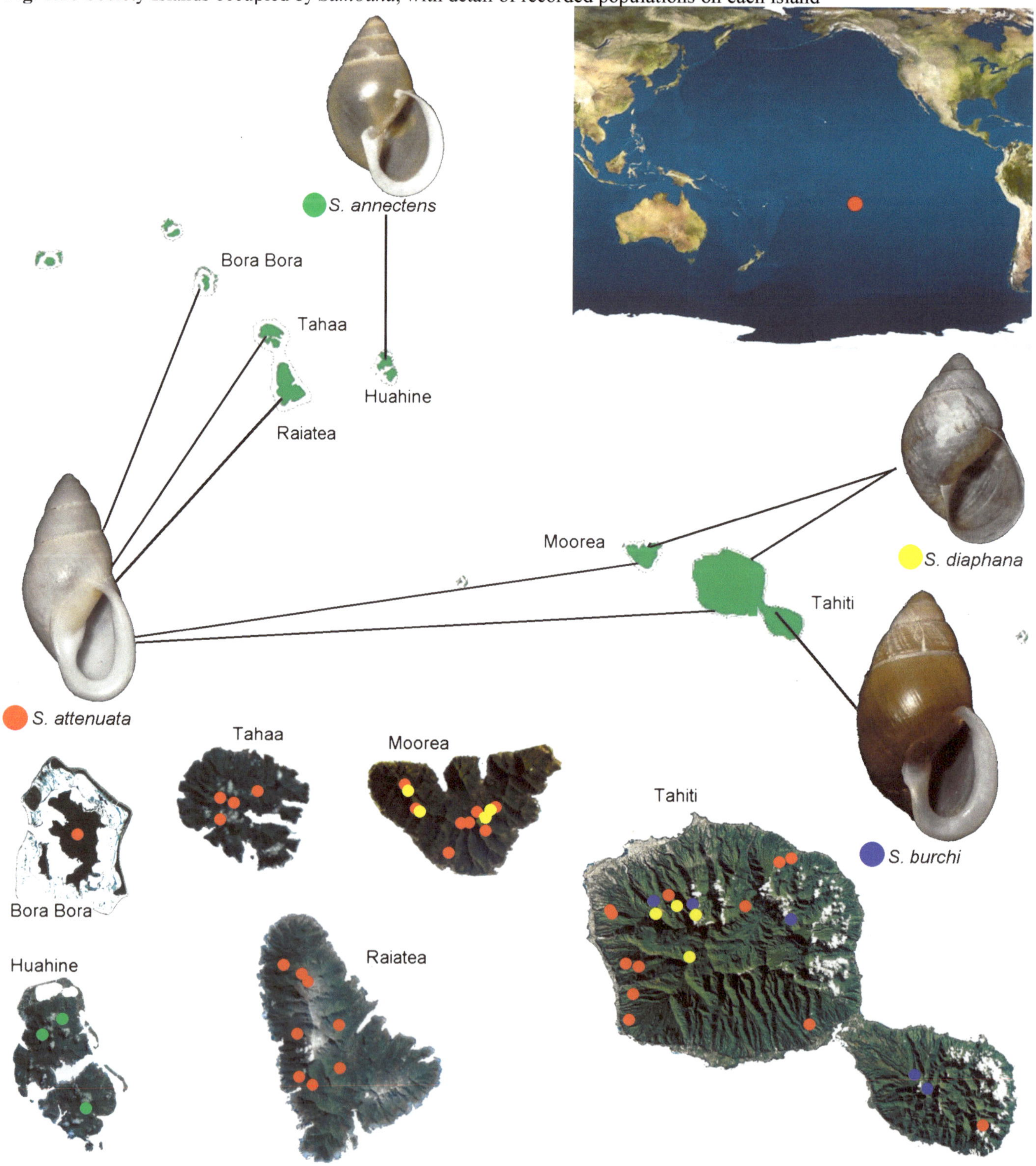

Samoana attenuata (Pease, 1865)

Partula carteriensis (non Quoy & Gaimard) Reeve 1850: pl4. Schmeltz 1869: 72. Martens 1877: 279. Garrett 1884: 70. Hartman 1885: 216

Partula gracilis Pease, 1864: 675 (nomen nudum)

Partula attenuata Pease, 1865: 672

Partula gracilis Pease, 1866: 197

Partula gracilis Pease, Pease 1867: 81, 1871b: 473. Paetel 1873: 104, 1883: 153, 1889: 267. Binney 1876: 244. Pfeiffer 1868: 156, 1877: 198. Gloyne 1878: 337

Partula attenuata Pease, Pfeiffer 1868: 156, 1877: 195. Pease 1871b: 473. Schmeltz 1874: 92. Gloyne 1878: 337. Hartman 1881a: 179, 1881b: 11, 1885: 216, 1892: 75. Paetel 1883: 152, 1889: 266. Garrett 1884: 70. Smith 1903: 451. Pilsbry 1909: 198

Partula gracilior 'Pease', Hartman 1881a: 183 (as syn of *gracilis*, nomen nudum)

Partula solitaria Crampton, 1932: 28

Partula solitaria Crampton, Clarke & Murray 1969: 31

Samoana solitaria (Crampton) Kondo 1968: 75, 1973: 19

Samoana attenuata (Pease), Kondo 1955: 91, 1968: 75, 1973: 20. Kilias 1969: 246. Kondo & Burch 1983: 105. Johnson *et al.* 1986: 98. Richardson 1990: 69. Coote & Loève 2003: 96. Lee *et al.* 2008: 616, 2009: 12

Dextral. Narrowly elongate torpedo shaped (most islands) or elongately ovate (Moorea). Apex sharp. Openly perforate. Surface shiny, faintly marked with growth lines and numerous minute, slightly wavy spiral striae. 5-5½ slightly convex whorls, the last compressed below the periphery, and strongly convex, somewhat swollen, at the base. Aperture nearly vertical, more than half the total length. Lip thick, expanded (except on Moorea), basal and outer margins thickened within except near upper end of lip, which is thin and slightly expanded. Columellar margin dilated, bifurcate above. Parietal callus a transparent film.

Pale yellowish-corneous, sometimes with a faint reddish tint towards the apex; suture transparent margined, or margined with a white line.

Body light brown. Mantle uniform on Tahiti, elsewhere light with dark spots and irregular streaks showing through shell.

Anatomy: Penis $^1/_3$ oviduct length, large, subcylindrical, often curved. Caecum distinctly separated from penis, thin and transluscent. Vas deferens subterminal. Vestigial retractor present. Caecum ornamented with few papillae internally, and irregular ridges. Secondary penial pilaster absent.

Type specimens:

Partula attenuata Pease, 1865 – Syntypes MCZ 24901; ZMB/Moll 75622, 74333.

Partula gracilis Pease, 1866 – Syntypes AMNH 42200; ANSP 59935; MCZ 24905, 25115; SMF 164972.

Partula solitaria Crampton, 1932 – Holotype ANSP 290231.

Distribution: Tahiti (Mt. Aorai, Papenoo, Tiitauiri, Fareteuira, Haapupuni, Pirai, Ururoa, Tiarei), Moorea (Faatoai, Uufau, Mt. Tohiea Belvedere), Raiatea (Tooroa ['Toloa'], Marahi ['Hapai'], Faaroa, Avera, Averarahi, Vaiaau, Miri Miri ['Miti Miti Aute Rahi']), Tahaa (Ruutia, Mt. Ohiri, Haamene valley, Mt. Furauti ['Mt. Purauti']) and Bora Bora (near Patio). A reference to *S. attenuata* from Huahine by Kondo (1973) is an error for *P. annectens.* Recorded from above 100 m but more abundant at mid-altitudes. As noted by Garrett (1884) "Owing to its peculiar habit of living on the foliage near the tops of trees, it easily escapes observation." Most specimens were collected around 3 m above ground.

Fig. 6.27 Society islands *Samoana.* After Kondo 1973 (*S. attenuata* – Tahiti; *S. burchi*; *S. annectens*; *S. diaphana*) and Kondo 1955 (*S. attenuata* – Tahaa), *S. attenuata* Moorea (ZSL 662). Scale bar 5 mm.

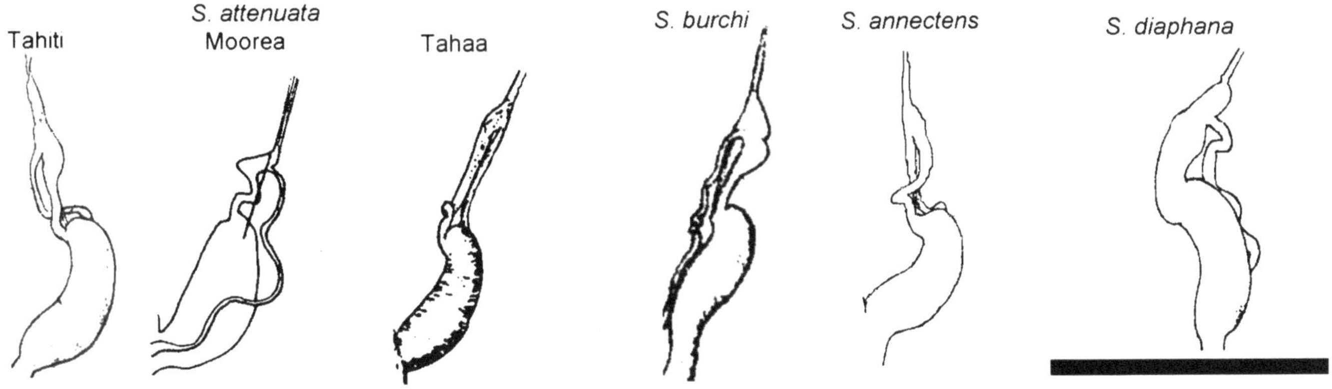

Note: The presence of embryos in 7 subadult shells from Purauti, Tahaa (Crampton MS) shows that reproductive maturity may occur before the lip is fully developed (length 12.2-14.9 mm).

Samoana annectens (Pease, 1865)

Bulimus annectens Pease, 1865: 671

Bulimus annectens Pease: Pfeiffer 1868: 48

Partula annectens Pease: Pease 1871b: 473. Pfeiffer 1877: 62. Hartman 1881a: 179; 1881b: 12, 1885: 218, 1892: 98. Pfeiffer & Clessin 1881: 303. Paetel 1883: 152, 1889: 266. Garrett 1884: 66. Smith 1903: 457. Pilsbry 1909: 254.

Samoana annectens (Pease), Kondo 1955: 92, 1968: 75, 1973: 19. Johnson *et al.* 1986: 98. Richardson 1990: 69

Samoana attenuata (Pease), Kondo 1973: 20 (error)

Dextral. Oblong-ovate, very thin, pellucid. Umbilicus open. Apex acute. Shiny with fine growth-lines, and very close, minutely crimped, spiral striae. 4½-5 convex whorls, body whorl compressed or somewhat flattened below the periphery. Aperture oblique, ovate. Lip thin, reflected. Columellar margin dilated and bifurcate above. Parietal callus indistinct.

Slightly yellowish corneous, sometimes with a pale pink-brown tint at the apex, suture bordered with an opaque white line. Lip white.

Body recorded by Garrett (1884) as: "The animal varies from pale luteous-yellow to light brownish yellow. The soft parts, as seen through the transparent shell, are mottled with slate-colored spots. The foot is about the same length as the shell, and the ocular peduncles are very long and slender." This species was photographed in 2005 (Fig. 6.32c).

Fig. 6.28 *Samoana attenuata*: a – *attenuata* type (MCZ 24901); b – *solitaria* type (ANSP 290231); c – *gracilis* type (ANSP 59935: '*gracilior*' Tahiti, ex Pease); d – specimen (UMZC 75283) (photo: b-c – F. Borrero)

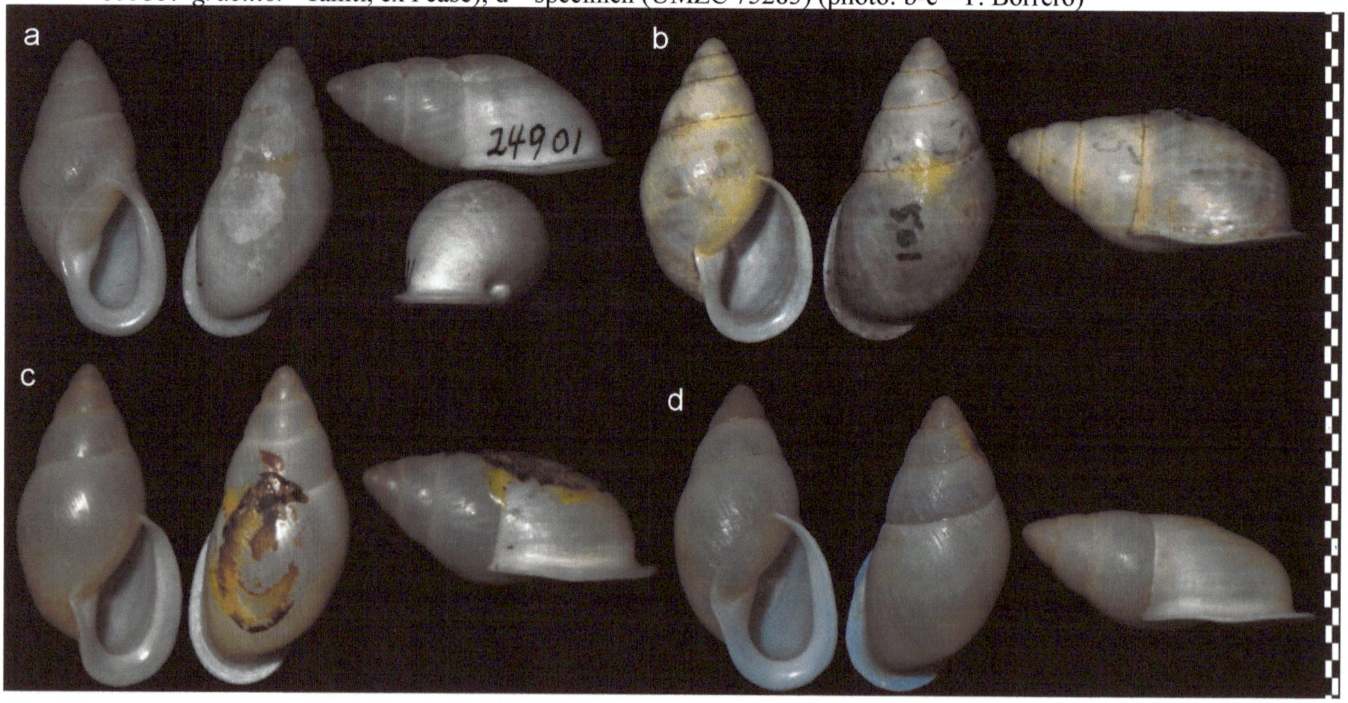

Fig. 6.29 *Samoana annectens*: a & b – types (MCZ 26352), b – Garrett specimen (ANSP 10105)

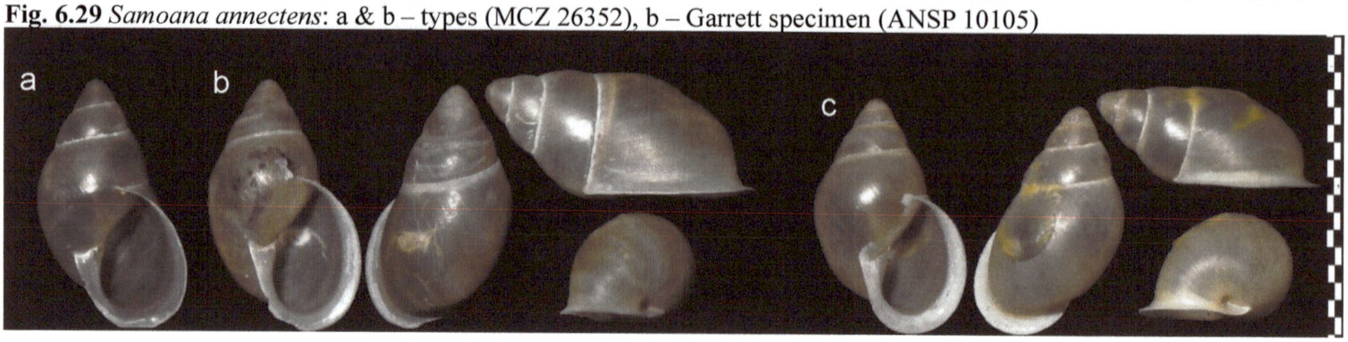

Anatomy: As *S. attenuata.*
Type specimens: Syntypes MCZ 26352.
Distribution: Huahine, recorded by Garrett (1884) only from Tepua valley, later recorded from Fare at 400-500 ft. by the Mangareva Expedition, in Fitii valley by J. Murray and M. Johnson in 1980 and Mauhuti by J. and E. Murray in 1987 (Murray field notes). Arboreal.

Samoana burchi Kondo, 1973

Samoana burchi Kondo, 1973: 20

Samoana burchi Kondo, Kondo & Burch 1983: 105. Johnson *et al.* 1986: 98. Richardson 1990:70. Lee *et al.* 2009: 12

Dextral. Ovate-conic. Thick. Umbilicus narrow. 5¼-5½ slightly convex whorls, apex obtuse. Suture slightly impressed. Aperture elongately elliptical, angled. Lip very thick, reflexed outwardly and with an internal thickened ledge. Parietal callus thin and transparent.
Dark brown with white suture. Body whorl with very narrow, ill-defined dark brown peripheral band. Lip white.
Body light brown, black marks on the mantle are faintly visible though the shell as large, irregular patches.
Anatomy: As *S. attenuata.*
Type specimens: Holotype BPBM 9006; paratypes BPBM 206910-16 (27 specimens).
Distribution: Tahiti, recorded from above 985-1600 m (Hamuta, Haoma, Taravao, Mt. Atara and Mt. Aorai). Arboreal, found at 1-3 m above the ground (Kondo 1973).

Samoana diaphana (Crampton & Cooke, 1953)

Partula diaphana Crampton & Cooke, 1953: 148

Samoana diaphana (Crampton & Cooke), Kondo 1955: 93. 1968: 75,. 1973: 19. Clarke & Murray 1969: 31. Kondo & Burch 1983: 105. Johnson *et al.* 1986: 98. Richardson 1990: 72. Lee *et al.* 2009: 12. Coote & Loève 2003: 96

Dextral. Semi-globose, moderately ovately conic. Extremely thin, fragile, transparent. Umbilicus minute, rimate. Slightly shiny, growth lines faint, spiral striae fine and dense on all whorls and into aperture. Apex obtuse. Body whorl not compressed. 4–4½ flatly convex whorls, spire slightly convex. Suture weakly impressed, scarcely visible narrow margin on mid-dorsal region only. Aperture ovately pyriform. Lip thin, very slightly expanding but not thickened, narrows towards outer insertion. Columella slightly twisted, somewhat compressed over umbilicus. Parietal callus very thin, indistinct.
Pale brown corneous.
Body light brown, mantle irregularly maculated with large whitish and blackish bold ziz-zag marks which show through the shell clearly.
Anatomy: As *S. attenuata* except spematheca shorter and attaching nearer to genital atrium, the caecum unusually well developed for a *Samoana*, the constriction at the junction of the penis and penial caecum less distinct. Vas deferens terminal. Vestigial retractor not present. Kondo (1973) thought the inflated form of the caecum to be a temporary state but this was disproved by later dissection of a specimen from Tahiti (Kondo & Burch MS).
Type specimens: Holotype BPBM 136041 (=9973).
Distribution: Tahiti and Moorea, mainly at higher altitudes (above 600 m) on Mt. Aorai and Mt. Pihaaiateta and in Hamuta valley on Tahiti, and from Hotutea, Faatoai, Maatea and Uufau on Moorea.

Fig. 6.30 *Samoana burchi*: a – holotype (BPBM 9006), b – specimen (UMMZ)

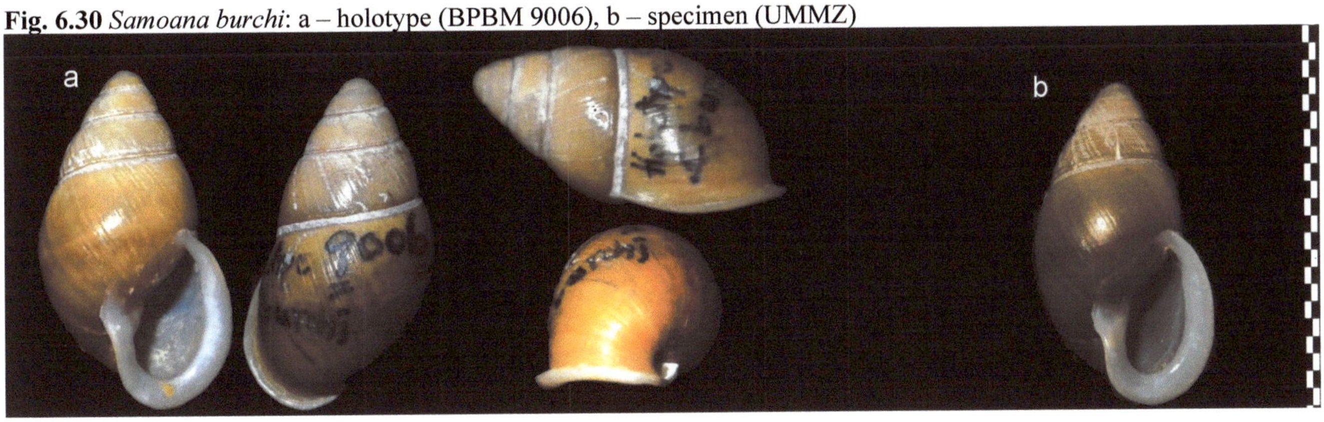

Fig. 6.31 *Samoana diaphana*: a – holotype (BPBM 136041); b – Moorea specimen (Clarke & Murray – NHMUK); c-e – Tahiti specimens (c – BPBM 206773; d – NHMUK 20140798 Mt. Aorai; e – JG, Pic Rouge)

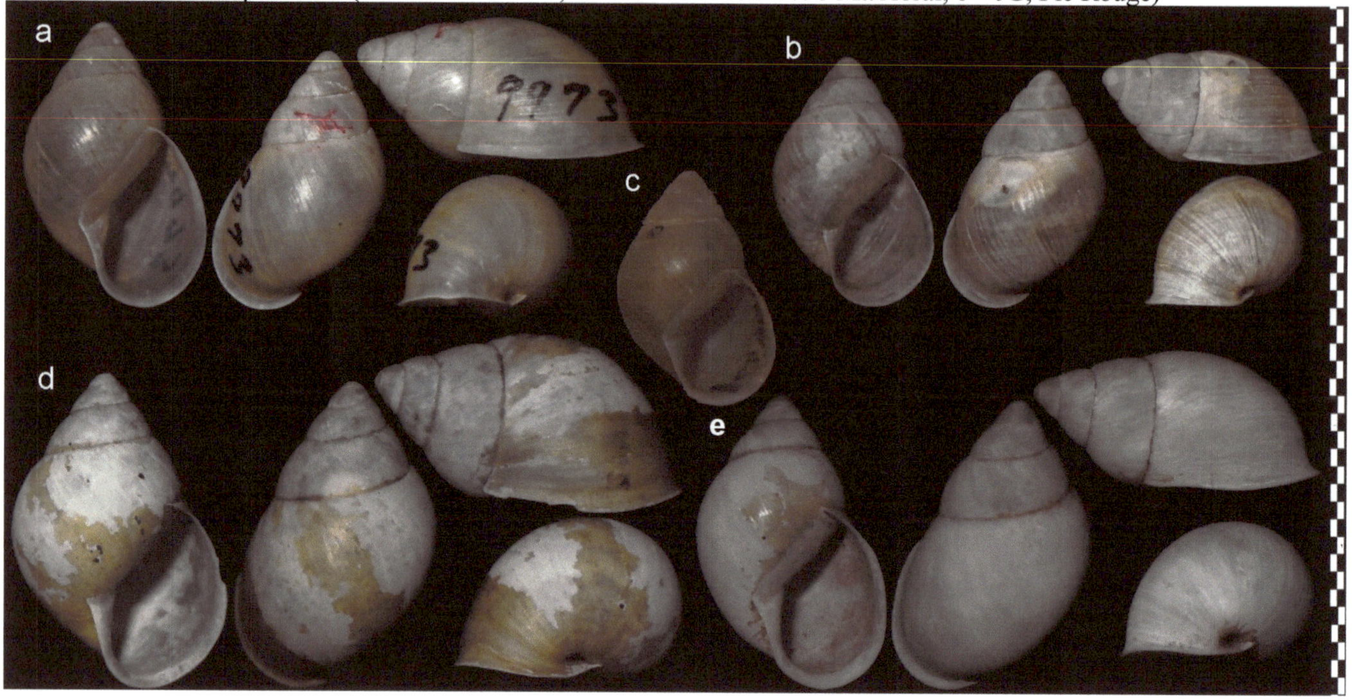

Fig. 6.32 Live *Samoana*: a – *S. fragilis*; b – *S. attenuata*; c – *S. annectens*; d – *S. decussatula*; e – *S. burchi*; f – *S. diaphana*; g – *S. ganymedes*; h – *S. gonochila*; i – *S. bellula*; *j* - *S. conica* (photos: a – G.C. Fiedler, b, e & f – T. Coote, c – F. Jacq, d – J.-Y. Meyer, g-i – D. Clarke, j - R. Cowie)

7. ***Palaopartula*** Pilsbry, 1909

Palaopartula Pilsbry, 1909: 307 (*Partula* section)
Palaeopartula Pilsbry, Richardson 1990: 6 (error)
Partula (*Palaopartula* Pilsbry, 1909) Schileyko 1999: 271
Palaopartula Pilsbry, Slapcinsky & Krauss 2016: 32

Dextral. Elongately conical. Spire elongate. Thin but strong. 5-6 moderately convex whorls. Protoconch with raised spiral striae, adult shell crossed by raised wavy spiral striae and coarse growth lines, giving a distinct granular appearance. Body whorl compressed laterally, base saccate. Umbilicus deeply perforated. Aperture unusually long, lip broadly expanded. Penis moderately long, slender, tapering. Penial caecum usually small, but may be well developed. Penis and base of caecum internally ornamented with ridges in a grid-like arrangement. Vas deferens enters caecum subterminally.

Type species: *Partula thetis* Semper, 1865

Radula: 14 laterals, distinction between laterals and marginals indistinct.

Distribution: Palau islands. All *Palaopartula* are arboreal (Kondo 1955).

Notes: Pilsbry's (1909) *Partula* section *Palaopartula* was overlooked by most subsequent authors. The molecular phylogeny of Partulidae (Lee *et al.* 2014) clearly shows the Palau species to form a monophyletic sister group to *Samoana*, distinct from *Partula*. In addition the shells are unique and the radula is also distinctive within Partulidae. Three species are known.

Table 7.1 *Palaopartula* shell measurements

	N	L			W			L aperture			W/L		
		range	mean	sd	range	mean	sd	range	mean	sd	range	mean	sd
P. calypso	38	25.7-31.0	27.65	1.227	11.7-15.4	13.59	0.886	9.8-14.1	11.79	0.914	37.8-48.8	43.08	2.673
P. thetis	106	25.0-30.3	27.03	0.951	12.7-16.0	13.47	0.470	13.6-17.0	14.84	0.566	50.0-54.3	49.83	1.682
P. leucothoe	101	17.8-22.0	20.02	0.570	9.9-13.0	10.88	0.355	9.2-13.9	11.32	0.329	52.1-57.1	54.33	1.575

Fig. 7.1 *Palaopartula* range and anatomy (anatomy after Kondo 1955)

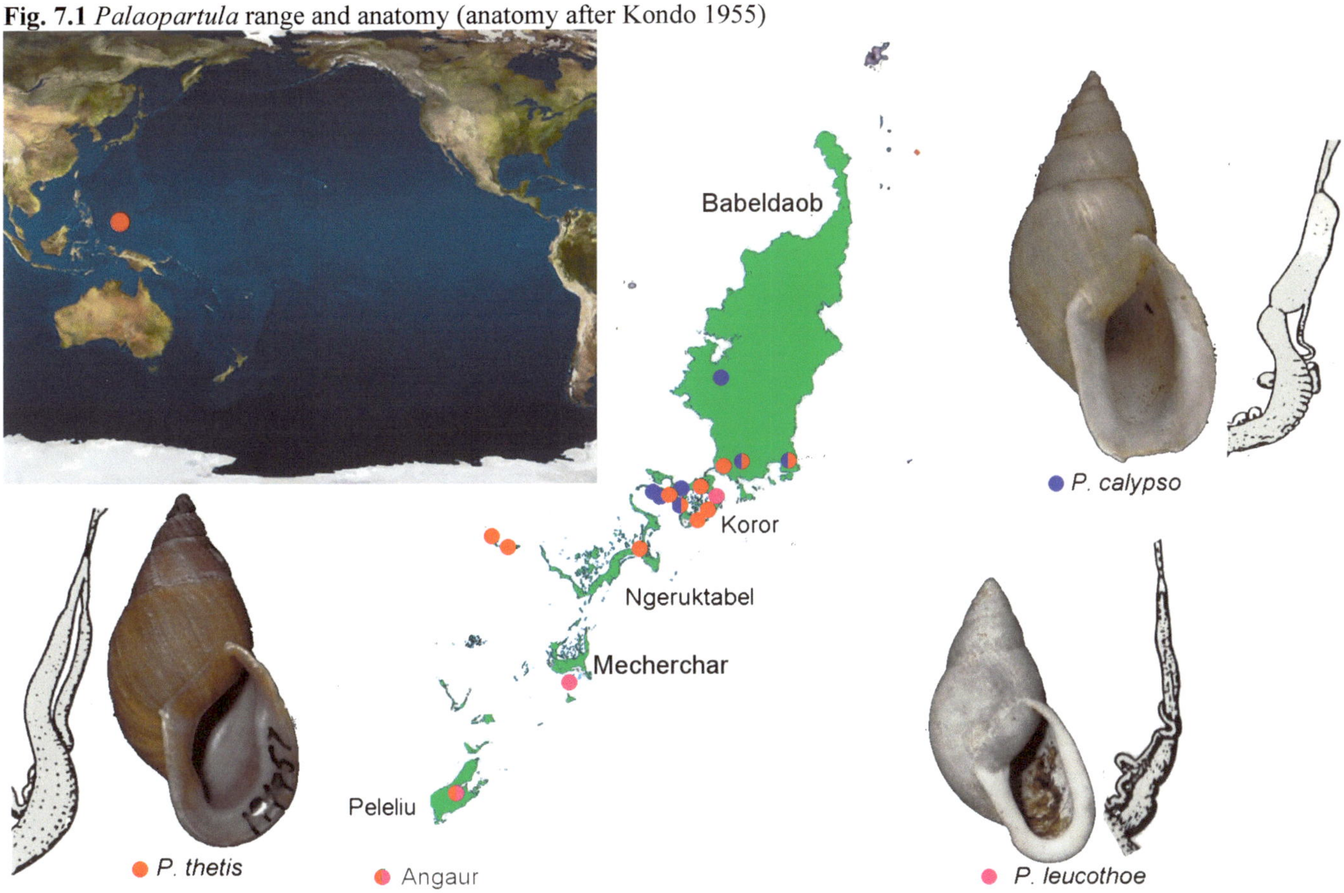

Palaopartula calypso (O. Semper, 1865)

Partula calypso O. Semper, 1865: 418

Partula calypso O. Semper, Pfeiffer 1868: 154, 1877: 194. Pease 1871b: 473. Hartman 1881a: 180, 1881b: 13, 1885: 218, 1892: 98. Pfeiffer & Clessin 1881: 300. Paetel 1889: 266. Smith 1903: 462. Pilsbry 1909: 307. Kondo 1955: 207, 1968: 77. Kilias 1969: 246. Richardson 1990: 15

Palaopartula calypso (O. Semper), Slapcinsky & Krauss 2016: 37

Dextral. Elongately ovate. Spire elongate, very acute. Thin but strong. Umbilicus open. 6 almost flat whorls. Protoconch with fine raised spiral striae, continuing onto shell where they are crossed by coarse growth lines, giving a distinctly granular appearance. Body whorl broadly compressed. Longitudinally somewhat wrinkled, with wavy spiral striae. Suture raised, striolate where crossed by growth lines. Columella nearly straight, oblique, with a slight internal nodular bulge. Aperture slightly oblique, oblong. Lip broadly expanded, especially over the umbilicus. Parietal callus thin, distinct only near columella. Whitish to light brown, spire darker, rosy towards the apex. Lip white.

Anatomy: Penis moderately elongate, ½ oviduct length. Caecum as long as penis. Penis broad at base, with a constriction between penis and caecum. Penis internally ornamented with 6-8 diagonal pilasters, terminating in knobs at constriction with caecum. Caecum with dense papillae in zig-zag crevices, apex tuberculate. Vas deferens enters caecum near base.

Type specimens: Lectotype ZMB Moll-39125a; paralectotypes ZMB Moll-39125b; ZMH 6925. Lectotype designation by Kilias (1969).

Distribution: Babeldaob ('Babelthaup'), Koror, Ngermalk/Malakal islands (records from Kondo 1955, Ó Foighil & Rundell 2012a, FMNH). Semper (1865) recorded the species from Peleliu, this may be an error. Partulids thought to belong to this species recorded on Mecherchar and Angaur island in 2003-7 (Rundell 2010) are probably *P. thetis*. Only four specimens in BPBM (compared to 170 *P. thetis* – Kondo 1955).

Palaopartula thetis (O. Semper, 1865)

Partula thetis O. Semper, 1865: 419

Partula thetis O. Semper, Pfeiffer 1868: 155, 1877: 194. Pease 1871b: 473. Pfeiffer & Clessin 1881: 300. Paetel 1889: 268. Pilsbry 1909: 307. Thiele 1931: 659. Kondo 1955: 209, 1968: 77. Zilch 1959: 461

Palaopartula thetis (O. Semper), Slapcinsky & Krauss 2016: 37

Dextral. Conically pyramidal, sometimes elongate. Moderately thick. Spire conic, apex pointed but mamillate. Umbilicus open, compressed. 5½-6 moderately convex whorls, ornamented with deep wavy spiral striae. Suture somewhat threadlike, variably striolate. Body whorl convex, regularly rounded, somewhat ascending in front. Columella vertical, slightly convex, with an internal basal nodular bulge. Aperture oblong-rounded, almost vertical, sometimes subauriculate. Lip broadly expanded, thin and without internal thickening, tapering rapidly from upper third to insertion. Parietal callus thin but distinct, straight.

Fig. 7.2 *Palaopartula calypso* paralectotypes: a – ZMH 6925; b – ZMB 39125

Light yellowish-brown with reddish on spire to dark brown with spire blackish. Some individuals have greenish-brown streaks on the spire and white streaks on the body whorl.
Body brown, mantle in preserved specimens spotted with white (ZMH 9828).
Anatomy: Penis moderate, ½ oviduct length. Slender with broad stalk, tapering. Penis with 10-15 diagonal pilasters and horizontal ridges forming a grid-like pattern, extending to middle of caecum, upper ⅓ smooth. Vas deferens nearly terminal, retractors short. Egg calcified.
Type specimens: Syntypes BPBM 87546; ZMH 6927.
Distribution: Widespread ("distributed universally over the numerous Palau islands" – Kondo 1955); specifically recorded from Babelthaup, Koror, Ngermalk, Auluptagel/Aurapushekaru, Ngeruktabel, Ulong islands (east and west), Peleliu and Angaur. The records from Peleliu are very sparse; Semper's (1865) syntypes were apparently from that island and there is a single specimen in BPBM (159790). There is also only one specimen from Angaur (BPBM 158199). Partulids on Mecherchar and Angaur island in 2003-7 and identified as *P. calypso* (Rundell 2010) may be *P. thetis*. One record from Kayangel island (ANSP 191958) is very isolated from all other records and may be questionable. Records from Kondo (1955), Ó Foighil & Rundell (2012b), BPBM and ANSP. Arboreal, in *Pandanus* axils (Slapcinsky & Kraus 2016).

Palaopartula leucothoe (O. Semper, 1865)

Partula leucothoe O. Semper, 1865: 417

Partula leucothoe O. Semper, Pfeiffer 1868: 155, 1877: 194. Schmeltz 1869: 71. Pease 1871b: 473. Pfeiffer & Clessin 1881: 300. Paetel 1889: 267. Pilsbry 1909: 306. Kondo 1955: 211, 1968: 77

Partula leucothoea Smith 1902: 479 (as syn. of *calypso*)

Palaopartula leucothoe (O. Semper), Slapcinsky & Krauss 2016: 37

Dextral. Conic-pyramidal, thin. Umbilicus open, compressed. Spire pyramidal, rather acute. 5½ almost flat whorls ornamented with dense, rippled spiral striae. Body whorl somewhat ascending in front. Slightly depressed below suture. Columella oblique, slightly concave. Aperture oblique, oblong, slightly compressed, margins subparallel. Lip thin, broadly expanded. Parietal callus thin, indistinct.
White, subhyaline. Lip white
Anatomy: Penis moderately long, ⅓ oviduct length, broad at base, tapering. Penis with 7-8 pilasters, terminating at neck. Caecum with 4-5 ridges running entire length.
Type specimens: Syntypes ZMH 6926.
Distribution: Peleliu (ANSP 170817, 294471, BPBM 159776), Kltaltebechel islet off Mecherchar (Ó Foighil & Rundell

Fig . 7.3 *Palaopartula thetis*: a – syntype (ZMH 6927); b & c – ANSP 191957

2012c) and Angaur (ANSP 294472). Crampton (MS) also listed measured a specimen apparently from Koror.
Note: Crampton (MS) labelled the Angaur specimens (ANSP 294472) *leucochroa*, noting it a possible new species. He did not explain this and there are no reasons apparent for separating this population.

Fig. 7.4 *Palaopartula leucothoe*: a – syntype (ZMH 6926); b – Crampton's 'leucochroa' specimen (ANSP 294472)

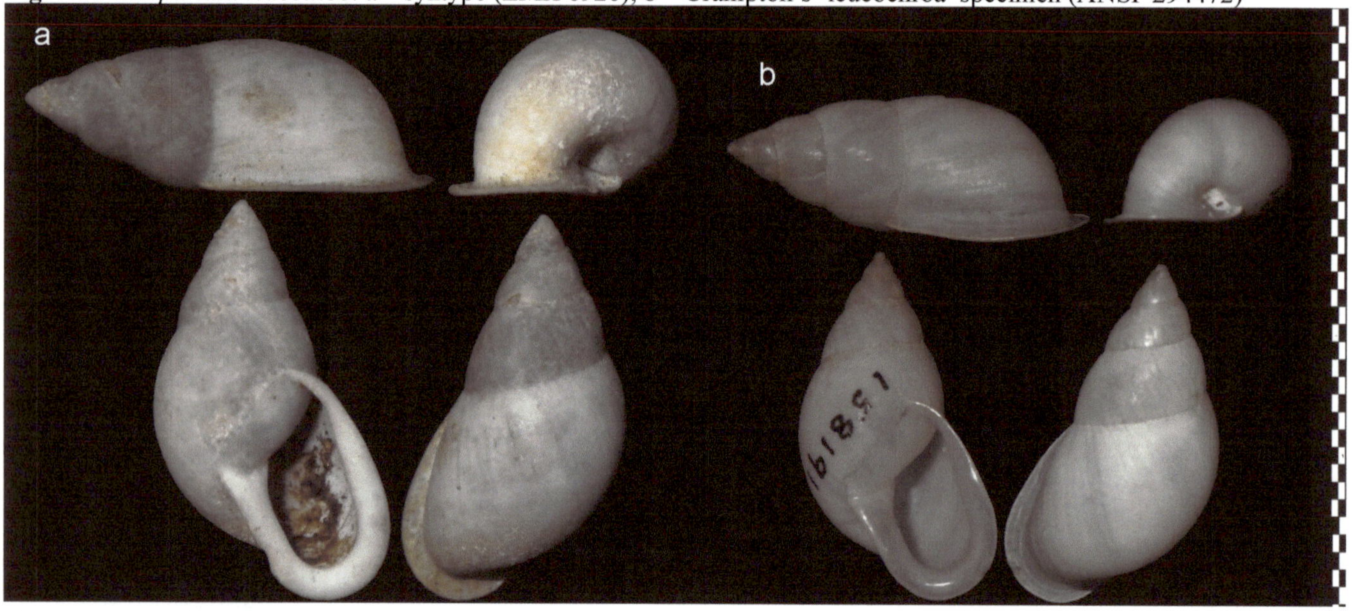

Fig. 7.5 Live *Palaopartula*: *P. thetis*, juvenile *P. calypso* and adult *P. calypso* (photos: R. Brewer)

8. ***Sphendone*** Slapcinsky & Krauss, 2016

Sphendone Slapcinsky & Krauss, 2016: 34

A monotypic genus from the Palau islands. Its phylogenetic placement is ambiguous; molecular phylogenies placing it with *Palaopartula* and *Samoana*, or between these and *Eua* (J. Slapcinsky pers. comm.). Diagnosis as for the type species.

Sphendone insolita Slapcinsky & Krauss, 2016

Sphendone insolita Slapcinsky & Krauss, 2016: 34

Normally dextral (2.5% sinistral, n=79). Obese-pupoid. 4½-5 moderately convex whorls with very faint, spiral striae, becoming more conspicuous on last two whorls. Umbilicus deeply perforated. Aperture vertical, base evenly rounded. Parietal callus distinct, raised at edges. Lip broadly expanded.

Brown, with indistinctly paler subsutural band on spire. Penultimate whorl paler or straw yellow-brown, with a narrow indistinct darker-brown subsutral band. Body whorl darker. Aperture brown internally. Rarely entirely white (4%, n=79).

Penis moderately long, robust. Penial caecum equal to penis in length. Penis and base of caecum internally ornamented with 10 longitudinal pilasters. coalescing to fomr a ridge between penis and caecum. Caecum with 10 thin, wiedly separated pilaster. Vas deferens and penial retractor apical.

Radula: 10 laterals.

Distribution: Palau islands (Ngeruktabel and Mecharchar). Terrestrial, found under rocks.

Type specimens: Holotype UF 42857, paratypes UF 271885-6, 248689-90, 332693.

Table 8.1 *Sphendone* shell measurements (from Slapcinsky & Kraus 2016)

	N	L			W			L aperture			W/L
		range	mean	sd	range	mean	sd	range	mean	sd	
S. insolita	56	18.3-22.7	21.10	0.800	12.9-14.9	14.00	0.400	10.5-21.1	11.30	0.400	63.6

Fig. 8.1 *Sphendone insolita* range, shell and anatomy (photos – J. Slapcinsky)

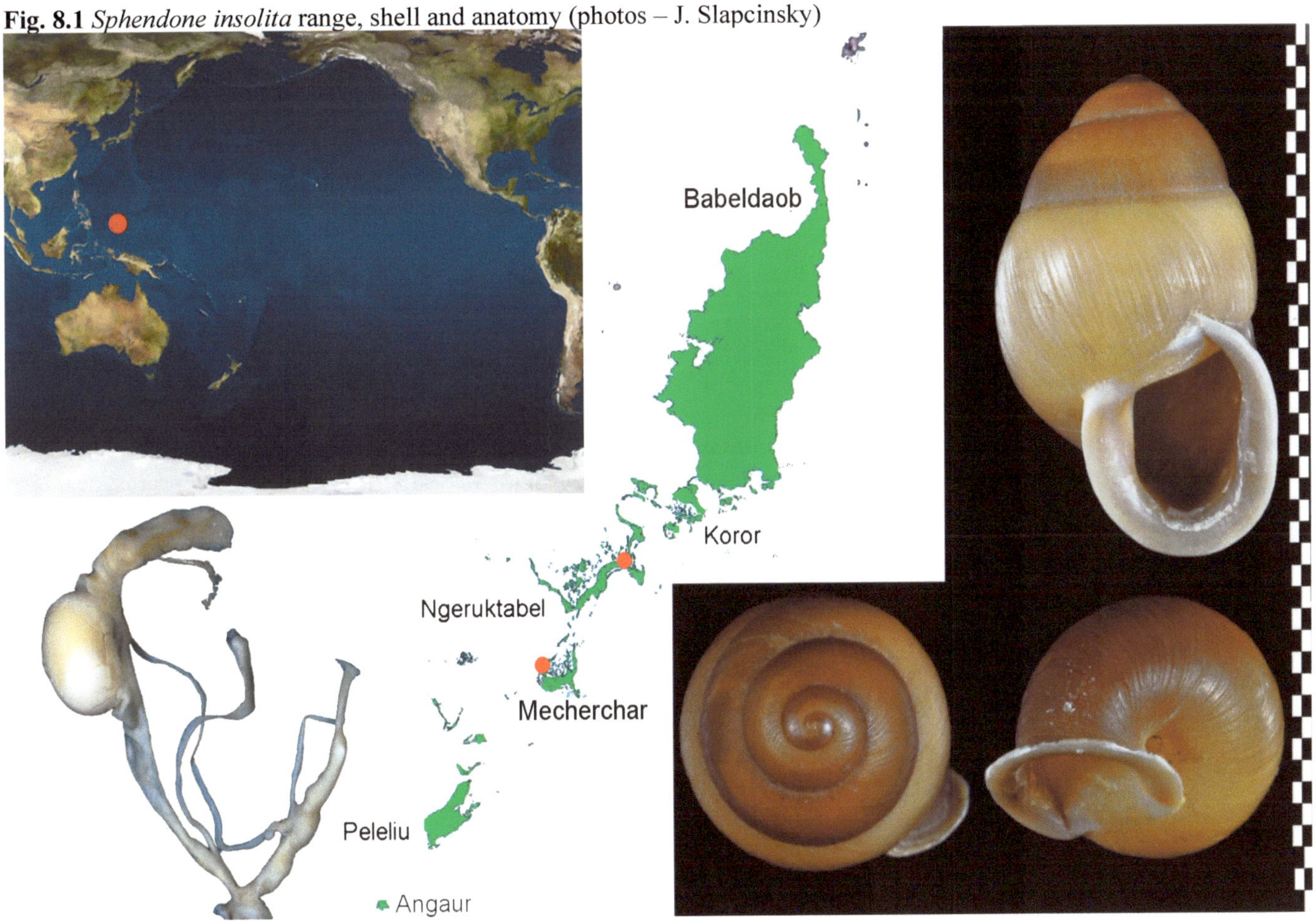

9. ***Partula*** Férussac, 1821

Partula Férussac, 1821: 65
Partulus Beck, 1837: 57 (error)
Partula Férussac, Herrmannsen 1852: 204. Hartman 1881a: 171, 1885: 206. Smith 1903: 422. Pilsbry 1909: 155. Solem 1959: 64. Richardson 1990: 6
Nenia Hartman, 1881b: 7 (*Partula* subgenus) [preoccupied]
Astræa Hartman, 1881b: 7 (*Partula* subgenus) [preoccupied]
Clytia Hartman, 1881b: 8 (*Partula* subgenus) [preoccupied]
Ilia Hartman, 1881b: 8 (*Partula* subgenus) [preoccupied]
Œnone Hartman, 1881b: 9 (*Partula* subgenus) [preoccupied]
Helema Hartman, 1881b: 9 (*Partula* subgenus) [preoccupied]
Pasithea Hartman, 1881b: 10 (*Partula* subgenus) [preoccupied]
Matata Hartman, 1881b: 14 (*Partula* subgenus) [preoccupied]
Sterope Hartman, 1881b: 14 (*Partula* subgenus) [preoccupied]
Harmonia Hartman, 1881b: 13 (*Partula* subgenus) [preoccupied]
Echo Hartman, 1881b: 11 (*Partula* subgenus) [preoccupied]
Takombaua Pilsbry, 1909: 165 (*Partula* section)
Leptopartula Pilsbry, 1909: 165 (*Partula* subgenus) [preoccupied]
Melanesica Pilsbry, 1909: 166 (*Partula* subgenus)
Carolinella Pilsbry, 1909: 166 (*Partula* subgenus)
Marianella Pilsbry, 1909: 166 (*Partula* subgenus)
Marianna Pilsbry, 1909: 313 (error)
Thacombaua Pilsbry, Thiele, 1931: 659 (misspelling)
Rennellia Clench, 1941: 20 (*Partula* section)
Melanisica Pilsbry, Richardson 1990: 6 (misspelling)
Takombana Pilsbry, Richardson 1990: 6 (misspelling)
Eua (*Thakombaua*) Pilsbry, Schileyko 1999: 264
Partula Férussac, Schileyko 1999: 267
Partula (*Partula*) Schileyko 1999: 268
Partula (*Melanesica*) Pilsbry, Schileyko 1999: 268
Partula (*Carolinella*) Pilsbry, Schileyko 1999: 269
Partula (*Marianella*) Pilsbry, Schileyko 1999: 269
Partula (*Leptopartula*) Pilsbry, Schileyko 1999: 270
Partula (*Rennellia*) Clench, Schileyko 1999: 271
Rennellia Clench, Schileyko 1999: 271

Shell dextral or sinistral, globular to elongate-conical. Thin to solid, 4-6 moderately to highly convex whorls. Protoconch spirally striated, sculpture of spiral incised lines, most prominent on spire. Aperture ovate, slightly oblique, lip usually expanded and reflexed. Parietal tooth sometimes present. Penis short or long, very varied in form. Penial caecum present and usually distinct. Penis ornamented internally with longitudinal fleshy ridges but no large pilasters, caecum containing rows of irregular tubercles, sometimes on ridges. Vas deferens enters caecum subterminally. Penial retractor inserted near apex of caecum. Egg capsule is calcified in most species.
Radula: 5-14 laterals, usually 8-11.
Type species: *Helix faba* Gmelin, 1792.
Distribution: New Guinea, Solomons, Vanuatu, Carolines, Marianas, Fiji, Cooks, Australs, Marquesas and Societies. Also reported from the Talaud islands in Indonesia, but probably in error.
Notes: *Partula* is the most diverse of the Partulidae genera, with 264 named taxa, of which 77 are valid species. Pilsbry (1909) divided the genus into eight subgenera, of which *Samoana* and *Palaopartula* have been raised to separate genera (and *Samoana* contains *Marquesana* Pilsbry, 1909). The remaining subgenera within *Partula* are: *Partula* s.s., *Leptopartula* Pilsbry, 1909, *Thakomeua* Pilsbry, 1909, *Melanesica* Pilsbry, 1909, *Carolinella* Pilsbry, 1909 and *Marianella* Pilsbry, 1909. Of these, *Leptopartula* is a polyphyletic grouping of thin-shelled *Partula*. The remaining groups are monophyletic but of limited taxonomic value and are not used here.

9.1 Melanesia *Partula*

39 species have been described from New Guinea, the Solomon islands and Vanuatu, these are reduced to 6 here.

Table 9.1.1 Melanesian *Partula* shell measurements

	N	L			W			L aperture			W/L		
		range	mean	sd	range	mean	sd	range	mean	sd	range	mean	sd
grisea	36	15.5-19.9	17.25	1.358	7.0-10.3	8.766	0.781	7.0-10.0	8.16	0.893	47.6-55.1	50.62	2.251
similaris	120	13.5-20.3	19.04	0.555	7.1-11.7	10.88	0.368	7.0-11.1	10.27	0.368	53.3-61.5	57.20	2.532
flexuosa	16	14.7-19.9	16.65	0.979	8.2-10.6	9.26	0.391	7.1-8.6	7.94	0.506	48.4-61.6	55.76	2.775
micans	23	13.0-17.4	15.35	0.813	7.0-10.6	8.48	0.430	7.0-8.9	7.30	0.341	48.4-61.6	55.12	2.081
cramptoni	23	21.5-28.8	23.80	0.686	12.8-15.0	13.80	0.463	9.6-13.5	12.92	0.547	52.0-63.3	58.63	1.718
vanikorensis	25	18.9-27.7	21.84	1.208	10.0-14.6	13.06	1.187	9.1-10.7	10.20	0.553	54.5-65.3	59.76	3.382
radiosa	2	19.8-23.9	21.85	-	11.4-13.2	12.30	-	9.7-11.8	10.75	-	55.2-57.8	56.50	-
pacifica sinistral	1	-	17.19	-	-	14.60	-	-	12.00	-	-	65.00	-
pacifica dextral	336	21.0-28.9	24.25	2.591	10.9-15.0	13.53	0.593	9.5-13.0	12.35	0.737	51.1-62.1	57.03	1.888
pyramis	25	21.7-25.0	22.48	1.003	11.7-14.0	12.00	0.435	11.5-13.0	11.56	0.547	50.2-57.5	52.03	1.805
auraniana	55	17.5-23.8	18.56	0.594	10.8-12.3	11.54	0.602	7.6-11.9	8.68	0.707	50.4-65.2	61.78	2.599

Fig. 9.1.1 New Guinea and Solomons *Partula*. Islands: 1 – Manus, 2 – Mussau, 3 – New Hannover, 4 – New Ireland, 5 – New Britain, 6 – Trobriands, 7 – d'Entrecasteaux, 8 – Woodlark, 9 – Laughlans, 10 – Louisiades, 11 – Bougainville, 12 – Choiseul, 13 – Ontong Java, 14 – Sta. Isabel, 15 – Shortlands, 16 – New Georgias, 17 – Guadalcanal, 18 – San Cristobal, 19 – Bellona, 20 – Renell, 21 – Sta. Cruz

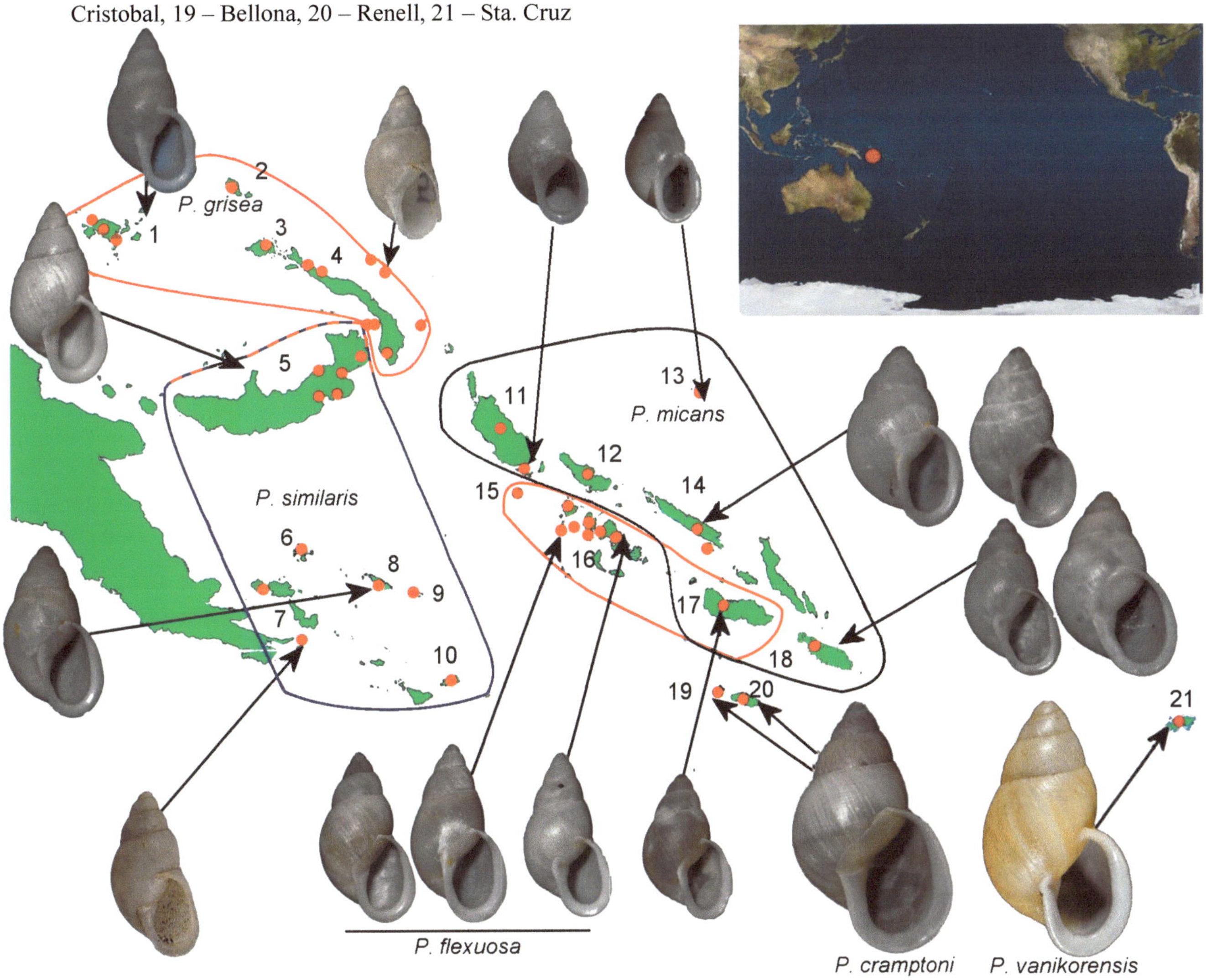

New Guinea

Two similar species are present on the western peninsula of New Guinea and the islands off its west coast. These are adaptable species found on several islands and may have been spread by human activity.

Partula grisea (Lesson, 1830)

Bulimus grisea Lesson, 1830: 325
Helix carteriensis Quoy & Gaimard, 1832: 117
Bulimus carteriensis Quoy & Gaimard, Deshayes & Milne-Edwards 1839: 370
Bulimus griseus Lesson, Pfeiffer 1841: 82
Partula carteriensis Quoy & Gaimard, Pfeiffer 1842: 112, 1848: 68, 1853a: 460, 1856: 162, 1859: 509, 1868: 157, 1877: 198. Albers 1850: 186. Morelet 1853: 372. Adams 1855: 145. Schmeltz 1869: 72. Paetel 1873: 104, 1883: 152, 1889: 266. Kobelt 1880: 16. Hartman 1881a: 180, 1885: 216, 1892: 75. Pfeiffer & Clessin 1881: 301. Smith 1903: 451. Pilsbry 1909: 298. Thiele 1928: 124. Rensch 1937: 580. Kondo 1955: 186, 1968: 76. Richardson 1990: 16
Partula grisea Lesson, Pfeiffer 1842: 115, 1848: 68, 1853a: 450, 1859: 510, 1868: 159, 1877: 199. Catlow & Reeve 1845: 163. Adams 1855: 145. Paetel & Schaufuss 1869: 83. Paetel 1873: 104, 1883: 153, 1889: 267. Hartman 1881a: 181, 1881b: 11, 1885: 217. Pfeiffer & Clessin 1881: 301. Brazier 1883: 105. Smith 1903: 453. Pilsbry 1909: 304. Kondo 1968: 76. Richardson 1990: 31
Partulus griseus Lesson, Albers 1857: 99
Sterope carterensis Quoy & Gaimard, Hartman 1881b: 14
Partula hartmanni E.A. Smith, 1884: 265
Partula hartmanni E.A. Smith, Hartman 1885: 216, 1892: 75. Paetel 1889: 267. Smith 1903: 457. Pilsbry 1909: 301. Kondo 1955: 188
Partula obliterata Pilsbry, 1909: 300
Placostylus dorseyi Dall, 1910: 219
Partula gerardensis Soos, 1911: 355
Partula mathildae C.R. Boetgger, 1918: 296
Partula paravicini Clench, 1933: 24
Partula obliterata Pilsbry, Rensch 1937: 580
Partula gerardensis Soos, Rensch 1937: 580
Partula paravicini Clench, Rensch 1937: 580
Partula (*Melanesica*) *flexuosa dorseyi* Dall, Haas 1939: 103
Partula dorseyi (Dall), Kondo 1955: 187, 1968: 76. Richardson 1990: 22

Fig. 9.1.2 New Guinea and Solomons *Partula* anatomy (except *P. vanikorensis*, see Fig. 9.1.9): a-d – *P. grisea*; e – *P. similaris*; f-i – *P. micans*; j – *P. cramptoni* (Rennell island). a – Nanua (*carteriensis*); b – Rambyuto (*hartmanni* NHMUK 1884.6.12.224); c – Rambyuto (*hartmanni*); d – Malie (*dorseyi*) & Guadalcanal BPBM 194055 (*flexuosa*); e – New Britain (UF 366509); f – Kulambangara (*coxi*); g – Guadalcanal (*micans*); h – Sta. Isabel (NHMUK 2184) and Ontong Java (NHMUK 1957.11.8.83,86-90); i – Shortlands (UMMZ). a, c-d, f-g, j after Kondo (1955). Scale bar 5 mm.

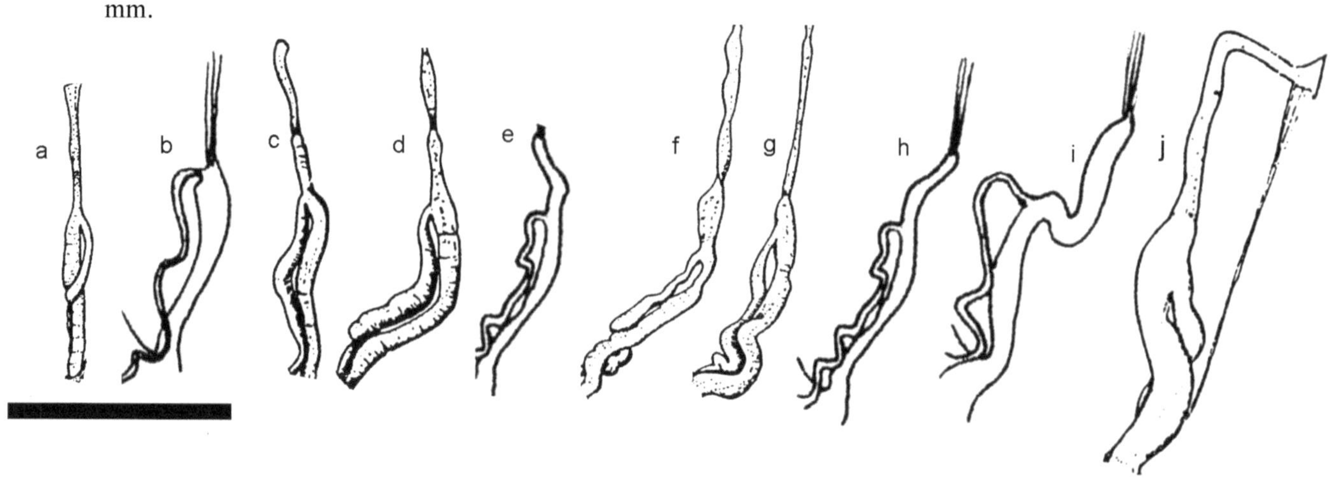

Dextral. Elongate, apex acute. Thin, translucent and glossy. Umbilicus narrow. 4½-5 slightly convex whorls with transverse growth lines and variably distinct fine spiral striae. Body whorl slightly inflated, compressed laterally, base saccate. Aperture oval, inflected. Lip narrowly reflexed, internally thickened, thin near the posterior angle, the outer margin slightly sinuous. Grey to pale horn with a pinkish tip. Base varies from whitish to brown, the spire usually darker. Lip and aperture white.
Anatomy: Penis short, ½ oviduct length, simple. Caecum distinct but short. Internally ornamented with minute tubercles, very weakly developed pilasters in penis. Penial caecum narrow at apex, vas deferens subapical.
Type specimens:
Bulimus grisea Lesson, 1830 – Apparently lost.
Helix carteriensis Quoy & Gaimard, 1832 – Syntypes MNHN-IM-2000-27658.
Partula hartmanni E.A. Smith, 1884 – Holotype and paratypes NHMUK 1886.12.169-172 (dry – holotype & 3 paratypes), 1884.6.12.224-232 (spirit paratypes).
Partula obliterata Pilsbry, 1909 – Syntypes ANSP 10119.
Placostylus dorseyi Dall, 1910 – Holotype FMNH 11254.
Partula gerardensis Soos, 1911 – Holotype HNHM, destroyed in World War II.
Partula mathildae C.R. Boetgger, 1918 – Lectotype SMF 5942; paralectotypes SMF 5943, 90465; ZMB Moll-75984 (2 paratypes ex SMF).
Partula paravicini Clench, 1933 – Holotype MCZ 59854; paratypes ANSP 160346, MCZ 59855.
Distribution: New Guinea islands – Admiralty islands (Manus, Rambyuto), Mussau (St. Matthias), Bismarck archipelago (Anair, Lihir, Mahur, Malie, New Hanover, New Ireland). *P. dorseyi* was recorded from the Lihir group of the Bismarck archipelago (type locality Anair ['Amir'], also Malie). *P. hartmanni* was recorded from the Admiralty islands (Rambyuto and two islets off the north-west coast of Manus island: 'Wild' and 'Pigeon' Islands). *P. obliterata* was of uncertain origin. *P. paravicinii* from St. Matthias island. The type locality for *P. grisea* is not certain; but Pilsbry thought that La Coquille only landed in the west of New Guinea, however, he was mistaken and La Coquille called at New Ireland, York and New Britain. Ó Foighil *et al.* (2011) propose that both *P. similaris* and *P. grisea* (as *P. carteriensis*) were introduced to the New Guinea islands in prehistoric times, possibly from the Solomon islands.
Note: Kondo (1955) thought that *P. hartmanni* was distinct as its eggs were calcified whereas *P. carteriensis* were uncalcified,

Fig. 9.1.3 New Guinea and Bismarck archipelago *Partula* types: a-g *P. grisea*; h-k *P. similaris*. a – *P. grisea* original figure (Lesson 1830); b – *P. carteriensis* MNHN IM-2000-2765; c – *P. hartmanni* NHMUK 188.6.12.169; d – *obliterata* ANSP 10119; e – *dorseyi* FMNH 11254; f – *mathildae* ZMB 75984; g – *paravicinii* ANSP 160346; h – *occidentalis* QM MO2811; i – *kubaryi* CMNH 624247; j – *similaris* NHMUK 96.11.30.12; k – *woodlarkiana* NHMUK 96.11.30.10; l – Manus *grisea* umbilical view BPBM 1966-264 (not type) (photos: b – M. Caballer; e – J. Gerber; h – D. Potter; i – C. Sturm)

however this reflected the acidity of the preservative rather than the actual egg-shell (Kondo MS), and he did not list *P. hartmanni* as a separate species in 1968.

Partula similaris Hartman, 1886

Partula similaris Hartman, 1886: 30
Partula woodlarkiana Hartman, 1886: 33
Partula kubaryi Hartman, 1888: 284
Partula kubaryi Hartman, Pilsbry 1909: 299. Leschke 1912: 107
Partula similaris Hartman, Hedley 1892: 97. Smith 1903: 454. Pilsbry 1909: 302. Soos 1911: 355. Iredale 1941: 64. Kondo 1968: 76. Richardson 1990: 56
Partula occidentalis Hedley, 1892: 98
Partula woodlarkiana Hartman, Hedley 1892: 98. Smith 1903: 454. Iredale 1941: 64
Partula occidentalis Hedley, Pilsbry 1909: 303. Iredale 1941: 64. Kondo 1968: 76. Richardson 1990: 43

Dextral. Elongately ovate-conical, apex acute. Thin, translucent and glossy. Umbilicus narrow. 4½-5 convex whorls with transverse growth lines and variably distinct fine spiral striae. Body whorl slightly inflated, compressed laterally, base saccate. Aperture rounded oval, inflected. Lip narrowly reflexed, internally thickened, thin near the posterior angle, the outer margin slightly sinuous. Parietal callus shiny, not strongly margined.

Whitish to pale brown, the spire usually darker. Lip white.

Anatomy: Penis shortened *turneri*-type with very weak pilasters. Penial caecum narrow at apex, vas deferens subapical.

Type specimens:
Partula similaris Hartman, 1886 – Holotype NHMUK 96.11.30.1.
Partula woodlarkiana Hartman, 1886 – Holotype NHMUK 96.11.30.10.
Partula kubaryi Hartman, 1888 – Holotype CMNH 62.4247.
Partula occidentalis Hedley, 1892 – Syntypes QM MO2811(=990826).

Distribution: New Guinea south east peninsula (East Cape) and islands – Bismarcks (Lolobau, New Britain, Watom/Vuatom), d'Entrecasteaux (Boiaboiawaga islet, Goodenough, Samarai), Trobriand (Kiriwina), Louisiade archipelago (Rossel/Yela), Woodlark (Muyua), Laughlan Islands.

The population on East Cape peninsula of Papua New Guinea (Clench 1936) is the only large island partulid population. It is found at low altitudes, but is not recorded from high forest sites. It is locally common on Boiaboiawaga islet and around the village of Guasapa on Muyua/Woodlark (Ó Foighil *et al.* 2011). The species may be locally abundant in small pockets around some villages, such as Marmar and Ulege villages on New Britain, but is not present in native forest (Ó Foighil *et al.* 2011).

Note: Includes *similaris*, *woodlarkiana* and *occidentalis* on the basis of their close genetic relationship (Ó Foighil *et al.* 2011). *kubaryi* was recorded from New Britain (Bismarck Archipelago), *occidentalis* from Samarai, *similaris* from the Louisiade group, *woodlarkiana* from Woodlark island, *obliterata* was of uncertain origin. Ó Foighil *et al.* (2011) propose that both *P. similaris* and *P. grisea* (as *P. carteriensis*) were introduced to the New Guinea islands in prehistoric times, possibly from the Solomon islands.

Richardson (1990) listed *P. kubaryi* as a synonym of *P. carteriensis* but the shell is closer to *P. similaris*, to which genetic group the New Britain population belongs.

Solomon islands

A large number of species have been described from the Solomon islands, but only four are likely to be valid. Most are closely related to the New Guinea species.

Partula flexuosa Hartman, 1885

Partula flexuosa Hartman, 1885: 204
Partula incurvum Hartman 1886: 31
Partula regularis Hartman, 1886: 31
Partula minor Hartman, 1886: 31
Partula minima Hartman, Hartman 1892: 99 (misspelling)
Partula hastula Hartman, 1886: 33
Partula incurva Hartman, Hartman 1886: 31. Smith 1903: 454. Pilsbry 1909: 291. Clench 1941: 20. Kondo 1968: 76. Richardson 1990: 36
Partula flexuosa Hartman, Hartman 1892: 75. Smith 1903: 453. Pilsbry 1909: 291. Clench 1941: 20. Kondo 1968: 76. Richardson 1990: 27

Partula hastula Hartman, Hartman 1892: 75. Smith 1903: 453. Pilsbry 1909: 291. Clench 1941: 20. Solem 1959: 74. Kondo 1968: 76. Richardson 1990: 32

Partula regularis Hartman, Hartman 1892: 75. Smith 1903: 454. Pilsbry 1909: 292. Rensch & Rensch 1935: 84. Clench 1941: 20. Kondo 1955: 185, 1968: 76. Clench & Turner 1972: 216. Richardson 1990: 52

Partula laevigata (non Pfeiffer), Smith 1903: 452 (misidentification). Pilsbry 1909: 294 (as syn. of *hollandiana*)

Partula minor Hartman, Smith 1903: 455. Pilsbry 1909: 287, 1910: 320. Solem 1959: 72. Kondo 1968: 77. Richardson 1990: 40

Partula hollandiana Pilsbry, 1909: 293

Partula hollandiana Pilsbry, Clench 1941: 20. Solem 1959: 74. Richardson 1990: 34

Dextral. Elongately ovate to ovate-conic. Thin, surface dull or moderately shiny. Umbilicus open, deep and moderately wide. Spire elongate with 4½-5½ convex whorls with distinct, close spiral striae. Suture well impressed. An inconspicuous ridge just above the suture of the penultimate whorl. Sculpture of regular spiral striae, present on all whorls but fading slightly on upper part of body whorl. Body whorl flattened laterally near aperture, convex, saccate at base. Aperture slightly oblique, ovate, small. Lip well expanded, thickened within. Columellar margin dilated above, and slightly grooved where it joins the body whorl. Parietal callus transparent.

Pale grey to grey-brown, sometimes with a yellowish brown medial band on the body whorl, with faint traces of lighter and darker spiral bands. Aperture white or bluish inside, lip white.

Anatomy: As *P. grisea*.

Type specimen:

Partula flexuosa Hartman, 1885 – Holotype CMNH 62.4238.

Partula incurvum Hartman 1886 – Holotype CMNH 62.4240.

Partula hastula Hartman, 1886 – Syntypes CMNH 62.4239.

Partula regularis Hartman, 1886 – Holotype CMNH 62.4241.

Partula minor Hartman, 1886 – Syntypes CMNH 62.4243.

Partula hollandiana Pilsbry, 1909 – Holotype CMNH 62.4237.

Distribution: Vella Lavella, Simbo (Eddystone), Ghizo, New Georgia, Roviana (Rubiana), Vangunu, St. George (off Santa Isabel) and Gaudalcanal. *P. flexuosa* has been recorded from all these islands, *P. hastula* only from Simbo (given as Erromango, New Hebrides by Hartman [1886], corrected to Simbo by Solem [1959] on the basis of a note in Hartman's handwriting), *P. incurva* from Rubiana and New Georgia, *P. regularis* from Guadalcanal. The *P. hollandia* specimen was of unknown origin.

Fig. 9.1.4 *Partula flexuosa*: a – *flexuosa* holotype CMNH 62.4238; b – *hastula* syntype CMNH 62.4239; c – *incurva* holotype CMNH 62.4240; d – *regularis* holotype CMNH 62.4241; e – *hollandiana* holotype CMNH 62.4237; f – *minor* syntype CMNH 62.4243; g – Vella Levella specimen (BPBM 1966-168) (photos: a-f – C. Sturm)

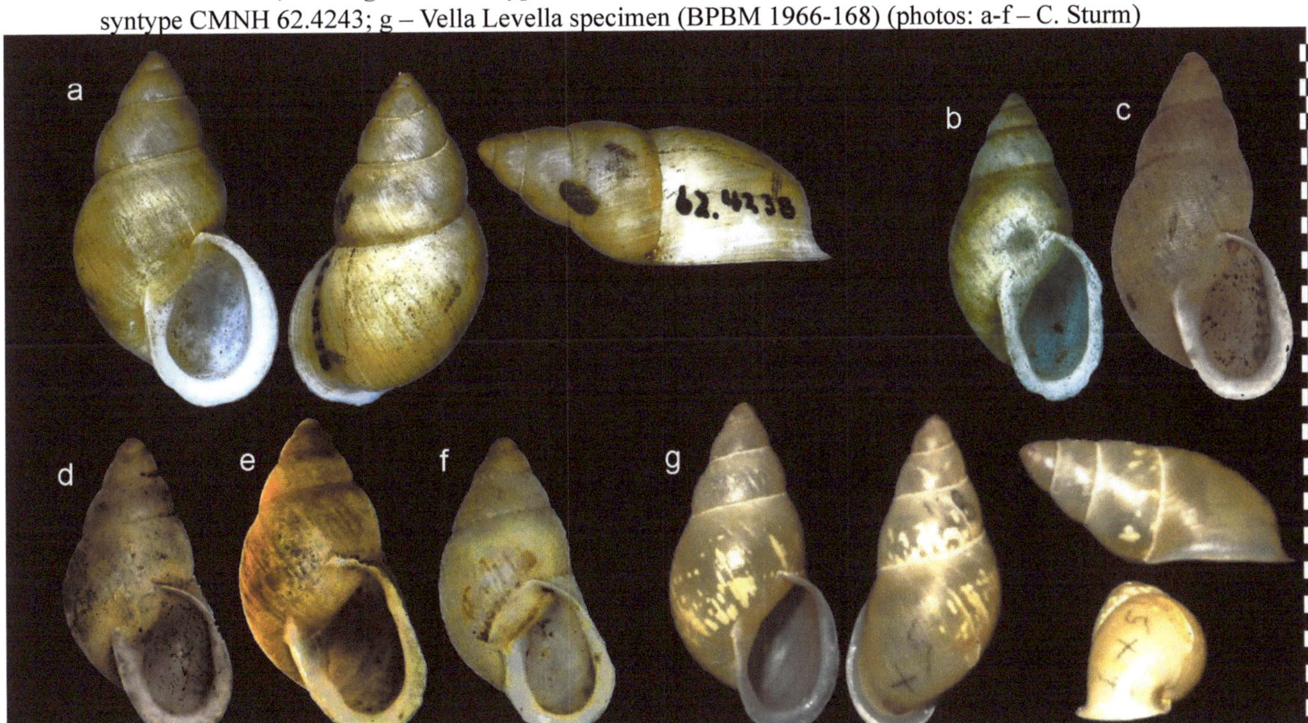

The types of *P. minor* are labelled 'Erromango' but this appears to have been in error as suspected by Solem (1959).
Note: This taxon could be synonymous with *P. similaris*.

Partula micans Pfeiffer, 1853

Partula micans Pfeiffer, 1853a: 451

Partula micans Pfeiffer, Pfeiffer 1854c: 138, 1855b: 276, 1856: 162, 1859: 513, 1868: 160, 1877: 206. Adams 1855: 145. Paetel & Schaufuss 1869: 83. Paetel 1873: 104, 1883: 153, 1889: 267. Kobelt 1880: 18. Hartman 1881a: 181, 1881b: 11, 1885: 216, 1892: 75. Pfeiffer & Clessin 1881: 303. Smith 1903: 452. Pilsbry 1909: 295. Clench 1941: 19. Kondo 1955: 184, 1968: 76. Richardson 1990: 40

Partula cinerea Albers, 1857: 98

Partula cinerea Albers, Pfeiffer 1859a: 510, 1877: 199. Kobelt 1880: 18. Hartman 1881a: 180, 1881b: 14, 1885: 216, 1892: 75. Pfeiffer & Clessin 1881: 301. Smith 1885: 595. Paetel 1889: 266. Smith 1903: 478. Pilsbry 1909: 296. Kilias 1969: 246. Richardson 1990: 17

Partula coxi 'Angas', Cox 1868: 46. Paetel 1883: 152 (all nomen nudum) Paetel 1889: 266. Richardson 1990: 19.

Partula pellucida Pease, 1871b: 457

Partula pellucida Pease, Pfeiffer 1877: 199. Kobelt 1880: 18. Hartman 1881a: 185, 1885: 217, 1886: 35, 1892: 76. Pfeiffer & Clessin 1881: 301. Paetel 1889: 268. Smith 1903: 456. Pilsbry 1909: 297. Clench 1941: 20. Kondo 1968: 76. Richardson 1990: 50

Partula coxi Hartman, 1885: 217

Partula perlucens Hartman 1886: 31

Partula perlucens Hartman, Smith 1903: 457. Pilsbry 1909: 293. Clench 1941: 20

Partula coxi Hartman, Hartman 1886: 32. Smith 1903: 455. Pilsbry 1909: 296. Clench 1941: 20. Kondo 1955: 184, 1968: 76

Partula milleri Solem, 1959: 72

Partula milleri Solem, Kondo 1968: 77. Richardson 1990: 40

Dextral. Oblong-conic to ovately conic. Thin, translucent and slightly glossy. Distinctly decussated. Umbilicus open but very narrow, deeply rimate. 4½-5 convex whorls ornamented with dense growth lines and spiral striae. Striae fading on body whorl. Spire acutely conic. Suture impressed. Body whorl as long as spire, impressed just below the suture and strongly convex above the periphery and at base. Body whorl may be compressed laterally behind the aperture. Columella lightly arcuate, margin

Fig. 9.1.5 *Partula micans*: a – *micans* syntype (NHMUK); b – *perlucens* holotype (NHMUK 96.11.30.15); c – *pellucida* syntype (NHMUK 96.11.30.5); d – *coxi* holotype (CMNH 62.4245); e – *cinerea* lectotype (ZMB 101545); f-g – Ontong Java (NHMUK 1957.11.8.86a & b); h – *coxi* specimen (NHMUK 78.11.30.1); i – Santa Isabel (NHMUK 2184); j – Shortlands (UMMZ 'NILAP1'); k – *milleri* holotype (USNM 619738) (photo: d – C. Sturm)

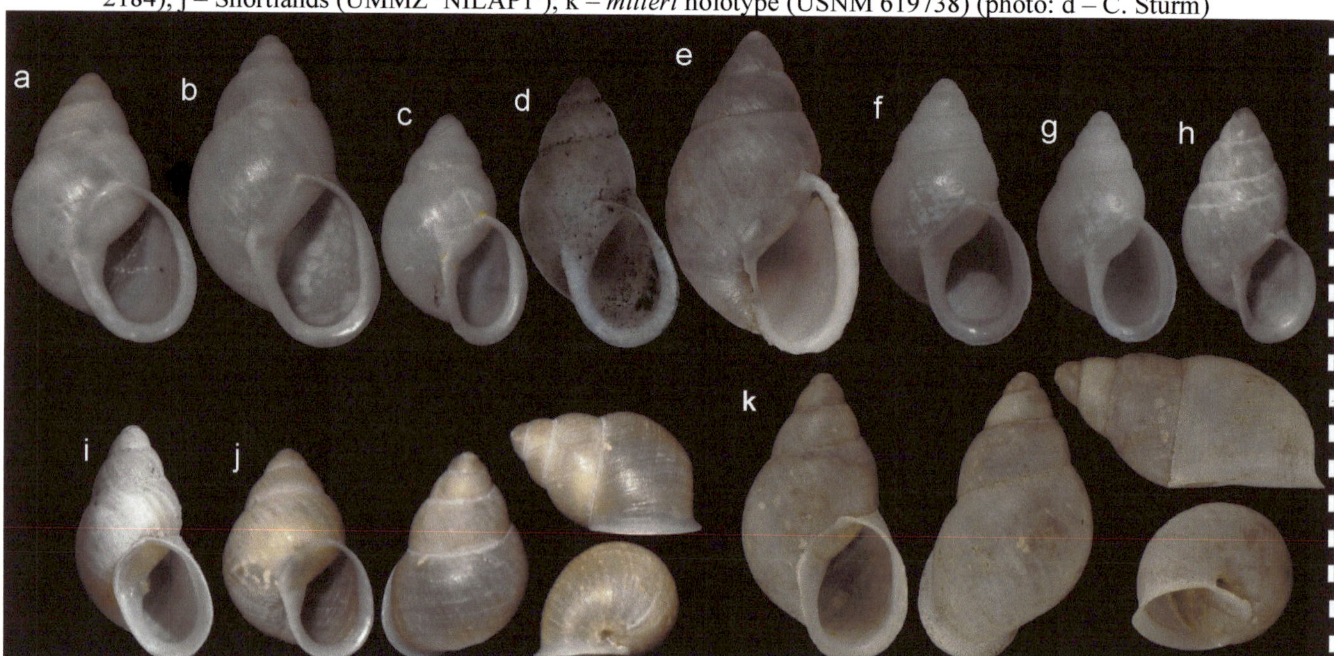

dilated, spreading. Aperture ovate to oblong-oval, slightly oblique. Lip expanded, acute, margins converging, very slightly thickened within. Parietal callus thick, slightly opaque, strongly bordered, giving the aperture a very distinctive appearance. White to pale corneous. White sutural line, sometimes a thin white subsutural line on body whorl. Lip whitish.
Anatomy: Penis short, ⅓-½ oviduct length. Caecum distinct but short. Penis with faint narrow pilasters, caecum with 4 rows of rugae and scattered papillae. Ornamentation may be much reduced and indistinct. Vas deferens subapical.
Type specimens:
Partula micans Pfeiffer, 1853 – Syntypes ANSP 151448 (labelled 'paratypes'); NHMUK un-numbered.
Partula cinerea Albers, 1857 – Lectotype ZMB Moll-101545a; paralectotype ZMB Moll-101545b. Lectotype designation by Killias (1968).
Partula pellucida Pease, 1871 – Syntypes MCZ 94837; NHMUK 96.11.30.5. The former is probably the CMNH specimen transferred to MCZ, catalogued incorrectly as 'neotype'.
Partula coxi Hartman, 1885 – Holotype CMNH 62.4245.
Partula perlucens Hartman, 1886 – Holotype NHMUK 96.11.30.1.
Partula milleri Solem, 1959 – Holotype USNM 619738. 14 juveniles in the same lot are assumed to be paratypes, but were not mentioned in the original description.
Distribution: Choiseul, Santa Isabel ('Ysabel'), Kolambangara (St. George group), Ontong Java, Shortland Islands, Guadalcanal, Ulawa, Three Sisters, Ugi (=Bio, Golfe), Treasury islands. *coxi* recorded from Santa Isabel, Kolambangaraa and Ontong Java; *pellucida* from Santa Isabel and Guadalcanal; *perlucens* from Ugi; *cinerea* from Treasury island (according to Dell 1955). A single record from Espiritu Santo in Vanuatu is a temporary introduction (the *P. milleri* sample, collected from the south side of Pallikula Bay by W.B. Miller). This locality was an American military base in World War II and it is probable that *Partula* were accidentally transported on the considerable material that was moved between the Solomons and Pallikula Bay. There are no subsequent records.
Note: A note with a specimen of *P. cinerea* in NHMUK (96.7.13.35.8) donated by C.M. Cooke says 'intermediate and also related to P. similaris Hart. + woodlarkiana Hrt. which is much more closely related if not P. occidentalis Hedley. If not this species then a new species C.M.C.'
P. milleri has not previously been synonymised with *P. micans* although this has long been suspected (Kondo MS) and even in the original description Solem (1959) thought it close to Solomons species. It can be identified as *P. micans* by the distinctive appearance of the aperture caused by the strongly bordered callus.

Partula cramptoni Clench, 1941

Partula cramptoni Clench, 1941: 183

Partula cramptoni Clench, Clench 1958: 180, 1968: 173. Zilch 1959: 455. Kondo 1955: 183, 1968: 76. Richardson 1990: 20

Dextral. Ovately conical. Thick, shiny but often worn dull. Umbilicus rimate, partially covered by columella. 5 slightly convex whorls ornamented with fine spiral striae and fine, irregular growth lines, apical two whorls finely rugose. Striae absent from body whorl. Suture impressed, not indented. Aperture subovate, angled. Lip expanded. Columella thickened.
Light brown with an off-white spire to brownish green with a blackish brown spire. Body whorl deep brown behind the lip, usually forming a red or brown band in the aperture.
Anatomy: Penis short, ⅓ oviduct length. Caecum broad, subcylindrical with slight median bulge. Vas deferens subapical.
Type specimens: Holotype AMNH 79014; paratypes AMNH 79008; BPBM 9166, 9159; MCZ 93021, 106381-2.
Distribution: Renell and Bellona.
Note: Clench (1941) placed *P. cramptoni* in a separate section '*Rennellia'*.

Partula vanikorensis (Quoy & Gaimard, 1832)

Helix Vanikorensis Quoy & Gaimard, 1832: 115

Partulus vanikorensis Beck 1837: 57

Bulimus vanikorensis Quoy & Gaimard, Deshayes & Milne-Edwards 1839: 371. Pfeiffer 1842: 124, 1848: 71. Catlow & Reeve 1845: 162

Partula vanikorensis Quoy & Gaimard, Gould 1852: 85. Pfeiffer 1853a: 446, 1856: 162, 1856: 271, 1859: 507, 1868: 156, 1877: 195. Adams 1855: 145. Paetel 1873: 104, 1883: 153, 1889: 268. Semper 1874: 158. Hartman 1881a: 185, 1881b: 10, 1885: 212. Garrett 1884: 47. Smith 1903: 443. Pilsbry 1909: 289. Solem 1959: 73. Kondo 1968: 76. Richardson 1990: 64

Dextral. Ovate-conic or regularly ovoid, solid. Umbilicus open, oval and shallow, partially covered. 5-6 convex whorls, longitudinally and transversely striate. Suture linear. Aperture oval, a little contracted. Lip wide, reflexed but only slightly thickened internally. Columellar margin dilated, parietal callus distinct.

Fawn to brown or white, longitudinal darker bands visible on lighter individuals, mainly on body whorl. Lip white, sometimes violet.

Body uniform yellowish fawn (Quoy & Gaimard 1832).

Anatomy: Penis short, slender; caecum moderately long, not distinctly separated from penis. Internal ornamentation reduced to low ridges and papillae.

Type specimens: Syntypes MNHN IM-2000-27661.

Distribution: Santa Cruz group (Vanikoro). Kondo MS suggested that Iredale's (1927) *Partula* from on Santa Cruz Island (Nendö) may have been *P. auraniana* but subsequent material from the island is all of *P. vanikorensis.* Fragments of a large *Partula* from Lomlom island in the Santa Cruz group (BPBM 1972-1) may belong to this species. Terrestrial.

Fig. 9.1.6 *Partula cramptoni*: a – holotype AMNH 79014; b – specimen BPBM 1966-305

Fig. 9.1.7 *Partula vanikorensis*: a – cotype MNHN IM-2000-27661; b-d – Santa Cruz specimens (b – BPBM 1978-22; c-d – BPBM 1966-71) (photo: a – M. Caballer)

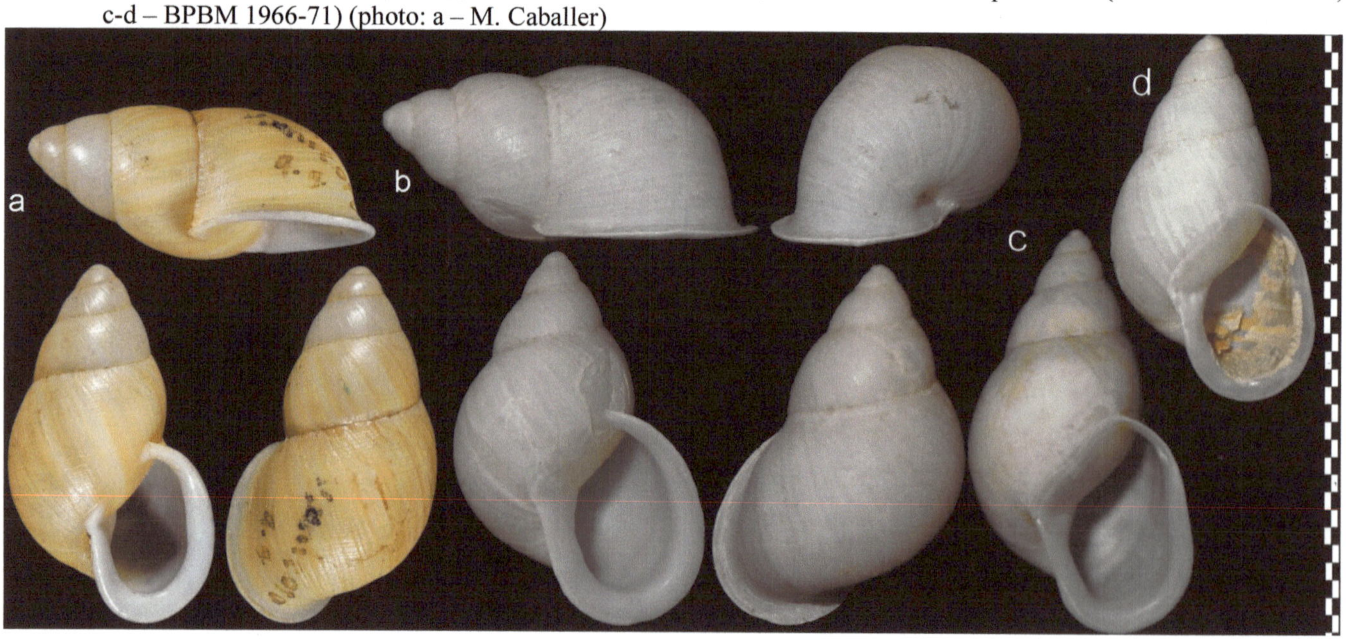

Note: Solem (1959) thought that one of the two MCZ specimens (MCZ 24844, ex NHMUK) was a mislabelled *Partula otaheitana sinistrorsa*, the other (MCZ 192357) was similar to *Samoana alabastrina.* He thought that photographs of the labelled types (MNHN) were mislabelled as they did not match Quoy & Gaimard's (1832) original figures. However, this is not the case. This very variable species most closely resembles Vanuatu's *P. auraniana*, which is also closer geographically than any species of the Solomon islands.

Vanuatu

Solem (1959) divided the Vanuatu *Partula* into three superspecies (*macgillivrayi*, *pyramis* and *auraniana*). In at least the first two cases these can best be regarded as species rather than superspecies, Solem (1959) had noted that the groups of species related to *macgillivrayi* and *pyramis* were regarded by Hartman as belonging to only two species, although neither he nor Hartman synonymised the species within them. *P. milleri* is removed from the list of Vanuatu species, being a temporary introduction of *P. micans* (see p. 77). Here the number of Vanuatu species is reduced from 12 to 4.

Partula radiosa (Pfeiffer, 1854)

Bulimus radiosus Pfeiffer, 1854a: 58
Bulimus radiosus Pfeiffer, Pfeiffer 1856: 159, 1859: 496, 1868: 143, 1877: 176
Buliminus (*Rhachis*) *radiosus* (Pfeiffer), Kobelt 1902: 987
Partula radiosa (Pfeiffer), Solem 1959: 67. Richardson 1990: 52

Dextral. Elongately conic. Moderately thick. 5 ½ moderately convex whorls. Umbilicus open, moderately narrow. Suture indistinctly raised. Aperture vertical. Columella dilated moderately. Lip moderately expanded, not reflexed, only very slightly and partially thickened internally. Parietal callus very thin.
Uniform pale yellow corneous, possibly strigated. Lip white.
Anatomy: Unknown; one adult specimen (NHMUK) contained a single calcified egg capsule.
Distribution: Type locality unknown. One specimen from Nonanakao on Erromango island, collected from 'virgin Kauri forest' (11.viii.1971: NHMUK jar No. 20). A specimen from Efate labelled 'minor' (BPBM 8694) may be this species.

Fig 9.1.8 Distribution of *Partula* species in Vanuatu

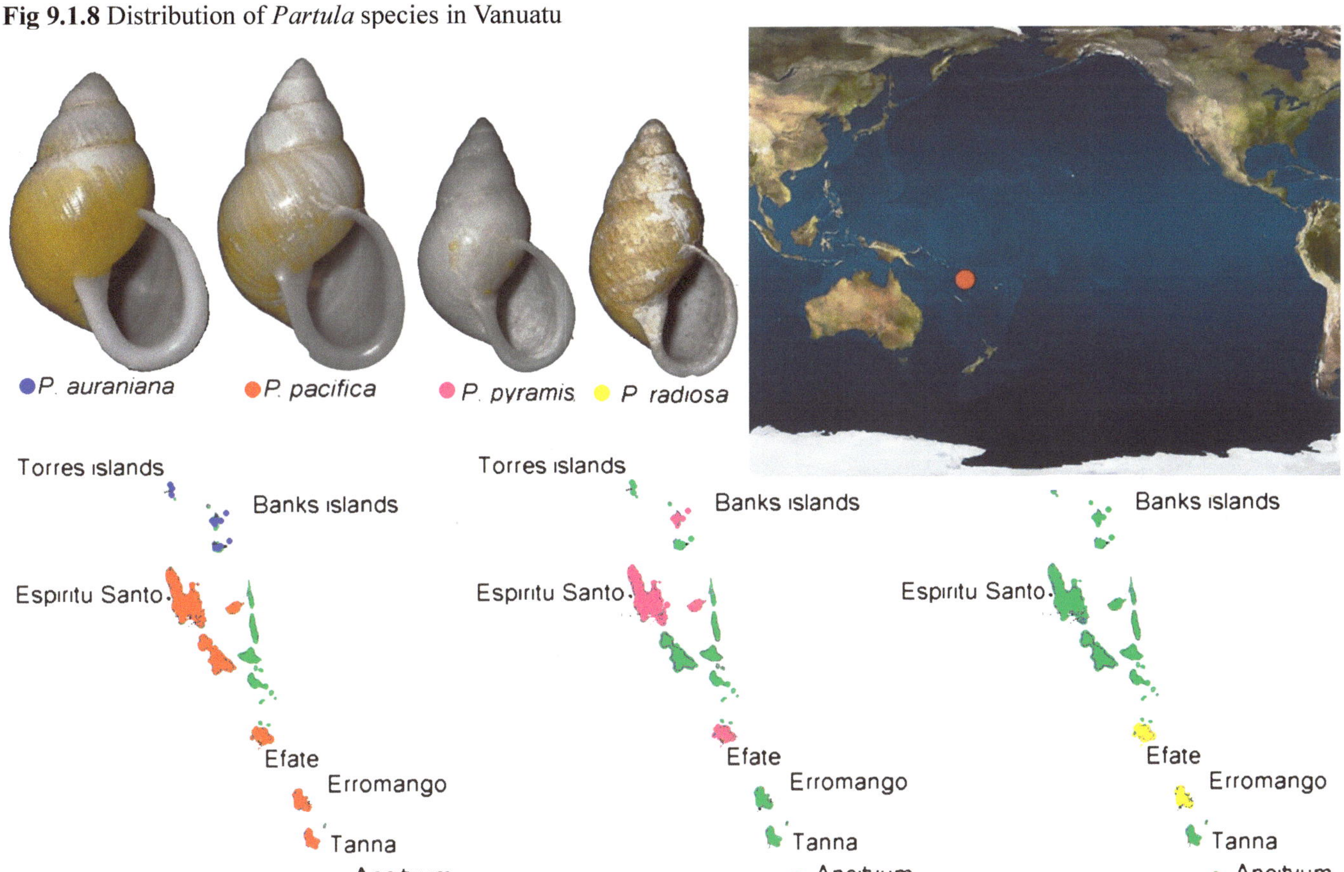

Type specimens: Holotype subadult and two paratype juveniles NHMUK un-numbered.
Note: Solem (1959) thought it similar to *P. macgillivrayi*, differing only in the presence of a supra-sutural cord in *P. radiosa*. However, the form of the shell is distinct and it is retained here as a separate species.

Partula pacifica Pfeiffer, 1854

Partula pacifica Pfeiffer, 1854d: 154

Partula macgillivrayi Pfeiffer, 1855a: 977

Partula macgillivrayi Pfeiffer, Pfeiffer 1856: 162, 1857: 61, 1859a: 508, 1868: 157. Paetel & Schaufuss 1869: 83. Paetel 1873: 104, 1883: 153, 1889: 267. Kobelt 1880: 19. Hartman 1881a: 180, 1881b: 13, 1885: 222, 1892: 99, 1897: 44. Pfeiffer & Clessin 1881: 301. Smith 1903: 470. Pilsbry 1909: 278. Solem 1959: 65. Richardson 1990: 39

Partula pacifica Pfeiffer, Pfeiffer 1856: 162, 1859a: 509, 1868: 158, 1877: 198. Hartman 1881a: 185, 1881b: 10, 1885: 208. Garrett 1884: 47. Paetel 1889: 268. Smith 1903: 480. Pilsbry 1909: 258

Partula turneri Pfeiffer, 1860: 140

Partula turneri Pfeiffer, Pfeiffer 1861: 16, 1868: 159, 1877: 202. Kobelt 1880: 19. Pfeiffer & Clessin 1881: 302. Paetel 1889: 218. Smith 1903: 466. Pilsbry 1909: 277. Thiele 1931: 659. Clench 1941: 19. Iredale 1941: 64. Kondo 1955: 181, 1968: 76. Solem 1959: 67. Zilch 1959: 454. Richardson 1990: 62

Partula caledonica Pfeiffer, 1862: 389

Partula caledonica Pfeiffer, Pfeiffer 1868: 157. Brazier 1871: 585. Crosse 1871: 184. Paetel 1873: 104, 1883: 152. Hartman 1881a: 180, 1881b: 13, 1885: 222, 1892: 99. Pfeiffer & Clessin 1881: 301. Crosse 1894: 172. Smith 1903: 465. Pilsbry 1909: 258. Solem 1959: 68. Richardson 1990: 14

Fig. 9.1.9 Anatomy of Vanuatu and Santa Cruz *Partula*: a – *vanikorensis* Santa Cruz (BPBM 189717, 1978-22); b-g – *pacifica* (b-e – *turneri* Tanna: b – ANSP A1762c; c-d – FMNH 173349; e – Kondo 1955; f – *macgillivrayi* Aore BPBM 1967-23; g-h – *caledonica* Efate FMNH 175201); i – *auraniana* Vanua Lava (FMNH 109348); j-l – *pyramis* (i – *albescens* Aore, Kondo 1955; j-k – *pyramis* Malekula, FMNH 175215). Scale bar 5 mm.

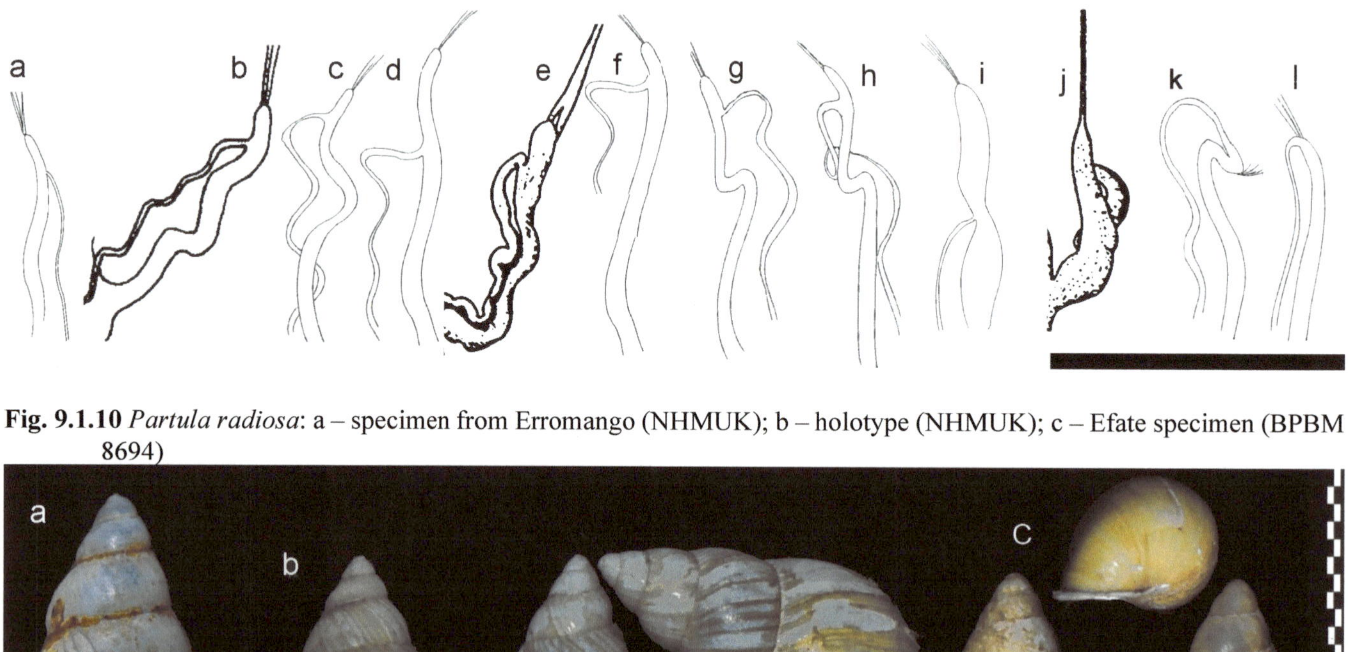

Fig. 9.1.10 *Partula radiosa*: a – specimen from Erromango (NHMUK); b – holotype (NHMUK); c – Efate specimen (BPBM 8694)

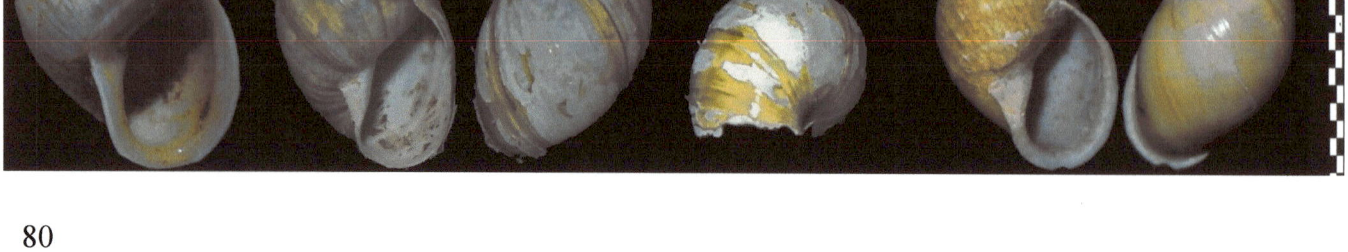

Partula pfeifferi Crosse, 1871: 184 (unjustified replacement name)
Partula brazieri Pease, 1872: 27
Partula pfeifferi Crosse, Brazier 1871: 585. Kobelt 1880: 19. Paetel 1889: 268. Crosse 1871: 184, 1894: 172
Partula brazieri Pease, Pease 1871a: 27. Pfeiffer 1877: 94. Garrett 1887: 135. Paetel 1889: 266. Smith 1903: 466. Pilsbry 1909: 271. Richardson 1990: 13
Partula turneri perstrigata Pilsbry, 1909: 277
Partula artensis 'Montrouzier', Pilsbry 1909: 278 (as syn. of *caledonica*, nomen nudum)
Partula (*Melanesica*) *turneri perstrigata* Pilsbry, Pilsbry & Cooke 1934: 15, Solem 1959: 67

Normally dextral (sinistral specimens from Efate and Aoba). Elongately ovate-conical to oblong-acuminate. Moderately thick, glossy. Umbilicus open, deeply rimate. 5-5½ convex whorls with wavy striae, fading in middle of body whorl. No suprasutural cord. Body whorl impressed in middle, saccate at base. Columella simple, slightly arcuate, flat, slightly grooved or subplicate. Columellar margins spreading. Aperture oblong, slightly oblique. Lip expanded, margins converging, thickened within. Off-white to pale buff with pale yellowish-green or darker chestnut growth striae. Lip white.

Fig. 9.1.11 *Partula pacifica*: a-g – types: a – *turneri* (NHMUK); b – *pacifica* (NHMUK); c – *perstriga* (ANSP 39848); d-e – *macgillivrayi* (NHMUK); f – *caledonica* (NHMUK); g – *brazieri* (ANSP 59846); h & i – Efate *caledonica* showing colour variation (FMNH 175201); j – Tanna *turneri* (FMNH 175349); k-m – Aoba *macgillivrayi* (BPBM 1967-38: k – abnormal individual, l – typical, m – sinistral)

Anatomy: Penis short-vermiform, ¼-⅓ oviduct length. Few reduced papillae in caecum and five strong pilasters in penis. Length of caecum varies greatly within the same population (10-30% of total penis length in Tanna – FMNH 175349). Vas deferens subterminal. Retractor sometimes split at penial insertion.
Type specimens:
Partula pacifica Pfeiffer, 1854 – Holotype NHMUK (un-numbered).
Partula macgillivrayi Pfeiffer, 1855 – Syntypes NHMUK (2 adults, 1 subadult, Cuming collection).
Partula turneri Pfeiffer, 1860 – Syntypes NHMUK (label annotated 'macgillivrayi, var=', Cuming collection).
Partula caledonica Pfeiffer, 1862 – Syntypes NHMUK (3 specimens, Cuming collection).
Partula brazieri Pease, 1872 – Holotype ANSP 59846; paratype ANSP 463934.
Partula turneri perstrigata Pilsbry, 1909 – Lectotype ANSP 59848; paralectotype ANSP 69174, 463718. Lectotype designation by Baker (1963).
Distribution: Erromango, Tanna, Aneityum, Efate ('Sandwich', 'Vate'), Aore, Aoba, Malakula, 'Yatalo' and Espiritu Santo. *P. caledonica* was also reported from Vanua Lavu, Banks Islands (Brazier) but not supported by specimens. *turneri* was described from Tanna, *macgillivrayi* from Aneityum and *caledonica* from Efate. Brazier reported the single specimen of *P. brazieri* from Tutuila but the locality was questioned by Garrett and Hartman. Specimens of this variable species (e.g. BPBM) have been variously misidentified as *P. albescens* and *P. pyramis.*
Note: *macgillivrayi* has a wider, more reflexed columella, more truncate base, more prominent sculpture, slightly more rounded whorls than do other *P. pacifica* forms. *turneri* has light markings and lacks a sutural cord. *perstriga* is smallest.

Partula pyramis Hartman, 1886

Partula pyramis Hartman, 1886: 34
Partula proxima Hartman, 1886: 34
Partula eximia Hartman, 1886: 35
Partula carnicolor Hartman, 1888: 250
Partula albescens Hartman, 1888: 251
Partula albescens Hartman, Hartman 1892: 99. Smith 1903: 467. Pilsbry 1909: 282. Solem 1959: 70. Richardson 1990: 10
Partula pyramis Hartman, Hartman 1892: 99. Smith 1903: 471. Sykes 1903: 198. Pilsbry 1909: 281. Solem 1959: 69. Kondo 1968: 76. Richardson 1990: 51
Partula carnicola Hartman, 1892: 99 (misspelling)
Partula carnicolor Hartman, Smith 1903: 466. Pilsbry 1909: 286. Solem 1959: 71. Richardson 1990: 16
Partula proxima Hartman, Hartman 1892: 99. Smith 1903: 467. Pilsbry 1909: 287. Solem 1959: 69. Richardson 1990: 51
Partula eximia Hartman, Hartman 1892: 99, 1897: 44. Smith 1903: 468. Pilsbry 1909: 280. Solem 1959: 70

Dextral. Elongately ovate. Thin to solid. Umbilicus moderately wide, deeply rimate. Spire long, slender. 5-5½ convex whorls, last two whorls very strongly convex. Body whorl laterally compressed, base saccate. Sculpture with widely spaced spiral lines on penultimate whorl, fading on body whorl except at base. Aperture ovate, subvertical to oblique, lip rather narrowly reflexed, thickened within, columella dilated, slightly nodose in *proxima*, distinctly grooved at junction with body whorl in *carnicolor*. Parietal callus thin, transparent, thick and opaque at ends. Lip expanded, concave.
Plain or bluish milky to pale buff, usually marked with narrow yellowish-green streaks on the last whorl. *carnicolor* has a darker apex marked with a poorly defined medial brown band. Lip white, aperture flesh-tinted inside in *carnicolor*.
Anatomy: Similar to *P. pacifica* but penis shorter, may be stouter. Caecum very small and pointed or absent. Penial pilasters 5-6, narrow, 2-3 entering caecum, papillae reduced.
Type specimens: Specimens labelled holotype and paratype by Solem (1959) are lectotype and paralectotype.
Partula pyramis Hartman, 1886 – Lectotype and paralectotype CMNH 62.4305. Described from three specimens.
Partula proxima Hartman, 1886 – Syntypes AM (un-numbered); NHMUK 94.11.30.3-4.
Partula eximia Hartman, 1886 – Holotype CMNH 62.4292, not located.
Partula albescens Hartman, 1888 – Holotype and paratypes CMNH 62.4290.
Partula carnicolor Hartman, 1888 – Lectotype and paralectotype CMNH 62.4289.
Distribution: Aneityum, Banks group (Vanu Lava), Espiritu Santo group (Espiritu Santo, Aore, Sitova, Tutuba) and Efate (Vate). *albescens* from the Espiritu Santo group; *pyramis* from Efate; *proxima* from the Banks Group (Vanua Lava); *eximia* from Aneityum; *carnicolor* from Aorea island. Within the Espiritu Santo group specimens from Tutuba shorter and broader and with narrower lip than Aore, smaller and with a narrower lip on Sitova Island than on Aura Island. A specimen from Port Olry, Espiritu Santo (Bouchet *et al.* 2007) appears to be this species.

Note: *eximia* is more pyramidal than typical *P. pyramis*, with a slightly oblique aperture; *proxima* has a narrower and more compressed umbilicus, only indistinct spiral striae and a more oblique aperture. Solem (1959) confused his figures, labelling a type of *P. pyramis* and the *P. eximia* holotype as *P. eximia* CMNH 62.4292. The true type is his fig. 6 (see Fig. 9.1.12e). Richardson (1990) listed *P. eximia* as a synonym of *P. macgillivrayi* (=*P. pacifica*).

Partula auraniana Hartman, 1888

Partula auraniana Hartman, 1888: 250

Partula fraterna Hartman, 1888: 250

Partula paterna Hartman, Hartman 1892: 99 (misspelling)

Partula auraniana Hartman, Hartman 1892: 99. Smith 1903: 468. Sykes 1903: 198. Ancey 1905: 44. Pilsbry 1909: 284. Solem 1959: 70, 1973: 168. Kondo 1968: 77. Richardson 1990: 11

Partula fraterna Hartman, Smith 1903: 468. Sykes 1903: 198. Pilsbry 1909: 286. Solem 1959: 71. Richardson 1990: 27

Dextral. Ovately conic. Thin or thick. Umbilicus widely, deeply rimate. Apex obtuse. 4½-5½ convex whorls, spiral striae fading on body whorl, still distinct on base. Suture impressed. Body whorl strongly convex, slightly compressed laterally. Aperture almost vertical, ovate. Lip moderately reflexed, internally thickened. Columella vertical, heavily calloused, tapering. Parietal callus thin, transparent, thickened at ends.

Body whorl pale brown. Periostracum thin, greenish yellow with some faint darker streaks. Spire rosy but normally decorticated to very pale brown to white, whitish below the suture. Aperture faintly flesh-tinted within. Lip white.

Anatomy: Penis stout, moderate in length. Caecum slightly longer than penis. Vas deferens subterminal.

Type specimens: Specimens labelled holotype and paratype by Solem (1959) are lectotype and paralectotype.

Partula auraniana Hartman, 1888 – Lectotype and paralectotype CMNH 62.4295.

Partula fraterna Hartman, 1888 – Lectotype and paralectotype CMNH 62.4294.

Fig. 9.1.12 *Partula pyramis* types: a – *proxima* (NHMUK 94.11.30.3); b – *carnicolor* (CMNH 62.4289); c – *pyramis* (CMNH 62.4305); d – *albescens* (CMNH 62.4290); e-f – *eximia* (e – from Solem 1959; f – from Hartman 1886) (photo: b, c-d – C. Sturm)

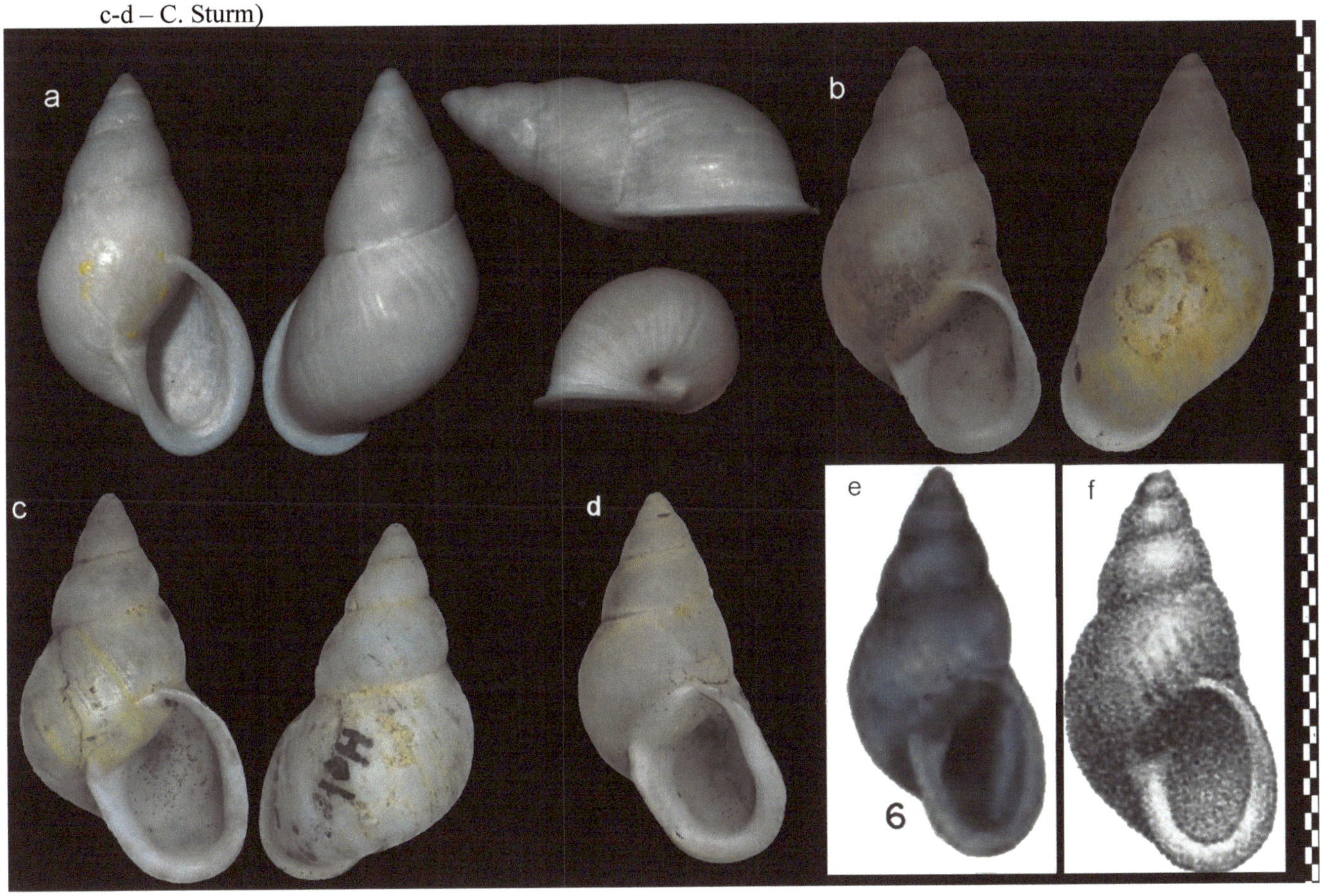

Distribution: Aore (='Aura' or 'Arag'), Banks group (Vanua Lava, Gaua, Mota Lava [Valua], Tikopa, Ravenga), Torres group (Bukabuka, Hiw, Metoma, Tegua, Lo, Toga). Sykes (1903) thought those from Lo a separate race.
Note: Solem (1959) thought that *P. carnicolor* intergraded with *P. auraniana* and *P. fraterna*, all three forming the *P. auraniana* superspecies. Kondo (MS) considered *P. auraniana* and *P. fraterna* to be synonyms. The two forms differ slightly in shape, *P. fraterna* in being larger, with a more acute spire and flatter body whorl. *P. fraterna* was recorded from Aore, Vanua Lava, Ravenga and Gaua islands.

Fig. 9.1.13 *Partula auraniana*: a – Santa Cruz specimen (BPBM 1966-306); b – Vanua Lava specimen (BPBM 206730); c-d – colour varieties (NHMUK un-numbered); e – paratype (CMNH 62.4295); f – *P. fraterna* lectotype (CMNH 62.4294); g – '*P. fraterna*' specimen (NHMUK 1902.12.1.194); h – atypical '*P. fraterna*' specimen (NHMUK 76.5.26.76) (photos: f – C. Sturm)

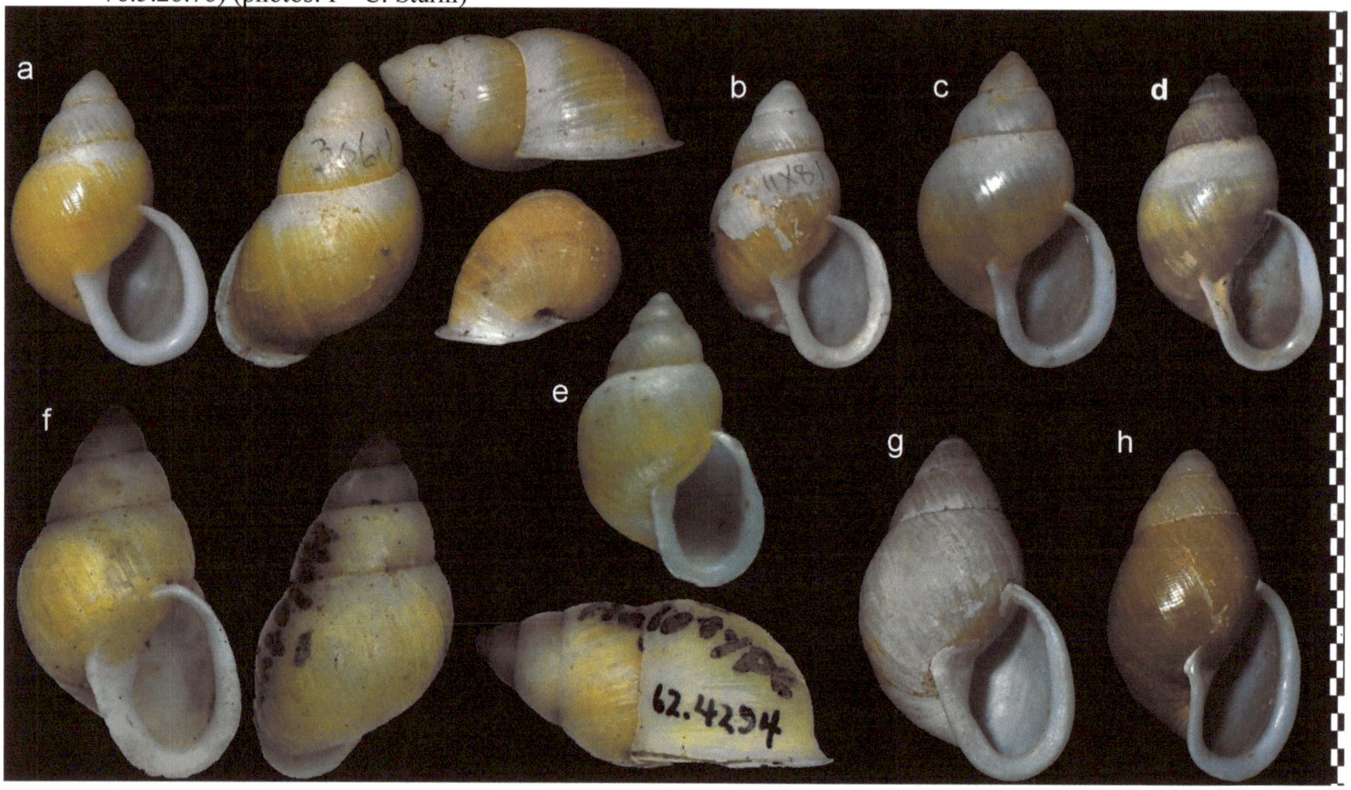

Fig. 9.1.14 *Partula flexuosa* of Guadalcanal, Solomons (photo: Simon's Specimen Shells Ltd.) and two views of *P. auraniana* of Torres island, Vanuatu (photos: O. Gargominy)

9.2 Micronesia

Partula are present on the higher islands of Micronesia: the Marianas and the Caroline islands. 12 species have been described from the islands, of which 9 are recognised here.

Marianas islands

6 species have been recorded from 9 islands in the Marianas group, one is known only from subfossils.

9.2.1 *Partula* of the Marianas islands, submarine banks shown in blue

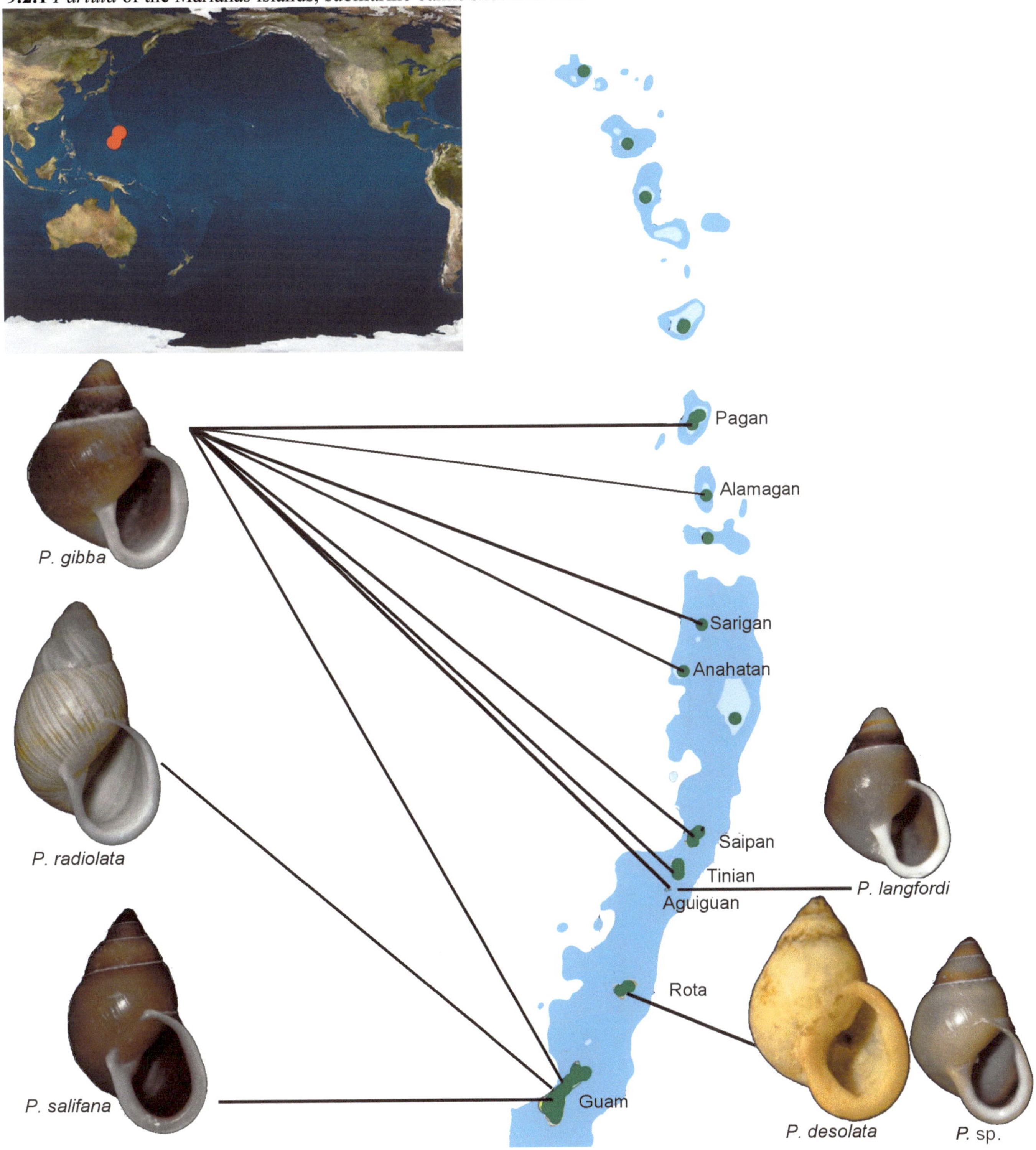

Partula gibba Férussac, 1821

Partula gibba Férussac, 1821: 66

Partula gibba Férussac, Quoy & Gaimard 1824: 485, 1832: 113. Gray 1825: 415. Sowerby 1825: 40. Poitiez & Michaud 1838: 497. Anton 1839: 40. Jay 1939: 57. Reeve 1842: pl3 f5, 1850: 15. Catlow & Reeve 1845: 163. Pfeiffer 1853a: 453, 1856: 162, 1856: 269, 1859: 513, 1868: 160, 1877: 208. Albers 1850: 186. Gray 1850: 116. Adams 1855: 145. Chenu 1859: 434. Paetel & Schaufuss 1869: 83. Pease 1871b: 473. Tapperone-Canefri 1874: 84. Hartman 1881a: 182, 1881b: 13, 1885: 211, 1892: 75. Paetel 1873: 104, 1883: 153, 1889: 267. Smith 1903: 439. Pilsbry 1909: 166, 1910: 315. Crampton 1925: 60. Thiele 1931: 658. Haltenorth & Jaeckel 1941: 53. Zilch 1959: 463. Kondo 1955: 199, 1968: 77. Richardson 1990: 29. Bauman 1996: 21

Helix gibba Quoy & Gaimard 1832: 113

Partulus gibbus Férussac, Beck 1837: 57

Bulimus gibbus Quoy, Deshayes & Milne-Edwards 1839: 371. Pfeiffer 1841: 25, 1842: 115, 1848: 69. Férussac & Deshayes 1851: 122

Table 9.2.1 Marianas *Partula* shell measurements

	N	L			W			L aperture			W/L		
		range	mean	sd	range	mean	sd	range	mean	sd	range	mean	sd
gibba sinistral	117	15.5-19.8	16.50	0.699	11.2-16.0	13.70	0.821	8.2-10.9	8.90	0.552	70.2-84.5	83.00	0.992
gibba dextral	5020	14.1-20.2	17.04	0.754	9.3-13.9	11.92	0.569	8.3-11.3	9.70	0.425	61.5-80.5	69.90	2.362
radiolata	1416	13.5-20.5	16.62	1.051	8.3-12.1	10.19	0.626	7.7-11.3	9.30	0.605	55.5-69.5	61.22	2.215
salifana	105	17.1-19.0	18.05	0.920	10.5-11.7	11.01	0.344	10.5-11.7	10.20	0.337	57.5-62.5	60.85	1.908
langfordi	73	12.9-15.0	13.60	0.240	9.1-11.4	10.00	0.250	-	8.00	-	69.8-76.0	72.06	1.011
sp. (Rota)	6	15.4-18.4	16.77	1.078	10.2-12.2	11.11	0.800	8.6-10.7	9.53	0.725	52.6-58.6	56.83	2.208
desolata	4	17.2-19.5	18.70	1.300	13.8-14.8	14.4	0.513	9.8-12.5	11.30	1.17	70.0-91.0	80.5	-

9.2.2 Anatomy of Marianas *Partula*: a-e – *P. gibba*; f-h – *P.* sp. (Rota); i-j – *P. radiolata*; k-m – *P. salifana*; n – *P. langfordi*. a-c, i, k after Kondo (1955), d-e – Guam (ANSP A987), f-h – Rota (f – elongate [BPBM 188957]; g – typical [BPBM 188959]; h – typical (FLMNH 45844); j – Guam (BPBM 152805); l-m – Guam (BPBM 152860); n – Kondo (1970). Scale bar 5 mm.

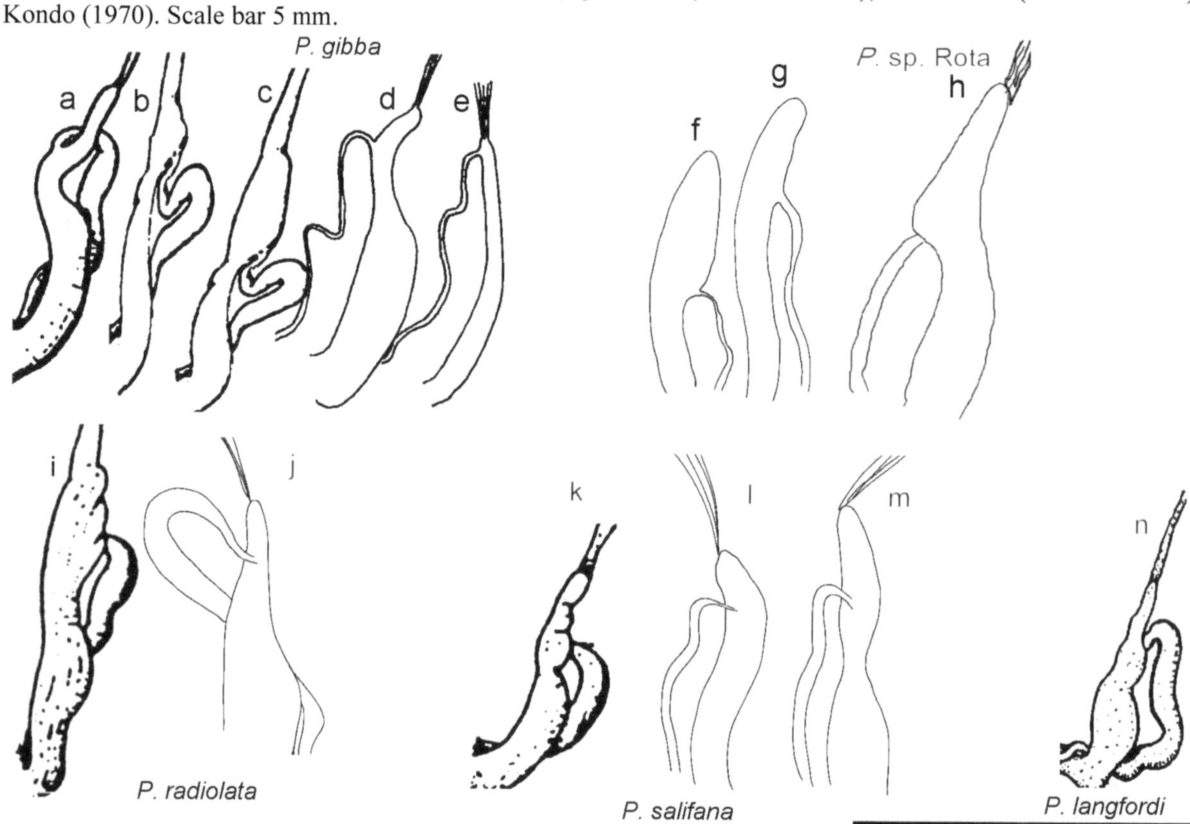

Partula mastersi Pfeiffer, 1853a: 453
Partula mastersi Pfeiffer, Pfeiffer 1855b: 269, 1857e: 110, 1859a: 513, 1868: 160, 1877: 208. Paetel & Schaufuss 1869: 83. Pease 1871b: 473. Paetel 1883: 153, 1889: 267
Partula bicolor Pease, 1872: 26
Partula bicolor Pease, Pease 1872: 473. Pfeiffer 1877: 201. Hartman 1881a: 180, 1881b: 13, 1885: 211, 1892: 75. Paetel 1883: 152, 1889: 266. Smith 1903: 439
Partula gibba bicolor Pease, Pilsbry 1909: 315

Dextral, rarely sinistral (12% in one site on Saipan, n=1,736 [Kondo 1970], 3% of an 1888 sample from Guam, n=220 and an isolated shell from 2013 [Kerr & Fielder 2015]). Ovately conic, spire acute. Solid, shiny. Umbilicus open, narrow. 4-4½ moderately convex whorls, weak growth lines and distinct spiral striae. Body whorl gibbous, variably swollen, base convex. Suture slightly adpressed. Aperture oblique, elongately ovate, outer margin straight sided. Lip reflexed, expanded and heavily thickened internally. Columellar lip dilated. Callus thin, shiny, most apparent at columella insertion.

Usually white or flesh-coloured with spire dull scarlet and body whorl pale yellowish brown. Suture white. Lip and aperture variably coloured, white in pale shells, flesh-tinted in dark shells. Some individuals reddish, with a brown body whorl; or white with a white spire; or castaneous-white and translucent; dull brown with a purple spire and flesh-coloured lip; lilac, the spire darker; light reddish brown, with bicoloured whorls. Colour patterns vary geographically.

Anatomy: Penis short, $^{1}/_{3}$ oviduct length. Caecum small, internally papillose, penis with distinct pilasters. Vas deferens

9.2.3 *Partula gibba*: a – type (MNHN 2000.27236); b – *bicolor* type (CMNH 62.4162); c – *mastersi* type (NHMUK); d – *mastersi* β (NHMUK); e-i – colour variation (ANSP); j – FLMNH 458444; k – sinistral, Saipan (BPBM 21537); l – Aguiguan (BPBM) (photos: a – P. Maestrati, b – C. Strum)

9.2.4 Crampton's (1925) *P. gibba* figures (plates 12-14 compiled, edited for position). Pl. 12-13 Guam, pl. 14 Saipan

subterminal. Kondo (1955) described an anatomical variant from Dedeo-Yigo with an enlarged caecum (Fig. 9.2.2c), similar to *P. salifana*, this was more strongly ornamented internally. He suspected that the two species hybridised, giving rise to the anatomical variant and to the difficulty of separating the species in the Mt. Lamlam population.

<u>Type specimens</u>:

Partula gibba Férussac, 1821 – Syntypes MNHN-IM-2000-272236.

Partula mastersi Pfeiffer, 1853 – Syntypes NHMUK 20140019-20 (var β), 20140017-18, all 'MC'.

Partula bicolor Pease, 1872 – MCZ 25345 listed as type. CMNH 62.4162 'topotypes'; described from 3 specimens sent by Brazier to Pease (Smith 1903) from a larger series. CMNH 62.4162 comprises 5 specimens. BPBM 87463 is labelled 'type', from SMF as part of the Möllendorff type material, but this cannot be a type.

<u>Distribution</u>: Guam, Aguiguan, Tinian, Saipan, Anatahan, Sarigan, Alamagan and Pagan islands. Specimens from Rota may not be this species (see. p. 90).

9.2.5 *Partula gibba* colour variation (data and colour forms from Crampton 1925)

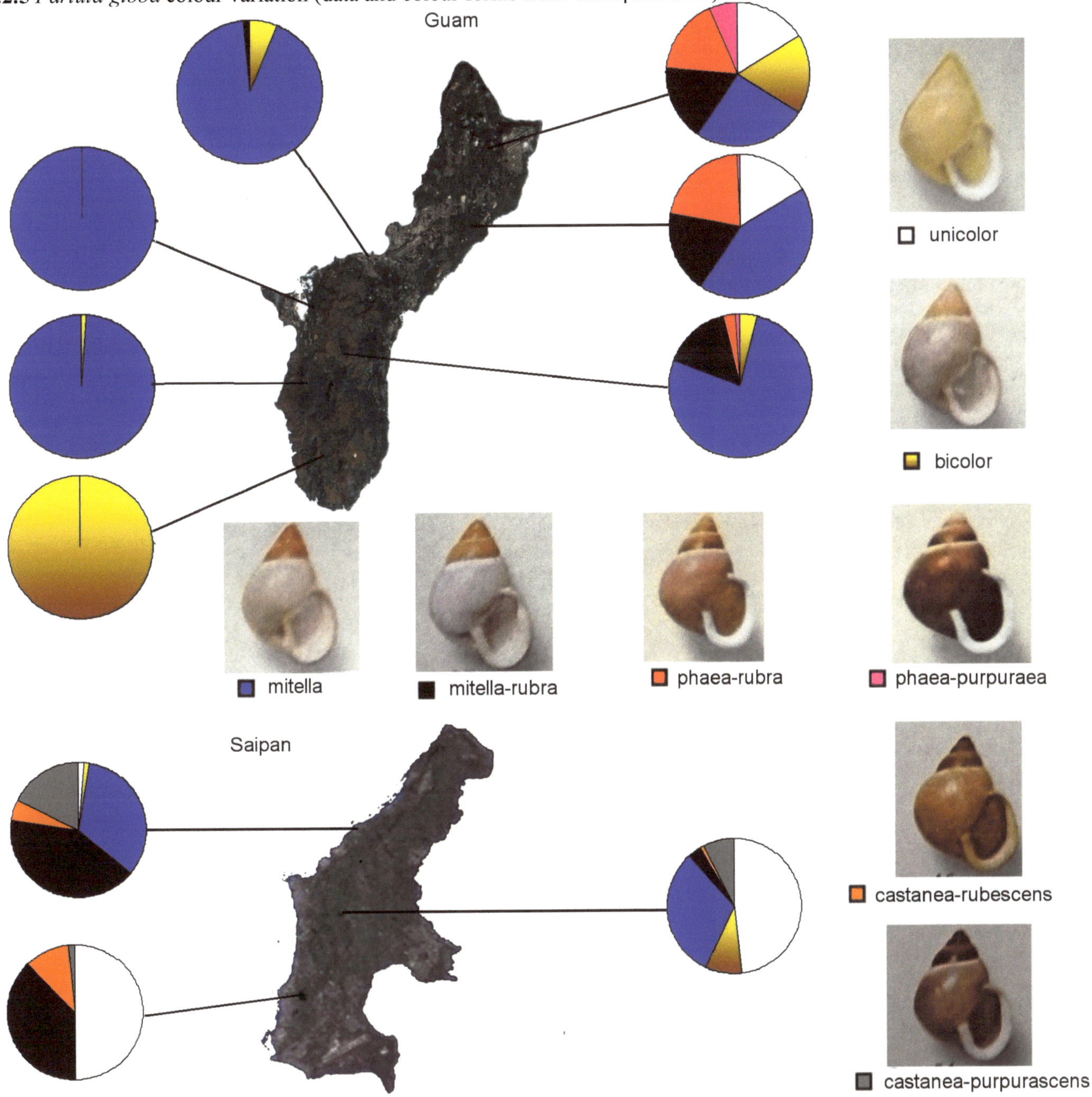

Partula langfordi Kondo, 1970

Partula new species Kondo 1968: 77

Partula langfordi Kondo, 1970: 75

Ovate conical, moderately thin. Umbilicus open. 5 slightly convex whorls, apex obtuse. Fine spiral striae distinct on spire. Aperture oblong ovate. Lip thickened and expanded. Columella with a faint sub-medial swelling, margin slightly obscuring umbilicus. Parietal callus thin, indistinct.

Buff superimposed with maroon, a maroon band on 2nd and 3rd whorls, band expanding apically over spire. Lip white.

Anatomy: Penis short, similar to *P. gibba*; caecum slightly expanded where vas deferens enters; vas deferens inflated.

Type specimens: Holotype BPBM 213183.

Distribution: Aguiguan island, at around 650 m.

Partula desolata Bauman & Kerr, 2013

Partula sp. cf./aff. *gibba* Bauman 1996: 21

Partula desolata Bauman & Kerr, 2013: 5

Ovate conical, solid. Umbilicus partially obscured, deep. 4½ nearly flat whorls with fine spiral striae. Suture emarginated, impressed. Aperture slightly oblique, moderately auriculate. Columellar margin straight, flared at insertion with body whorl. Lip expanded, thickened. Parietal callus thin, with a small nodule.

Original colour not known; light brown, darker near sutures.

Type specimens: Holotype BPBM 252143-A; paratypes FLMNH 449332-3; UGI 3001.

Distribution: Rota – late Holocene subfossils from 1,000 years ago, possibly to 400 years ago (Bauman & Kerr 2013).

***Partula* sp. (Rota)**

Ovately conic, spire acute. Umbilicus open, partially obscured by reflected lip. 4½ very slightly convex whorls with fine spiral striae on spire. Body whorl very slightly angulated. Suture raised. Aperture oblique, oblong, outer margin straight sided. Lip reflexed, expanded, thickened internally. Columella angled. Parietal callus thin, shiny.

Colour patterns are as in *P. gibba*, with forms unicolor, bicolor, phaea-purpuraea being recognisable. In most individuals the inner edge of the aperture is dark purple-brown. Suture clearly marginated white.

Anatomy: Penis short, similar to *P. gibba*; caecum distinctly expanded, vas deferens at base of caecum.

Distribution: Rota island. This species is currently being described (M. Hadfield & D. Sischo pers. comm.).

Partula salifana Crampton, 1925

Partula salifana Crampton, 1925: 25

Partula salifana Crampton, Kondo 1955: 201, 1968: 77. Richardson 1990: 56

Ovate conic, thick. Umbilicus open, slightly flattened. 5-5¼ whorls, slightly impressed below suture. Spiral striae on protoconch, fading on spire. Aperture elongate. Lip expanded, flattened, narrowing as it approaches body whorl. Callus distinct, slightly thickened at margin.

Chestnut to yellowish, apex often purple due to decortication. Lip white to yellow brown or purple. Aperture purplish.

Anatomy: Penis short, ¼ oviduct length, similar to *P. gibba*. Caecum broad. Penis ornamented with 6 pilasters.

Type specimens: Syntypes ANSP 290457, 290487, A1434A; BPBM 92311; DMNH 081968; MCZ 179487.

Distribution: Guam – Mount Alifan and two adjacent peaks. Arboreal, especially on *Pandanus* sp. (Crampton 1925).

9.2.6 a – *Partula langfordi* type (BPBM 313183); b – *P. desolata* type (BPBM 252143-A, from Bauman & Kerr 2013)

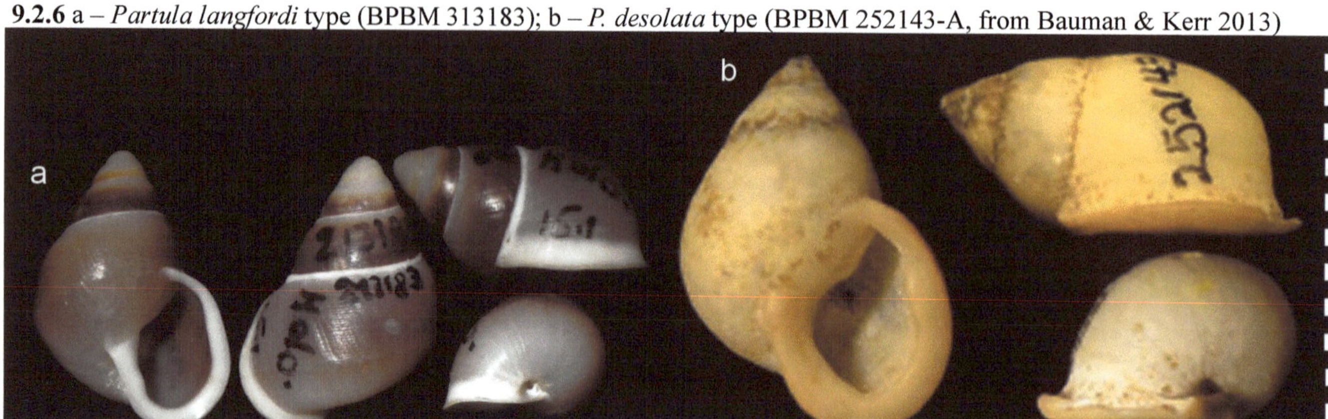

9.2.7 *Partula* sp. Rota: a-b – UF 458524; c – colour variation (BPBM 215324); d – elongate (BPBM 188951)

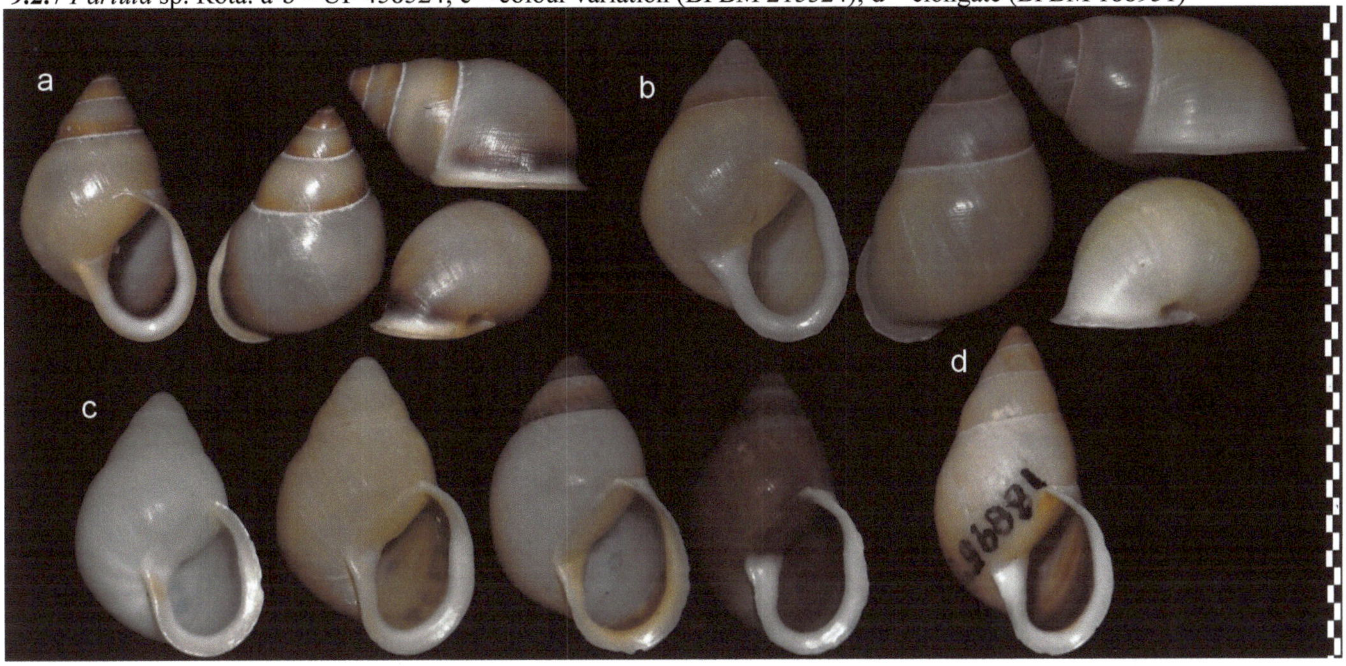

9.2.8 *Partula salifana*: a – type (ANSP 290457); b-e – colour variation (b-c: ANSP 455268; d-e: ANSP 344247)

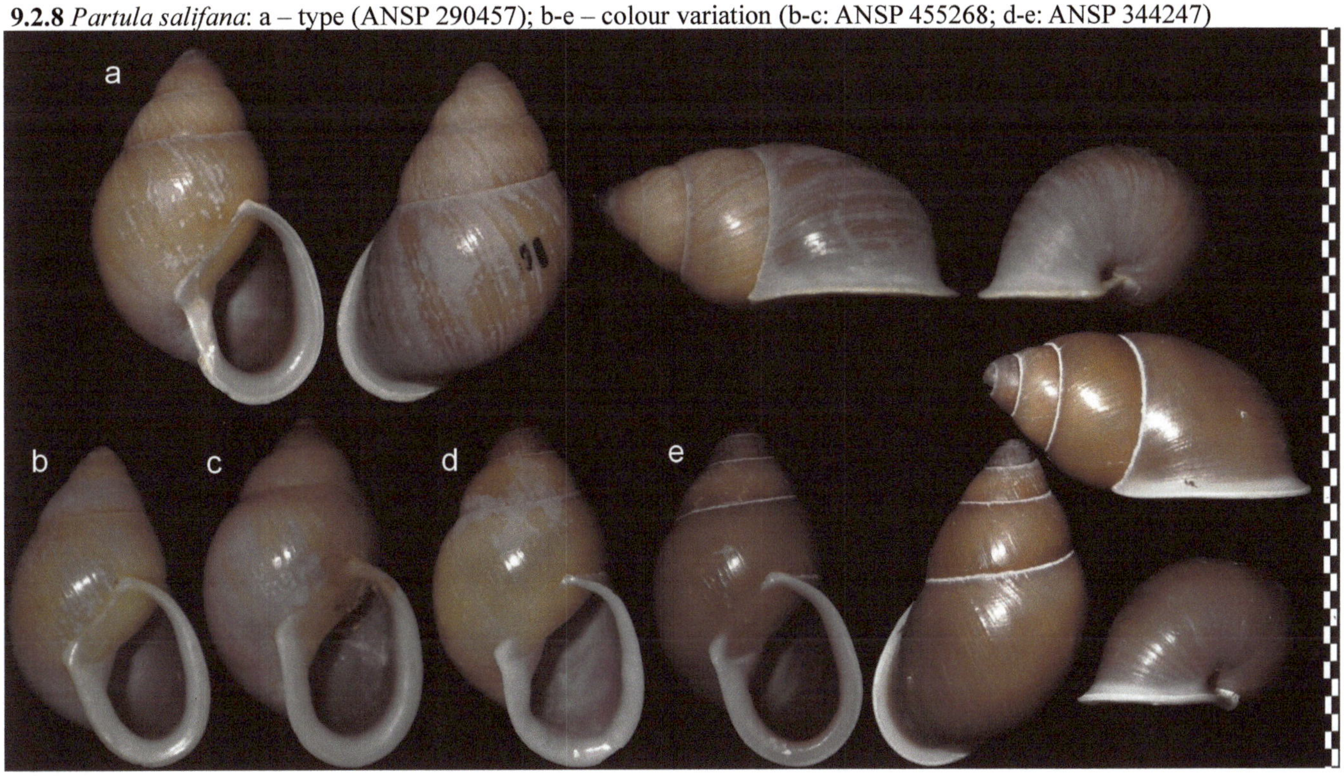

9.2.9 Crampton's (1925) *P. salifana* figures (edited for position)

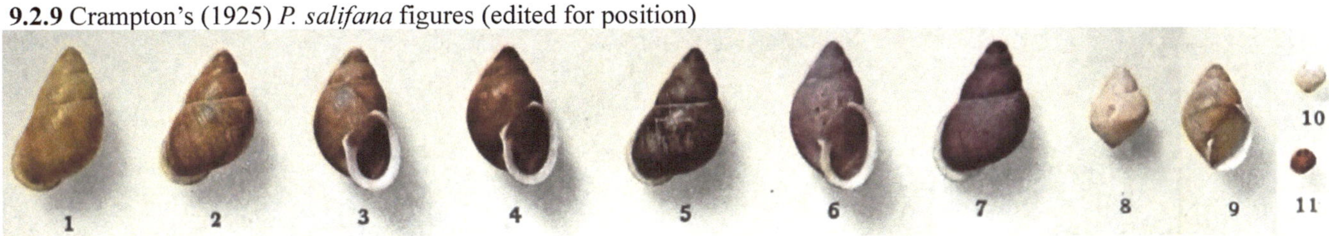

Partula radiolata (Pfeiffer, 1846)

Bulimus radiolatus Pfeiffer, 1846: 39

Partula radiolata (Pfeiffer), Pfeiffer 1846: 56, 1848: 69, 1853a: 450, 1856: 162, 1856: 270, 1859: 509, 1868: 158, 1877: 198. Albers 1850: 186. Reeve 1850: no 6. Adams 1855: 145. Pease 1871b: 473. Paetel 1873: 104, 1883: 153, 1889: 268. Kobelt 1880: 16. Hartman 1881a: 186, 1881b: 13, 1885: 223, 1892: 99. Pfeiffer & Clessin 1881: 301. Smith 1903: 439. Pilsbry 1909: 316. Crampton 1925: 34. Rensch 1937: 582. Kondo 1955: 202, 1968: 77. Richardson 1990: 52

Partula radiolata rushi Pilsbry, 1909: 318

Oblong-tapering. Umbilicus very narrow. Thin but strong. Spire obtuse, 4½-5 convex whorls. Faint impressed lines. Aperture obliquely oval. Lip thin, expanded, right margin somewhat straightened, columellar margin dilated above, spreading over umbilicus. Lip slightly thickened within.

Pale yellow with darker axial rays and brown lines. Lip white. Var. *rushi* sometimes has a brown band above the suture on the spire. Crampton (1925) defined colour varieties as pallida (22.3%, n=1,535), flava (53.0%), fulva (7.7%), strigata (10.1%), strigata-helix (1.9%), bicolorata (5.0%). Colours vary geographically.

Anatomy: Penis short, ¼ oviduct length. Caecum small, curved. Strongly ornamented internally with numerous papillae.

Distribution: Guam. Originally described from 'New Ireland' in error. *rushi* collected from the area of Port San Luis d'Apra. Historically also on the islet of Cabras (Crampton 1925).

Type specimens:

Bulimus radiolatus Pfeiffer, 1846 – Syntypes NHMUK 20140003 (var b), 20140002, all Cuming collection.

Partula radiolata v. *rushi* – Syntype ANSP 84841 (a single shell). Pilsbry (1909) referred to 'examples'; location of other specimens unknown.

Note: *rushi* form darker, spire dark brown, last 1½ whorls striped cream and dark brown, crossed by a brown peripheral band, aperture usually brown inside.

9.2.10 *Partula radiolata*: a – *rushi* syntype (ANSP 84841); b – *radiolata* syntype (NHMUK 200140002); c – *radiolata* var. b syntype (NHMUK 200140003); d-h – colour varieties (ANSP 191975); i – umbilical view (ZSL 700)

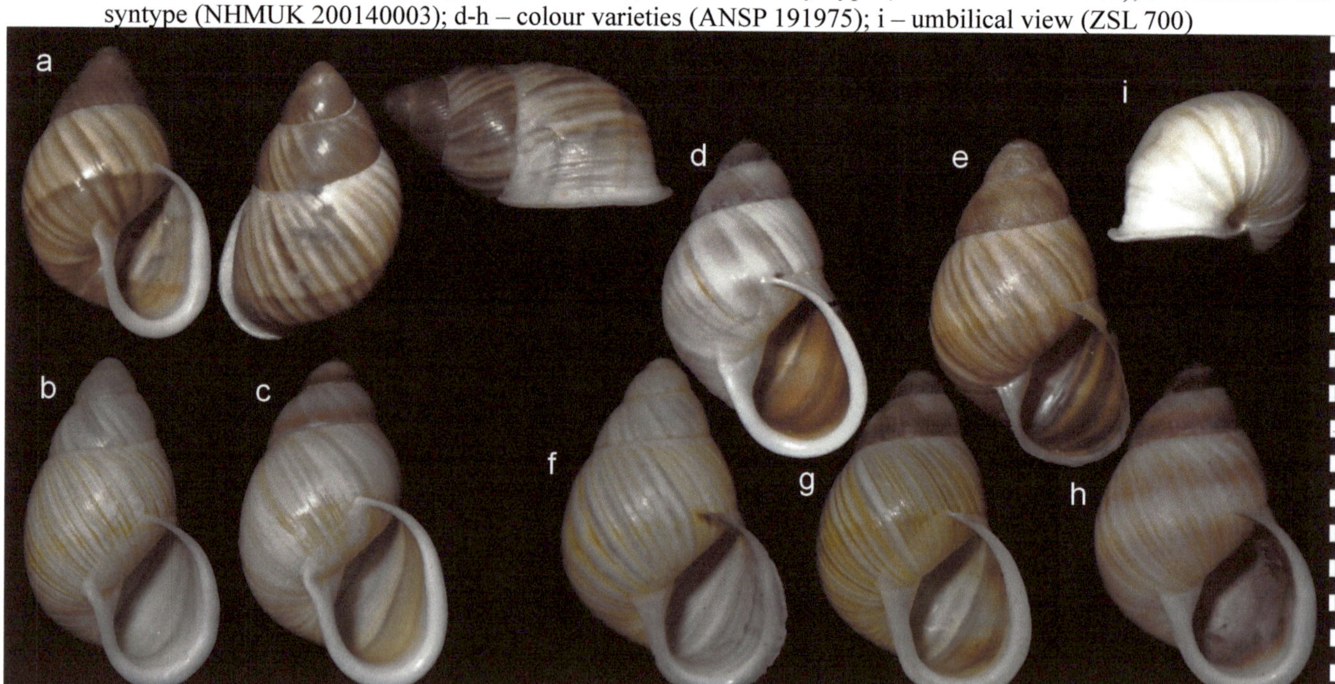

9.2.11 Crampton's (1925) *P. radiolata* figures (edited for position)

9.2.12 *Partula radiolata* colour patterns (data and colour forms from Crampton 1925)

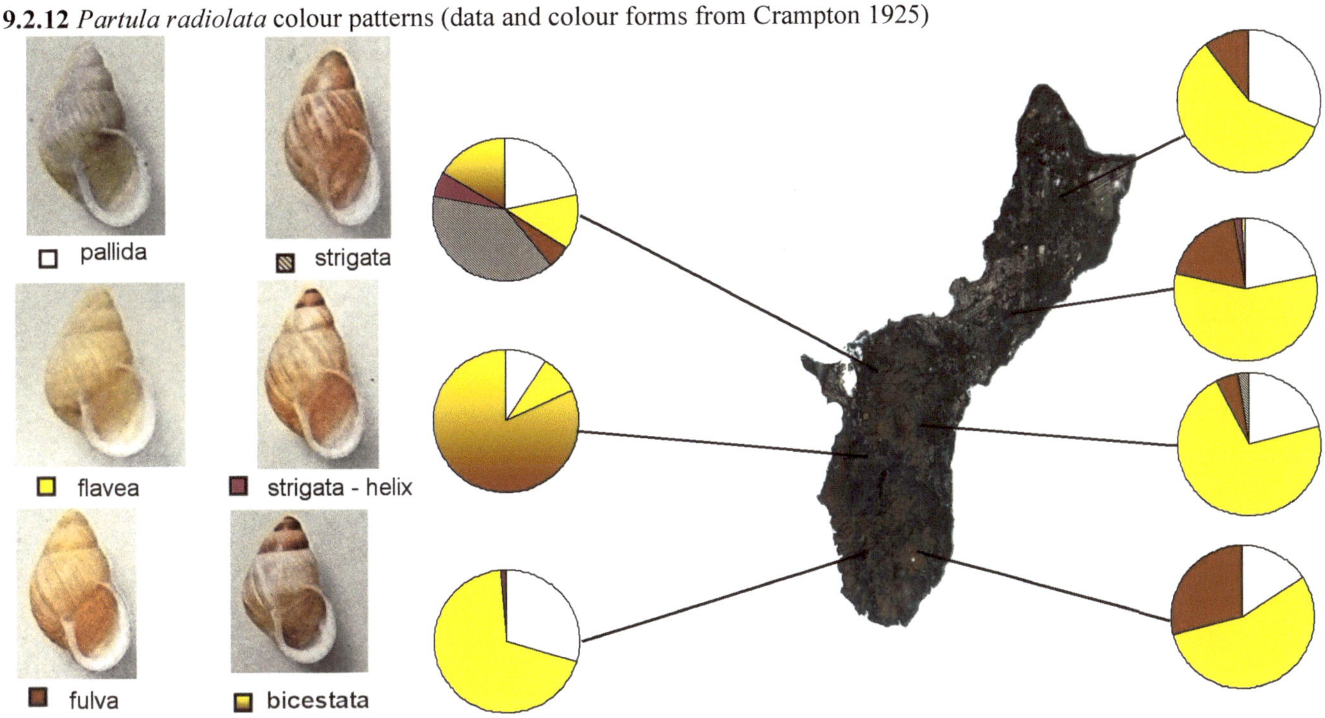

Caroline islands

Only two Caroline islands are high enough to support the habitats needed by partulids; three species are known from Kosrae and Pohnpei (Ponape).

Table 9.2.2 Measurements of Caroline islands *Partula*

	N	L			W			L aperture			L/W		
		range	mean	sd.	range	mean	sd.	range	mean	sd.	range	mean	sd.
rufa	82	18.9-24.4	22.44	0.283	12.7-16.5	14.98	0.917	9.9-14.1	12.24	0.802	58.5-73.5	66.67	2.867
guamensis	111	24.6-27.2	26.09	1.106	16.0-18.5	17.02	0.694	11.0-13.5	12.61	0.707	61.3-68.1	65.31	2.567
emersoni dextral	195	19.1-22.0	20.15	0.826	9.2-10.5	10.10	0.995	9.5-10.4	10.10	0.338	43.1-52.8	49.50	1.618
emersoni sinistral	4	19.4-20.3	19.68	0.419	10.5-11.0	10.80	0.245	9.2-9.8	9.45	0.252	53.8-56.7	54.90	1.283

9.2.13 Distribution of Caroline islands *Partula*. The arrow marks Temwen islet.

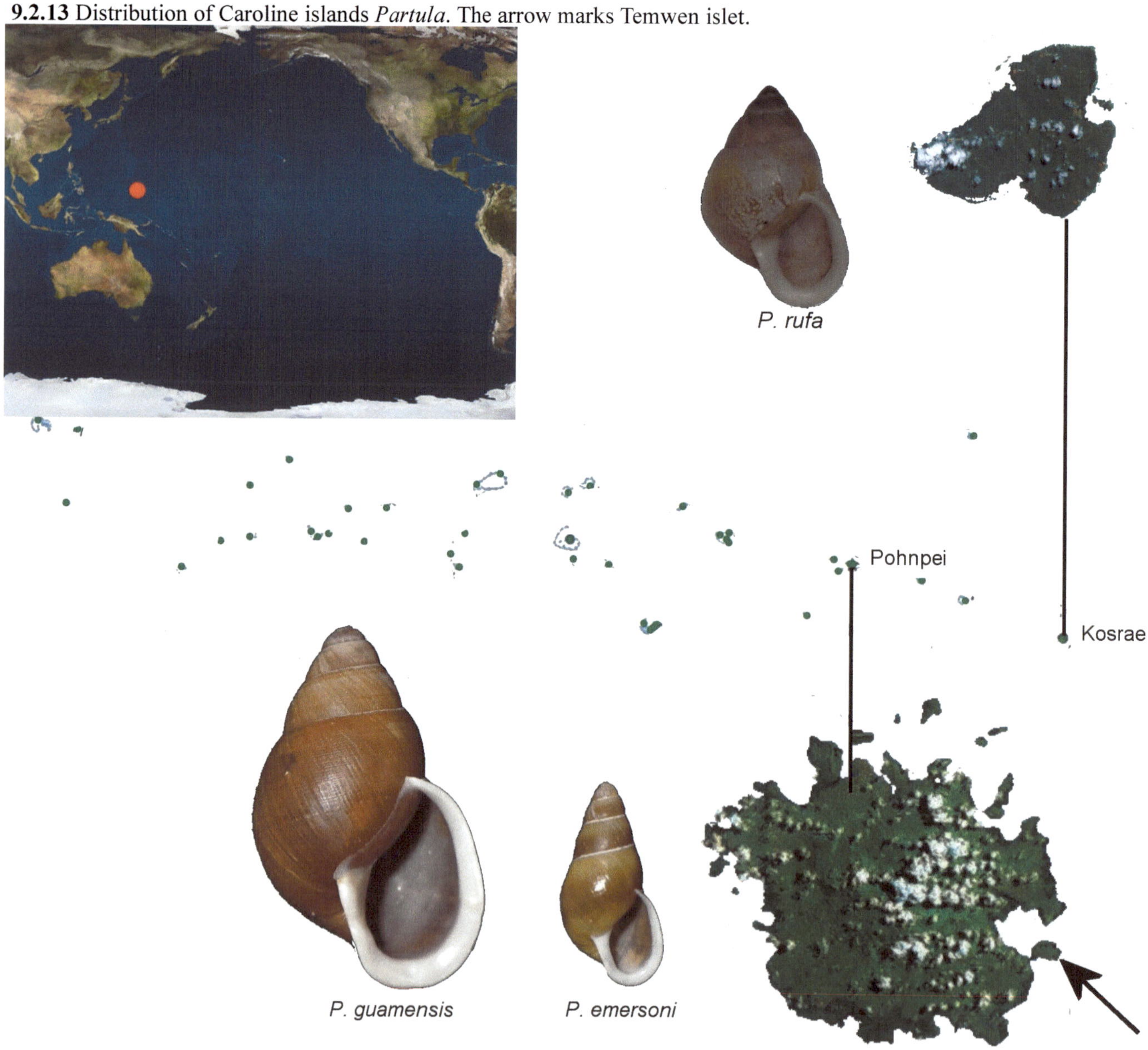

Partula emersoni Pilsbry, 1913

Partula emersoni Pilsbry, 1913: 67

Partula emersoni Pilsbry, Kondo 1955: 194, 1968: 77. Richardson 1990: 22

Normally dextral (0.5% sinistral, n=195; Crampton MS). Elongate conical. Thin, glossy. Umbilicus open, narrow, half covered. 5½ moderately convex whorls. Body whorl somewhat flattened above periphery, very convex beneath. Early whorls spirally striate, fading on penultimate whorl. Body whorl lacking striae but distinctly malleate. Suture moderately impressed, an inconspicuous cord immediately above, partly covering suture in intermediate whorls. Aperture elongately ovate. Lip expanded, reflected and slightly thickened within.

Shells may be either olive brown with a white or uniform apex, or banded with a darker subperipheral band of varying intensity. The lower border of the band is edged whitish. Lip whitish. Uniform shells are slightly commoner than banded shells (58%, n=198; Crampton MS) are found in similar proportions. Of the uniform shells only 40% have the white apex. According to Crampton (MS) colour patterns vary geographically, with the uniform shells predominating in Kupornjo ('Kupwuriso') (96%) and common at Chaptakai (Sapwtakai) (26%) while banded shells were found in all areas but dominated only in Jokaj ('Sokehs', probably referring to Kondo's Paipalap site) and Tolan Kamar (one of three sites with the same name, probably on Kupwuriso).

Anatomy: Penis moderately long (40% of oviduct length), simple. Caecum narrow, as long as penis. 5-6 weakly developed pilasters in penis and rows of distinct papillae in caecum.

Type specimens: Holotype BPBM 101001.

Distribution: Pohnpei. Localised. Arboreal, erroneously described as terrestrial by Kondo (1955). The origin of this species was initilly confused but was confirmed by specimens collected by Kondo (Pelep & Hadfield 2011). In 1936 Kondo collected the species only from the areas of Kupwuriso (from 30-660 m), Paipalap and Sapwtakai in the north-western half of the island.

9.2.14 Anatomy of Caroline islands *Partula* (after Kondo 1955). Scale bar 5 mm.

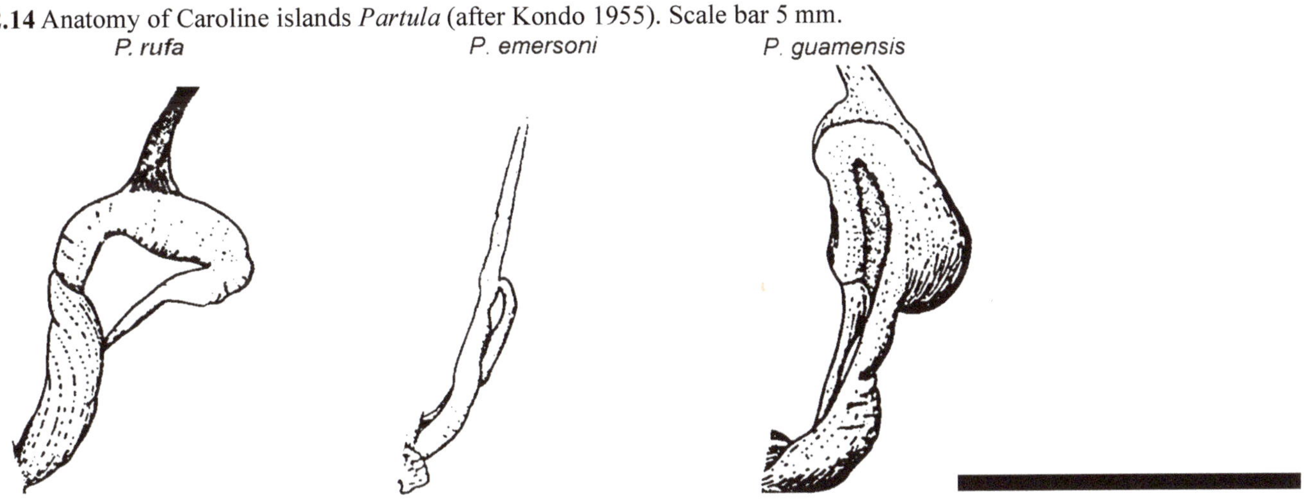

9.2.15 *Partula emersoni*: a – holotype (BPBM 101001); b – typical colour pattern (ANSP 204826); c – sinistral specimen (BPBM 157967)

Partula guamensis (Pfeiffer, 1846)

Bulimus guamensis Pfeiffer, 1846: 113

Bulimus guamensis Pfeiffer, Pfeiffer 1846: 54, 1848: 73

Partula guamensis Pfeiffer, Reeve 1850: no 4, 1859: pl1 f4. Férussac & Deshayes 1850: 124. Pfeiffer 1853a: 446, 1856: 162, 1859: 511, 1868: 159, 1877: 199. Adams 1855: 145. Pease 1871b: 400, 473. Martens 1880: 146, 1881: 12. Hartman 1881a: 180, 1881b: 13, 1892: 98, 1885: 221. Paetel 1883: 153, 1889: 266. Möllendorff 1900: 112. Smith 1903: 468. Pilsbry 1909: 308. Thiele 1931: 659. Kondo 1955: 192, 1968: 77. Zilch 1959: 462. Richardson 1990: 32

Partula brumalis Reeve, 1850: pl1 f12

Partula brumalis Reeve, Pfeiffer 1853a: 446, 1859: 511. Frauenfeld 1869: 874. Pease 1871b: 473. Martens 1881: 95. Paetel 1889: 266

Partula Ponapensis Hartman, 1885: 221 (substitute for *guamensis* Pfeiffer)

Partula rufa, sub-sp. *montana* Möllendorff, 1900: 112 (unjustified substitute for *guamensis* Pfeiffer)

Partula rufa, sub-sp. *grandis* Möllendorff, 1900: 112

Ovate conic. Umbilicus open, rimate. Solid, very delicately decussate. Spire short, obtuse. 5¼-6 convex whorls. Protoconch with dense spiral striae, extending on all whorls, variably developed on the body whorl. Body whorl swollen. Columella nearly straight. Aperture wide, semioval. Lip margins remote, right margin expanded, thickened internally. Columellar margin broadly reflexed, spreading.

Dark chestnut brown, sometimes with a slightly darker peripheral band. Aperture usually bluish or dull purplish inside,

9.2.16 *Partula guamensis*: a – *guamensis* type (NHMUK 71.4.12.29); b – *brumalis* type (NHMUK un-numbered); c – neonate (ANSP 191984); d – adult (ANSP 191984)

sometimes white. Lip white. Möllendorff's (1900) *grandis* was described with three colour varieties: dark purple-brown with a violet lip, a pale chestnut form with a white lip (form castanea) and a rare pale greenish yellow form (flavescens).

Anatomy: Penis very large, ¾ length of oviduct. Caecum expanded, recurved, with four distinct sections: funnel-shaped entrance to vas deferens, followed by a constriction, an enlarged main part and a narrow base. Interior of penis with 10 pilasters, caecum with at least 18 rows of tubercles. A second variety has a simpler internal structure, Kondo (1955) considered this to be a result of ontogeny.

Distribution: Pohnpei and Temwen islet. On Pohnpei recorded from throughout the island. Terrestrial – all live specimens collected by Kondo in 1936 were from leaves on the ground (Pelep & Hadfield 2011).

Type specimens:

Bulimus guamensis Pfeiffer, 1846 – Syntypes probably NHMUK 71.4.12.29, labelled 'Caroline islands, Ascension island', the largest of these matches the type description in size.

Partula brumalis Reeve, 1850 – Syntypes NHMUK un-numbered (Cuming collection).

Partula rufa grandis Möllendorff, 1900 – Not located.

Note: Möllendorff's *grandis* was larger than typical forms (30.5 × 19 mm), less solid, with weaker spiral sculpture, a more expanded lip. From Chaptakai, 800 ft, Mt. Nanalaut 1,900 ft.

Partula rufa Lesson, 1831

Partula rufa Lesson, 1831: 324

Partulus rufus Lesson, Beck 1837: 58

Bulimus (*Partula*) *rufus* Lesson, Pfeiffer 1842: 121, 1848: 229

Partula rufa Lesson, Pfeiffer 1848: 229, 1853a: 449, 1855a: 145, 1859: 512, 1868: 160, 1877: 205. Adams 1855: 145. Martens 1881: 94. Hartman 1881a: 186, 1881b: 10, 1885: 221, 1886: 30, 1892: 98. Pfeiffer & Clessin 1881: 302. Paetel 1889: 268. Möllendorff 1900: 112. Smith 1903: 469. Pilsbry 1909: 311. Richardson 1990: 55

Partula martensiana Pilsbry, 1909: 310

Partula martensiana higginsi Pilsbry 1924: 17

Partula martensiana Pilsbry, Kondo 1955: 190, 1968: 77. Richardson 1990: 40

Ovate-conic, solid and strong, opaque. Glossy. Umbilicus open, narrow and compressed. Conic, obtuse spire. 5 convex whorls ornamented with fine spiral striae, obsolete or very faint on body whorl. Aperture almost vertical. Lip only very slightly expanded, strongly thickened within. Columellar margin dilated above.

Yellowish to brown, spire reddish to purplish brown, periostracum often flaky. Aperture flesh-pink within. Lip white.

Anatomy: Penis robust, as long as oviduct. Caecum as long as penis. Penis with 8 branching pilasters, crowded in neck, reduced to papillae in caecum, separated by isolated tubercles. Ribbing extends into funnel-like opening of vas deferens.

9.2.17 *Partula rufa*: a – *martensiana* holotype (CMNH 62.4298); b – *higginsi* holotype (ANSP 128508); c – Kondo specimen (BPBM 155947a) (photos: a – C. Sturm; c – F. Borrero)

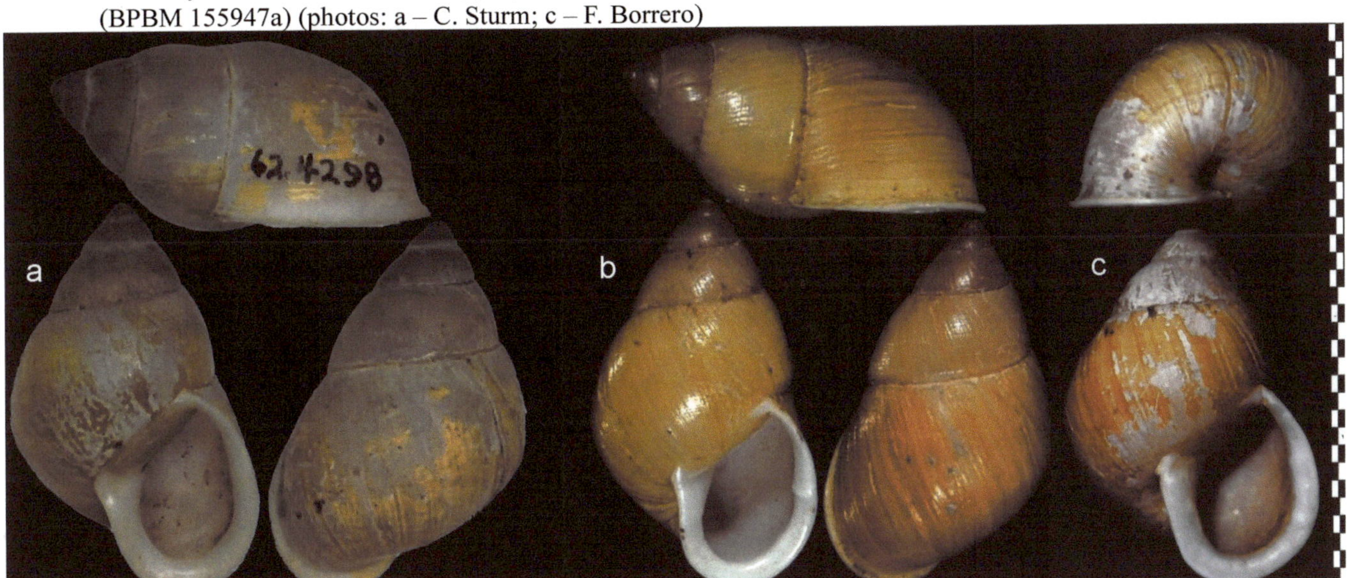

<u>Type specimens</u>:
Partula rufa Lesson, 1841 – Lost.
Partula martensiana Pilsbry, 1909 – Holotype CMNH 62.4298.
Partula maretnsiana higginsi Pilsbry, 1924 – Holotype ANSP 128508.
<u>Distribution</u>: Kosrae ('Ualan'). Terrestrial.
<u>Notes</u>: *P. rufa* has been a disputed species due to the early loss of the type specimen (in 1886 Hartman already referred to it as a 'lost' species). Hartman (1886) identified three specimens collected by Brazier on Kosrae as *P. rufa*. Pilsbry (1909) considered *P. rufa* to be smaller and narrower than the Kosrae species, which he named *P. martensiana*. However, he converted Lesson's measurements in lignes to millimetres incorrectly. The correctly converted measurements of 20.5 × 10 mm are only a little smaller than the 21.5-22.3 × 14.0-14.1 mm of his *P. martensiana*.
P. martensiana higginsi differed from the typical form in being narrower, the penultimate whorl distinctly angular above the suture, a narrower umbilicus and the presence of a tiny labial nodule. The first two features are within the range of variation of *P. rufa* and the labial structure is clearly an abnormality, as noted by Crampton (in Pilsbry 1924).

9.2.18 Living *Partula* on Guam (a-c), Rota (d) and Pohnpe (e): a – *P. gibba*; c – *P. radiolata*; d – *P*. sp.; e – juvenile *P. emersoni* (photos: a-c – G.C. Fiedler; d – D. Sischo; e – B. Holland)

9.3 Central Polynesia

Six species of *Partula* are scattered across central Polynesia, from the Fiji islands, Wallis and Futuna, and the Cook islands. Of these, the Fiji islands are usually considered part of Melanesia but are included in the Polynesia-Micronesia biodiversity hotspot (Myers *et al.* 2000).

Fiji

Three species are present in the central Polynesian Fiji islands, one on Rotuma and two on the smaller islands of central Fiji.

Table 9.3.1 Central Polynesian *Partula* shell measurements

	N	L			W			L aperture			W/L		
		range	mean	sd	range	mean	sd	range	mean	sd	range	mean	sd
P. lirata	324	14.7-22.3	19.22	1.224	8.5-11.7	10.21	0.595	7.8-11.1	9.80	0.875	47.5-63.5	53.22	2.273
P. lanceolata	73	17.7-21.6	19.71	0.841	10.1-11.9	10.88	0.406	9.5-11.5	10.34	0.365	50.5-60.5	55.24	1.820
P. leefei	209	11.4-14.8	13.40	0.685	7.3-9.3	8.45	0.342	6.5-8.3	7.51	0.318	58.5-68.5	62.83	1.698

9.3.1 Fijian *Partula* distribution. Islands: 1 – Rotuma, 2 – Taveuni, 3 – Laucala, 4 – Kanacea, 5 – Vanua Balavu, 6 – Avea, 7 – Cikobia-i-Lau, 8 – Munia, 9 – Mago, 10 – Tuvuca, 11 – Cicia, 12 – Naiau, 13 – Oneata

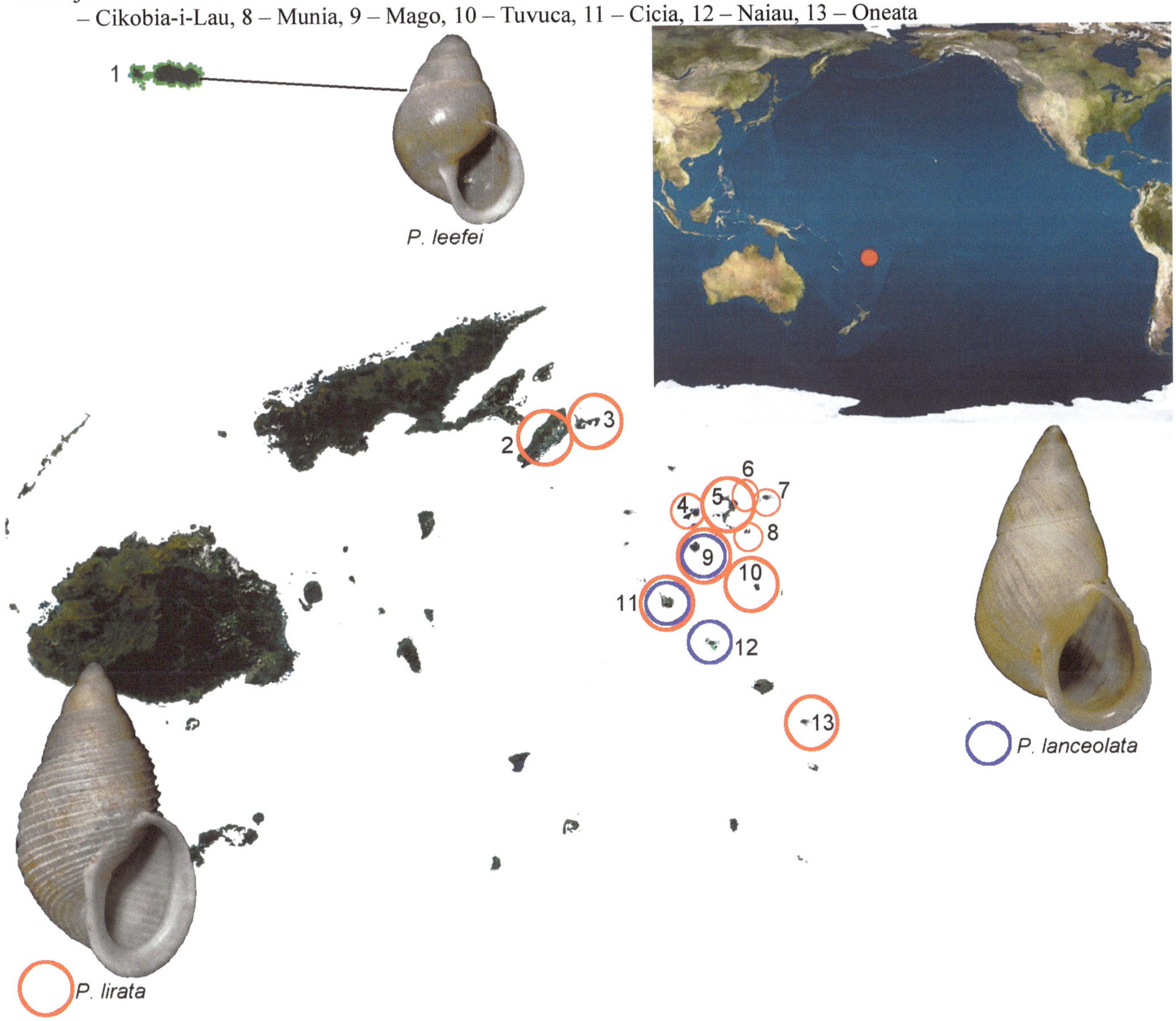

Partula lirata Mousson, 1865

Partula lyrata Pfeiffer, 1859: 158 (nomen nudum)

Partula lirata Mousson, 1865: 196

Partula lirata Mousson, Crosse 1865: 432. Heynemann 1867: 148. Pfeiffer 1868: 158, 1877: 198. Paetel & Schaufuss 1869: 83. Schmeltz 1869: 71, 1874: 91. Mousson 1870: 126. Paetel 1873: 104, 1883: 153, 1889: 267. Semper 1874: 15. Hartman 1881a: 183, 1881b: 14. Garrett 1887: 187. Smith 1903: 451. Pilsbry 1909: 275. Thiele 1931: 659. Pilsbry & Cooke 1934: 14. Zilch 1959: 453. Kondo 1968: 76. Richardson 1990: 37

Partula lyrata Mousson (misspelling), Heynemann 1867: pl1 f1. Mousson 1870: 126. Pfeiffer & Clessin 1881: 301. Hartman 1885: 216, 1892: 76. Richardson 1990: 39

Eua (*Thakombaua*) *lirata* Schileyko 1999: 264

Dextral. Elongate ovate-conic. Thin, surface dull. Umbilicus narrow, rimate. 5½ whorls. Protoconch of almost 3 whorls, the first half whorl smooth, following half with very faint spiral punctate-striate. Striae on rest of protoconch sparse, mainly on upper part. Spire with spiral striae developing into distinct, low ridges, increase in prominence. On the body whorl there are 8-12 strong spiral ridges above the periphery, separated by small, unequal ridges. The base is crossed by numerous small ridges. Spiral ridges reach the lip margin, making it crenulated. Body whorl laterally compressed, base convex. Aperture ovate, oblique. Lip narrowly expanded, internally thickened. Columella dilated above. Parietal callus distinct but transparent, usually with a deeply recessed nodular denticle.

White, pale yellowish, pale brown or dark liver brown with narrow paler streaks.

Anatomy: Penis short, $^{1}/_{7}$ length of oviduct, constricted at junction of penis and the variably developed caecum or tapering. Internally ornamented with a series of ridges which continue into caecum. Schileyko (1999) considered this a species of *Eua*, but it lacks the characteristic *Eua* penial pilaster. Kondo (1955) figured a specimen with an unusually inflated vas deferens, resembling a fully developed epiphallus, in all other specimens the vas deferens is broad but not inflated to this degree.

Distribution: Lau islands: Vanua Balavu (Vanua Mbalavu), Avea, Kanacea ('Kanathea'), Oneata, Laucala ('Lauthala'), Taveuni, Cikobia-i-Lau ('Thikombia-i-Lau', 'Cikambia'), Mago ('Mango'), Tuvuca ('Tuvutha'), Munia islet and Cicia ('Thithia'). Arboreal, recorded near the shore (Garrett). The record from Viti Levu ('Viti') (Schileyko 1999, specimen in Leiden) is probably an error. A record from Naiau (Brodie *et al.* 2016) is based on specimens (BPBM) of *P. lanceolata.*

Type specimens: Syntypes UZH 518343, MNHN-IM-2000-26716.

Notes: Kondo (1955) mentioned that "should the researcher go to the island of Vanua Mbalavu he would find a colony of smooth-to-lirate form a short distance north of Mvana village". These specimens (BPBM 179704 – Fig. 9.3.3 g-i) are typical *P. lirata* despite the variable sculpture.

9.3.2 Fijian *Partula* anatomy: a-i – *P. lirata* , j-q – *P. lanceolata* (*m-n* – cf. '*sarissa*', o-q – '*sarissa'*), r – *P. leefei.* a, d-f – Vanua Mbalavu, b, g-n – Mango, c – Cikobia-i-Lau, g – Naiau, h – Oneata, o-q – Cicia, r – Rotuma. Sources: a, r – Kondo 1955; b, j – Pilsbry & Cooke 1934; c – ZSL 839; d-f – FMNH 168151; g – BPBM 180218 & 78924; h – BPBM 180142 & 78936; i – BPBM 180068; k – ANSP A1763e; l – BPBM 788942; m – BPBM 78943; n – BPBM 78940; o-q – FMNH 168256. Scale bar 5 mm.

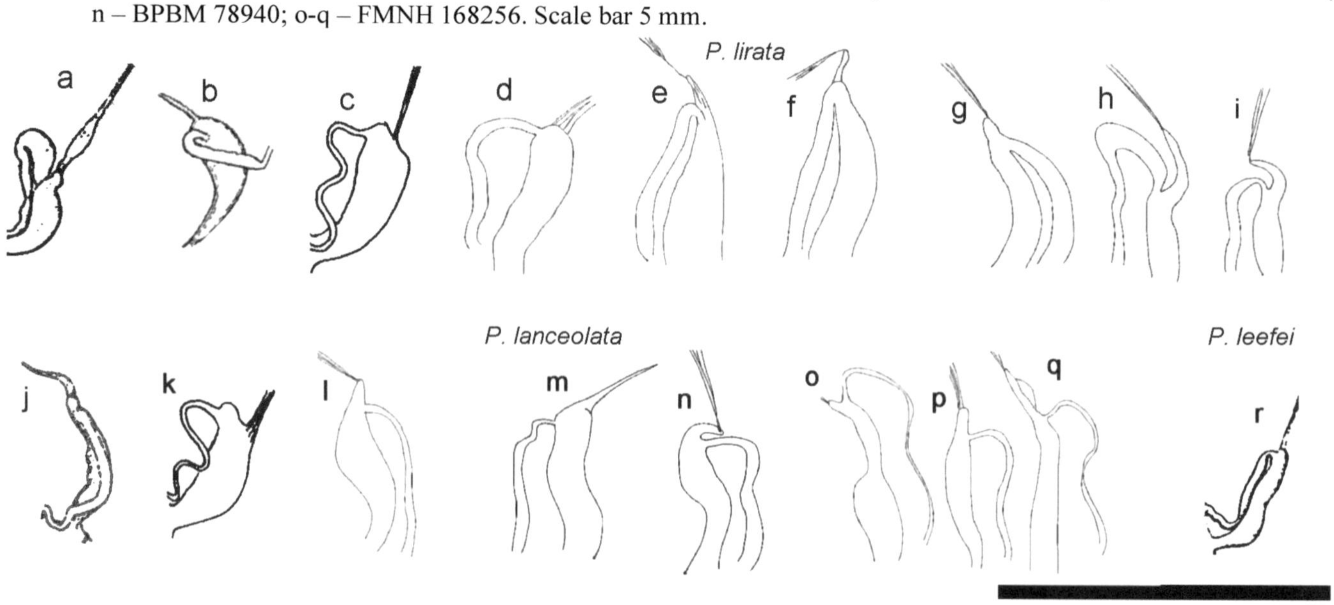

Partula lanceolata Cooke & Crampton, 1930

Partula lanceolata Cooke & Crampton, 1930: 5

Partula (*Thakombaua*) *lancelolata* Cooke & Crampton, Pilsbry & Cooke 1934: 15

Partula lanceolata Cooke & Crampton, Kondo 1955: 178, 1968: 76. Richardson 1990: 36

Dextral. Elongately ovate-conic. Thin but strong. Umbilicus flatly rimate. 5½ straight sided to slightly convex whorls usually lacking distinct sculpture other than faint growth lines. In some specimens (Crampton's '*sarissa*') 1-3 low spiral ridges may be present on the spire and incised spiral striae may also be visible. Suture faintly marginated. Aperture subovate, very simple. Parietal callus thin, shining. Lip thin and narrow, slopes inwardly from margin with slight inner thickening. Columella slender, usually only slightly dilated above. Parietal tubercle weakly developed in most shells, absent in others.

Off-white to light greyish brown. Apex sometimes with a faint trace of pale buff or fawn colour.

Anatomy: Similar to *P. lirata*; penis usually more elongate and the caecum more developed, but variable.

Type specimens: Syntypes ANSP 161288 (&A1763E), 295209, 339338; BPBM 10852, 78940-2; CMNH 62.39800; MCZ 63766.

Distribution: Lau islands: Mago ('Mango'), Cicia ('Thithia') and Naiau ('Nayau'). On Mago recorded from 250-500 ft., on Naiau from near the shore. Naiau specimens (BPBM) were listed as *P. lirata* by Brodie *et al*. (2016).

Note: Crampton (MS) recognised a third species '*Partula sarissa*' as distinct on the basis of its white colour and more developed sculpture than typical *P. lanceolata*. However, this cannot be reliably separated from that species and is not accepted here. It was first recorded from Naiau (ANSP 339328, 295211 ex BPBM [labelled 'Kaiau', probably in error], BPBM 78911) by Edwin Bryan, Jr. in 1928, in 1934, and then by Kondo in 1938. Noted not to be found in abundance and not sympatric with *P. lirata*. It was found only in a forested region on shrubs and trees and dead leaves at altitudes from 30 feet near the shore to 500 feet inland. 19 adults and 29 immatures were collected, embryonic young with shells were found in two gravid adults.

9.3.3 *Partula lirata*: a – type MNHN 2000.26716; b – type UZH 518534; c – specimen (NHMUK); d-f – colour variations collected by Garrett (ANSP 10115); g-i – variation in sculpture from Vanua Balavu (BPBM 179704); j – Cicia form (BPBM 78924) (photo: a – P. Maestrati; b – UZH)

Partula leefei E.A. Smith, 1897

Partula leefei Smith, 1897: 20

Partula leefei Smith, Pilsbry 1909: 274. Kondo 1968: 76. Richardson 1990: 36

Dextral. Ovate. Umbilicus open. Apex rounded. 5 moderately convex whorls ornamented with fine spiral striae, almost obsolete on body whorl. Apex flattened. Aperture ovate. Lip expanded and reflexed, slightly thickened, narrowed at upper margin. Columella broadly triangular. Parietal callus weakly developed or virtually absent, outer margin perfectly straight. Yellow-olive, always with a whitish subsutural zone of variable width. Some individuals have an additional indistinct band of matt-white or brown. Spire paler and area around umbilicus usually whitish. Aperture brownish inside. Lip white.

Anatomy: Penis short, $^{1}/_{7}$ oviduct length. Caecum short. Interior papillose in caecum, weakly pilastered in penis.

Type specimens: Holotype and 7 paratypes NHMUK 96.4.2.35-44.

Distribution: Rotuma, only recorded from the west of the island (Itutiu district) (Brodie *et al.* 2012, 2014) and the islet of Uea.

9.3.4 *Partula lanceolata* and '*sarissa*': a – *P. lanceolata* type (Mago: ANSP 161288); b – Crampton's (MS) proposed type of '*sarissa*' (Naiau: ANSP 29511); c-d – '*sarissa*' morphotypes from Cicia (FMNH 168256) (photo: a – F. Borrero)

9.3.5 *Partula leefei*: a – holotype (NHMUK 96.4.2.35); b – collected with the type series J.S. Gardiner & R.B. Leefe in 1896 (UMZC 75305); c – specimen (NHMUK un-numbered)

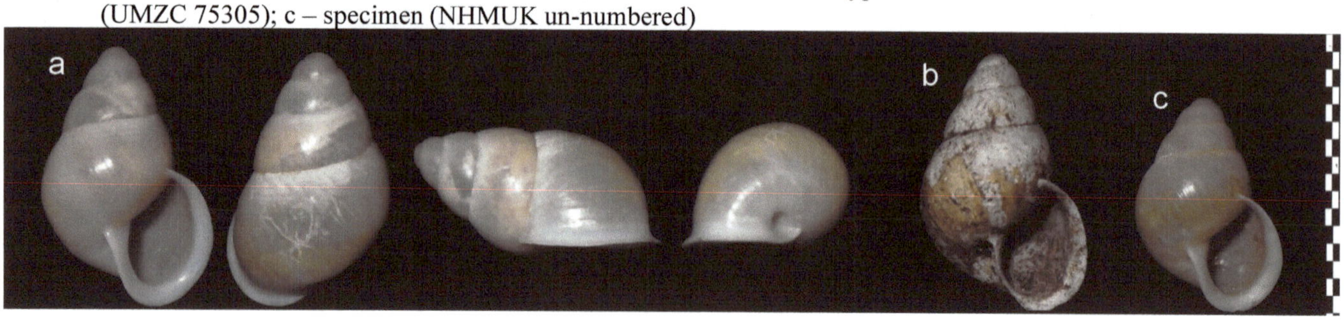

Wallis and Futuna

Only one species of *Partula* has been described from Wallis and Futuna, although there has been uncertainty over which name applies to it. Crampton identified a further two species as distinct and left an uncompleted manuscript and notes describing them as new species (Crampton MS dated 10/29/1949 and October 1952).

Partula obesa Pease, 1868

Partula obesa Pease, 1868a: 223

Partula subgonochila Mousson, 1871: 14

Partula obesa Pease, Pfeiffer 1877: 201. Hartman 1881a: 184, 1881b: 13, 1885: 221, 1892: 98. Pfeiffer & Clessin 1881: 302. Paetel 1889: 267. Smith 1903: 469. Pilsbry 1909: 260. Richardson 1990: 43

Partula subgonochila Mousson, Schmeltz 1874: 92. Pfeiffer 1877: 203. Pfeiffer & Clessin 1881: 302. Paetel 1883: 153. Smith 1903: 465. Pilsbry 1909: 273. Kondo 1955: 179, 1968: 76

Partula sub-gonocheila Mousson, Hartman 1881a: 188 (misspelling), 1881b: 14, 1885: 219, 1892: 98.

Table 9.3.2 Wallis and Futuna *Partula* shell measurements

	N	L			W			L aperture			W/L		
		range	mean	sd	range	mean	sd	range	mean	sd	range	mean	sd
obesa dextral	209	14.2-18.1	15.59	0.753	8.9-10.7	9.65	0.149	7.5-9.5	8.37	0.310	56.6-67.5	62.04	1.912
obesa sinistral	6	14.1-16.9	15.46	0.720	9.5-10.7	10.23	0.384	7.5-8.7	8.32	0.382	52.9-62.1	62.40	2.141

Fig. 9.3.6 Islands of Wallis and Futuna occupied by *Partula*

Fig. 9.3.7 *Partula obesa* anatomy, all '*subgonochila*' from Alofi except where stated. a – Kondo 1955; b – BPBM 115471 ('*subgonochila*' Futuna), BPBM 1974-20 ('*subgonochila*' Futuna), BPBM 1974-21 ('*subgonochila*' Futuna), BPBM 1974-25 ('*subgonochila*' & '*cyclochila*' Futuna), BPBM 115487 ('*orthochila*'); c – BPBM 115481 ('*orthochila*'); d-e – BPBM 11502; f – BPBM 115488 ('*subgonochila*', Futuna); g – BPBM 115511 ('*orthochila*'); h – BPBM 115487; i – BPBM 115481. Scale bar 5 mm.

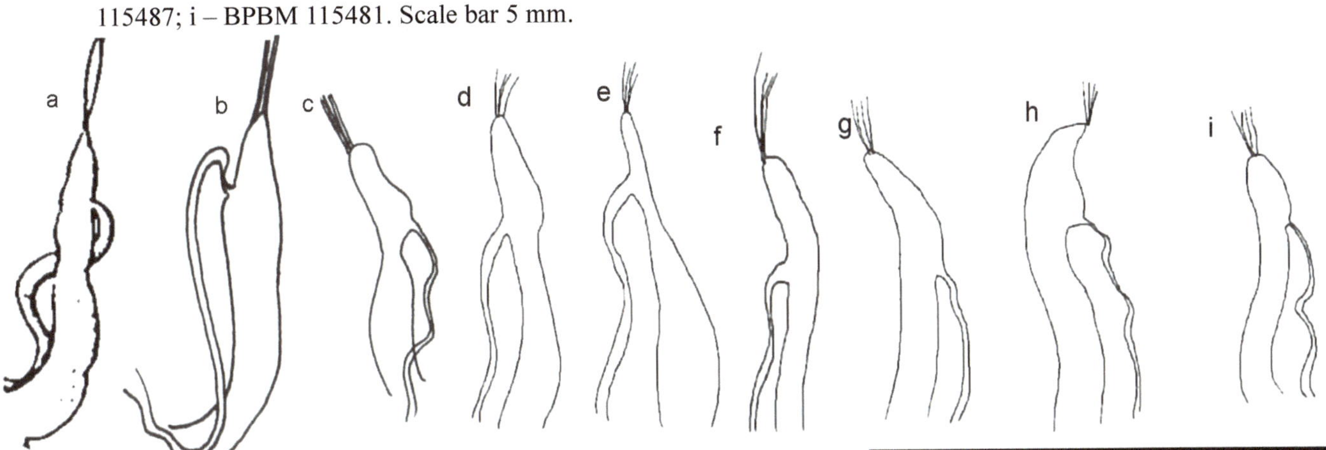

Usually dextral (total 96%, Futuna 96%, Alofi 100%, n=245). Ovately conic, moderately thin to thick. Umbilicus subperforate (Alofi) or open but narrow (Futuna). Finely, widely decussate, with fine but distinct spiral striae, usually fading on the body whorl. Apex obtuse (Futuna) or acute (Alofi). Spire convexly-conic, regular, the summit rather obtuse; suture not impressed, submarginate, marginate on last whorl (Alofi). 4½-5 slightly convex whorls, 5 on Alofi; the last moderately convex, not as long as the spire, not ascending in front. Aperture rather small, subvertical (making a 5° angle with the axis), oblong, obliquely truncate above. Lip expanded, angularly reflexed, somewhat labiate within, the margins not approaching; right margin evenly curved; columellar margin vertical, wide (narrow on Alofi), spreading, not folded within. In specimens from Futuna there is a low, broad ridge at the summit of the columella. Parietal callus thin, smooth and not distinctly margined.
Diaphanous greenish, lighter below suture, to medium brown. Adults often extensively decorticated, especially in a subsutural band. Lip whitish.
Anatomy: Penis short, ¼ oviduct length, caecum small but distinct. Interior ornament reduced.
Type specimens:
Partula obesa Pease, 1868 – Not located.
Partula subgonochila Mousson, 1871 – Syntypes BPBM 1664; UZH 518544.
Distribution: Futuna (Hoorn) islands: Futuna and Alofi, from sea level to 280 m.
Note: Listed as a synonym of *Bulimus gonochila* Pfeiffer, 1847 by Pilsbry (1909) and Richardson (1990), however, the type of *P. gonochila* is distinct (see p. 51).
Partula obesa Pease, 1868 is known only from the original description and figure, the type specimen having been lost early (Hartman 1881). There is a specimen of '*Partula obesa*' in NHMUK (un-numbered), labelled MacAndrew collection but it does not resemble Pease's original figure. Pease's description is generic and the only distinctive aspect is the colour: "Greenish yellow, with a broad white subsutural band. Lip white", a combination that is only found in *P. subgonochila* and *P. leefei*. The size (20 × 14 mm) matches the former, as does the figure (cf. Fig. 9.3.8 and 9.3.9g). The origin of the type specimen was unknown according to Pease (1868). Hartman (1881) cited "Islands Fortuna and Vavao, Gräff" but in MS stated that the single specimen was of unknown origin (Smith 1903).
Crampton (MS) considered two forms to be species separate from *P. subgonochila*: *'cyclochila'* differed in being sinistral, thicker and stouter, with a more pronounced suture, thicker and more reflexed lip and uniform columella; and *'orthochila'* differed in being fully perforate, wider, with more convex whorls and a less acute spire. The aperture is not truncate above and vertical rather than subvertical, the lip thinner and columella lacking an internal ridge. '*cyclochila*' was based on 6 adult and 7 subadult dead shells from Futuna collected by E.G. Burrows in 1932 ('brought from the mountains') (ANSP un-numbered ex BPBM 115437 'cotype'; BPBM 9721; ANSP un-numbered ex BPBM 115439; BPBM 115456, 115469, 115478). '*orthochila*' was based on 7 specimens from Alofi (BPBM 115531 & 115533, with specimens from both lots in ANSP, and BPBM 115511). Other specimens are present in BPBM. Both this and the typical form were sympatric in at least one population (BPBM 115502).
A further distinct form can be recognised from an extensive sample from Futuna (BPBM 1974-25). This has the extreme version of the *P. subgonochila* aperture with a distinct ridge at the base of the columella (Fig. 9.3.10 m-n). None of these forms differ in any consistent way in reproductive anatomy, each population containing several anatomical minor variants.

Fig. 9.3.8 Pease's (1868) *Partula obesa* figure (left) and live snails on Alofi (photo: J.-Y. Meyer)

Fig. 9.3.9 Wallis and Futuna *Partula*: a – '*subgonochila*' from Alofi (ANSP 295214 ex BPBM 115502); b – '*orthochila*' (ANSP ex BPBM 115531); c – '*cyclochila*' (ANSP ex BPBM 115437); d – extreme '*subgonochila*' form (BPBM 1974-25); e-i – variation in one population (BPBM 1974-25); j – *subgonochila* syntype (UZH 518544); k – specimen (ANSP 295215); l – aperture of '*orthochila*' (BPBM 115481); m-n – aperture of '*subgonochila*' forms (BPBM 1974-25)

Cook islands

The Cook islands support two *Partula* species, one of which (*P. hyalina*) is an introduced species (see p. 135). The endemic species is covered below.

Partula assimilis Pease, 1868

Partula assimilis Pease, 1868a: 230

Partula assimilis Pease, Pease 1870: 401, 1871b: 473. Schmeltz 1874: 92. Pfeiffer 1877: 197. Hartman 1881a: 179, 1881b: 14, 1892: 98. Pfeiffer & Clessin 1881: 301. Paetel 1883: 152, 1889: 266. Smith 1903: 464. Pilsbry 1909: 261. Kondo 1955: 175, 1968: 76

Table 9.3.3 Cook Islands *Partula* shell measurements

	N	L			W			L aperture			W/L		
		range	mean	sd	range	mean	sd	range	mean	sd	range	mean	sd
assimilis	25	16.1-18.5	16.63	0.943	9.0-11.0	10.08	0.425	8.3-10.0	8.64	0.382	52.9-62.1	60.62	2.141
hyalina	102	14.8-16.7	15.26	0.781	8.0-10.1	8.55	0.376	8.1-9.7	9.81	0.399	52.8-58.4	55.78	1.811

9.3.10 Cook islands *Partula*. *P. assimilis* reproductive anatomy shown (scale bar for anatomy 5 mm).

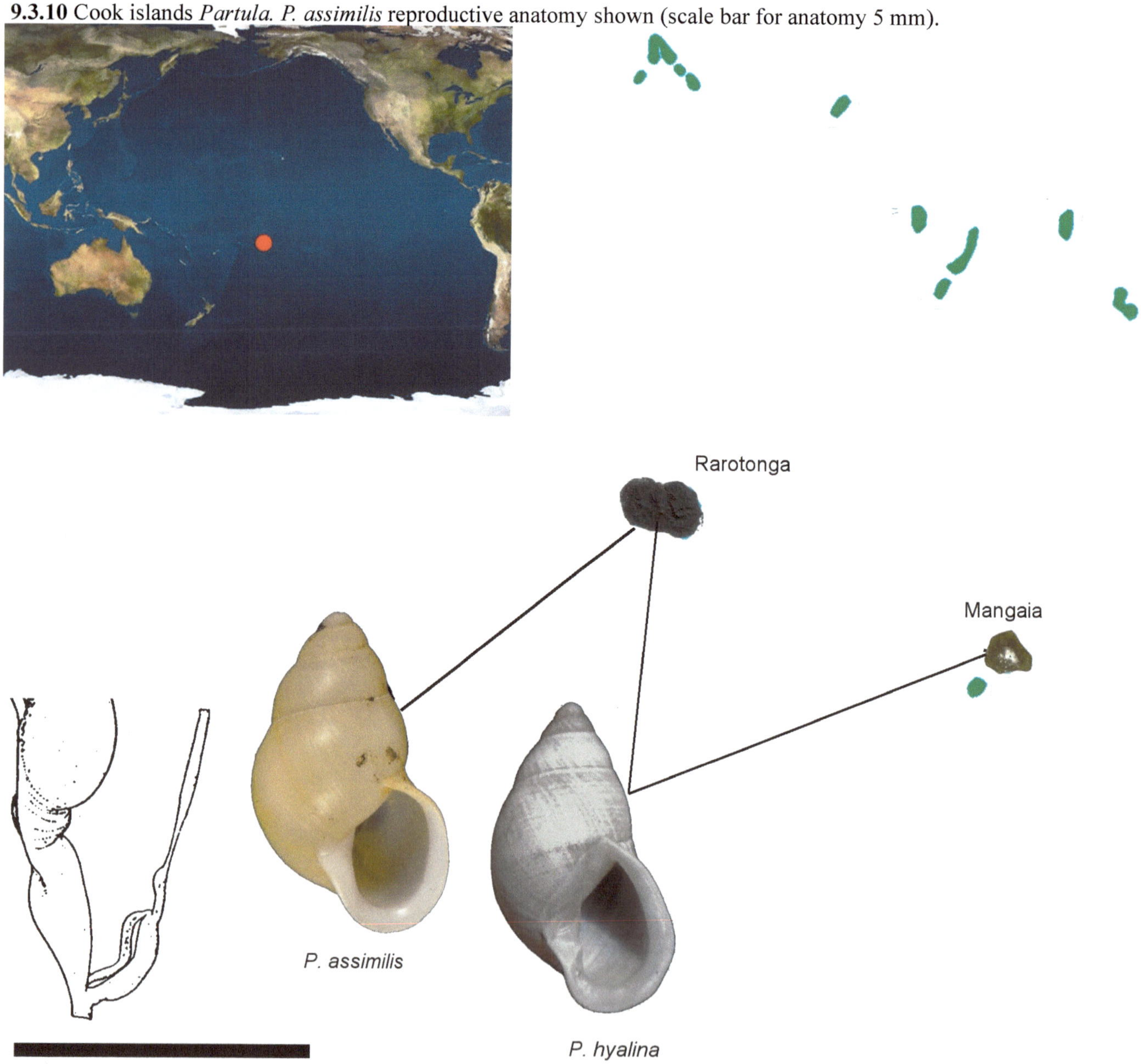

Partula assimilis var. *virgulata* Pease, 1870: 401

Partula cookiana 'Mouss.' Paetel 1873: 104. Schmeltz 1874: 92. Paetel 1883: 152, 1889: 266 (all nomen nudum)

Partula assimilis Pease β *virgulata* Pfeiffer, 1877: 197.

Partula virgulata Pease, Hartman 1881a: 189, 1881b: 14, 1885: 219, 1892: 98. Garrett 1884: 80. Smith 1903: 465. Richardson 1990: 67

Ovate-conic. Thin but strong. Umbilicus open, narrow. 4½-5 convex whorls ornamented with faint spiral striae, indistinct or obsolete on body whorl. Protoconch of 2½ whorls ornamented with subpunctate spiral lines. Body whorl convex. Aperture slightly oblique. Lip reflexed and somewhat thickened within. Columellar lip dilated above. Callus very thin.

Pale corneous, spire sometimes pale brown. Ornamented with indistinct, minuscule pale brown streaks. Aperture and lip white or flesh-tinted.

Anatomy: Penis short, ¼ oviduct length, broad. Caecum well developed. Interior sparsely papillose, pilasters in penis reduced. Vas deferens enters at base of caecum.

Type specimens:

Partula assimilis Pease, 1867 – Labelled paratypes SMF 4951, 164977. No types defined but Pease (1867) had 'large numbers' and apparent syntypes with labels indicating a Pease and Rarotonga origin are included in the collections of AM (C.132563), MCZ (24906, 25137), NHMUK (un-numbered), USNM (408811, 42508, 77853). Some in USNM with the same data ex. Garrett (USNM 76955, 77920) would be topotypes. Of these USNM 408811 are mislabelled *P. suturalis* and *P. varia.*

Partula assimilis var *virgulata* Pease, 1870 – Syntypes MCZ 24939; NHMUK 1934.6.22.11-15 (ex Clench, MCZ).

Distribution: Rarotonga (inland forests, including primary and secondary habitats – Brook 2010).

Note: Var *virgulata* is relatively short, streaked light red.

Richardson (1990) listed *P. assimilis* and *P. cookiana* as synonyms of *P. varia.*

Fig. 9.3.11 *Partula assimilis* types: a-b – *assimilis* (a – MCZ 24906; b – NHMUK; d-e – USNM 77853; f-g – USNM 42508); c – *virgulata* (NHMUK 1934.6.22.11); h-i – mislabelled *P. suturalis* (h) & *P. varia* (i) (USNM 408811)

9.4 Society Islands

The greatest diversity of all partulids lies in the Society islands, with 51 *Partula* species. The six islands occupied by partulids are treated individually, passing from east to west.

Fig. 9.4.1 Map of Society Island *Partula* with number of species on each island (note some species are present on more than one island)

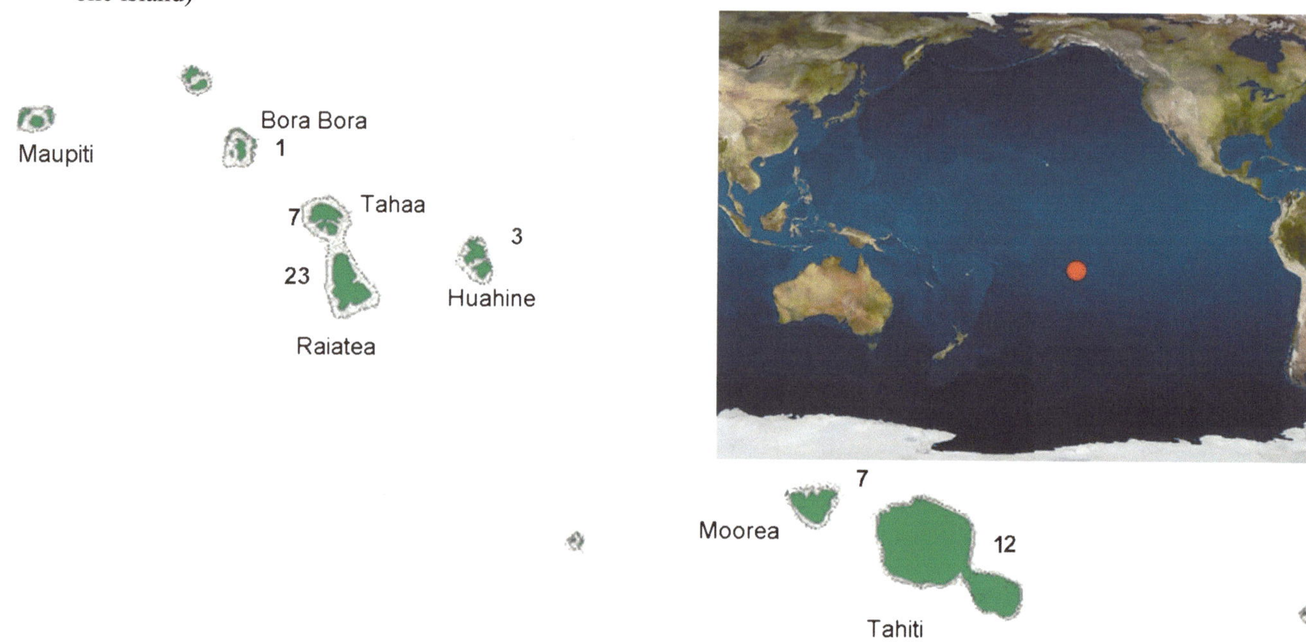

Fig. 9.4.2 Shells of Tahiti *Partula*: a – *P. affinis*; b – *P. compressa*; c – *P. diminuta*; d – *P. nodosa*; e – *P. incrassa*; f – *P. clara*; g – *P. hyalina*; h – *P. laevigata*; i – *cytherea*; j – *P. jackieburchi*; k – *P. otaheitana*; l – *P. producta*

Tahiti

17 species have been described from specimens reportedly collected on Tahiti. These were reviewed by Crampton (1916) who recognised 7 as valid species. Subsequently two further species were described (Cooke & Crampton 1930; Kondo 1981) but as noted by Kondo (1981) and Kondo & Burch (1980) the taxonomy of the species remained confused. 12 species are recognised here, of which *P. incrassa* is covered in the Moorean section (p. 151).

Partula otaheitana (Bruguière, 1792)

Helix perversa, etc., Chemnitz, 1786: 108 (unavailable work)
Bulimus otaheitanus Bruguière, 1792: 347
Helix otaheitana Bruguière, Bosc 1801: 112. Dillwyn 1817: 935. Wood 1818: pl34 f110
Partula otaheitana Bruguière, Férussac 1821: 66. Gray 1825: 415. Sowerby 1825: 40. Wood 1828: 24. Anton 1839: 40. Jay 1839: 57. Pfeiffer 1841: 25, 1842: 119, 1853b: 448, 1856: 162, 1856: 263, 1868: 160, 1877: 204. Reeve 1842: pl175 f16, 1849: pl3 f136. Catlow & Reeve 1845: 163. Gould 1846: pl7 f91. Albers 1850: 187. Férussac & Deshayes 1851: 126. Adams 1855: 145. Chenu 1859: 434. Pease 1864: 675. Frauenfeld 1869: 874. Paetel & Schaufuss 1869: 83. Schmetlz 1869: 71, 1874: 97. Paetel 1873: 104, 1883: 153, 1889: 267. Semper 1874: 158. Gloyne 1878: 337. Hartman 1881a: 184, 1881b: 9, 1885: 9, 1892: 75. Garrett 1884: 46. Smith 1903: 442. Mayer 1902: 122. Pilsbry 1909: 183. Crampton 1916: 113, 1924a: 558, 1925: 8. Kondo 1955: 153, 1968: 76. Kilias 1969: 247. Richardson *et al.* 1979: 327. Emberton 1982: 31. Kondo & Burch 1983: 105. Richardson 1990: 43
Partula lineata Lesson, 1831: 324
Partulus torosus Beck, 1837: 57 (nomen nudum)
Partulus otaheitanus Bruguière, Beck 1837: 58
Bulimus Otaheitanus Bruguière, Lamarck *et al.* 1839: 371

Table 9.4.1 Tahiti *Partula* shell measurements: D- dextral, S – sinistral, bold – species totals

	N	L			W			L aperture			W/L		
		range	mean	sd	range	mean	sd	range	mean	sd	range	mean	sd
***otaheitana* D**	3586	14.8-21.7	18.27	0.857	8.6-11.9	10.45	0.440	6.3-9.9	9.48	0.518	50.6-67.3	56.77	2.356
***otaheitana* S**	9960	14.4-22.0	18.50	0.993	8.2-14.0	10.55	0.513	8.0-11.0	9.56	0.491	56.2-68.2	61.88	2.984
o. otaheitana D	1490	14.8-21.1	18.43	0.875	8.6-12.4	10.64	0.459	6.3-11.4	9.61	0.424	56.8-66.7	58.17	1.945
o. otaheitana S	2487	16.9-20.9	18.22	0.902	9.0-13.3	10.67	0.474	8.0-11.4	9.480	0.466	51.6-62.9	59.18	2.143
o. rubescens S	1425	17.8-21.6	20.61	1.024	9.8-14.0	11.58	0.555	8.5-10.5	10.44	0.585	46.6-66.4	58.20	2.132
o. sinistrorsa D	772	15.4-21.0	17.95	0.830	8.9-10.8	10.12	0.410	7.8-10.1	9.23	0.630	48.5-59.2	55.07	2.420
o. sinistrorsa S	2799	15.1-22.0	18.71	1.140	8.1-11.9	10.39	0.530	7.6-11.7	9.53	0.510	50.0-60.5	56.23	2.250
o. crassa D	5	18.9-20.0	18.81	0.451	9.9-11.9	10.64	0.852	9.0-9.8	9.32	0.516	50.6-67.3	57.84	2.510
o. crassa S	2721	14.4-20.9	17.28	0.908	8.2-12.1	10.01	0.511	9.0-9.8	9.13	0.438	50.0-66.8	58.08	2.227
jackieburchi	121	15.7-20.8	18.43	1.091	9.9-14.0	11.39	0.820	8.0-10.7	9.31	0.955	54.5-57.5	56.00	1.055
***producta* D**	272	17.7-21.4	19.31	0.880	9.8-11.7	10.62	0.453	8.4-10.8	9.83	0.456	53.8-57.5	55.00	2.444
***producta* S**	3405	15.6-21.0	16.38	0.718	8.9-11.7	9.20	0.378	87.9-10.5	8.37	0.399	50.6-60.1	51.65	2.013
laevigata	9	19.8-21.7	20.27	0.690	11.5-13.2	12.55	0.556	9.9-11.6	10.72	0.740	55.6-64.6	61.15	3.424
***cytherea* D**	8	16.4-21.7	20.39	1.760	10.4-12.6	11.61	0.810	8.5-10.7	9.94	0.883	56.2-63.5	60.04	2.640
***cytherea* S**	59	16.5-23.7	19.08	1.035	10.5-14.3	12.17	0.767	8.6-11.9	10.02	0.632	55.6-74.7	63.73	3.964
***affinis* D**	5154	15.3-19.2	16.62	0.756	8.6-11.5	9.77	0.393	8.0-10.0	8.81	0.422	50.8-64.7	58.73	2.114
***affinis* S**	39	16.0-19.0	16.81	0.683	9.6-11.3	10.16	0.462	8.4-10.0	8.90	0.442	55.1-63.7	60.60	1.832
compressa	47	16.6-21.8	18.37	1.153	8.9-12.5	10.55	0.710	8.3-12.0	9.39	0.672	48.8-65.0	60.97	2.155
***nodosa* D**	1095	12.9-18.4	15.13	1.010	7.7-11.1	9.29	0.520	7.6-9.5	8.22	0.506	52.4-70.5	61.39	2.220
***nodosa* S**	655	14.1-17.9	15.55	0.741	8.6-10.6	9.34	0.406	7.6-8.9	8.15	0.403	57.8-65.6	60.04	2.424
diminuta	209	12.6-15.7	14.16	0.634	8.1-10.3	9.10	0.345	6.6-8.2	7.63	0.391	59.5-70.5	64.28	2.207
i. incrassa	63	17.4-18.3	15.45	0.973	7.1-9.6	8.61	0.505	50.5-62.5	8.38	0.523	50.0-62.0	55.79	2.250
clara	553	12.9-17.5	15.12	0.710	7.3-9.9	8.75	0.382	7.3-10.5	8.52	0.417	50.5-66.5	57.80	2.157
hyalina	1028	12.6-17.7	15.31	0.778	7.7-10.1	8.74	0.413	7.4-10.2	8.92	0.475	51.5-63.5	57.12	2.062

Partula tahulana Anton, 1839: 40 (nomen nudum)
Partula otaheitana var. *minor* Grateloup 1840a: 426; 1840b: 167
Partulus torosus Beck, Pfeiffer 1842: 123, 1848: 72 (all nomen nudum)
Partula lineata Lesson, Pfeiffer 1842: 117, 1848: 72, 1853b: 449, 1856: 162, 1859: 508, 1868: 156, 1877: 195. Albers 1850: 187. Reeve 1850: 7. Adams 1855: 145. Pease 1864: 675, 1871b: 473. Paetel & Schaufuss 1869: 83. Schmeltz 1869: 72, 1874: 92. Hartman 1881a: 183, 1881b: 10, 1885: 213, 1892: 75. Garrett 1884: 50. Smith 1903: 446. Pilsbry 1909: 156. Crampton 1925: 12. Richardson 1990: 36
Bulimus (*Partula*) *isabellinus* Pfeiffer, 1846: 39
Bulimus (*Partula*) *isabellina* Pfeiffer, Pfeiffer 1846: 55, 1848: 70
Bulimus (*Partula*) *amabilis* Pfeiffer, 1846: 38
Bulimus (*Partula*) *amabilis* Pfeiffer, Pfeiffer 1848: 71
Bulimus otaheitanus Bruguière, Pfeiffer 1848: 71
Partula reeveana Pfeiffer, 1848: pl65 f10
Partula reeveana Pfeiffer, Pfeiffer 1853b: 447, 1854c: 137, 1856: 162, 1856: 275, 1859: 511, 1868: 159, 1877: 201. Adams 1855: 145. Kobelt 1880: 18. Hartman 1881a: 185, 1881b: 10, 1885: 212. Garrett 1884: 47. Paetel 1889: 268. Smith 1903: 443
Partula amabilis Pfeiffer, Reeve 1851: pl2 f8a. Pfeiffer 1853b: 448, 1856: 162, 1859: 507, 1868: 156, 1877: 195. Adams 1855: 145. Pease 1864: 673, 1871b: 473. Paetel & Schaufuss 1869: 83. Schmeltz 1869: 71. Gloyne 1878: 337. Hartman 1881a: 179, 1881b: 10, 1885: 212. Garrett 1884: 47. Smith 1903: 441
Partula isabellina Pfeiffer, Reeve 1851: pl2 f8b. Pfeiffer 1853b: 448, 1859: 507, 1868: 156, 1877: 194. Adams 1855: 145. Paetel 1873: 104, 1883: 153, 1889: 267. Hartman 1881a: 183, 1881b: 10, 1885: 212. Garrett 1884: 47. Smith 1903: 443. Pilsbry 1909: 184
Partula rubescens Reeve, 1851: pl3 f12
Partula taheitana 'Bruguière, Gould, 1852: pl84 f91 (ex errore). Pease 1871b: 473. Schmeltz 1874: 92. Hartman 1881a: 185, 1881b: 10, 1885: 212. Garrett 1884: 47
Partula rubescens Reeve, Pfeiffer 1853b: 446, 1855b: 271, 1856: 162, 1859: 507, 1868: 156, 1877: 195. Adams 1855: 145. Pease 1864: 675, 1871b: 473. Gloyne 1878: 337. Hartman 1881a: 185, 1881b: 10, 1885: 212. Garrett 1884: 47. Smith 1903: 441
Partula stenostoma Pfeiffer, 1855a: 97
Partula nitens Pfeiffer, 1855b: 293
Partula nitens Pfeiffer, Pfeiffer 1856: 162, 1859: 513, 1868: 160, 1877: 206. Hartman 1881a: 184, 1881b: 10, 1885: 214. Paetel 1889: 267. Smith 1903: 445. Pilsbry 1909: 238

9.4.3 Anatomy of Tahiti *Partula* species, after Kondo (1955) except *P. jackieburchi* (Kondo 1981) and *P. compressa* (BPBM 145548). Scale bar 5 mm.

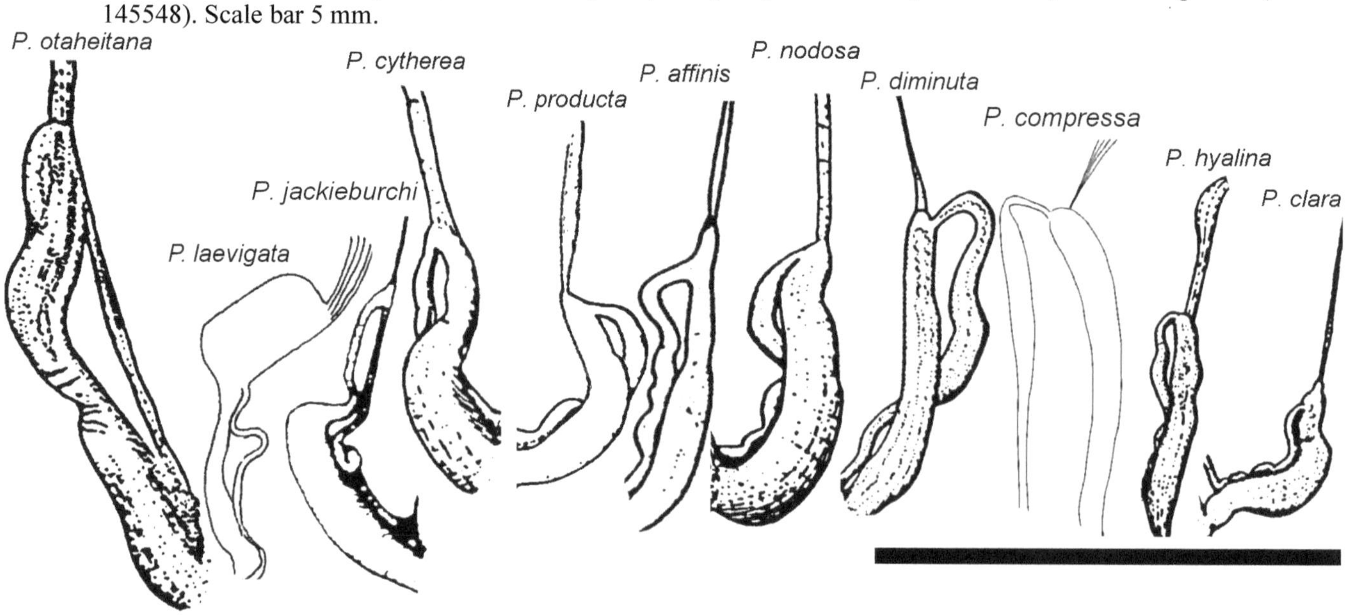

Partula stenostoma Pfeiffer, Pfeiffer 1856: 162, 1856: 61, 1859: 507, 1868: 156, 1877: 95. Hartman 1881a: 187, 1881b: 10, 1885: 215. Garrett 1884: 50. Paetel 1889: 268. Smith 1903: 481. Pilsbry 1909: 192. Richardson 1990: 56

Partula lignaria Pease, 1865: 671

Partula lignaria Pease, Pfeiffer 1868: 160, 1877: 205. Pease 1871b: 473. Paetel 1873: 104, 1883: 153, 1889: 267. Schmeltz 1874: 92. Gloyne 1878: 338. Hartman 1881a: 183, 1881b: 10, 1885: 212, 1892: 75. Garrett 1884: 49 (as var. of *otaheitana*). Smith 1903: 443

Partula rufa Carpenter (non Lesson) 1865: 675 (nomen nudum). Hartman 1885: 10

Partula sinistrorsa 'Pease', Carpenter in Pease, 1865: 675 (as syn. of *rubescens*). Paetel 1873: 104. Schmeltz 1874: 92. Pfeiffer 1877: 209. Hartman 1881a: 185, 1881b: 10, 1885: 213 (all nomen nudum). Garrett 1884: 49 (as var. of *otaheitana*)

Partula turricula 'Pease' Carpenter in Pease, 1865: 675 (as syn. of *rubescens*, nomen nudum)

Partula crassa 'Pease', Carpenter in Pease, 1865: 675 (as syn. of *otaheitana*). Schmeltz 1874: 92. Pfeiffer 1877: 209. Gloyne 1878: 337. Hartman 1881a: 181, 188b: 10, 1885: 213. Paetel 1883: 152, 1889: 266. Garrett 1884: 49. Smith 1903: 442. Pilsbry 1909: 191 (as var. of *otaheitana*) (all nomen nudum)

Partula perversa 'Pease', Carpenter in Pease, 1865: 675 (as syn. of *otaheitana*, nomen nudum). Smith 1903: 442

Partula tahitana Bruguière, Pease 1871b: 473 (ex errore). Schmeltz 1874: 92

Partula turricula Pease, 1872: 196

Partula dentata Paetel, 1873: 104, 1883: 153 (all nomen nudum)

Partula sinistralis 'Pease', Paetel 1873: 104, 1889: 268. Pfeiffer 1877: 209. Hartman 1881a: 185, 1881b: 10, 1885: 213 (all nomen nudum). Garrett 1884: 49 (as var. of *otaheitana*)

Partula lineata Lesson, Hartman 1881a: 183. Pilsbry 1909: 312

Partula turricula Pease, Hartman 1881a: 223, 1881b: 10, 1885: 213. Garrett 1884: 47. Pilsbry 1909: 283. Solem 1959: 75

Partula brevicula 'Pease', Hartman 1881a: 180, 1881b: 10. Garrett 1884: 49. Smith 1903: 49. Pilsbry 1909: 185 (all nomen nudum)

Partula turricula Pease, Smith 1903: 442. Pilsbry 1909: 283

Partula otaheitana lignaria Pease, Pilsbry 1909: 186. Crampton 1916: 286. Kondo & Burch 1983: 105

Partula otaheitana dubia 'Pease.', Pilsbry 1909: 188

Partula otaheitana amabilis Pfeiffer, Pilsbry 1909: 189. Crampton 1916: 130. Kondo 1955: 156, 1968: 76. Kondo & Burch 1983: 105

Partula otaheitana rubescens Reeve, Pilsbry 1909: 190. Crampton 1916: 149. Kondo 1955: 158. Kondo & Burch 1983: 105

Partula othaeitana sinistrorsa 'Pease' Garrett Pilsbry 1909: 190. Crampton 1916: 219, 1925: 8. Kondo 1955: 157. Kondo 1968: 76. Kondo & Burch 1983: 105

Partula otaheitana sinistrorsa confluens Pilsbry, 1909: 192 (variety)

Partula otaheitana otaheitana Bruguière, Crampton, 1916: 119. Kondo 1955: 155. Kondo & Burch 1983: 105

Partula otaheitana sinistralis Pease, Crampton 1916: 250, 1825: 8. Kondo & Burch 1983: 105

Partula otaheitana crassa Pease, Crampton 1916: 268. Kondo & Burch 1983: 105

Partula otaheitana crassa occidentalis Crampton, 1916: 274 (variety)

Partula otaheitana crassa aberrans Crampton, 1916: 282 (variety)

Sinistral or dextral, varying geographically. Ovately to slenderly conical. Shell moderately thin, strong. Subperforate. 4½-5 moderately convex whorls. Sculpture of fine spiral striae usually fading on body whorl. Suture impressed. Aperture elongate, usually slightly compressed. Columella straight or slightly twisted. Parietal callus shiny, thin but well defined. A deeply recessed parietal tooth may be present, variably developed; elongate, inward-running when fully developed. Lip expanded. Colour highly variable: from white with yellowish streaks to yellow, red or dark brown. Spire usually with a red tint. Cestata banding patterns may be present and numbers of bands vary. Lip usually coloured dorsally, white ventrally, aperture pale.

Anatomy: Penis large, robust, ½-¾ oviduct length, development varying geographically in different subspecies. Short in north-western populations with a well developed caecum not separated from the penis. Southern populations have a long penis and caecum, separated by a notable constriction. Eastern populations have a vermiform penis, without a distinct caecum. Internally the penis is ornamented with several branching pilasters which run into the caecum as ridges of papillae.

Type specimens:

Bulimus otaheitanus Bruguière, 1792 – Not located. Supposed paratypes ZMB 101546 (3 specimens ex Coll. Dunker, labelled "This teste Cumings is the type of *P. otaheitana* Brug. Tahiti", probably meaning typical, and not nomenclatural

Fig. 9.4.4 *Partula otaheitana* type specimens: a – *amabilis* one of Reeve's (1851) figures of Pfeiffer's types and the possible types NMW.1955.158; b – *P. lineata* (after Lesson 1831); c – *crassa* (UMZC specimen – Pease duplicate, not a type); d – *brevicula* (ANSP 59558); e – *confluens* (ANSP 2630); f – *stenostoma* (NHMUK); g – *isabellinus* (NHMUK); h – *lignaria* (ANSP 2631); i – *perversa* supposed type (MCZ 24997); j – *reeveana* (NHMUK); k – *rubescens* (NHMUK); l-m – *sinistralis* (l – syntype BPBM 5761; m – Pease duplicate ANSP 294795); n – *sinistrorsa* (MCZ 24997); o-p – *turricula* (AMNH 42056) (photos: a – H. Wood; c – F. Borrero)

types), MCZ 24997 (Pease), FLMNH 195882 'paratypes'; SMF 164966 'syntype'. None are actual types.

Partula lineata Lesson, 1831 – Not located.

Bulimus (*Partula*) *isabellinus* Pfeiffer, 1846 – Holotype NHMUK un-numbered.

Bulimus (*Partula*) *amabilis* Pfeiffer, 1846 – Possible syntypes NMW.1955.158 labelled 'I. of Annaa' ex Melvill-Tomlin collection. Reeve (1851) figured Pfeiffer's *amabilis* from the Cuming collection supposedly collected in the 'Annaa or Chain islands' (Pfeiffer 1848) in the Tuamotus by Captain Rohr (Reeve 1851). NMW specimens are the only ones labelled 'Annaa' to resemble Reeve's figures. Other specimens in ZMH ('Tuamoto Archipel: Anna Atoll') are dextral, as are NHMUK '*amabilis*' specimens (labelled 'P. reeveana. Pfr. Solomon Islds.') (see p. 118).

Partula reeveana Pfeiffer, 1848 – Syntypes NHMUK un-numbered.

Partula rubescens Reeve, 1851 – Syntypes NHMUK un-numbered.

Partula stenostoma Pfeiffer, 1855 – Syntypes NHMUK un-numbered.

Partula nitens Pfeiffer, 1855 – Not located, described from an unknown locality from Cuming specimens.

Partula lignaria Pease, 1865 – Syntypes AMNH 42487; ANSP 2631; MCZ 25058, 25136, 25142; SMF164966.

Partula turricula Pease, 1872 – Syntypes AMNH 42056; ANSP 59926. ANSP 59926a listed as holotype by Baker (1963). These sets of specimens include dextral as well as sinistral specimens, whereas Pease (1872) only mentioned sinistrals, this suggests that the dextrals were added to the lots subsequently (and so would not be types).

Partula sinistrorsa Garrett, 1884 – Lectotype ANSP 59548; paralectotype ANSP 2628, 463797; MCZ 24997, 297864. Lectotype designation by Baker (1963).

Partula crassa Garrett, 1884 – Lectotype ANSP 59921a; paralectotype ANSP 59921; MCZ 25078. Lectotype designation by Baker (1963).

Partula perversa – Supposed syntypes of nomen nudum MCZ 25317 & 24997.

Partula sinistralis Garrett, 1884 – No types defined, described in passing by Garrett. Types would be Garrett specimens: BPBM 5761.

Partula brevicula – Supposed lectotype of nomen nudum ANSP 59558 (according to Baker 1963); paralectotypes ANSP 45189, 59558; MCZ 24997.

Partula otaheitana sinistrorsa confluens Pilsbry, 1909 – Supposed syntypes of variety ANSP 2630, 463921.

Distribution: Recorded from almost all valleys and mountain ridges. Arboreal.

Subspecies: Seven subspecies were recognised by Crampton (1916), of which *P. otaheitana affinis* is best considered a separate species (see p. 124). Although Crampton gave precise limits to the ranges of the subspecies the distinctions between them are not clear. Kondo (1955) accepted Crampton's subspecies and defined distinct anatomical forms for the four subspecies that

Fig. 9.4.5 Crampton's (1916) *Partula otaheitana* figures (edited for position)

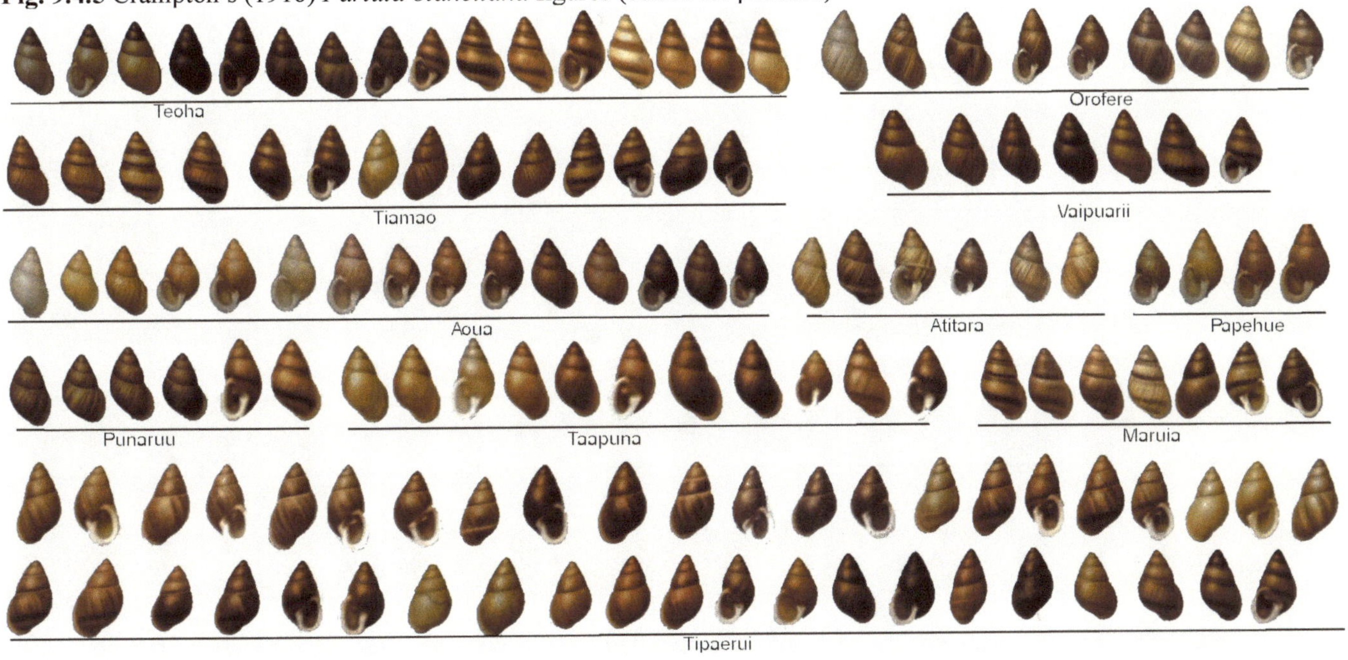

Fig. 9.4.5 (cont.)

Fautaua
Hamuta
Pirai
Tuauru
Mahaena
Pohaitara
Maruapoo
Ururoa
Ahonu
Papenoo
Papeiha
Paraura
Haavini
Vaiaaia
Aiurua
Vaitupa
Oopu
Apirimaue
Faarumai
Tautira
Aiavaro
Tiarei
Vavii
Vaihiria
Vaipoe
Opiriroa
Tenaire
Otuna
Moaroa
Atehi

he dissected. Subsequent more extensive dissections (Kondo & Burch MS and present study) identify four anatomical forms which overlap or are linked by intermediates.

Partula otaheitana otaheitana Bruguière, 1792 – penis short, base inflated, caecum broad, tapering, separated from penis by constriction. Similar to *P. o. crassa*. Includes *P. o. lignaria* Pease, 1864 and *P. o. amabilis* Pfeiffer, 1846.

Partula otaheitana crassa Garrett, 1884 – Intermediate between *P. o. otaheitana* and *P. o. sinistrorsa.* Crampton (1916) described two varieties (*occidentalis* – Aoua and Papehue valleys, and *aberrans* – Taapuna valley).

Partula otaheitana sinistrorsa Garrett, 1884 – penis long, caecum usually separated by a constriction. Includes *P. o. sinistralis* Garrett, 1884. Similar to *P. o. crassa.*

Partula otaheitana rubescens Reeve, 1850 – penis vermiform, constriction between penis and caecum indistinct. This form is anatomically distinct from other subspecies.

Notes: *Partula lineata* Lesson, 1831 was described from a specimen collected on the Coquille expedition, apparently in 'Oualan' (Kosrae) in the Caroline islands in 1822. The type specimen is lost and the species is known only from the original description on illustration. It does not resemble any known western Pacific *Partula* and has been suggested to be a misidentified *P. suturalis* (Garrett 1884), however, the Coquille did not visit Moorea but landed at Tahiti, Bora Bora, 'Eua, Bougainville, New Britain, Rotuma and Kosrae (islands occupied by partulids, listed in chronological order of their visit). Pilsbry (1909) thought the locality to be correct and that it was likely to be an overlooked species. He noted that "The nodule on the parietal wall, shown in Lesson's figure but not mentioned in the description, is a feature appearing also in species of the Society, Fiji and Marianne groups." *P. lineata* appears to be attributable to *P. otaheitana.*

P. stenostoma is an enigmatic taxon, considered to be similar to Tahitian species by Pilsbry (1909) but Hartman (1881) thought it Moorean and Smith (1902) considered it to be *P. suturalis vexillum*. However, it is a closer match to *P. otaheitana* and is here considered a synonym.

Grateloup's (1840a, b) *P. otaheitana* var. *minor* is indistinct, lying within the range of all *P. otaheitana* populations. His specimens have not been located.

Fig. 9.4.6 *P. otaheitana* anatomical variaiton compared to *P. jackieburchi* and *P. laevigata*: a – Fautaua; b – Mt. Marau (ZSL 519 and 1116, MNHN Th118); c – Pohaitara (after Kondo 1955); d – Punaruu (ANSP A730); e – Teihomono Plateau (ZSL 680); f – Vavii; g – Papenoo; h – Ahonu (ANSP A782), Vaipoe (ANSP A824), 'Vavii to Oofu' (ANSP 318706); i – Vairaharaha (ANSP A835); j – Vaitunamea; k – Temarua; l – Pueu; m – Faone; n – Teihomono Plateau (ZSL 695). a, c, f-g, l – after Kondo (1955), j-k – after Kondo & Burch (1983); m – after Kondo (1981). Scale bar 5 mm.

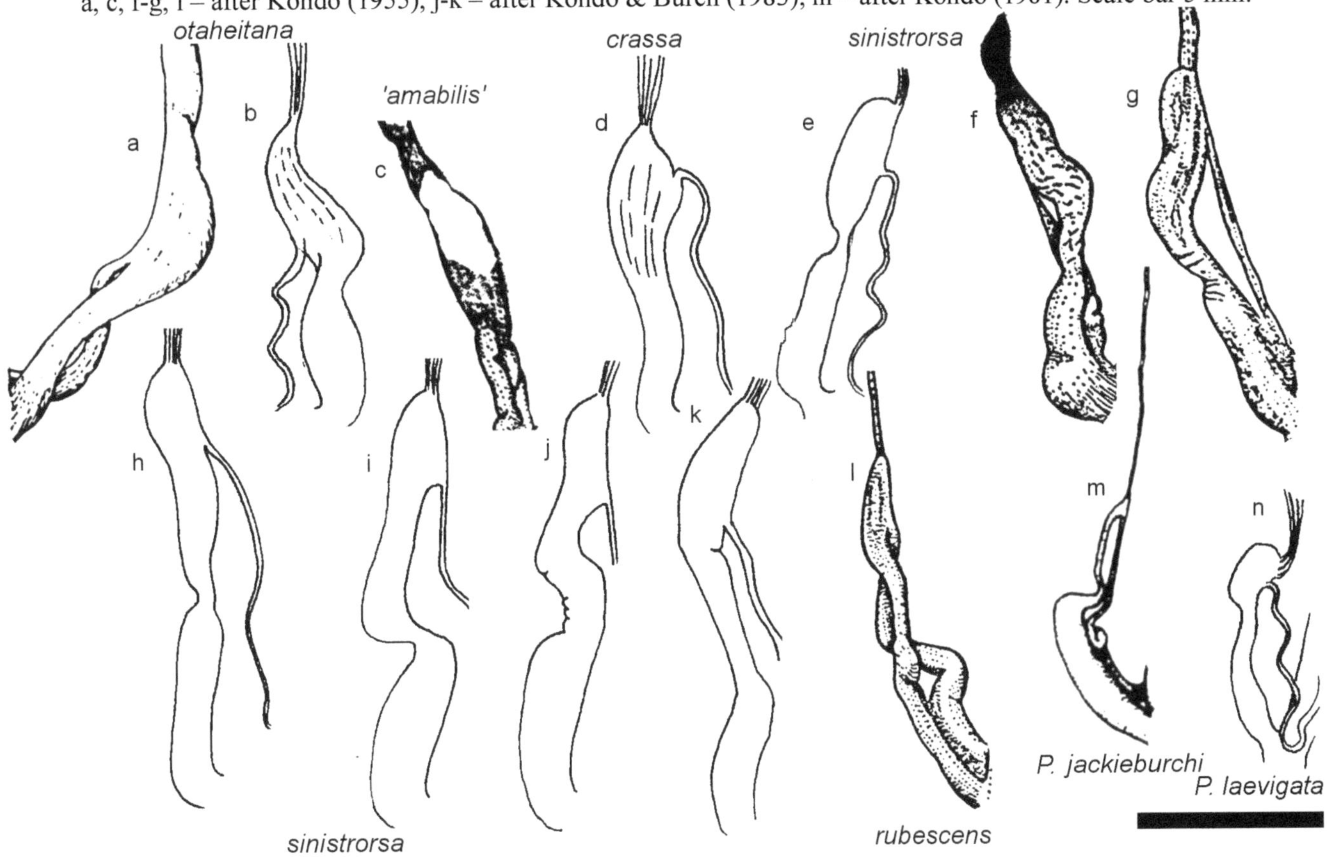

Fig. 9.4.7 Distribution of *P. otaheitana* anatomical forms: left – distribution of subspecies recognised here; right – Crampton's (1916) map of subspecies ranges. Dissected populations marked with colour-coded points. Boundaries between subspecies are drawn in yellow. Populations on the boundary between *rubescens* and *sinistrorsa* containing a mixture of the two morphotypes are shown as split green and blue points.

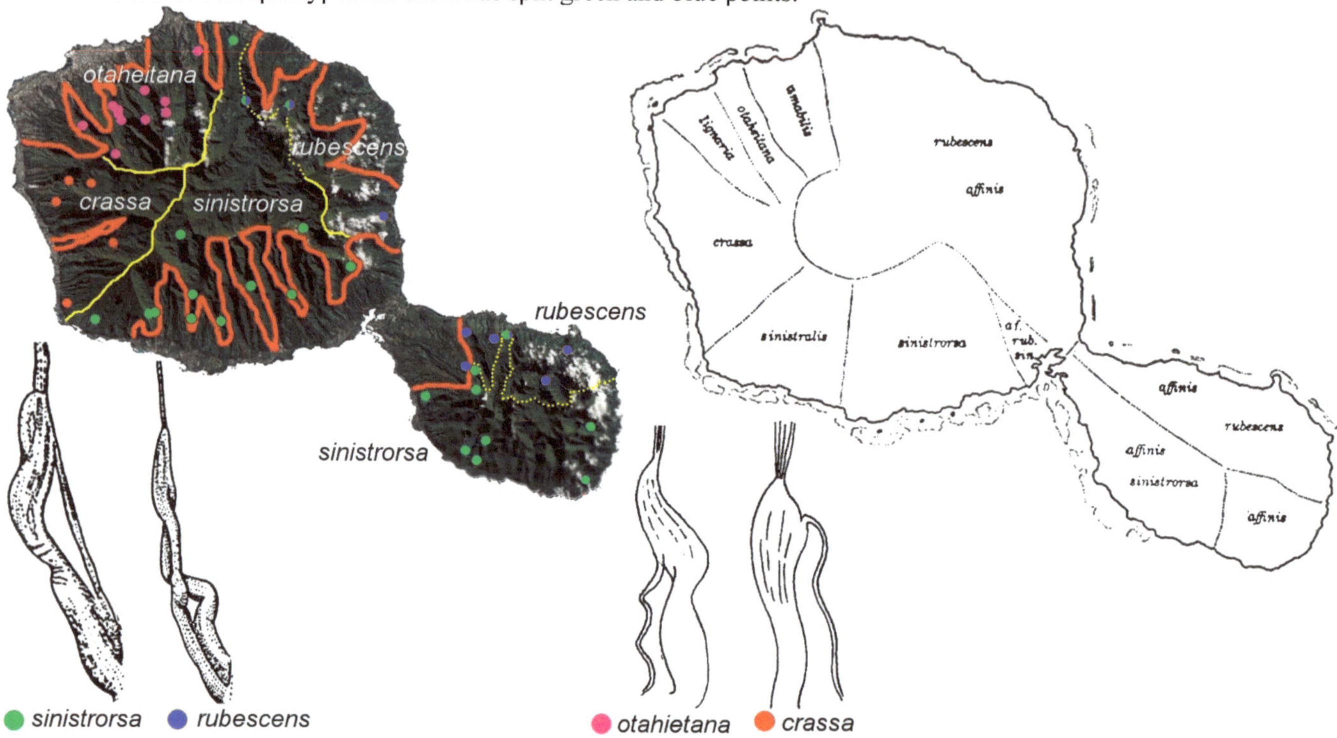

Fig. 9.4.8 *Partula otaheitana* geographical variation in colour pattern (bar charts) and chirality (pie charts). Mountain samples collected by T. Coote in top right. Shaded valleys contain populations of mixed chirality.

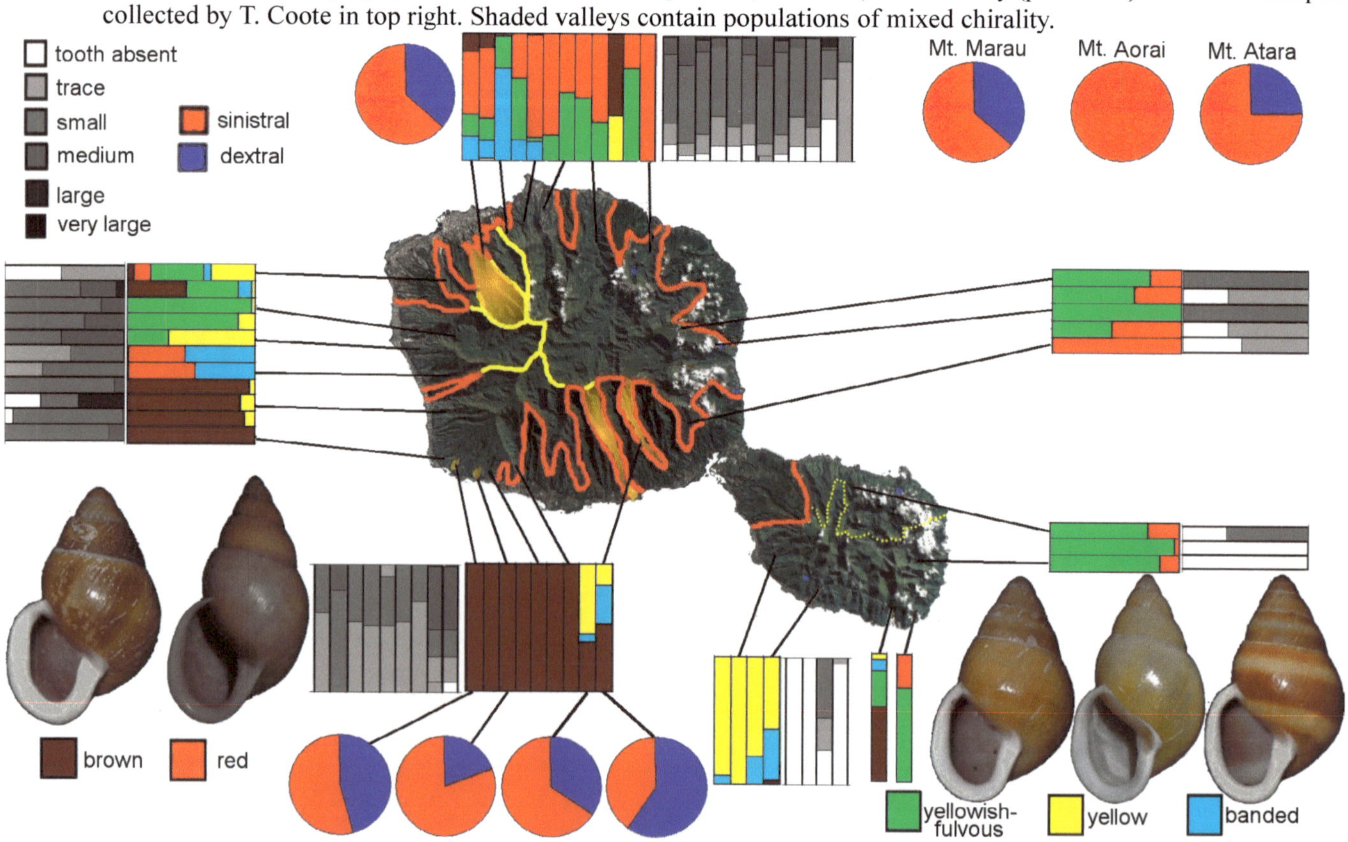

Partula laevigata Pfeiffer, 1856

*Partula laevigata P*feiffer, 1856: 162

Partula laevigata Pfeiffer, Pfeiffer 1857b: 334, 1859a: 511, 1868: 160, 1877: 203, 1881: 302. Schmeltz 1869: 72. Hartman 1881a: 183, 1881b: 14, 1885: 217, 1892: 76. Paetel 1889: 267. Smith 1903: 452. Pilsbry, 1909: 294

Partula newcombianum Hartman, 1886: 34

Partula newcombianum Hartman, Hartman 1892: 98. Smith 1903: 464

Partula newcombiana Hartman, Pilsbry 1909: 305. Fulton 1917: 238. Kondo 1968: 77. Richardson 1990: 43

Partula neweenitiarum Hartman 1897: 44 (misspelling)

Dextral. Ovately conic, spire usually acute. Umbilicus open, narrow. 5½ moderately convex whorls, body whorl inflated, frequently compressed, giving a distinctitly shouldered profile. Sculpture of fine spiral striae, prominent on spire, usually fading on body whorl. Aperture oblong-ovate, lip expanded, thickened within. Compression of body whorl often resulting in a distinctly sinuous lip edge. Columella slanted, expanding at insertion. Parietal callus thin, shiny, indistinct except at ends. A small nodular palatal tooth is present but variably developed.

Off-white with light brown strigations, spier rose tinted. Lip white, aperture variably tinted with a slate shade. Sometimes unbanded but frequently crossed by three dark spiral bands. These may be narrow, with only a peripheral band being dsitinct, or broad, appearing as a dark shell with two pale bands. Frequently bands are obscured by a darker brown shading.

Anatomy: Penis short, slender, with a distinctly inflated caecum. Internally ornamented with 3-4 low fleshy ridge. Vas deferens at base of caecum, moderately inflated.

Type specimens:

Partula laevigata Pfeiffer, 1857 – Holotype NHMUK un-numbered.

Partula newcombiana Hartman, 1886 – Holotype with neonate NHMUK 1919.12.31.55. Type originally registered as CMNH 62.4336; thought by Hartman to have been lost in a shipwreck (Smith 1903). However, Smith was correct in thinking that it had actually been returned as it was in the Layard material obtained by NHMUK in 1911 (Fulton 1917).

Distribution: Recorded from Teihomono plateau and Mt. Atara/Taravao plateau (800-1050 m). Arboreal on low vegetation (T. Coote pers. comm.).

Fig. 9.4.9 *Partula laevigata*: a – *laevigata* syntype (NHMUK); b – *newcombiana* holotype (NHMUK 1919.12.31.55); c-f – *'Partula amabilis'* (c-d – NHMUK; e-f – 'ZMH 6996); h – Teihomono specimen (ZSL 666)

Notes: This is an overlooked species, Pfeiffer's specimens being of unknown origin and Hartman's *P. newcombiana* having a disputed provenance. *P. newcombiana* was supposedly the westernmost *Partula*, reputedly having been collected by Layard on Salibabu, one of the Talaud islands off the east coast of Sulawesi. Palaeontological records do not include any partulids, although they record zonitids and helicarionids (Onoe *et al.* 2009). Layard collected widely but does not seem to have collected in the Talauds himself, although he could have obtained a shell in exchange with someone who did. Hartman (1886) recorded "I received this shell from Mr. Layard, through Mr. A. Garrett. In outline and general appearance it resembles some varieties of *P. varia*." The resemblance to *P. varia* is only superficial and it is most similar to *P. laevigata* and specimens labelled *P. otaheitana amabilis* (NHMUK and ZMH 6996). The NHMUK specimens are of unknown origin (dubiously labelled 'Solomon Is') and the ZMH specimens appear to be incorrectly labelled 'Anaa Atoll', apparently copying Reeve's erroneous locality in the descriptions of *amabilis*. These '*amabilis*' specimens correspond to the extant populations on Tahiti.
The extant populations were initally referred to dextral forms of *P. otaheitana*, however they do not match reversed shells of that species. Although shells of opposite coil are not exact mirror images they are very close (Fig. 9.4.11). Furthermore, the Mt. Atara partulids include *P. laevigata* and both sinistral and dextral *P. otaheitana.* In these *P. laevigata* and *P. otaheitana* never share the same colour patterns. Dissections of specimens from the Teihomono plateau also show that there are distinct anatomical forms corresponding to *P. otaheitana* and *P. laevigata*. This species is futher distinguished from *P. otaheitana* by having a nodular parietal denticle (Fig. 9.4.11f), as is found in most other dentate species, whereas *P. otaheitana*'s denticle takes the form of a short inward-running ridge (Fig. 9.4.11g).

Fig. 9.4.10 *Partula laevigata* variation and anatomy: a – Mt. Atara partulids: *P. otaheitana* (left - four yellow, sinistral and right middle, two brown dextrals), *P. laevigata* (top right, two dextral banded) and *Samoana burchi* (bottom right, one); b-d – anatomy from Teihomono plateau (scale bar 5 mm): b-c – *P. laevigata* (ZSL 666, 695), d – *P. otaheitana* (ZSL 680). e – *P. laevigata* Mt. Atara. Photos: T. Coote.

Fig. 9.4.11 *P. laevigata* distribution and shells in comparison to sympatric *P. otaheitana* (right): a – Teihomono; b – Taravao. c-i – comparison of dextral shells (left) with revesed images of sinistral shells (right): c-d – *P. laevigata* (left) and *P. otaheitana* (right): c – Teihomono; d – Mt. Atara; e – *P. suturalis* dextral and reversed sinistral (NHMUK Clarke & Murray 525); f-n – parietal denticles: f – *P. laevigata* Teihomono; g – *P. otaheitana* Teihomono; h – *P. laevigata* Mt. Atara; i – *P. cytherea* Mt. Marau; j – *P. compressa* Pic Rouge; k – *P. producta* Orofero; l – *P. diminuta* Pirai, m – *P. nodosa* Orofero, n – *P. affinis* Tipaerui

Partula jackieburchi (Kondo, 1981)

Samoana jackieburchi Kondo, 1981 (for 1980): 25

Samoana jackieburchi Kondo, Kondo & Burch 1983: 105. Johnson *et al.* 1986: 97. Richardson 1990: 74

Partula jackieburchi (Kondo) Johnson *et al.* 1993a: 43

Sinistral. Elongately conic. Subperforate. 5½ straight-sided whorls, body whorl convex. Sculpture of fine spiral striae, prominent on spire, usually fading on body whorl. Aperture oblong-ovate, lip expanded. Columella simple, slightly twisted in Mt. Aorai population. Parietal callus thin, shiny, well defined. Palatal tooth variably developed; if present it forms a tiny swelling, only well developed in Tiarei valley.

Lemon-corneous, spire reddish, tip reddest.

Anatomy: Penis conical, cylindrical, short (¼ oviduct length except in Faatautia valley where it is 1/3 and in Faone, Tautira and on Mt. Aorai where it reaches ½), usually strongly curved into a 'C' shape. Internally ornamented with 3-4 fleshy ridges, one of which branches twice into inverted Y-shaped spurs (Fautautia, Aivaro), these may be reduced and indistinct or extensively anastomosed (Aivaro). Caecum present as simple tube indistinct from the vas deferens, or terminally expanded (Faone and Mt Aorai, one specimen Tautira), translucent. Vestigial retractor present in Faone and Aivaro. Little variation has been found in the series dissected (37 by Kondo, 9 in the present study).

Type specimens: Holotype and paratypes BPBM 9007 ex 1970-131.4. Other paratypes BPBM 86690-8, 86700-3, 86745-51, 86756, 86881, 137018-9, 1970-133, 1970-143, 1970-189, 1970-190B, 1970-214.

Distribution: Recorded from Mt. Aorai (north slope and ridge 5600-6000 ft, 1934), Tiarei (360 m, 1970), Faatautia (100 m, 1970), Faone (40 m, 1970), Tautira (200-1500 ft, 1927) and Aivaro ('Aiavaro', 100-1500 ft, 1927) valleys. 31 additional specimens: Aiavaro (BPBM 86746, 86748, 86754-5), Faone (BPBM 1970-131), Tautira (BPBM 86880, 86882). On the ridge of Mt. Aorai collected on the ground by Elwood C. Zimmerman, Donald Anderson and Yoshio Kondo. Neighbouring colonies were *P. otaheitana*. Elsewhere arboreal on ferns, shrubs and tree trunks, including ginger, *Coffea*, *Hibiscus tiliaceus*, *Musa troglodytarum*, *Freycinetia* and *Cordyline terminalis*.

Notes: First collected in 1927 by L.H. MacDaniels (1927), then by the Mangarevan Expedition (1934) and by J.B. Burch, J.N. Cather, J.B. Burch Jr. and E. Oshima (1970) but not recognised until 1980 when apparent *Partula otaheitana* were being dissected to distinguish them from large *P. producta*. Kondo (1981) considered *jackieburchi* to be classifiable as a *Samoana*, based on its anatomy. It resembles *Samoana* in having a vestigial penial retractor attaching to the penis as well as the normal, well-developed branch attaching to the caecum. However, it lacked the penial pilasters found in all true *Samoana*. The branch of the penial retractor is also present in *P. otaheitana* (shown in Kondo's 1981 figure). Johnson *et al.* (1993a) demonstrated that *P. jackieburchi* was interfertile with *P. otaheitana*, although their crosses had only limited success and the hybrids they figured show a range of anatomies, none of which have been recorded from wild-collected specimens. This suggests that although the two species will cross in laboratory conditions wild hybridisation appears to be insignificant. In the valleys occupied by *P. jackieburchi*, Crampton (1916) noted that in the Faarumai, Tiarei and Mahena group of valleys *P. otaheitana* "is exceedingly rare or absent" although *P. affinis* were abundant. None were recorded in Farapa or Vahii. In Tiarei he found 15 *P. otaheitana* and this is the only valley where the two species have been recorded in sympatry. The 1970 sample from Tiarei (BPBM 1970-A190) contains just a single *P. otaheitana*. Johnson *et al.*'s (1993a) only sample with both species is from almost exactly the same point in Tiarei valley, again with only a single *P. otaheitana*. This suggests that *P. jackieburchi* is only present where *P. otaheitana* is absent or extremely rare.

The conchological differences between *P. otaheitana* and *P. jackieburchi* are minimal: *P. jackieburchi* is always slightly

Fig. 9.4.12 *Partula jackieburchi*: a – holotype (BPBM 9007); b-c – colour variants (b – BPBM 137018; c – BPBM 1970-189.1)

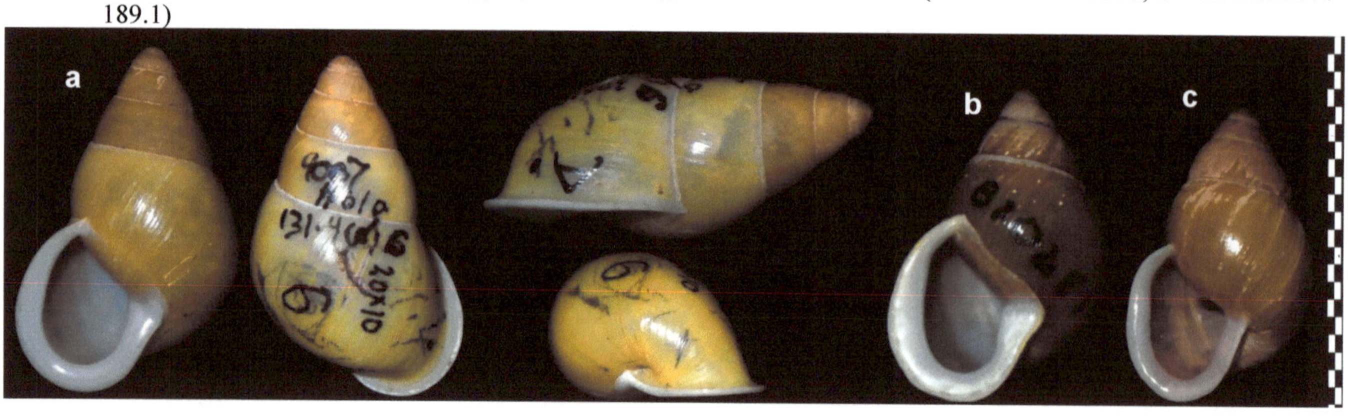

shorter (length compared to *P. otaheitana*: 89% Tiarei, 91% Tautira, 93% Aiavaro). The width tends to be much closer (97% Tiarei, 95% Tautira, 104% Aiavaro). For the species as a whole the differences are negligible (1%) due to the variability of different *P. otaheitana* populations. The aperture tends to be rounder in *P. jackieburchi*, the lip thicker, the columella evenly curved or straight rather than angled, and the sculpture more persistent.

Partula producta Pease, 1865

Partula producta Pease, 1865: 671

Partula producta Pease, Pfeiffer 1868: 156. Pease 1871b: 473. Schmeltz 1874: 92. Hartman 1881a: 185, 1881b: 10, 1885: 10, 1892: 75. Paetel 1883: 153, 1889: 268. Garrett 1884: 66. Smith 1903: 448. Pilsbry 1909: 192. Crampton 1916: 301. Kondo 1955: 163, 1968: 76

Partula affinis producta Kondo & Burch 1983: 101

Elongately ovate to ovately conical, solid. Umbilicus open, rimate. Glossy with fine, but weak spiral striae. 5-5½ moderately convex whorls, the last as long as spire. Body whorl roundly angulate at base, distinctly compressed laterally. Aperture oblong, nearly vertical. Lip narrowly expanded and reflexed, slanting, a little contracted in upper part, moderately thickened within. Columella subnodose. Parietal callus thin, transparent except at edges; a weak tubercular parietal tooth present in 2.3% of specimens (n=43) in the main range, well developed in all 4 specimens from an isolated population in Orofero valley.

Yellow to light brown. Typically banded, resembling lyra (three evenly spaced bands, sutural zone and umbilicus exhibit the ground colour, the basal band is obscured on the spire so only two are visible) and some cestata. Differs from *P. otaheitana sinistrorsa* in having a pale sutural band. Colour may vary so that there is a continuous gradation from pale to banded to dark. Aperture dark within. Lip white or pinkish.

Anatomy: Penis short, ½ oviductal length, narrow. Caecum elongate, may be pointed; vas deferens usually subapical (near base of caecum). Interior with three main smooth vertical ridges, and flanking subsidiary ridges. Caecum with tuberculate ridges.

Type specimens: Syntypes AMNH 42106; ANSP 59643, 295272.

Distribution: Southern Tahiti. Originally recorded from one valley by Garrett (1884): "This species only occurred to my notice in one valley, on the southwest coast of Tahiti [Faarahi valley], where it is abundant, lurking beneath decaying leaves and under heaps of loose stones." Elsewhere Garrett refers to "the district of Papieri", which Crampton (1916) considered

Fig. 9.4.13 *Partula jackieburchi* range: red & blue points – *P. jackieburchi* only; red – Kondo & Burch (1979), blue – Johnson *et al.* (1993a); purple – sympatric *P. jackieburchi* and *P. otaheitana.* Bars show tooth development and shells show typical examples, where two shells are shown left is *P. jackieburchi*, right is *P. otaheitana* from adjacent areas

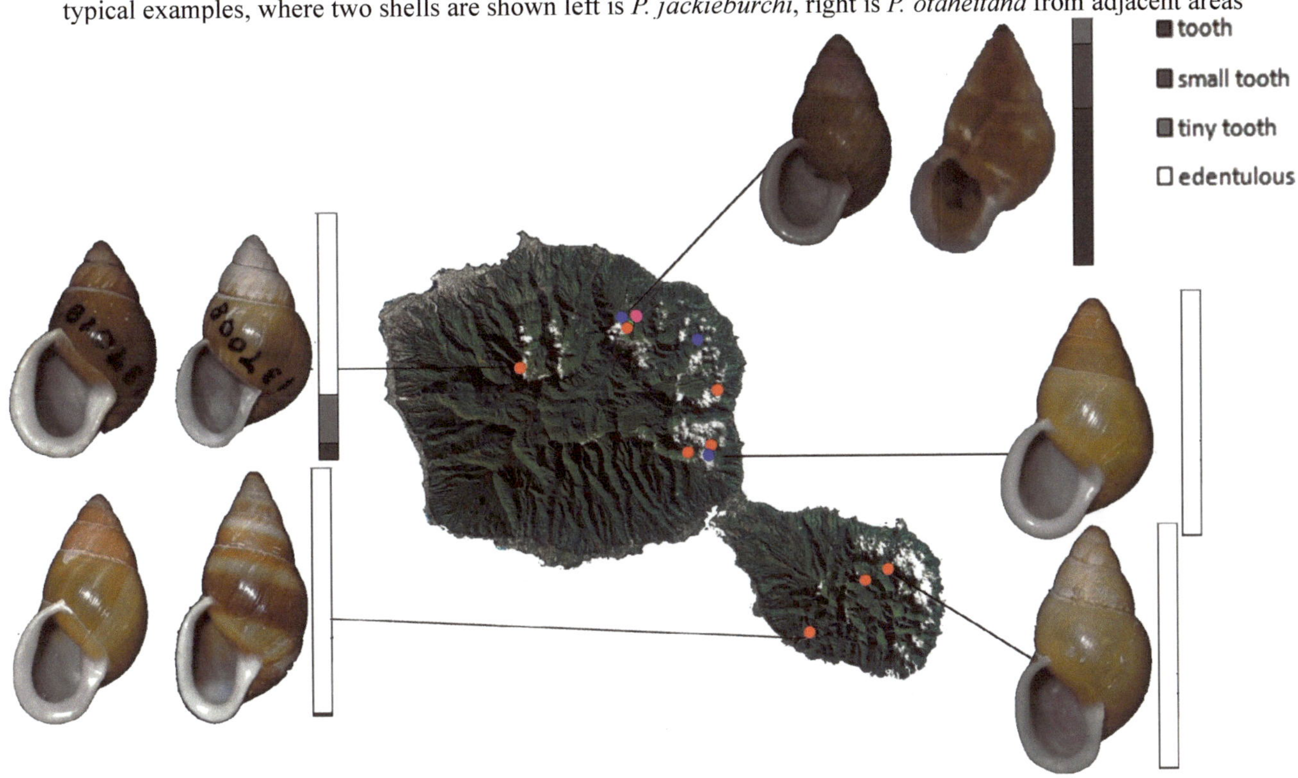

Fig. 9.4.14 Crampton's (1916) *Partula producta* figures (edited for position), Titaviri, Papeiti, Vaihiria, Temarua, Taharua, and Tereeia specimens misidentified by Crampton as *P. otaheitana*

Fig. 9.4.15 *Partula producta*: a – type (ANSP 59643); b-k – specimens: b-d – Garrett (ANSP 59928); e – dentate Orofero specimen (T. Coote); f-h – Crampton (ANSP 272443); i-j – Burch (UMMZ 117); k-l – sinistral, Taharuu valley (ZMH 9832) (photos: a – F. Borrero)

an error for the adjacent Faarahi. Crampton (1916) noted that it was "somewhat elusive", and that he had found it through searching deliberately for it in the region recorded by Garrett. Kondo & Burch (1983) identified the species (as '*Partula affinis producta*') from Faone, Apirimaue, Titaviri, Vaihiria, Vairaharaha, Faarahi, Vaitunamea, Taharua, Papeiti, Temarua and Tereehia. They concluded that Crampton had misidentified sinistral *P. producta* as *P. otaheitana sinistrorsa-sinistralis* with which it can only be distinguished anatomically. Although Garrett and Crampton collected *P. producta* exclusively from the ground, Burch's specimens (BPBM material and notes used in Kondo & Burch 1979) were collected both on the ground and on vegetation. As Garrett and Crampton had found, the Faarahi population was mostly terrestrial (64%, n=22; compared to 6% for *P. otaheitana*, n=34), as were the ones in Titaviri: "mostly ground collected, but some arboreal… found mainly under leaves". Others were mixed: "mostly arboreal but some on ground… on bamboo, coffee and banana" (Vaitunamea) or largely arboreal (87% in Tereehia, n=75). Kondo & Burch (1980) overlooked a population of *P. producta* from Oopu valley (BPBM 1970-107) which contains a mixture of sinistral *P. producta* and *P.*

Fig. 9.4.16 *Partula producta* geographical variation. Sampling points marked according to chirality: red – sinistral, blue – dextral, green – mixed (and shown in detail on right)

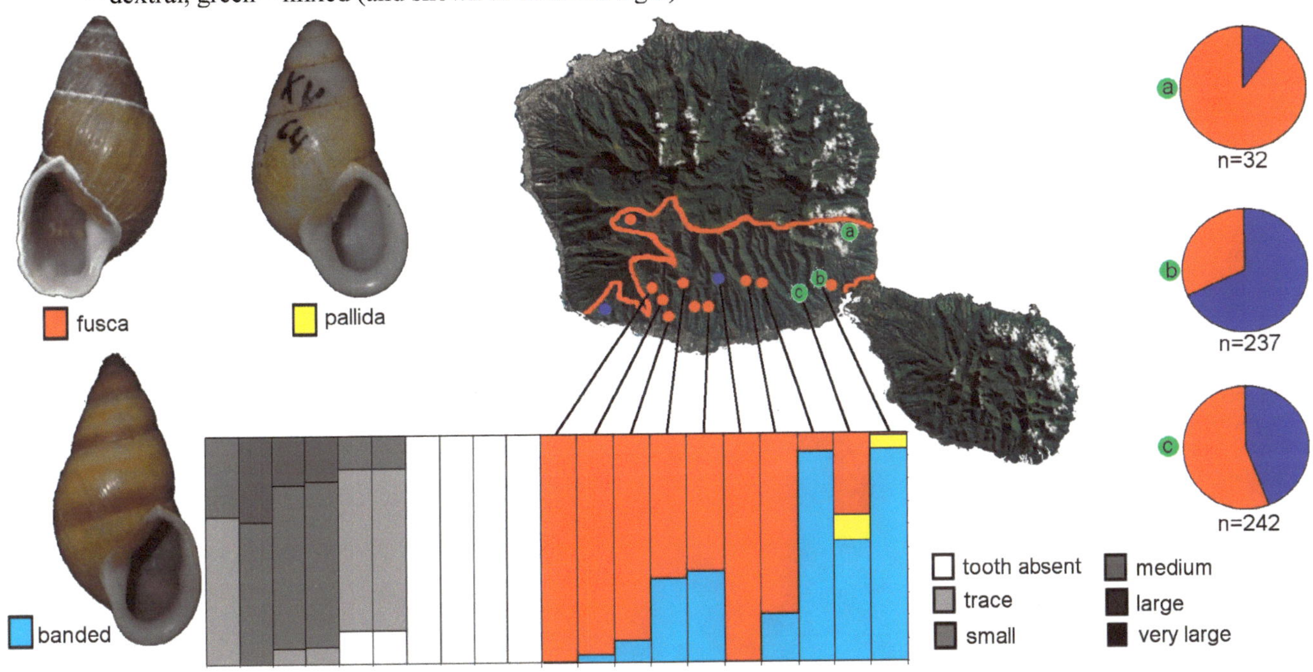

Fig. 9.4.17 Anatomy of *P. producta* (a-c), *P. nodosa* (d-f), *P. affinis* (g-k) and *P. compressa* (l-m): a – Faarahi (ANSP A864); b – Teoha (ANSP A745); c – Vaihiria (ANSP A754); d – Kondo 1955; e – *composita* Taapuna (ANSP A757); f – *laeva* Maruapoo (ANSP A770) & *intermedia* Aoua (ANSP A772); g – 'Vavii to Oofu' (ANSP 318706); h – unknown (NHMUK Cheeseman 1925); i-j – Faaroa (ZSL 'T6'); k – Papenoo ('*stolida*' BPBM 89584); l – Aorai (BPBM 145548) & Viriviriterai (ZSL 675); m – Orofena (BPBM 142356). Scale bar 5 mm.

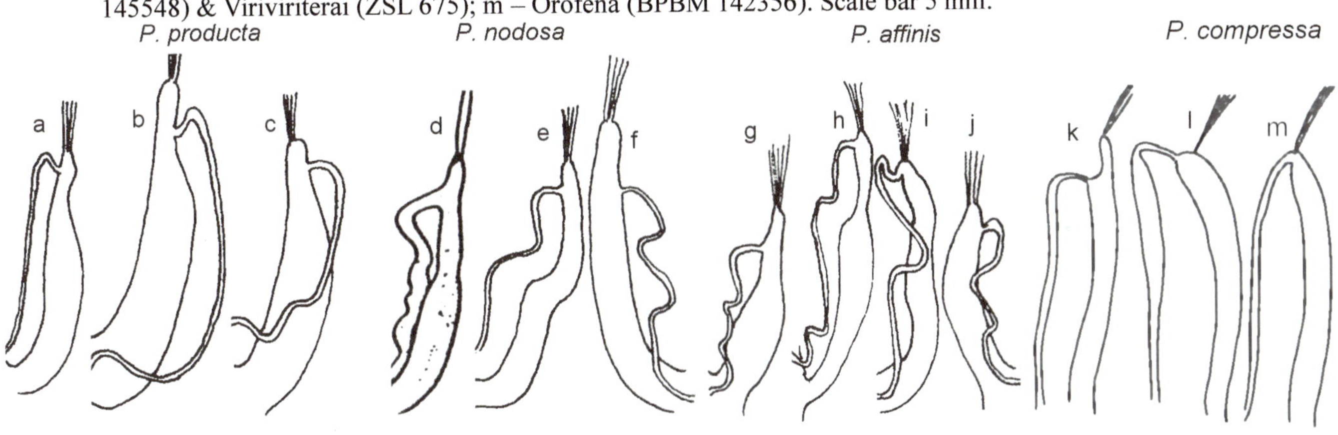

otaheitana. Sinistral specimens from Taharuu valley (Kondo & Burch 1983 and ZMH 9832) show little shell variation but include a small number of anatomical *P. o. sinistrorsa* as well as *P. producta.* In addition, T. Coote collected four dark dextral shells from Orofero valley, below the crest of Mt. Tahiti, 1300 m.

Partula affinis Pease, 1868

Partula affinis 'Pease', Carpenter in Pease, 1865: 675 (as syn. of *rufa*, ex errore, nomen nudum)

Partula affinis var. *dubia* 'Pease', Carpenter in Pease, 1865: 675 (as syn. of *varia*). Garrett 1884: 49 (nomen nudum)

Partula affinis Pease, 1868a: 224

Partula affinis Pease, Pease 1871b: 473. Schmeltz 1874: 92. Pfeiffer 1877: 204. Gloyne 1878: 338. Hartman 1881a: 179, 1881b: 10, 1885: 212. Garrett 1884: 47. Smith 1903: 444. Kondo 1968: 76. Kilias 1969: 246. Kondo & Burch 1979: 79. Emberton 1982: 31. Richardson 1990: 7. Johnson *et al.* 1993a: 43. Coote & Loève 2003: 95

Fig. 9.4.18 *Partula affinis*: a – *affinis* paralectotype (NHMUK 2044); b-e – *dubia* supposed types; f – E. Cheeseman specimen (NHMUK); g – UMMZ 159; h – UMZC 75351; i – ANSP 272775; j, m-n, q-s – ANSP 272432; k-l – UMZC 75295; o-p – ZSL 'T6'; t-u – Garrett *P. affinis* specimens misidentified as '*P. stolida*' (MCZ 50872)

Partula affinis var. *bacca* 'Pease', Smith 1903: 444 (nomen nudum)
Partula otaheitana affinis Pease, Pilsbry 1909: 187. Crampton 1916: 185. Kondo 1955: 161
Partula otaheitana affinis erythraea Crampton, 1916: 195
Partula affinis affinis Pease, Kondo & Burch 1983: 105

Usually dextral (99.5%, n=5,212: Crampton 1916), some populations largely sinistral. Elongately ovate. Umbilicus open, compressed. Shell solid, surface with irregular growth lines, fine spiral striae, indistinct on body whorl. Suture impressed, sometimes faintly marginated. Lip narrowly thickened. A thin callosity is sometimes present between lip insertion and columella. Columella dentate. A small, nodular parietal tooth sometimes present.

Light or dark brown, sometimes reddish. Uniform or banded, some with three dark red-brown bands in cestata, frenata and bisecta patterns. Crampton described a variety *erythraea* from Farapa valley, being plain yellow, overshaded by red (72%) or with reddish bands (24%), only 4% were plain whitish.

Body and mantle black.

Anatomy: Penis very short (shorter than buccal mass), 1/4-1/3 oviduct length, broad, may be club-shaped. Internally ornamented

Fig. 9.4.19 Crampton's (1916) *Partula affinis* figures (edited for position)

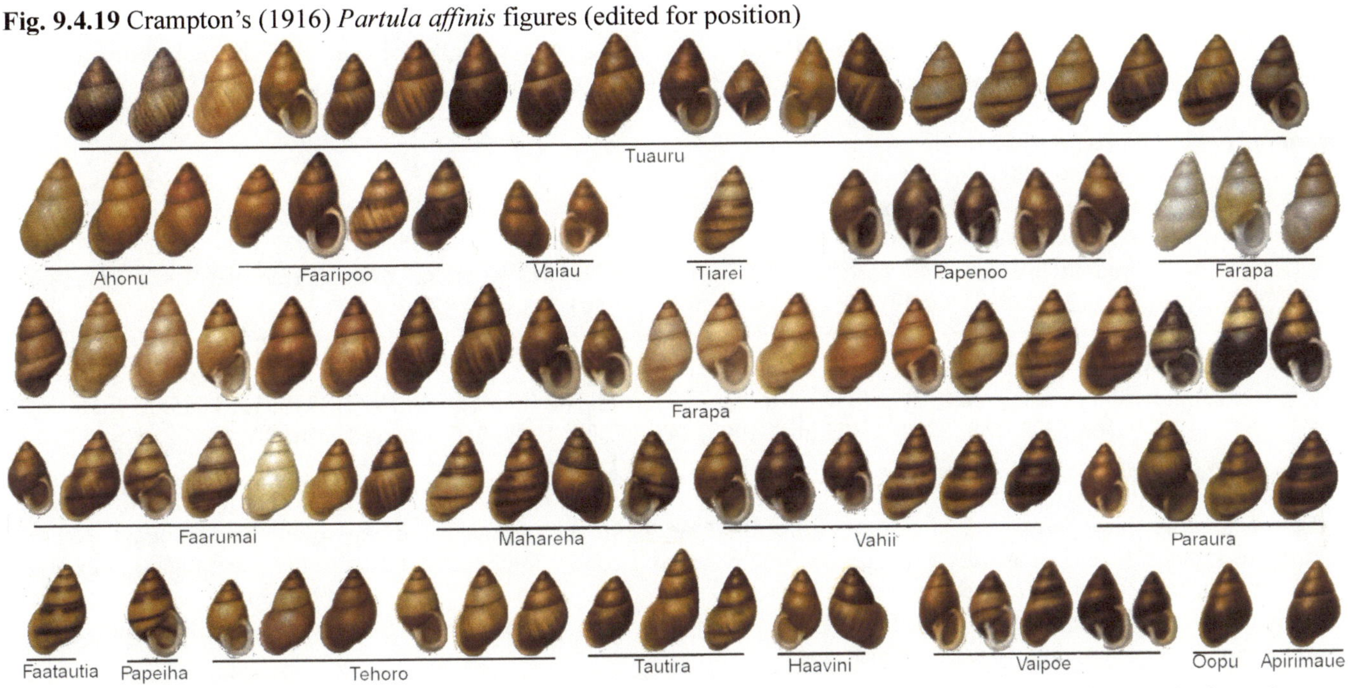

Fig. 9.4.20 *Partula affinis* geographical range. Sampling points marked by chirality: red – dextral, yellow >95% dextral, green >67% dextral, blue >60% sinistral. Sampling points figured in bar chart by region, in clock-wise order.

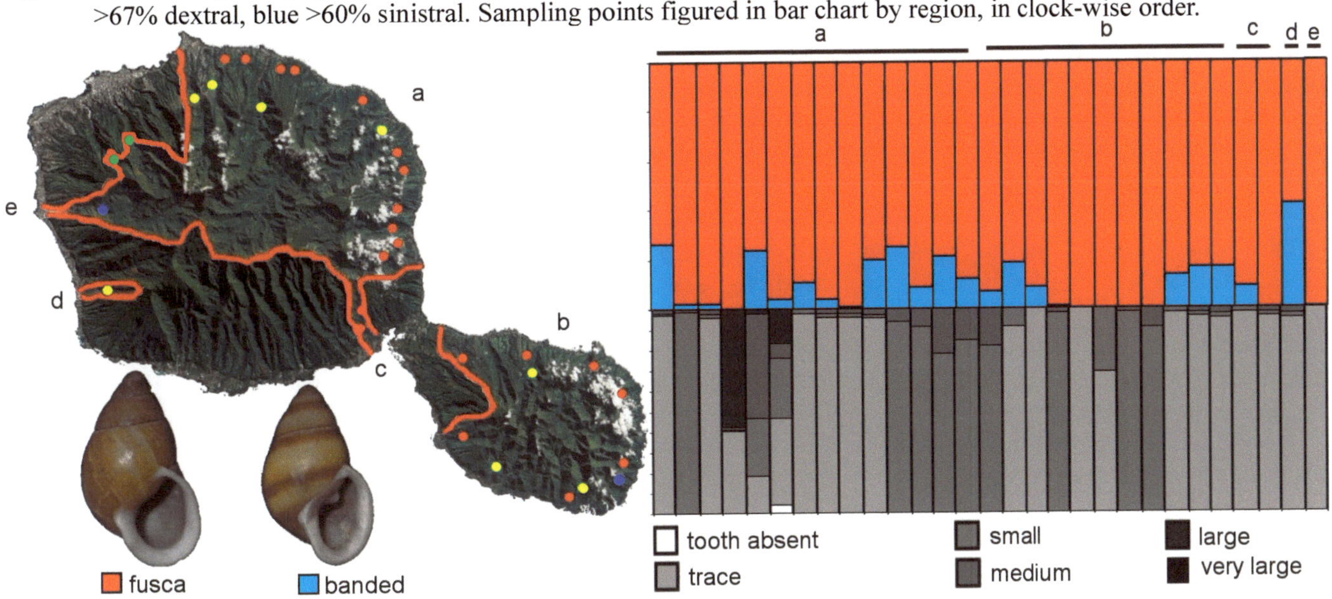

with 5-6 vertical ridges which run into the short caecum. Vas deferens usually slightly subterminal.
Type specimens:
Partula affinis Pease, 1868 – Lectotype ('holotype' – Baker 1963) ANSP 59560, paralectotypes: AMNH 118932; ANSP 2603, 5960-1, 463936; FLMNH 111860-62; MCZ 25313; NHMUK 2004; ZMB/Moll-74327, 75621.
Partula affinis var. *dubia* (nomen nudum) – Supposed types MCZ 297865, 25315.
Distribution: Recorded from almost all valleys in the north-eastern quarter of Tahiti-Nui and Tahiti-Iti, with isolated records in the south (Tuauru, Ahonu, Faaripoo, Papenoo, Farapa, Faarumai, Tiarei, Mahaena, Vahii, Paraurua, Faatautia, Papeiha, Utuufai, Faone, Tehoro, Haavini, Tautira, Aionifaa, Aiurua, Vaiau, Hototunu, Vaiaaia, Aiavaro, Vavii, Vaipoe, Haoma, Oopu, Apirimane).
Notes: Pease (1868b) considered this species allied to *P. otaheitana*, and it was treated as a subspecies by Crampton (1916). Lundman (1947) and Emberton (1982) observed that this arrangement was dubious as '*P. o. affinis*' and '*P. o. rubescens*' were sympatric and non-hybridising in northern valleys. There is also some confusion with *P. compressa* (as *P. stolida*). Crampton (1916) tentatively proposed that all three names referred to *P. affinis*: "Certain of the *affinis* shells, it is true, are much compressed in the last whorl, but such shells fall far short of the length given by Garrett. It may be that such examples are really abbreviated relatives of the shells named by Pease as *stolida*; if so then, the type so distinguished is only a modified *affinis*, for my own specimens certainly belong to the latter primary variety." The confusion arises from the labelling of some variants of *P. affinis* as *P. stolida* whereas examination of types shows that true *P. stolida* is a synonym of *P. compressa*.

Partula compressa Reeve, 1850

Partula compressa Pfeiffer, Reeve, 1850: pl4 f20
Bulimus compressus Pfeiffer, 1850: 75
Partula compressa Pfeiffer, Pfeiffer 1853b: 447, 1856: 162, 1859: 508, 1868: 157, 1877: 196. Adams 1855: 145. Frauenfeld 1869: 874. Pease 1871b: 473. Paetel 1873: 104. Schmeltz 1874: 81. Hartman 1881a: 181, 1881b: 13, 1885: 223, 1892: 99. Pfeiffer & Clessin 1881: 301. Garrett 1884: 112
Partula stolida 'Pease' Carpenter in Pease 1865: 675 (in syn. of *vanikorensis*, nomen nudum)
Partula stolida Pease, 1866a: 198. Garrett 1884: 70
Partula stolida Pease, Pease 1871d: 473. Pfeiffer 1877: 195. Hartman 1881a: 189, 1881b: 10, 1885: 213, 1892: 75. Paetel 1883: 153, 1889: 268. Smith 1903: 445. Pilsbry 1909: 257. Richardson 1990: 56
Partula compressa Reeve, Pilsbry 1909: 181. Richardson 1990: 19

Elongately ovate to ovately conical. Solid, glossy. Deeply rimate. Spire conic, obtuse, with 5-5½ nearly flat whorls. Body whorl as long as spire, narrowed and compressed at the base. Faint spiral striae on spire. Columella subnodose. Aperture oblong to subtriangular. Lip broadly expanded, a little reflexed, strongly thickened internally. Parietal callus thin except at ends, curved. Dentate (*stolida*) or edentulous (*compressa*).
Chestnut coloured with irregular darker streaks or uniformly dark brown (Viriviriterai). Suture marked with a white line. Lip and interior of aperture white, tinted brownish violet.
Anatomy: Penis short, ⅓ oviduct length, cylindrical, no distinct caecum. Vas deferens terminal, usually inflated.
Type specimens:
Bulimus compressus Pfeiffer, 1850 – Syntypes NHMUK (un-numbered).
Partula stolida Pease, 1866 – Syntypes MCZ 25311, 104508.
Distribution: Higher parts of Tahiti; recorded by Garrett (1884) from Papenoo, subsequently collected (1934) on Mt. Orofena (=Orohena) (5000-5400 ft) by Harold St. John, Mt. Aorai by D. Anderson and recent records from Pic Rouge, Mt. Marau, Mt. Aorai, Tefaiti and Orofero (all collected by T. Coote – 2001, 2004), and from low vegetation on Viriviriterai plateau (E. Loève 1996, 720-940 m; photographed in 2003 by J.-Y. Meyer at 920-1100 m, '*Partula* sp.' and '*P. cf. affinis*' in Meyer 2004).
Garrett (1884) writes: "I took a few examples of this ground species about two miles up Papenoo valley, on the northeast coast of Tahiti. They were all found lurking among the roots of ferns."
Notes: Although both Pfeiffer and Reeve considered Pfeiffer the author of the name *compressa*, Reeve's figure was published two months earlier than Pfeiffer's and so he must be designated as the authority.
This species has been confused with *P. affinis* due to early misidentifications or mislabelling. Pilsbry (1909) noted that specimens of *P. compressa* sent to him by Garrett in 1862 matched another specimen that Garrett had sent as *P. stolida*, and also that this resembled a compressed *P. affinis*. Garret seems to have mislabelled these; some of his '*P. stolida*' are *P. affinis* (e.g. MCZ 50872, USNM 76999). Further confusion was provided by Garrett's five specimens in USNM registered as *P. compressa* from Raiatea (USNM 76961) and some in MCZ which are misidentifications of *P. navigatoria radiata*, as noted by Hartman (1885) and Pilsbry (1909). Despite this Garrett (1884) was clear that *P. compressa/stolida* was not synonymous with *P. affinis*.

In the absence of any clear identification of the ill-defined original *P. stolida* Pilsbry thought this could not be excluded from *P. compressa*, "yet the statement that it is "dentate" probably indicates that what Pease originally had was a form closer to *P. affinis*." *P. stoldia* specimens are dentate, whereas *P. compressa* are not, otherwise the two are similar. Pilsbry (1909) left *P. compressa* as a valid species with uncertainty; Crampton (1916) failed to clarify the situation as he did not find

Fig. 9.4.21 *Partula compressa*: a-c – *compressa* types (NHMUK un-numbered); d-f – *stolida* types (MCZ 104508); g-h – '*stolida*' Pease collection MCZ 25311; i – '*stolida*' Garrett ANSP 59562; j-m – T. Coote collected specimens (j – Pic Rouge; k – Tefaiti; l – Mt. Aorai; m – Orofero); n-q – dissected specimens (n – Mt. Aorai BPBM 145548; o – Mt. Orofena BPBM 142356); p-q – Viriviriterai (ZSL 675)

find the species, despite Garrett's clear locality. Given its scarcity in collection, and the fact that there have been only six collections (Reeve, Garrett, St. John, Anderson, Loève and Coote), this failure is unremarkable.

St. John and Anderson specimens labelled *P. compressa* and *P. stolida* were set aside for further examination by C. Montague Cooke Jr. in 1935. In 1975 Kondo re-identified one as *Samoana burchi* (BPBM 145006) and the others as *P. affinis* based on dissection of one sample (BPBM 88984), the others (BPBM 142356, 145548) are true *P. compressa.*

Partula cytherea Cooke & Crampton, 1930

Partula cytherea Cooke & Crampton, 1930: 3

Partula cytherea Kondo 1955: 162, 1968: 76

Partula affinis cytherea Cooke & Crampton, Kondo & Burch 1983: 105

Usually sinistral (88%; n=67). Broadly ovate-conic. Umbilicus open. 5-5½ slightly convex whorls. Surface glossy, with weak growth lines and faint spiral striate, fading on body whorl. Aperture sub-auriform, oblique. Lip thickened, inwardly contracted, margin shouldered at its parietal insertion. Parietal callus well marked. Columella straight, slightly flaring at insertion. Parietal tooth nodular, almost always well developed (93%, n=67).

Orange-brown to olive-brown base, spire red brown, darkening to purple-brown at apex as a result of decortication.

Anatomy: Penis short (¼-½ oviduct length), constriction dividing caecum and penis; vas deferens apical or subterminal.

Type specimens: Syntypes ANSP 161301, 272446 (body preserved as BPBM 87053), BPBM 10851. Crampton & Cooke (1930) designated two as 'cotypes' (ANSP 161301, BPBM 10851) and BPBM 87053 as 'paratypes', this distinction is not valid as no single holotype was defined; all are syntypes. BPBM 87049, 87086, 89581 are from the same series.

Distribution: Known only from the upper slopes of Mt Orohena 500-4000 ft (33 specimens, 1927), Mt. Aorai (3 in 1930) and Mt. Marau (8 specimens, 1,400 m, 2005, Coote). Arboreal ('on ferns and shrubs, mostly on leaves of trees and lower plants' – MacDaniels 1927 specimen notes). Superficially similar specimens from Mt. Marau at 1,328 m are anatomically *P. otaheitana* (MNHN Th118). There do not appear to be sympatric *P. otaheitana* and *P. cytherea* populations.

Note: Cooke & Crampton (1930) considered this distinct from *P. affinis* on aperture angle and shape and wider umbilicus. Kondo (annotated thesis) noted "dissected: not a good species YK 29/III/1977" but later reversed his opinion (MS).

Fig. 9.4.22 *Partula cytherea*: a – type (ANSP 272446); b – umbilical view of specimen (BPBM 87053); c – Mt. Marau sinistral and dextral specimens (ZSL) (photo: a – F. Borrero)

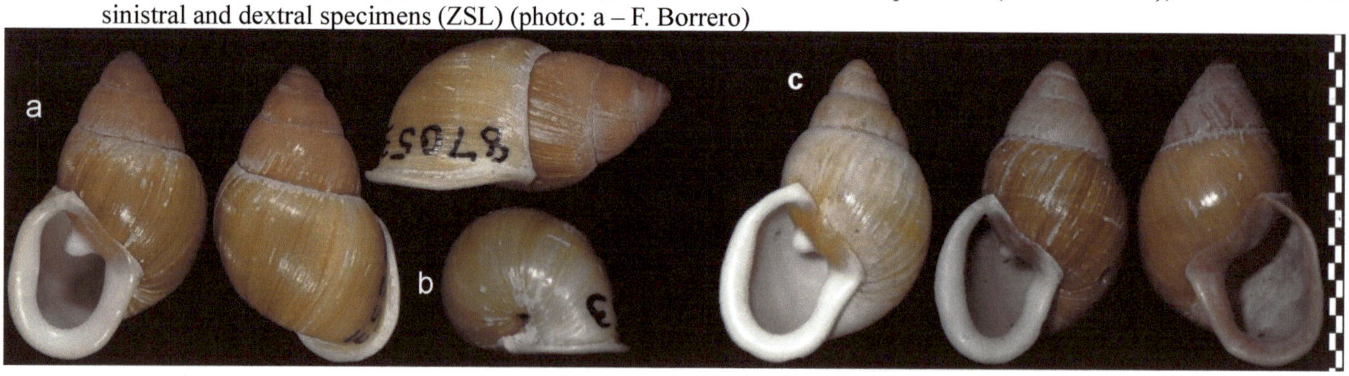

Fig. 9.4.23 Distribution of *Partula cytherea* with chirality variation (red – sinistral, yellow – dextral) and *P. compressa*

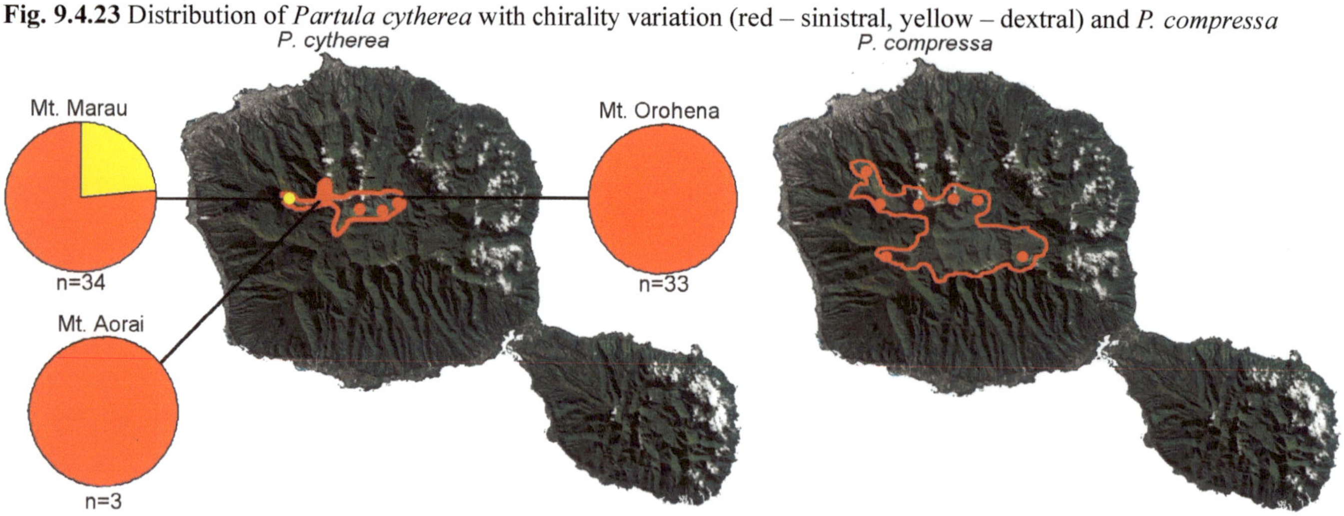

Partula nodosa Pfeiffer, 1853

Partula nodosa Pfeiffer, 1853a: 262

Partula nodosa Pfeiffer, Pfeiffer 1853b: 449, 1856: 266, 1856: 266, 1859a: 512, 1868: 160, 1877: 206. Adams 1855: 145. Schmeltz 1869: 72, 1874: 92. Pease 1871b: 473. Paetel 1873: 104, 1883: 153, 1889: 267. Hartman 1881a: 184, 1881b: 10, 1885: 10, 1892: 75. Garrett 1884: 65. Smith 1903: 445. Pilsbry 1909: 179. Crampton 1916: 87. Kondo 1968: 76

Partula trilineata 'Pease' Carpenter in Pease, 1865: 675 (as syn. of *nodosa*, nomen nudum)

Partula trilineata Pease 1866: 195; 1867: 81. Pfeiffer 1877: 195

Partula nodosa var. *trilineata* Pease 1871b: 473

Partula nodosa var. *sinistralis* Mayer 1902: 127

Partula nodosa var. *laeva* Pilsbry 1909: 195 (unjustified replacement for *sinistralis*)

Partula nodosa var. *composita* Crampton, 1916: 91

Partula nodosa var. *exigua* Crampton, 1916: 100

Partula nodosa var. *intermedia* Crampton, 1916: 101

Partula affinis nodosa Pfeiffer, Kondo & Burch 1983: 105

Usually dextral (60%, n=1,922: Crampton 1916). Ovately conical. Umbilicus open. 5½ straight sided whorls, the last as long as spire. Surface with widely spaced spiral striae. Columella distinctively curved and usually with a prominent nodule at mid-point. Aperture vertical or slightly angled, oblong and narrow. Lip only slightly expanded, internally thickened. Margins subparallel, right straightened. Parietal callus thin. Parietal denticle nodular, variably developed.

Brown, ornamented with a wide white band at the suture, and often some other pale bands. Crampton (1916) divided

Fig. 9.4.24 *Partula nodosa*: a-b – *nodosa* type (NHMUK 79.9.12.66); c – *trilineata* type (ANSP 59563); d – *laeva* (ANSP 276120); e – *intermedia* (ANSP 276121); f – *composita* (ANSP 276119); g – *exigua* (part of sample ANSP 272438); h – colour variety (UMZC 75341); i-l – *sinistralis* types (MCZ 25404)

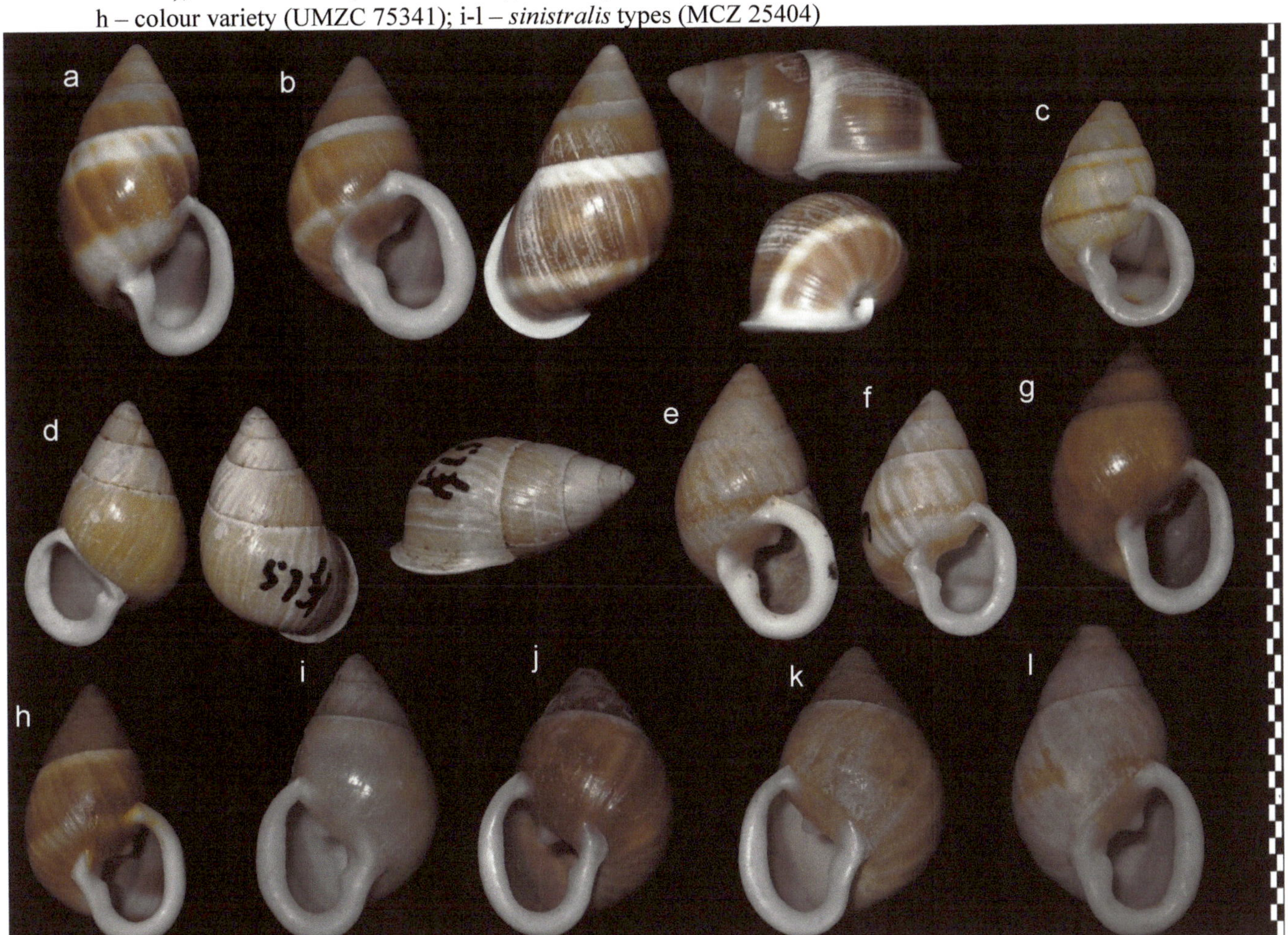

colour variation into four colour classes: I (Pilsbry's pallidor) – plain light; II ('trilineata) – pale to brown with 1-3 spiral bands, spire usually brown, suture margined with white; III (concrescens) – bands connected to form a broad band; IV (nodosa) – uniformly dark. These banding patterns all intergrade and are modifications of frenata.

Anatomy: Penis very short (shorter than buccal mass), ¼ oviduct length.

Geographical variation: The proportion of dextral shells in the population varied geographically: dextrals 98% Taapuna, 100% Punaruu, Papehue, Aoua, Orofere (only 18 individuals), Atehi (only 4), but 3.5% in Maruapoo (Crampton 1916). Four varieties have been described differing in chirality and colour pattern. Genetically *intermedia* and *composita* are not well defined, and there are no data for the other varieties (Lee *et al.* 2007b). Samples from most populations have been dissected and are anatomically identical (other than in being reversed due to different chirality). These supposed taxa probably reflect local genetic drift in different valleys rather than true subspecies and are not recognised here.

var. *sinistralis* Mayer, 1902 – Maruapoo valley. Mostly sinistral (97%, n=80).

var. *intermedia* Crampton, 1916 – Papehue, Aoua, Orofere valleys. Dextral, intermediate between *composita* and *exigua*.

var. *composita* Crampton, 1916 – Punaruu and Taapuna valleys. Chirality variable (155 sinistral: 148 dextral).

var. *exigua* Crampton, 1916 – Atehi valley. Only 4 specimens, all dextral.

Type specimens:

Partula nodosa Pfeiffer, 1853 – Syntypes ANSP 59565, NHMUK 79.9.12.66.

Partula trilineata Pease 1866 – Lectotype ANSP 59563; paralectotypes AMNH 42049; ANSP 2651, 59564; MCZ 24940, 25130. Lectotype designation by Baker (1963).

Partula nodosa var. *sinistralis* Mayer 1902 (and *laeva* Pilsbry, 1909) – Syntypes MCZ 25404-5.

Fig. 9.4.25 Crampton's (1916) *Partula nodosa* figures (edited for position)

Partula nodosa var. *composita* Crampton, 1916 – No types defined but any of Crampton's series would be types (ANSP 276117-9, 320751 and A757).

Partula nodosa var. *exigua* Crampton, 1916 – No types defined but any of Crampton's series would be types (ANSP 272438).

Partula nodosa var. *intermedia* Crampton, 1916 – No types defined but any of Crampton's series would be types (ANSP 276121 and A772).

Distribution: West side of Tahiti, recorded in seven valleys from Taapuna to Orofere. Garrett (1884) found it "restricted to a limited area about two miles up Punaavia valley on the west coast of Tahiti… It is entirely confined to the south side of the stream which flows through the valley, and circumscribed in a narrow area about three-fourths of a mile in length." (Garrett). Arboreal.

Note: *P. nodosa laeva* was proposed by Pilsbry (1909) as a replacement name for *P. nodosa sinistralis* Mayer, 1902 which he stated was preoccupied. However, there is no earlier use of *sinistralis* in combination with *nodosa*, hence his replacement name is unnecessary.

Partula diminuta C.B. Adams, 1851

Partula diminuta C.B. Adams, 1851: 125

Partula filosa Pfeiffer, 1853a: 262

Partula filosa Pfeiffer, Pfeiffer 1853b: 450; 1855a: 162, 1855b: 267, 1868: 160, 1877: 206. Adams 1855: 145. Pease 1864: 675. Hartman 1881a: 182, 1881b: 10, 1885: 10, 1892: 75. Paetel 1883: 153. 1889: 267. Garrett 1884: 64. Mayer 1902: 126. Smith 1903: 445. Pilsbry 1909: 179. Crampton 1916: 81. Kondo 1955: 165, 1968: 76

Partula diminuta Pfeiffer 1853b: 452, 1859: 513, 1868: 168, 1877: 206. Adams 1855: 145. Hartman 1881a: 181, 1881b: 10, 1885: 213. Garrett 1884: 47. Paetel 1889: 266. Smith 1903: 478. Pilsbry 1909: 259

Partula lineolata Pease, 1868a: 224

Partula lineolata Pease, Pease 1871b: 473. Schmeltz 1874: 92. Pfeiffer 1876: 206

Partula affinis filosa Pfeiffer, Kondo & Burch 1983: 105

Ovate-conical. Umbilicus open. Thick, surface with close spiral striae on all whorls, dull or slightly shiny. Spire obtuse. 5 flattened whorls, body whorl convex. Columella slightly plicate above. Aperture a little oblique, triangular-semioval. Lip slightly expanded, with a prominent thick callus within. Small parietal tooth present. A small tubercle is often present on the columella.

White (3%), dull yellowish with pink spire (6%), or red-brown with darker purplish apex (33%), brown (26%) or dark purple-brown with darker apex (32%). Numerous transverse white decorticated lines visible on all whorls. Lip white.

Anatomy: Penis short, $^1/_3$ oviduct length, tubular. Vas deferens apical (or rarely very slightly subterminal), broad.

Type specimens:

Partula diminuta C.B. Adams, 1851 – Holotype MCZ 279302 (labelled paralectotype but is the holotype). Originally in the Museum of Amherst College (Pilsbry 1909) but later transferred to MCZ.

Partula filosa Pfeiffer, 1853 – Syntypes NHMUK (un-numbered).

Partula lineolata Pease, 1867 – Lectotype ANSP 59933a (Baker 1963); paralectotypes AMNH 42156; ANSP 59933; MCZ 24933. Baker (1963) listed ANSP specimens as holotype and paratypes but this is in error and should be regarded as a lectotype designation.

Distribution: Only found in Pirae valley, commonest at lower elevations. Crampton (1916) recorded it from 375 ft upwards. By 1909 the lower parts had been cleared for agriculture. The upper limit was just above 550 ft. Described as "abundant on foliage" by Garrett (1884) and collected on "plantain (few), *Dracaena* (ti), ginger (opui), turmeric, and birds-nest fern (oaha)" by Crampton (1916), but Burch's 1970 specimens were mainly from the ground, some from near ground level to 2-3 ft, some up to 6 ft (BPBM 1970-199-2002).

Fig. 9.4.24 Crampton's (1916) *Partula diminuta* figures (edited for position)

Fig. 9.4.27 *Partula diminuta*: a-b – *filosa* syntypes (NHMUK); c-d – *lineolata* paralectotypes (AMNH 42156); e – *diminuta* holotype (MCZ 279302) f-l – colour forms (f-h – ANSP 295322; i-l – UMZC 75355; k – UMMZ 200).

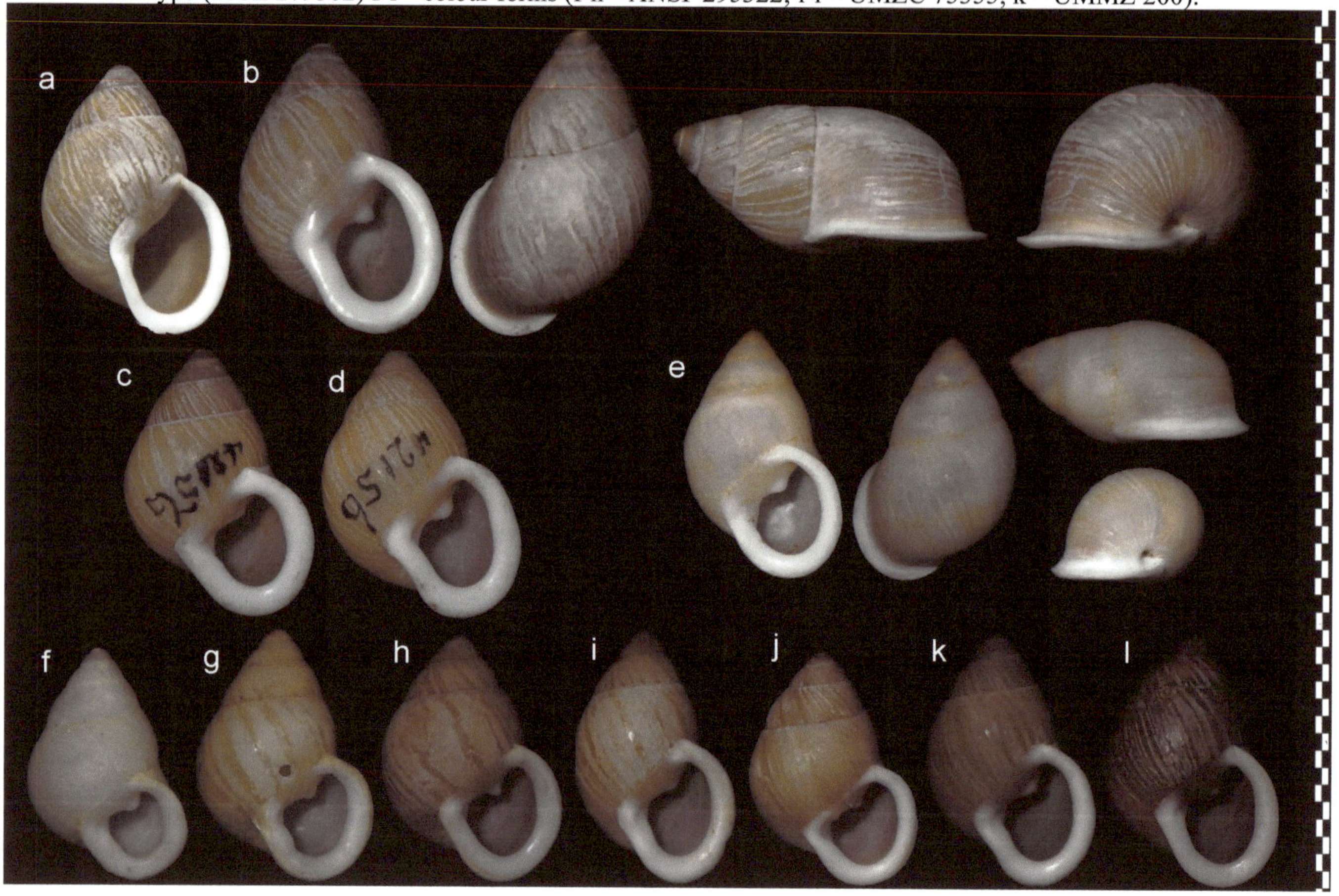

Fig. 9.4.28 Map of *P. nodosa* and *P. diminuta* ranges and variation, percentages given for the proportion of dextral shells, all other populations 100% dextral

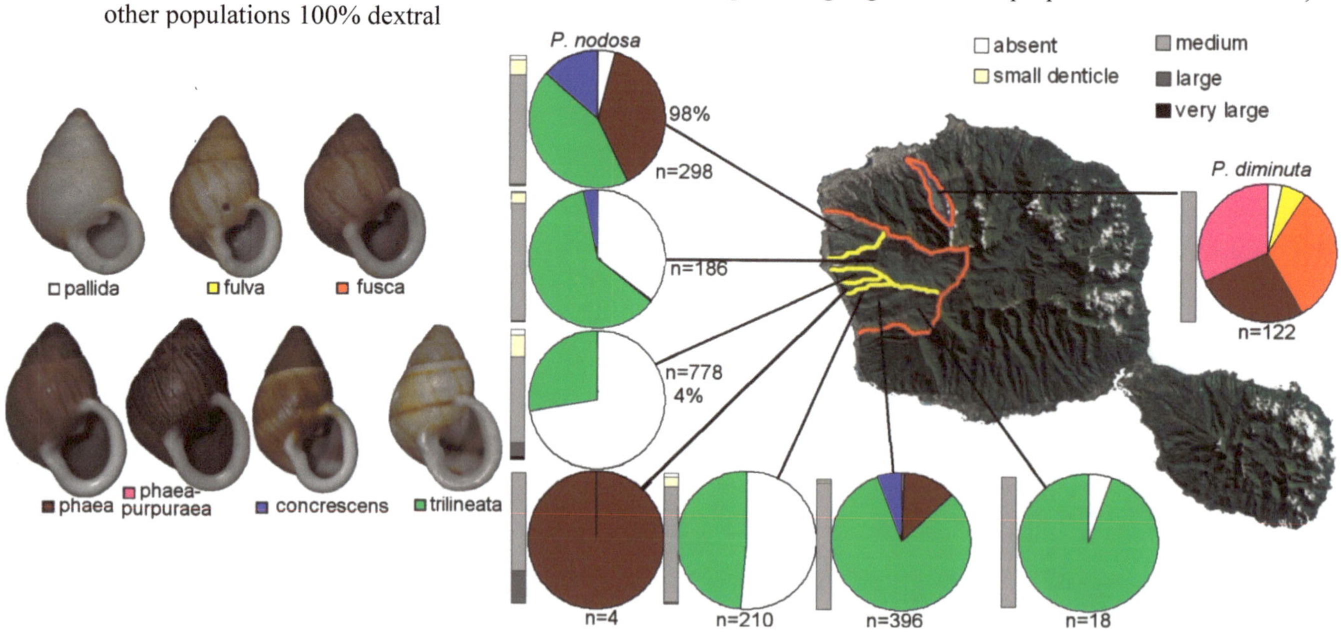

Partula clara Pease, 1865

Partula clara Pease, 1865: 671

Partula clara Pease, Pease 1871b: 473. Pfeiffer 1868: 159, 1877: 199. Hartman 1881a: 181, 1881b: 11, 1885: 216, 1892: 75. Paetel 1883: 152, 1889: 266. Garrett 1884: 56. Smith 1903: 445. Pilsbry 1909: 197. Crampton 1916: 57, 1925: 8. Kondo 1968: 76. Emberton 1982: 33. Kondo & Burch 1983: 104

Partula clara parva Crampton 1916: 62

Partula clara minor Crampton 1916: 65

Partula clara angusta Crampton 1916: 66

Partula clara prima Crampton 1916: 67

Oblong-ovate. Umbilicus open. Thin, translucent. Shining with fine growth lines and fine wavy spiral striae visible on all whorls. 4½ convex whorls. Body whorl compressed below periphery. Aperture ovate, slightly oblique. Lip reflexed, slightly thickened within except near ends. Columellar margin dilated above, extending over the umbilicus edge. Parietal callus very thin, shining.

Very pale brown, rarely dark brown; often strigated darker. Pale brown shells with red-brown bands)cestata) occur in some populations. Suture white margined. Aperture usually pale brown or pinkish inside, lip white.

Body light brown, mantle variably very light brown, whitish or light brown with white maculae (some Vaihiria individuals).

Anatomy: Penis short (shorter than buccal mass), ¼ oviduct length; slightly longer than in *P. hyalina*. Caecum usually separated from penis by a constriction, this is usually conspicuous but is occasionally very slight. Numerous pilasters in penis, restricted to basal ¾, only papillae in caecum. Vas deferens subterminal, usually medial on the caecum, except in Tereehia valley where it may be subapical (17%, n=12; Kondo & Burch MS, ANSP A890, BPBM 1970-178), this population rarely shows the constriction between penis and caecum (25%).

Type specimens:

Partula clara Pease, 1865 – Syntypes AMNH 42061; ANSP 59555; MCZ 24931, 25141.

Partula clara parva Crampton 1916 – No type defined, types would be ANSP 276077-9.

Partula clara minor Crampton 1916 – No type defined, types would be ANSP 276081-5 and A938C-D.

Partula clara angusta Crampton 1916 – No type defined, types would be ANSP 276086 and A890.

Partula clara prima Crampton 1916 – No type defined, types would be ANSP 276088-9, 276092-8 and A938E.

Distribution: Upper portions of the valleys of the west and southern parts of Tahiti-Nui. Recorded up to 600 m. Arboreal.

Notes: Crampton (1916) defined seven geographical forms to which he ascribed six names. Four are straightforward geographical variants: *parva* in the west (Punaruu to Papehue valleys), *minor* in the south-west (Orofere to Tiamao), *angusta* from one valley in the south-west (Tereehia), *prima* in the south (Otuna to Vaihiria, with a separate population labelled '*prima*?' from Papenoo). In addition he described *marmorata* from the north and east (Maara to Titaviri, and Tahiti-Iti) and

Fig. 9.4.29 *Partula clara*: a – syntype (AMNH 42061); b – *prima* (UMMZ 128); c – *prima* (Otuna ANSP 276088); d – *parva* (Panaruu ANSP 276077); e – *minor* (Orofere ANSP 276081); f – *angusta* (Tereehia ANSP 276086); g-h – Mt. Marau (ZSL); i-m – Crampton specimens (ANSP 272439)

incrassa from Tiapa valley. The latter is not a *P. clara* form, and is here recognised as a distinct species (see p. 153). Kondo & Burch (1983) demonstrated that *marmorata* is a form of *P. hyalina*. Crampton's *P. clara* varieties do not differ anatomically but do differ significantly in shell proportions: *angusta* and *minor* differ significantly in length, width and proportions (t=7.466, 3.795 and 4.419 respectively, $P<0.05$), *angusta* and *prima* differ in length (t=6.187, $P<0.05$).
Garrett (1884) regarded it as a rare species approaching extinction. Crampton (1916) thought that although *P. clara* was not common it had increased in abundance since Garrett and Meyer.

Fig. 9.4.30 Crampton's (1916) *Partula clara* and *P. hyalina* figures (edited for position). Rows 1-3 – *P. clara*, 4-5 *P. hyalina*

Fig. 9.4.31 *Partula clara* geographical variation showing Crampton's subspecies. Valleys with recorded populations shown as coloured dots (blue – dissected)

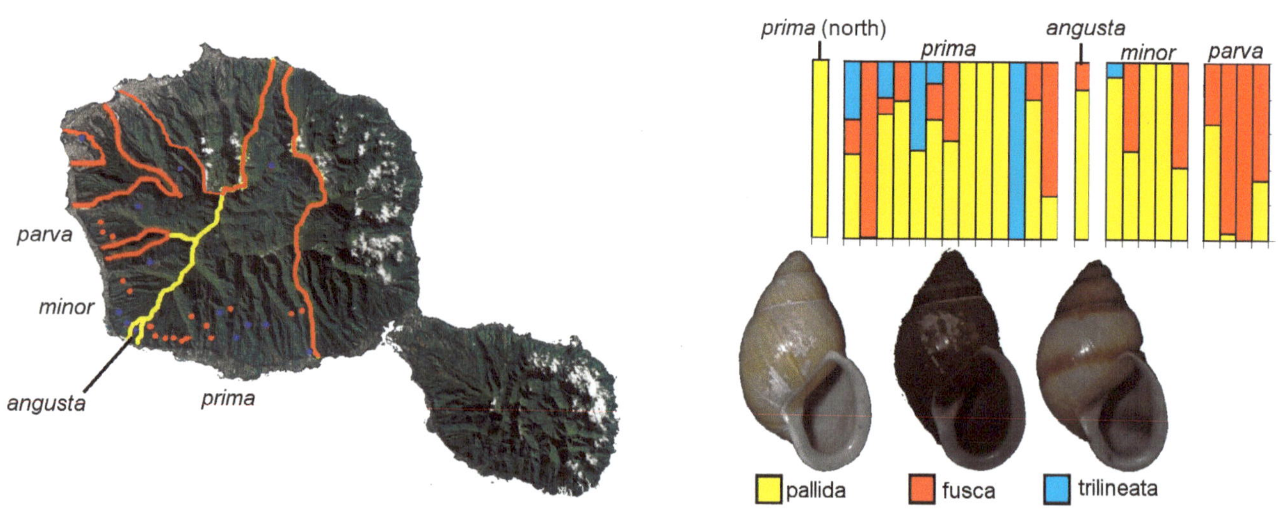

Partula hyalina Broderip, 1832

Partula hyalina Broderip, 1832: 32

Bulinus hyalinus Broderip, Deshayes & Milne-Edwards 1839: 371. Pfeiffer 1848: 67

Partula hyalina Broderip, Sowerby 1833: f 3, 1841: 5. Müller 1836: 32. Beck 1837: 57. Anton 1839: 40. Jay 1939: 57. Pfeiffer 1842: 116, 1848: 67, 1853b: 451, 1855b: 271, 1856: 162, 1859: 510, 1868: 159, 1877: 198. Reeve 1842: pl175 f1, 1849: pl 3 f 14. Catlow & Reeve 1845: 163. Albers 1850: 186. Shuttleworth 1854: 148. Adams 1855: 128. Pease 1864: 671, 1871b: 473. Frauenfeld 1869: 874. Paetel & Schaufuss 1869: 83. Schmeltz 1869: 72, 1874: 92. Paetel 1873: 104, 1883: 153, 1889: 267. Semper 1874: 158. Garrett 1879: 19, 1881: 396, 1884: 65. Kobelt 1880: 275. Hartman 1881a: 173, 1881b: 11, 1885: 216, 1892: 75. Pfeiffer & Clessin 1881: 301. Mayer 1902: 122. Smith 1903: 479. Pilsbry 1909: 180. Crampton 1916: 38. Böttger 1920: 295. Kondo 1955: 168, 1968: 76. Kilias 1969: 247. Emberton 1982: 33. Kondo & Burch 1983: 101. Richardson 1990: 34

Partula clara Pease, Kondo 1955: 166 (misidentification)

Partula clara marmorata Crampton 1916: 70

Elongately ovate. Thin but strong. Umbilicus open. Spire conic, apex acute. 5 very slightly convex whorls, body whorl convex, somewhat compressed laterally. Ornamented with fine spiral striae on all whorls. Aperture very oblique, ovate. Lip broadly reflexed, strongly thickened within. Columellar margin dilated above, extending over the umbilicus edge, and then receding at attachment to body whorl, giving a distinct angled outline, margin weakly nodose. Parietal callus thin.

White to very pale brown. Crampton (1916) noted it to be the only *Partula* 'universally devoid of colour'. The apex sometimes has a very faint reddish or brown tint. A light brown form (*marmorata*) exists in eastern populations, this is uniform except for a white lip, or may have a reddish sutural band at the apex.

Body pale grey-brown, mantle silvery grey outside, white with fine black mottling inside.

Anatomy: Penis short, ¼ oviduct length, no distinct caecum; vas deferens attached terminally.

Type specimens:

Partula hyalina Broderip, 1832 – Syntypes NHMUK (un-numbered); SMF 164967. These are the only types matching the type locality (NHMUK – 'Oheteroa, MC'; SMF – 'Rurutu, Austral islands, ex Cuming'). Supposed types of uncertain status are AMNH 42055 ('Austral islands') and NMB 1435a (no data). Supposed types with the wrong provenance data (and so unlikely to be types) are MCZ 297851 ('Society islands, Turnaco'), ZMUC 75633 ('Tahiti'). ZMB-MOL 74336 and FMNH 11357 (ex MCZ) are from Pease, and therefore not types.

Partula clara marmorata Crampton, 1916 – No type defined, types would be ANSP 276099-101, 276105.

Distribution: Recorded from most valleys on Tahiti. Introduced populations have been recorded in the Austral (Raivavae, Rimatara, Rurutu and Tubuai) and Cook islands (Aitutaki, Atiu, Mangaia, Mauke and Rarotonga). These are thought to be prehistoric introductions from around 30,000 years ago (Lee *et al.* 2007). The oldest historical records from the Australs are from 1828 (Rurutu 1828, Tubuai 1870s, Rimatara 1921, Raivavae 1922) and the Cooks from the 1870s (Mangaia

Fig. 9.4.32 *P. hyalina* (a & b) and *P. clara* (c-i) anatomical variation: a & d after Kondo (1955); b – various locations (Tahiti: ANSP 318706, A880, A937, A937A-B; MNHN Th40; NHMUK Cheeseman 1925; ZSL 539, 730, 751, 1124. Raivavae: MNHN Rv39); c – *angustata* possible hybrid Tereehia (ANSP A890); e & f – Papenoo (ZSL 1165); g – *angustata* Tereehia (ANSP A890); h – *parva* Tiamao (ANSP A938D) and *prima* Papeiti (ANSP A938E); i – *minor* Vaipuarii (ANSP A938C). Scale bar 5 mm.

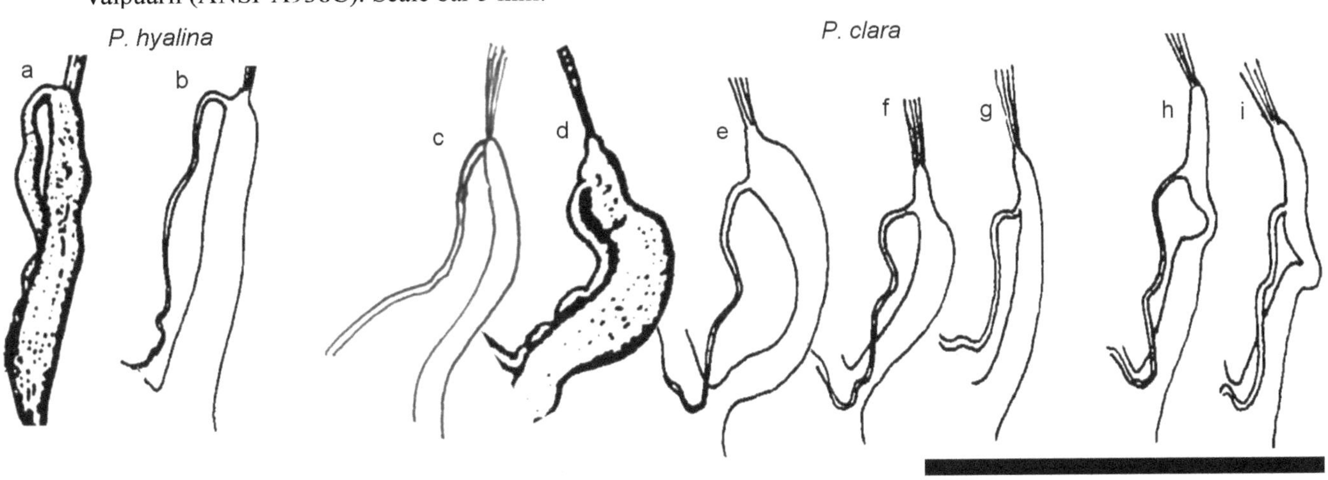

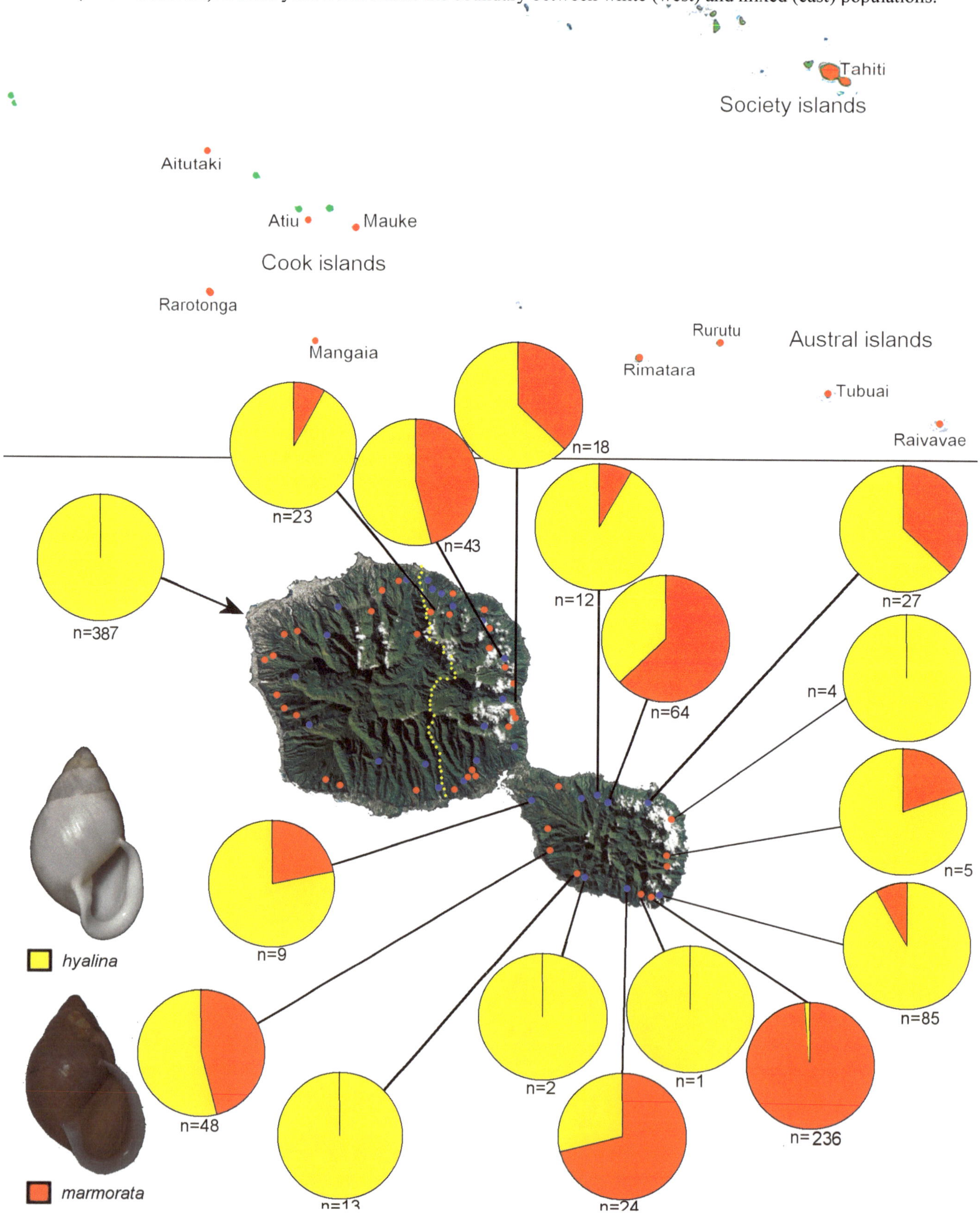

Fig. 9.4.33 *Partula hyalina* range and variation. Top: occupied islands (red) in Societies, Cooks and Australs. Bottom: distribution of *hyalina* and *marmorata* forms on Tahiti. Valleys with recorded populations shown as coloured dots (blue – dissected). Dotted yellow line marks the boundary between white (west) and mixed (east) populations.

1870s, Atiu and Mauke 1925). Arboreal; Crampton (1916) recorded it on *Musa* and *Colocasia* and noted that small snails were found on *Dracaena*. At this time it was most abundant in larger, drier valleys.

<u>Notes</u>: Kondo & Burch (1983) recognised that *P. clara marmorata* is form of *P. hyalina*, having a *hyalina* reproductive anatomy. Conchologically *marmorata* shells are difficult to distinguish from *P. clara* with confidence. Kondo & Burch (MS) considered a red-brown sutural line in the embryonic shells to be diagnostic of brown *P. hyalina* (Fig. 9.4.34), but this is not a constant feature.

Fig. 9.4.34 *Partula hyalina*: a-b – syntypes (NHMUK); c – Raivavae (MNHN Rv39); d – *marmorata* Titaviri (ANSP 276101); e – *marmorata* Vaitutaepuaa (MNHN Th40); f – *marmorata* Pierae (ZSL); g-i – apex colouration (g – *P. hyalina* Haavini; h – *marmorata* Haavini; i – *P. clara* Vaihiria)

Fig. 9.4.35 Live Tahitian *Partula*: a – *P. otaheitana*; b – *P. diminuta*; c – *P. nodosa*; d – *P. i. incrassa*; e – *P. cytherea* juveniles; f – *P. affinis*; g – *P. clara* & h – *P. hyalina*. Photos: a & f – J.-Y. Meyer; b – D. Clarke; c-e & g-h – T. Coote.

Moorean *Partula*

12 species have been described on Mooorea (of which one, *Partula solitaria*, is a synonym of *Samoana attenuata*). Seven are recognised as valid species here, although hybridisation is apparently extensive (see p. 245). These are the most studied of all *Partula*, with Crampton's (1932) monograph, the extensive genetic studies of Clarke, Murray and Johnson and the genetic analysis of Burch's samples (Lee *et al.* 2009).

Table 9.4.2 Measurements of Moorean *Partula* species: D – dextral, S – sinistral, bold – species totals

	N	L			W			L aperture			L/W		
		range	mean	sd.	range	mean	sd.	range	mean	sd.	range	mean	sd.
***suturalis* D**	7997	15.0-20.8	18.92	1.030	8.5-11.5	10.51	0.496	7.5-10.7	9.66	0.547	47.5-64.5	55.58	2.206
***suturalis* S**	4671	18.9-25.0	19.04	1.072	10.3-13.0	10.55	0.662	9.9-12.9	9.75	0.563	45.5-64.5	55.37	2.309
s. suturalis D	3290	15.1-20.8	17.68	0.852	8.5-11.5	9.99	0.419	7.5-10.7	9.15	0.612	46.5-57.5	54.95	2.222
s. vexillum D	4707	17.4-22.9	19.78	1.154	9.7-12.7	10.87	0.550	8.5-11.7	10.02	0.563	45.5-64.5	55.37	2.309
s. vexillum S	4772	18.9-23.5	19.05	1.072	10.3-13.0	10.55	0.662	9.9-12.9	9.75	0.367	47.5-64.5	54.06	2.025
aurantia	1380	15.1-21.1	18.48	0.839	8.5-11.3	10.00	0.421	7.7-10.7	9.40	0.367	47.5-61.5	54.06	2.025
***mirabilis* D**	4189	14.1-19.9	16.62	0.776	8.1-11.1	9.62	0.407	7.3-10.5	8.67	0.409	49.5-65.5	57.85	2.111
***mirabilis* S**	7	14.4-19.5	16.17	0.747	8.3-11.0	9.73	0.328	7.4-10.4	8.67	0.361	49.3-66.0	59.79	2.711
***taeniata* D**	28539	12.3-20.2	15.80	1.002	7.3-11.5	9.21	0.564	6.5-10.9	8.52	0.598	46.5-70.5	58.29	3.565
***taeniata* S**	1	-	13.90	-	-	9.70	-	-	7.50	-	-	70.00	-
incrassa	172	12.6-18.3	15.31	0.911	7.1-9.6	8.63	0.436	7.3-9.7	8.45	0.567	48.5-66.5	57.12	2.378
i. exigua	109	12.6-17.45	15.23	0.875	7.5-9.7	8.64	0.396	7.3-9.7	8.48	0.480	51.5-64.5	56.70	2.463
mooreana	5535	13.8-20.8	17.31	1.183	7.9-11.5	9.89	0.543	6.9-11.5	9.34	0.622	48.5-66.5	57.16	2.432
tohiveana	587	17.4-24.5	21.58	0.970	10.1-13.7	12.10	0.492	8.9-12.5	11.05	0.473	47.5-63.5	56.06	2.061

Fig. 9.4.36 Moorean *Partula* anatomy. After Kondo (1955), scale bar 5 mm.

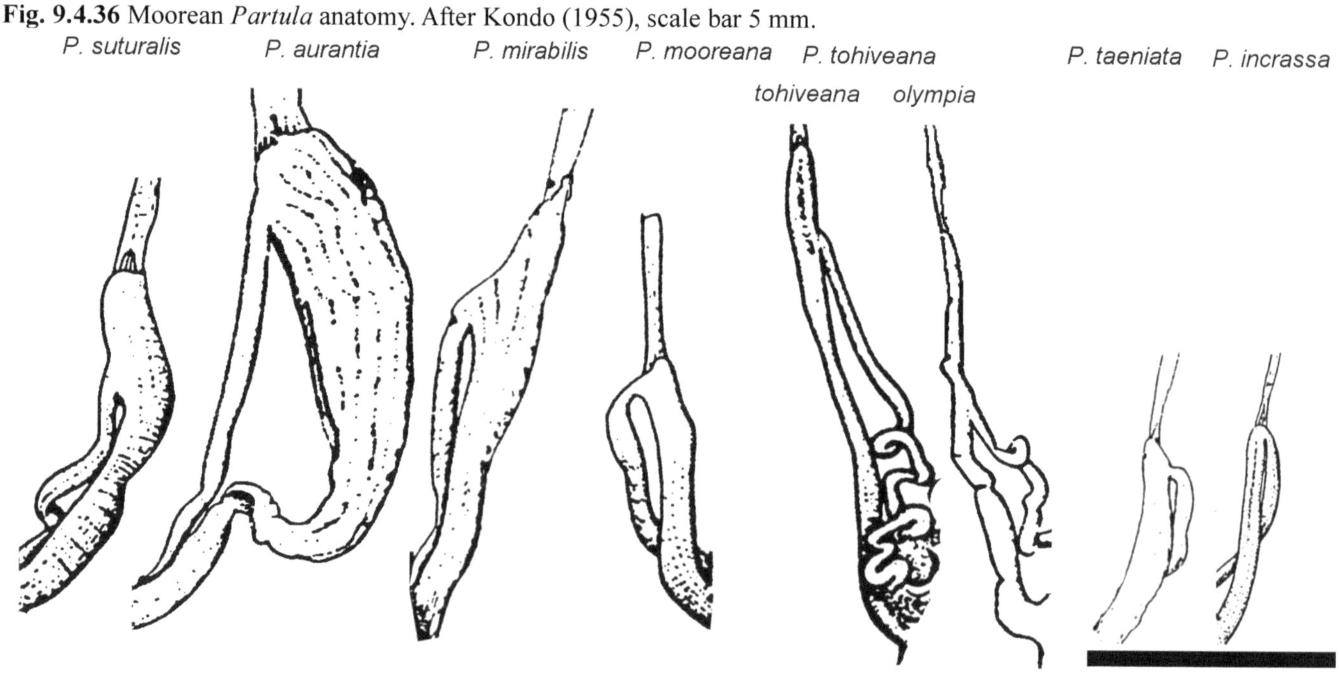

Partula suturalis Pfeiffer, 1855

Partula lineata Lesson, Reeve 1850: pl2 f7 (misidentification). Hartman 1881a: 189, 1885: 215. Garrett 1884: 50

Partula suturalis Pfeiffer, 1855: 98

Partula suturalis Pfeiffer, Pfeiffer 1856: 162, 1856: 62, 1859a: 508; 1859b: 62, 1868: 157, 1877: 196. Hartman 1881a: 179, 1881b: 10, 1885: 215, 1885: 10, 1892: 75. Garrett 1884: 50. Paetel 1889: 268. Smith 1903: 446. Pilsbry 1909: 201. Crampton 1924b: 558, 1925: 15, 1932: 91. Kondo 1968: 76. Clarke & Murray 1969: 31. Lipton & Murray 1979: 129. Murray & Clarke 1980: 101. Emberton 1982: 33. Johnson 1982: 145. Murray *et al.* 1982: 316. Clarke *et al.* 1984: 101. Richardson 1990: 57

Partula strigosa Pfeiffer, 1856: 244

Partula strigosa Pfeiffer, Pfeiffer 1857c: 384, 1859: 509, 1868: 159, 1877: 198. Kobelt 1880: 16. Hartman 1881a: 187, 1881b: 10, 1885: 215. Paetel 1883: 153, 1889: 268. Smith 1903: 447

Partula alternata 'Pease' Carpenter in Pease 1864: 675 (as syn. of *nodosa*, nomen nudum)

Partula nodosa Pease (non Pfeiffer), 1864: 67. Schmeltz 1874: 337. Gloyne 1878: 337. Garrett 1884: 50. Hartman 1885: 215

Partula vexillum Pease, 1865: 675 (nomen nudum)

Partula vexillum Pease, Pease 1866: 198, 1867: 81, 1871: 473. Pfeiffer 1877: 196. Gloyne 1878: 337. Hartman 1881a: 195, 1881b: 10, 1885: 215. Garrett 1884: 50. Paetel 1889: 268. Smith 1903: 446

Partula ornata 'Anthony', Smith 1903: 448 (nomen nudum)

Partula suturalis strigosa Pfeiffer, Pilsbry 1909: 204. Crampton 1932: 118. Kondo 1968: 76. Clarke & Murray 1969: 31

Partula suturalis vexillum Pease, Pilsbry 1909: 204. Crampton 1932: 100. Kondo 1955: 150, 1968: 76. Clarke & Murray 1969: 31

Partula dendroica Crampton, 1924a: 114

Partula dendroica Crampton, Crampton 1932: 47. Kondo 1955: 146

Dextral or sinistral, varying geographically. Oblong-conic. Umbilicus narrow. Spire with spiral striae, fading on body whorl, but may persist around the umbilicus, rarely on all whorls. 5-5½ slightly convex whorls, the last slightly shorter than the spire, somewhat swollen above, tapering towards the base. Suture slightly raised. Aperture slightly oblique, truncate-oblong. Lip thin, narrowly expanded, sinuated above, columellar margin dilated, reflexed and thickened within. Columella almost vertical, swollen or inflated just below the insertion. Parietal callus very thin; generally no parietal tooth, but when developed it is a low callous nodule, only forming a distinct tooth in *vexillum*.

Colour highly variable; from pale yellowish to black with a variable number of bands (0-6). Spire may be pinkish. Suture margined with a white line. Lip white. Of Crampton's 14,766 specimens (including '*P. dendroica*') 65% were frenata, 10% cingulata, with rare apex, fusca, strigata, phaea, atra, unilineata, lyra, zonata and bisecta forms.

Anatomy: Penis elongate, $^1/_3$ oviduct length, club- or bat-shaped. Caecum well developed, half as long as penis, inflated. Vas deferens submedian.

Type specimen:

Partula suturalis Pfeiffer, 1855 – Syntypes NHMUK un-numbered.

Partula strigosa Pfeiffer, 1857 – Syntypes NHMUK un-numbered.

Partula vexillum Pease 1866 – Lectotype ANSP 59646; paralectotypes ANSP 464353 MCZ 24814, 24886, 24936. Lectotype

Fig. 9.4.37 Moorean *Partula* species: a – *P. aurantia*; b – *P. suturalis*; c – *P. mirabilis*; d – *P. taeniata*; e – *P. incrassa*; f – *P. mooreana*; g – *P. tohiveana*

designation by Baker (1963). AMNH 42141 labelled 'type' is a Garrett specimen, not a true type.

Partula alternata (Pease, MS) Smith, 1903 – Syntypes MCZ 25069.

Partula dendroica Crampton, 1924 – ANSP 272252 labelled types but no types defined, described based on all specimens collected by Crampton prior to 1923. ANSP 272252 are the specimens illustrated in the original description. Other specimens collected by Crampton prior to 1923 can also be regarded as syntypes: ANSP 272252, 272757, 279072-6, 279070-1; BPBM 92317-21; NHMUK 1935.10.15.21-22.

Distribution: Moorea – widespread in the upper parts of valleys, rarely down to 40 m (Murray & Clarke 1980). The most arboreal Moorean *Partula*, found up to 15 m from the ground, often on *Hibiscus tiliaceus*.

Subspecies: Three subspecies have been described, there are no differences in reproductive anatomy and they do not form clear mitochondrial genetic groups. However, the south-eastern populations are visibly distinct from the others and two

Fig. 9.4.38 Geographical variation in *P. suturalis* anatomy. 10 snails dissected in each sampled valley; valleys shaded by dominant anatomical form (grey – unsampled valleys). Material collected by Clarke & Murray in 1962 (NHMUK).

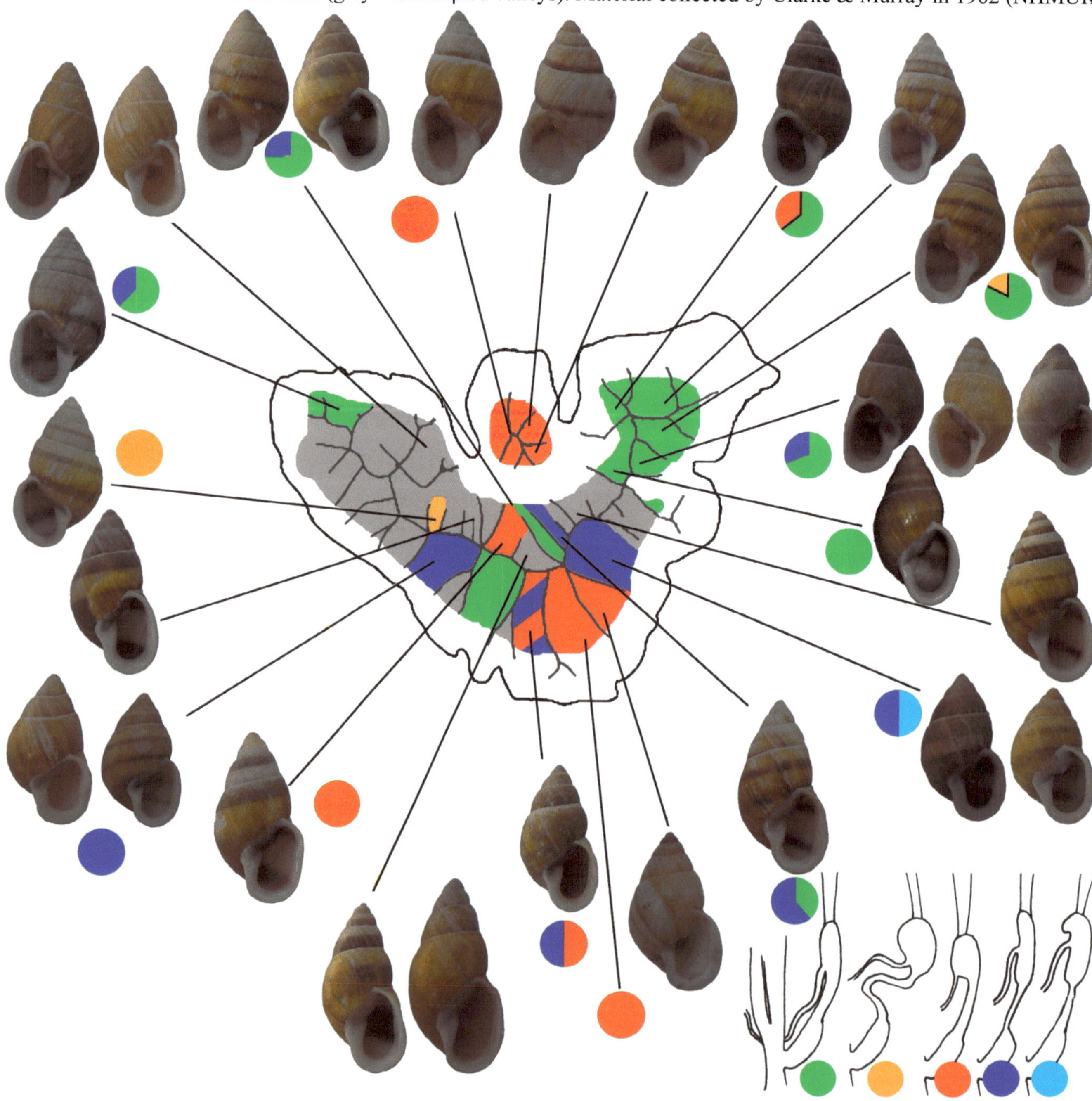

subspecies are retained here.

Partula suturalis suturalis (Pfeiffer, 1855) **comb. nov.** – south-eastern valleys from Maata to Atiraa (Murray & Clarke 1980). Usually dextral. Oblong-conic. Whorls nearly flat. Aperture slightly oblique, truncate-oval. Weak parietal tooth.

Partula suturalis vexillum (Pease, 1866) – all Moorean populations except for the south-eastern valleys. Usually dextral but chirality varying geographically. More elongate than *P. s. suturalis*, whorls more convex, parietal tooth more developed. Includes '*Partula suturalis dendroica*' of Mt. Rotui.

Fig. 9.4.39 *Partula suturalis*: a – *suturalis* syntype (NHMUK); b-c – *strigosa* syntypes (NHMUK); d – *vexillum* syntype (ANSP 59646); e – *dendroica* topotype (NHMUK); f-m, q-v – *vexillum* Crampton's illustrated variation (ANSP 272764); n-o – *dendroica* Crampton's illustrated variation (ANSP 272757)

Note: Pfeiffer's (1855) *P. suturalis* is indistinguishable from *P. strigosa*, accordingly '*P. suturalis strigosa*' should be corrected to *P. s. suturalis*.
Genetically there are two clear groups; a distinct *suturalis* clade related to *P. mirabilis* and a north to north-eastern group which mixes *suturalis, vexillum* and *'dendroica'* (Lee *et al.* 2009). This appears to indicate extensive gene flow in this clade with *suturalis* invading the *vexillum* clade from the eastern edge of its range, leading to a breakdown of the subspecies.
Garrett (1884) considered Lesson's *Partula lineata* to belong to this species, but subsequently (letter to Hartman, 1884) regarded it as a 'lost' species of unknown identity. Here it is considered a synonym of *P. otaheitana*.

Fig. 9.4.40 Crampton's (1932) *P. suturalis* figures (edited for position)

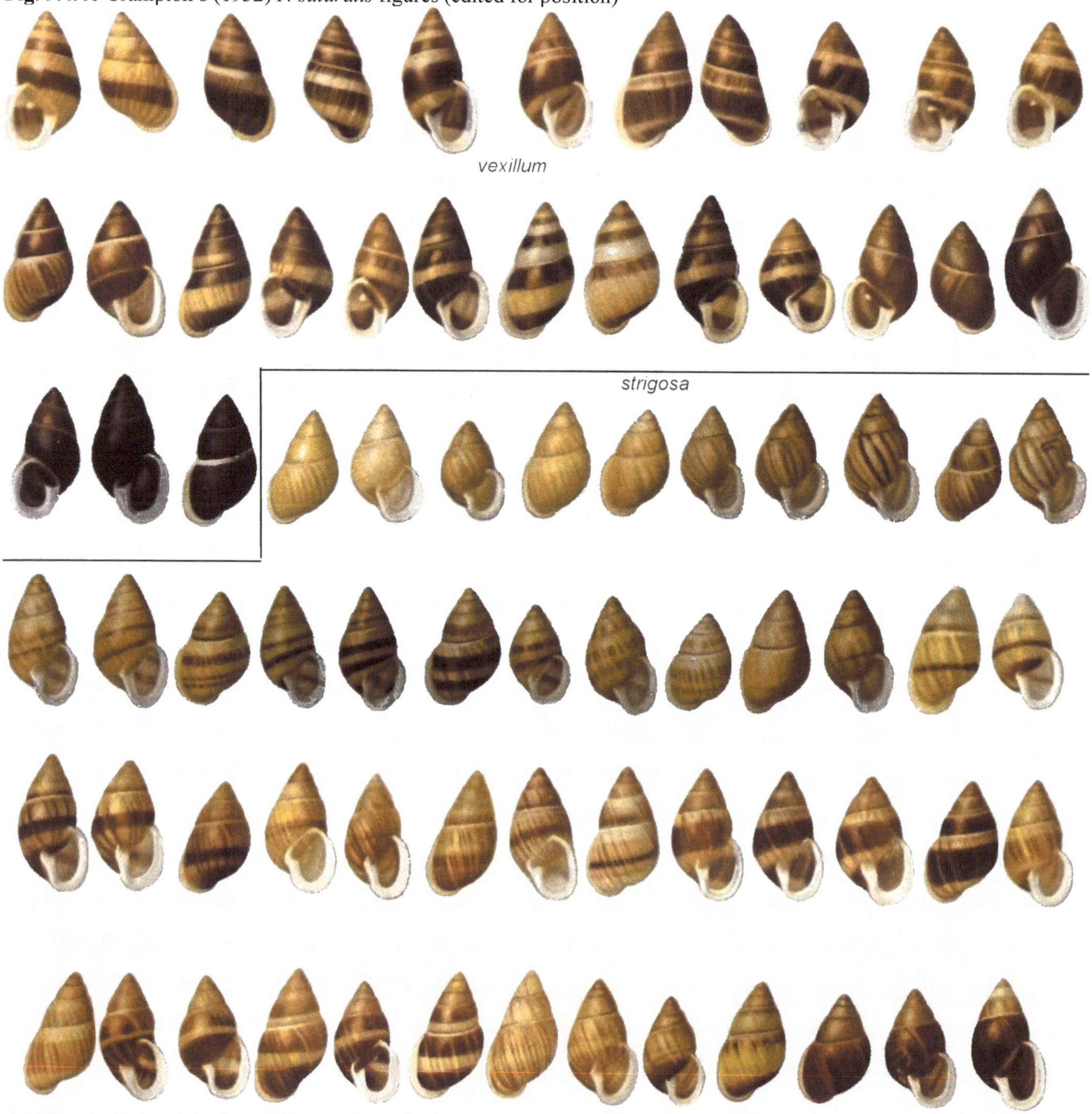

Fig. 9.4.40 (continued)

Partula aurantia Crampton, 1932

Partula aurantia Crampton, 1932: 54

Partula aurantia Crampton, Kondo 1955: 147, 1968: 76. Clare & Murray 1969: 31. Murray & Clarke 1980: 98. Clarke *et al.* 1984: 101. Richardson 1990: 11

Predominantly dextral (only 0.1% sinistral, n=1,600). Elongate-conic. Umbilicus open. 5-5½ slightly convex whorls. Spiral striae moderately distinct on spire, fading on body whorl, still visible around umbilicus (strong on all whorls in one specimen from Faamariri). Suture not impressed. Top of columella shallowly grooved. Aperture relatively small and

Fig. 9.4.41 *Partula suturalis* geographical variation. Valleys shaded according to chirality: red – >80% dextral, green – mixed, blue – >80% sinistral. The range of the subspecies *P. s. suturalis* and *P. s. vexillum* are marked. Based on Crampton (1932) data, n=15,049.

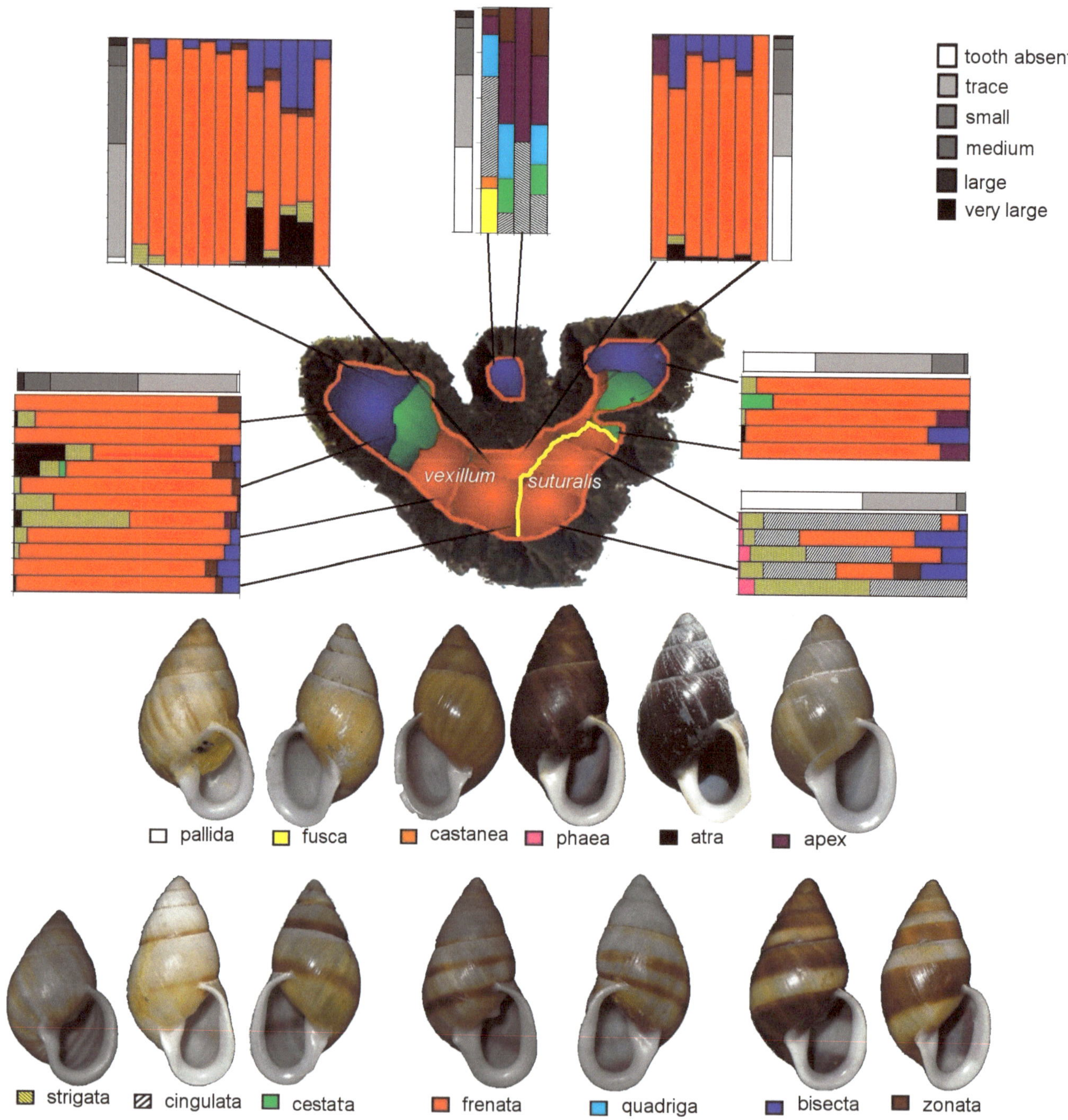

narrowly oblong, though highly variable. Lip bevelled, thickened internally. Columella vertical, plicate. Parietal callus thin, shining; parietal tooth generally absent or very weakly developed.
Commonly orange-brown, but may be banded or wholly dark brown. Crampton's 1,642 specimens were 65% lyra, 33% fusca and also phaea and zonata. Although not listed by Crampton, some shells are frenata.
Body: "it is interesting to note that the foot of *aurantia* presents a characteristic form, and it is also curiously pigmented along a zone a little within the margin, in a way that is unique." (Crampton 1932)

Fig. 9.4.42 *Partula aurantia*: a – syntype (NHMUK); b-c – Paraoroa valley (ANSP 279054); d-f – Crampton's illustrated series of colour forms (ANSP 272758)

Fig. 9.4.43 Crampton's (1932) *P. aurantia* figures (edited for position)

Anatomy: Penis very long, approximately oviduct length; caecum longer than penis (1.5-2 times penis length), club-shaped. Interior of penis with few pilasters, strongly rugose with 8-10 ridges in caecum, ridges visible externally. Penial retractor attaches subterminally. Vas deferens terminal.
Type specimens: Syntypes ANSP 272758, 279053-4, 279055-62; BPBM 92329-31, 9993; NHMUK 1935.10.15.9-13. Only BPBM 92329-31 are labelled types but Crampton did not define any specific type specimens and based his description on all specimens collected by him, thus all Crampton specimens should be regarded as syntypes.
Distribution: Northeastern Moorea: Paraoro, Paraoro valleys of the inner circle and Vaipohe and Faamaariri of the outer. The two areas being separated by a low ridge (330 m). Some association with *Freycinetia* sp. (Murray & Clarke 1980).
Garrett did not find this species, presumably he collected too low down in the valleys it occupied. Crampton (1932) found it to be restricted to high parts of valleys in 1907-19, but in the Paparoa valley in 1923 he found it in the valley floor "scarcely 25 feet above sea-level". He did not detect and change in the environment over this time and though this evidence of an expansion of the range of the species. Variations in abundance in other valleys he discounted as being due to sampling in different areas in different years. He considered it to be expanding from its origin "northern Paparoa, at mid-level, not very far from the low saddlelike divide between Paparoa and Paraoro to the north."

Partula mirabilis Crampton, 1924

Partula mirabilis Crampton, 1924a: 116

Partula mirabilis Crampton, Crampton 1925: 26, 1932: 64. Kondo 1955: 149, 1968: 76. Clarke & Murray 1969: 31. Murray & Clarke 1980: 108. Murray *et al.* 1982: 316. Richardson 1990: 41

Partula taeniata propinqua Crampton 1932: 171 (partim)

Partula mirabilis propinqua Murray & Clarke 1980: 108 (partim)

Dextral (sinistral in 0.26% of Crampton's specimens, n=4,631); ovate-conic to elongate-conic. Umbilicus narrow. 5 slightly convex whorls. Body-whorl flattened towards aperture. Suture of the last whorl impressed. Surface shiny with spiral striae, fewer on last whorls. In some areas the sculpture is indistinct (Mouaroa and Poua) or only distinct around the umbilicus (Fareaito, Faataofe, Tefeo and Opohuno) or spire (some from Maramu). Aperture narrowed, elongated, almost oblong in general outline. Lip sharply bevelled, thin and smooth. A thin, roughened callus present between the insertions of the lip. Parietal tooth almost always present, but rarely prominent.

Fig. 9.4.44 *Partula aurantia* geographical variation

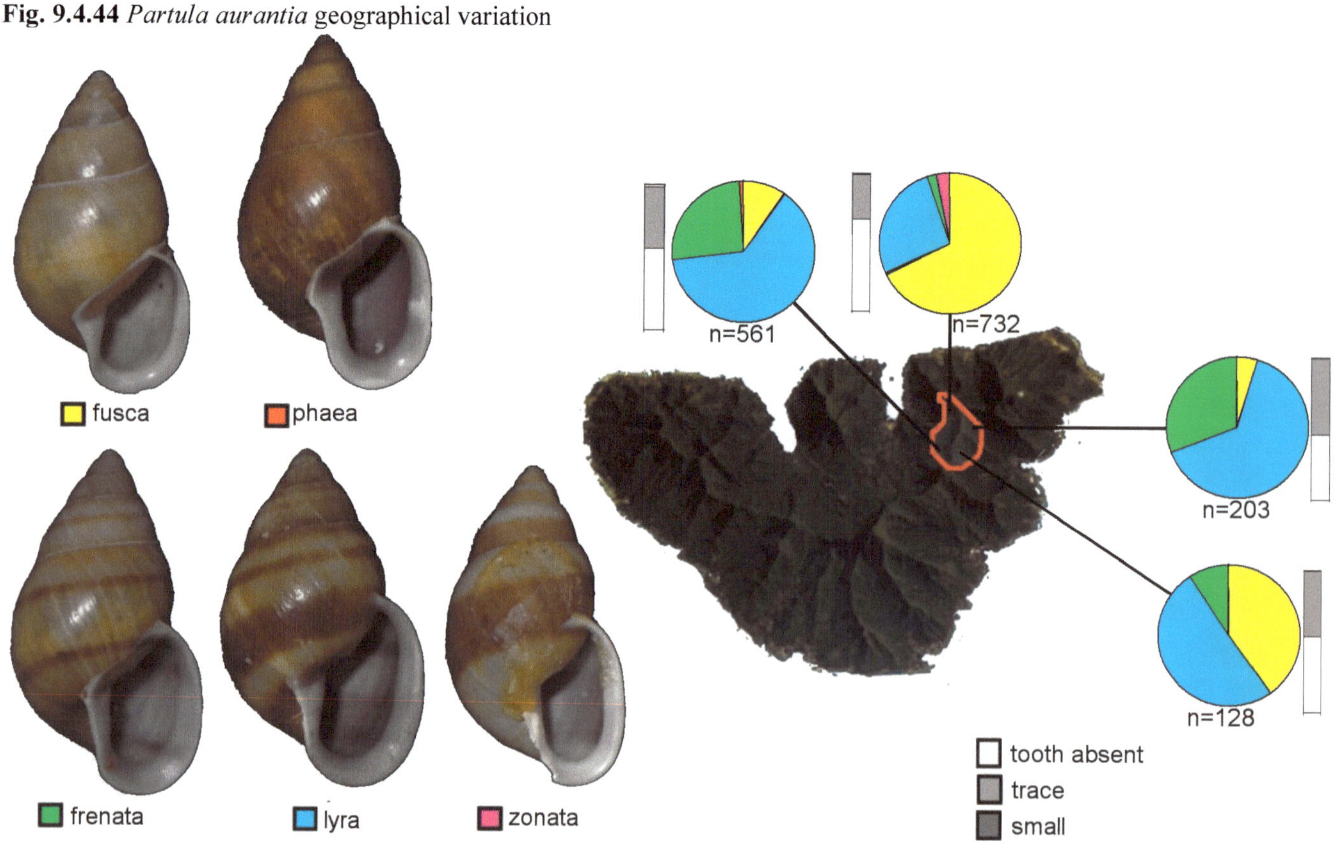

Basic shell colour light to dark brown with varying patterns. Crampton (1924a) recorded 31% of his 5,001 specimens as the lyra form, 24% phaea, 18% apex, 10% zonata. He also recorded pallida, fusca, mitella, fulva and bisecta, noting that this species showed more colour variation than any other Moorean species. There are also some cestata.

Anatomy: Penis long, muscular, ½ oviduct length. Caecum large, broad, ½ penis length, slightly constricted between penis and caecum.

Type specimens: Syntypes ANSP 268328, 268333-5, 268337-41, 269113-25, 269127-8, 279041-52; BPBM 92322-7; NHMUK 1935.10.15.1-8. Only BPBM 92322-7 are labelled as types (incorrectly as 'paratypes'), but Crampton did not designate specific types and based his description on all his specimens collected between 1907 and 1923.

Fig. 9.4.45 *Partula mirabilis*: a – syntype (NHMUK); b-h – Crampton's illustrated varieties (ANSP 272760)

Fig. 9.4.46 Crampton's (1932) *P. mirabilis* figures (edited for position)

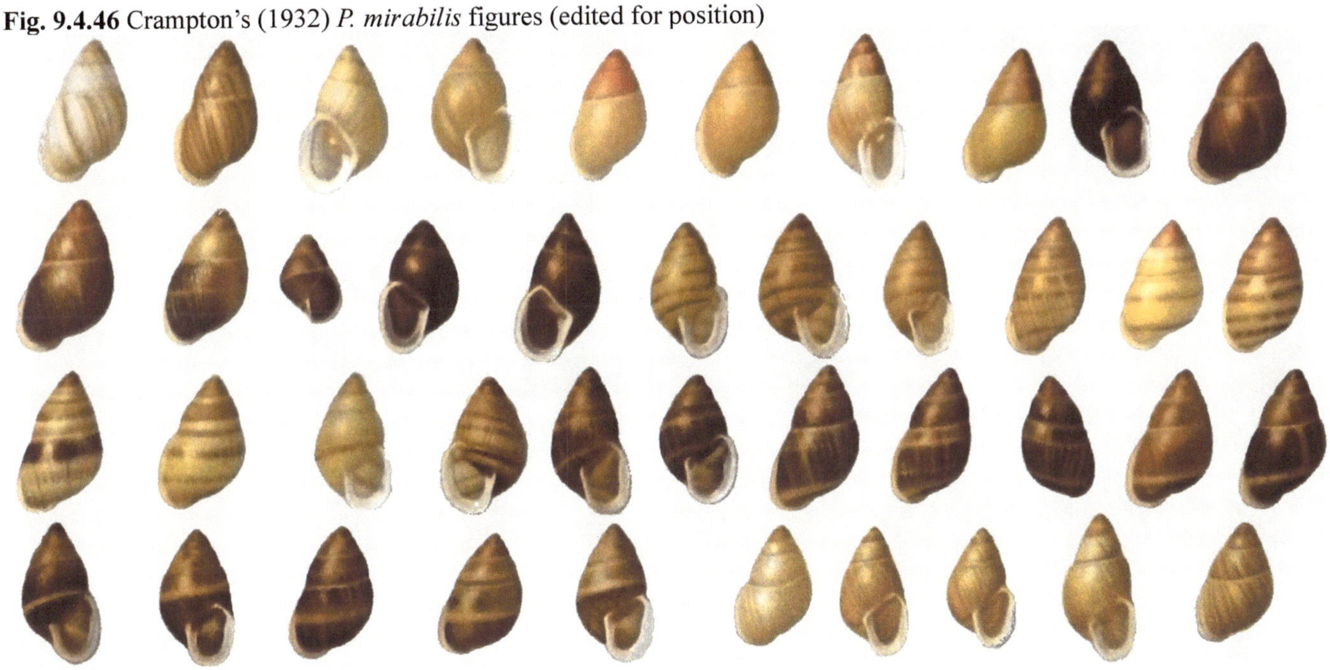

Distribution: Central mountains of Moorea, distributed over both sides of the main crater rim.
Notes: A population of partulids from the south side of Mt. Rotui has been referred to as *P. taeniata propinqua* (Crampton 1932) and *P. mirabilis propinqua* (Murray & Clarke 1980). This is a hybrid form, see p. 152.

Partula taeniata (Mørch, 1850)

Bulimus (*Partulus*) *tæniatus* Mørch, 1850: 29

Partula spadicea Reeve, 1850: pl4 f24

Partula taeniata (Mørch), Pfeiffer 1853b: 451, 1856: 162, 1859: 509, 1868: 157, 1877: 198. Adams 1855: 145. Carpenter 1864: 675. Hartman 1881a: 188, 1881b: 11, 1885: 215, 1892: 75. Garrett 1884: 72. Paetel 1889: 268. Smith 1903: 448. Pilsbry 1909: 205. Crampton 1925: 12, 1932: 130. Clarke & Murray 1969: 21. Lipton & Murray 1979: 129. Murray & Clarke 1980: 107, 1984: 343. Emberton 1982: 33. Murray *et al.* 1982: 316. Clarke *et al.* 1984: 101. Richardson 1990: 58

Partula spadicea Reeve, Pfeiffer 1853b: 451; 1856: 162, 1857f: 273, 1859: 509, 1868: 157, 1877: 198. Frauenfeld 1869: 874. Schmeltz 1869: 72. Paetel 1873: 104, 1883: 153, 1889: 268. Garrett 1884: 73

Partula erhelii Morelet, 1853: 371

Partula erhelii Morelet, Pfeiffer 1859a: 509, 1868: 157, 1877: 198. Pease 1871: 473. Paetel 1873: 104, 1883: 153, 1889: 266. Hartman 1881a: 182, 1881b: 11, 1885: 215. Smith 1903: 478. Pilsbry 1909: 212. Crampton 1925: 12. Richardson 1990: 22

Partula decussata Carpenter (non Pfeiffer), 1865: 675 (ex errore)

Partula lineata Carpenter (non Lesson) 1865: 676

Partula striolata 'Pease' Carpenter in Pease 1865: 675 (as syn. of *taeniata*, nomen nudum)

Partula simulans 'Pease' Carpenter in Pease 1865: 675 (as syn. of *taeniata*, nomen nudum)

Partula elongata Pease, 1866: 196

Partula striolata Pease, 1866: 197

Partula simulans Pease, 1866: 202

Partula striolata Pease, Pease 1867: 81, 1871: 473. Pfeiffer 1877: 203. Gloyne 1878: 338. Paetel 1883: 153, 1889: 268

Fig. 9.4.47 *Partula mirabilis* geographical variation, colour variation (main pie charts) and tooth development (bars) based on data from Crampton (1932), anatomical variation (small pie charts and penis shapes shown bottom right) from material collected by Clarke & Murray in 1962 (NHMUK), 10 specimens dissected at each point

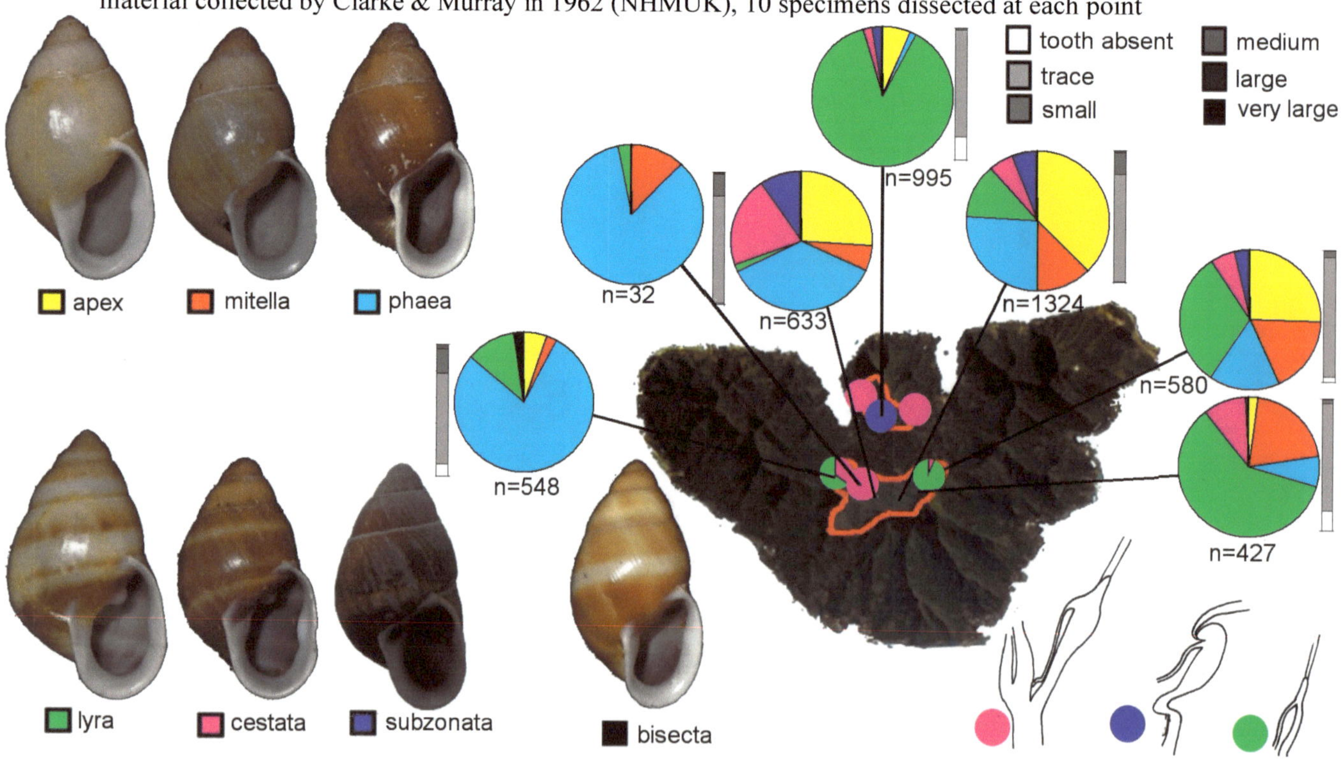

Partula simulans Pease, Pease 1867: 81. Schmeltz 1869: 72, 1874: 92. Paetel 1873: 104, 1883: 153, 1889: 268. Pfeiffer 1877: 206. Gloyne 1878: 338

Partula elongata Pease, Pease 1867: 81, 1871: 473. Schmeltz 1874: 72. Pfeiffer 1877: 196. Paetel 1883: 153, 1889: 266. Garrett 1884: 68

Partula nucleola 'Pease', Schmeltz 1869: 72, 1874: 92. Paetel 1873: 104. Pfeiffer 1877: 209 (all nomen nudum). Garrett 1884: 72 (form of *taeniata*)

Partula concinna Pease, 1872: 196

Partula gracilior 'Pease', Hartman 1881a: 183 (as syn. of *gracilis*, misidentified)

Partula concinna Pease, Hartman 1886: 35

Partula corneola Hartman, 1886: 32

Partula corneola Hartman, Hartman 1892: 75. Smith 1903: 454

Partula peraffinis 'Pease', Pfeiffer 1877: 197 (nomen nudum)

Partula taeniata spadicea Reeve, Pilsbry 1909: 207

Partula taeniata elongata Pease, Pilsbry 1909: 210. Crampton 1932: 139. Kondo 1955: 151, 1968: 76. Clarke & Murray 1969: 31

Partula taeniata simulans (Pease), Pilsbry 1909: 205. Crampton 1932: 132. Clarke & Murray 1969: 31

Partula taeniata striolata (Pease), Pilsbry 1909: 205. Crampton 1932: 132. Clarke & Murray 1969: 31. Clarke *et al.* 1984: 101

Partula taeniata spadicea (Reeve), Pilsbry 1909: 207. Crampton 1932: 131. Clarke & Murray 1969: 31

Patula taeniata nucleola 'Pease', Pilsbry 1909: 209. Crampton 1932: 160. Clarke & Murray 1969: 31

Partula taeniata propinqua Crampton, 1932: 170 (partim)

Partula mirabilis propinqua (Crampton) Clarke & Murray 1969: 31 (partim)

Almost always dextral, only rare aberrant sinistrals recorded. Narrowly ovate to elongately oval, or conical. Umbilicus narrow. Thin; slightly translucent. Surface dull or slightly shiny, with dense, wavy growth lines and fine spiral striae on all whorls. These may be indistinct but are always present and usually strongly marked. Suture appressed. 5 slightly convex whorls, the last about as long as the spire, obsoletely angular above and below, often compressed near mouth. Fold of the columella compressed, twisted, receding above abruptly, giving a nodular appearance. A parietal tooth is rarely present (1% according to Garrett [1884] but not recorded subsequently). Lip broadly expanded, reflexed.

Colour varies from white to dark brown, there may be darker striations or bands, which may be either dark or pale. Lip white. Of Crampton's 32,059 specimens he recorded 34% fusca, 27% fulva, 24% pallida, also phaea-purpurea, frenata, lyra and zonata.

Anatomy: Penis short (⅓-¼ oviduct length), slender, subcylindrical. Caecum short, indistinct (¼ penis length). Penis weakly papillate in upper ¼ with 5-6 strong pilasters at base, the main one intruding deep into caecum. Vas deferens subapical. Egg shell membranes uncalcified in populations sympatric with *P. incrassa*.

Type specimens:

Bulimus (*Partulus*) *tæniatus* Mørch, 1850 – Holotype ZMUC-GAS-1985. ZSM/Moll-20020055.10 is wrongly labelled a paratype.

Partula spadicea Reeve, 1850 – Holotype NHMUK un-numbered. Supposed syntype USNM 408590 is ex Pease collection as is unlikely to be a true type.

Partula corneola Hartman, 1886 – Holotype CMNH 62.4242.

Partula elongata Pease, 1866 – Lectotype ANSP 59533; paralectotypes AMNH 42037; ANSP 2626, 464475, 59533; MCZ 24899, 24921, 25127, 25294; SMF 164971 ('paratype'). Lectotype designation by Baker (1963).

Partula simulans Pease, 1866 – Lectotype ANSP 59941; paralectotypes AMNH 42090; ANSP 59538; MCZ 24881, 24955, 25116; NHMUK; SMF 4950. Lectotype designation by Baker (1963).

Partula striolata Pease 1866 – Lectotype ANSP 59648; paralectotypes AMNH 42134; ANSP 464474; MCZ 24910, 25293, 25303. Lectotype designation by Baker (1963).

Partula concinna Pease, 1871 – Holotype CMNH 62.4244.

Partula nucleola Garrett, 1884 – Lectotype ANSP 59531; paralectotypes ANSP 2667, 59532, 59943, 463795; MCZ 24897, 24937, 50854; NHMUK. Lectotype designation by Baker (1963).

Partula taeniata var. *multicolor* (label name) – Supposed syntypes of label name BPBM 87513; ZSM 110381.

Partula gracilior 'Pease' Hartman, 1881 – Supposed type of nomen nudum MCZ 25059.

Partula taeniata propinqua Crampton, 1932 – Syntypes ANSP 279042, 282832-3. No types designated, any Crampton specimens of this form from Vaihiaaiia, Matapoopoo and Tepatu would be types.

Distribution: All Moorean valleys. Arboreal on shrubs up to 5 m above ground (Murray *et al.* 1993). "occurs in prodigious numbers on the foliage of bushes" (Garrett 1884).

Geographical variation: Six subspecies were considered valid by Crampton (1932) (as 'varieties'):

spadicea Reeve, 1850 – Mt. Rotui; Johnson *et al.* (1993b) considered this synonymous with the sympatric *elongata*.

propinqua Crampton, 1932 – Matapoopoo and Tepatu; Johnson *et al.* (1993b) considered this to be a form of *P. mirabilis*; it is a hybrid of *P. taeniata* and *P. mirabilis* (see p. 152).

striolata Pease, 1866 – North-eastern Moorea, intergrades with *taeniata* and hybridises naturally with *P. incrassa*.

nucleola Garrett, 1884 – North-western Moorea. Typical form in Faatorai valley, eastwards resembling *elongata*.

elongata Pease, 1866 – Central and south Moorea, the most distinctive form.

simulans Pease, 1866 – Synonymous with *taeniata* Mørch, 1850; separate eastern and western populations. Eastern populations appear to be *striolata* and *elongata* hybrids and western populations *elongata* and *nucleola* hybrids.

Fig. 9.4.48 *P. taeniata*: a – *taeniata* holotype (ZMUC 1985); b – *elongata* lectotype (ANSP 59533); c – *striolata* (ANSP 59648); d – *simulans* paralectotype (NHMUK); e – *nucleola* paralectotype (NHMUK); f – *spadicea* holotype (NHMUK); g – *propinqua* syntype (ANSP 282833); h – *concinna* holotype (CMNH 62.4244); i – *corneola* holotype (CMNH 62.4242); j – *erhelii* holotype (NHMUK); k-s – Crampton specimens (ANSP 272772) (photos: h-i – C. Sturm)

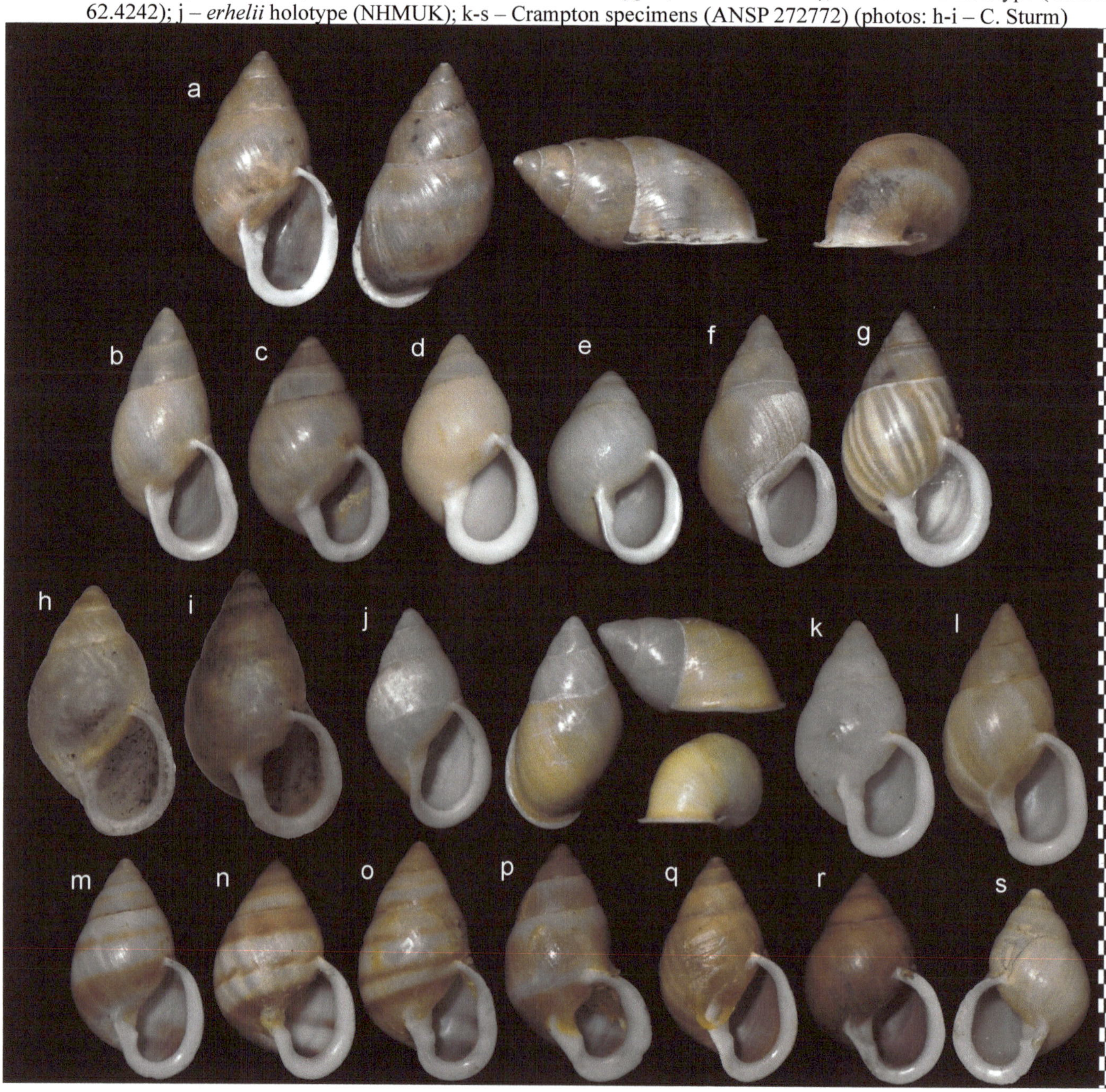

Fig. 9.4.49 Crampton's (1932) *P. taeniata* figures and subspecies designations (edited for position)

None of the anatomical forms are distinct, all integrade morphologically, all are interfertile (Murray & Clarke 1980) and the named subspecies are not separated in mitochondrial genes (Lee *et al.* 2009), accordingly the subspecies are not considered valid here, but rather represent clinal geographical variants.

Note: The origin of the single *P. erhelii* specimen is uncertain; Morelet (1853) considered it to be Moorean although it was given to him by a young surgeon, M. Erhel, who brought it from Tahiti. Its identity has also been disputed; it was tentatively identified as a form of *P. taeniata* by Hartman (1881a) but this was disputed by Pilsbry (1909). Crampton (1926) though it unidentifiable and of doubtful provenance. It is here considered to be a *P. taeniata* with an atypically twisted aperture.

P. mirabilis/taeniata propinqua – Crampton (1932) named a population from Tepatu valley on the south side of Mt. Rotui as *P. taeniata propinqua*, Murray & Clarke (1980) re-identified this population as a form of *P. mirabilis* but noted that it was also able to hybridise with *P. taeniata*. It is notable that Crampton (1932) identified specimens from the next valley, Matapoopoo as *P. mirabilis*, and also found *P. taeniata elongata* in Tepatu. Crampton's (1932) *P. taeniata propinqua* specimens are clearly forms of *P. taeniata*, contrary to Murray & Clarke's (1980) findings. Clarke & Murray (1969) stated that "some of the individuals described by Crampton as *taeniata propinqua* are in fact *mirabilis*. Other individuals described as *taeniata propinqua* resemble intermediates between *mirabilis* and *taeniata spadicea*. On the northern and western slopes of Rotui there appear to be two entities, corresponding to *spadicea* and *propinqua-mirabilis* which seem to intergrade in some places, but remain distinct in others." In describing their crosses of *propinqua* and *mirabilis* Murray & Clarke (1980) wrote: "these *mirabilis* from Tepatu [*propinqua*] and Matapoopoo [*mirabilis*] form fertile hybrids, although there is a hint of incipient isolation since three of the four matings did not show reciprocal crossing. The snails from Matapoopoo crossed freely with those from Maramu [*mirabilis*]. Thus it appears that there is a widespread entity, including the Tepatua '*propinqua*', that can be referred to *P. mirabilis*." Further, in crossing *taeniata* and *propinqua* "There is clear morphological and genetic evidence of hybridization, and the hybrids from at least two of the matings are fertile. In three of the matings there is non-reciprocal crossing, so that there seems to be some restriction of exchange between the two groups." Dissections of Murray and Clarke's Tepatu *P. mirabilis* (including some resembling *propinqua*) and Crampton's original specimens identified a range of anatomical forms between *P. taeniata* and *P. mirabilis*. *propinqua* appears to be a hybrid between *P. taeniata* and *P. mirabilis* as noted by Murray &Clarke (1980): "Thus morphologically and genetically the *P. mirabilis* from Tepatu form a bridge between *P. mirabilis* and *P. taeniata*."

Fig. 9.4.50 *Partula taeniata* variation showing the distribution of supposed subspecies. Crampton (1932) data; n=30,488

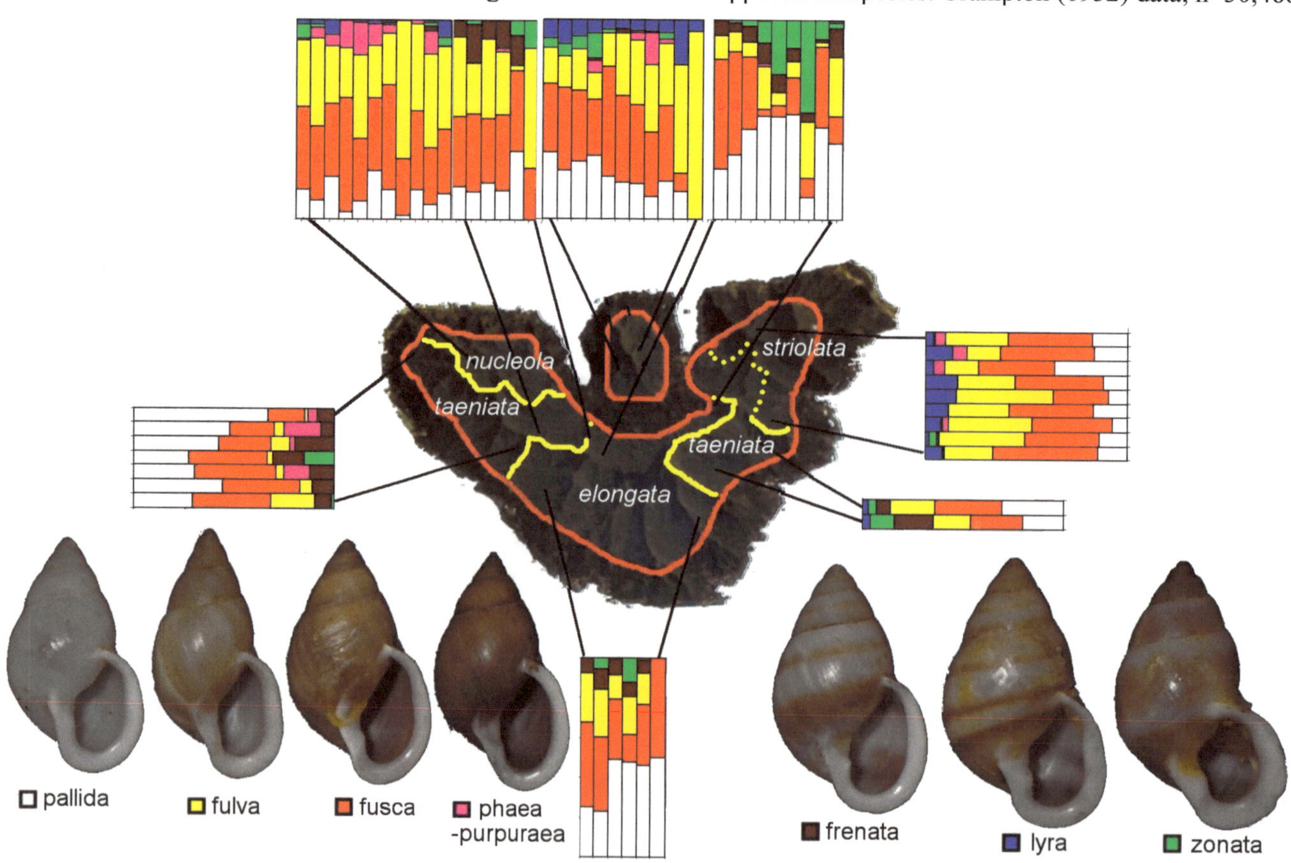

Partula incrassa (Crampton, 1916) **comb. nov.**

Partula clara incrassa Crampton, 1916: 63

Partula exigua Crampton 1932: 30

Partula exigua Crampton, Crampton 1932: 30. Kondo 1955: 142, 1968: 76. Clarke & Murray 1969: 31. Murray & Clarke 1980: 103. Clarke *et al*. 1984: 101. Richardson 1990: 23

Fusiform-conic. Thin and fragile. Umbilicus open but compressed. 5 slightly convex whorls with fine growth lines. Surface shiny with fine, dense, wavy spiral striae in most areas, or nearly smooth (Faamariri, Marehapa). Suture slightly impressed. Last whorl flattened near the lip. Columella usually more or less straight. Apertures elongately oval, almost straight and inclined backwards slightly. Lip very thin and narrow, very little internal thickening. Columellar insertion angled. Parietal callus thin but distinct. Edentulous.

White to light brown (Moorea 84% pallida, 11% fusca and 5% frenata, n=161 [Crampton 1932]; Tahiti 82% pallida, 6% fusca, 12% frenata, n=67 [Crampton 1916]). Bands where present (usually two banded frenata) may be lighter or darker than the base colour. Lip white, occasionally with faint brown tint in darker shells. Suture margined with a dull white line. Rarely on Tahiti the apex may be pinkish.

Fig. 9.4.51 *Partula mirabilis/taeniata propinqua*: a-g – Crampton specimens; h-i – Clarke & Murray specimens; j-p – Burch specimens. a-e – *mirabilis* (a-b – various ANSP 272760; c-e – Matapoopoo ANSP 279043); f-g – *taeniata propinqua* (f – Tepatu ANSP 282832; g – Matapoopoo ANSP 282833); h-i – *mirabilis propinqua* Matapoopoo 161-2 (NHMUK); j-p – *mirabilis propinqua* (Matapoopoo: j-l – BPBM 1970-273; m-p – UMMZ 1970-273); q – living *P. taeniata* from Opohunu Bay in 2015 (T. Coote)

Anatomy: Spermatheca short, white (yellowish in similar *P. taeniata*) and usually with a terminal bulb. Penis very short (1/3 oviduct length), slender, subcylindrical. Caecum at least ½ penis length. Vas deferens nearly apical. Egg membranes are calcified (unlike sympatric populations of *P. taeniata*).

Type specimens:

Partula clara incrassa Crampton, 1916 – Syntypes ANSP 276080. No types defined, types would be all Crampton specimens.

Partula exigua Crampton, 1932 – Syntypes ANSP 280867-77, 272753-6; BPBM 92328; NHMUK 1935.10.15-6. Only BPBM 92328 are labelled as types but Crampton did not define any specific types and all the material he collected was used in the type description, therefore all Crampton specimens are syntypes.

Distribution: Moorea and Tahiti. *P. i. incrassa* on Tahiti (Aoua valley only). *P. incrassa exigua* on Moorea, restricted to the east: Mt. Teaharoa (2000 ft), Paraoro, Paparoa, Maherapa, Faamaariri, Vaipohe and Puutu valleys. "infrequent, on vegetation of higher growth" (Pilsbry 1909). Intergrades with *P. taeniata* in Paparoa (Murray & Clarke 1980).

Subspecies: Two taxa are recognised as subspecies here based on distribution, they do not differ notably in morphology.

P. incrassa incrassa (Crampton, 1916) – Tahiti.

P. incrassa exigua (Crampton, 1932) **comb. nov.** – Moorea. More variable than nominate form.

Fig. 9.4.53 *P. taeniata* and *P. incrassa* anatomical variation: a – *elongata* Maramu (Kondo 1955); b – *simulans* Haapiti (ZSL 1209); c – *striolata* Faamariri (ANSP JJM S103P); d – *nucleola* Faatoai (ZSL 1251); e & f – various valleys (e – 2 specimens ANSP A939; f – 170 specimens, Clarke & Murray 1962, NHMUK); g-i – Crampton Tepatu *propinqua* (ANSP A756); j – Crampton's Matapoopoo *P. mirabilis* (ANSP A691); k – *exigua* Mt. Teaharoa (Kondo 1955); l – *incrassa* Atehi (ANSP A938A)

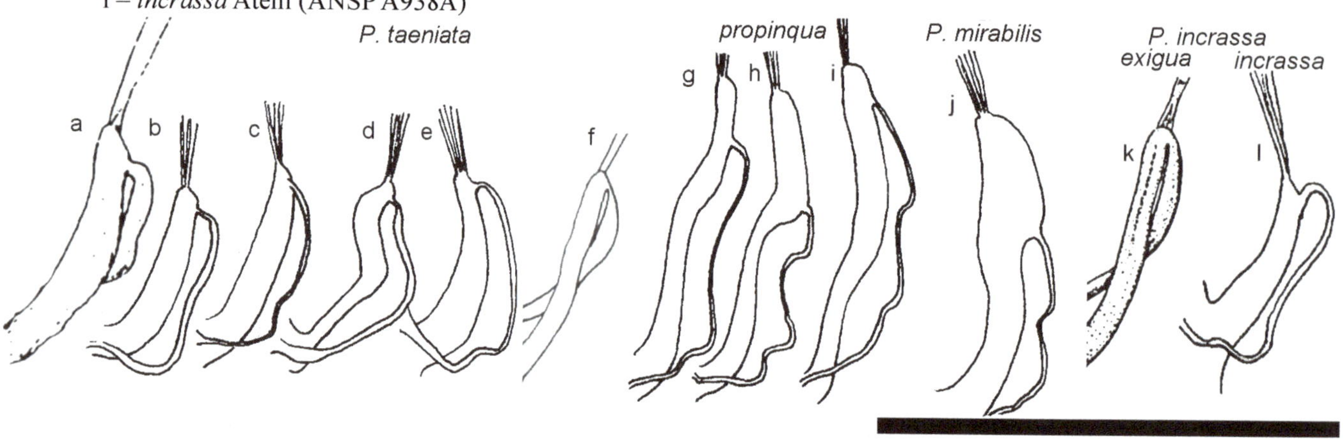

Fig . 9.4.54 *Partula incrassa*: a – *exigua* syntype (NHMUK); b-d – Clarke & Murray *exigua* specimens (NHMUK); e – Tahiti *incrassa* syntype (ANSP 276980)

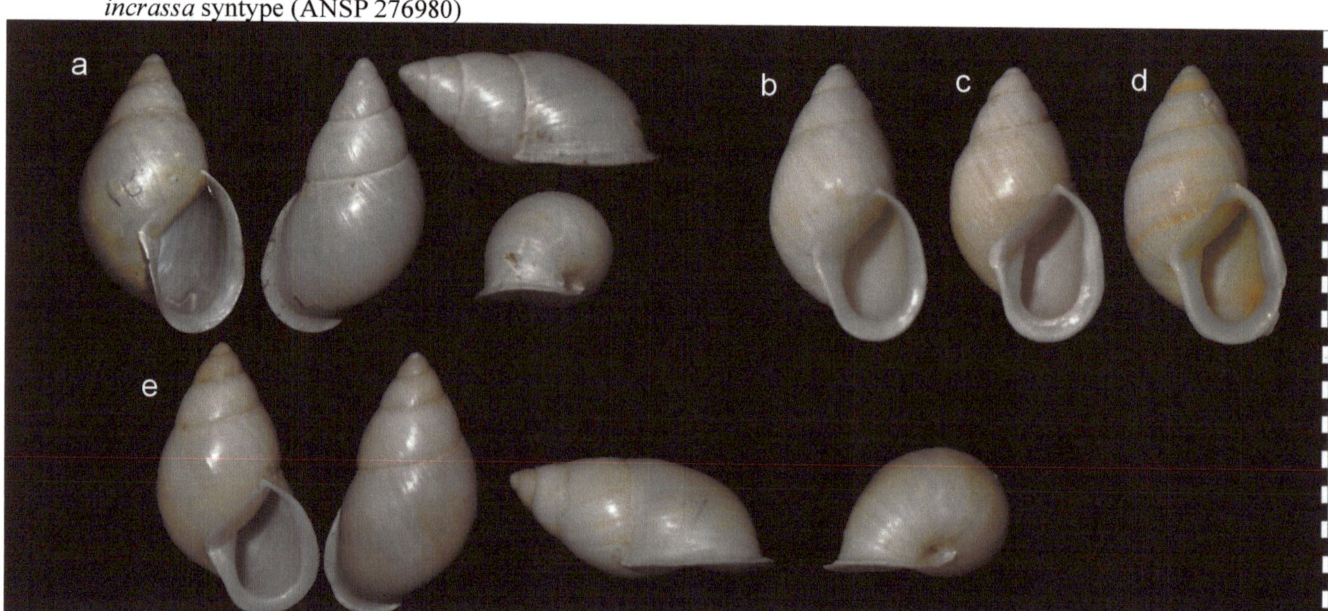

Fig. 9.4.55 Crampton's *P. incrassa* figures; row 1 *P. i. incrassa* (Crampton 1916), 2-3 *P. i. exigua* (Crampton 1932) (edited for position)

Fig. 9.4.56 *Partula incrassa* variation on Moorea and Tahiti

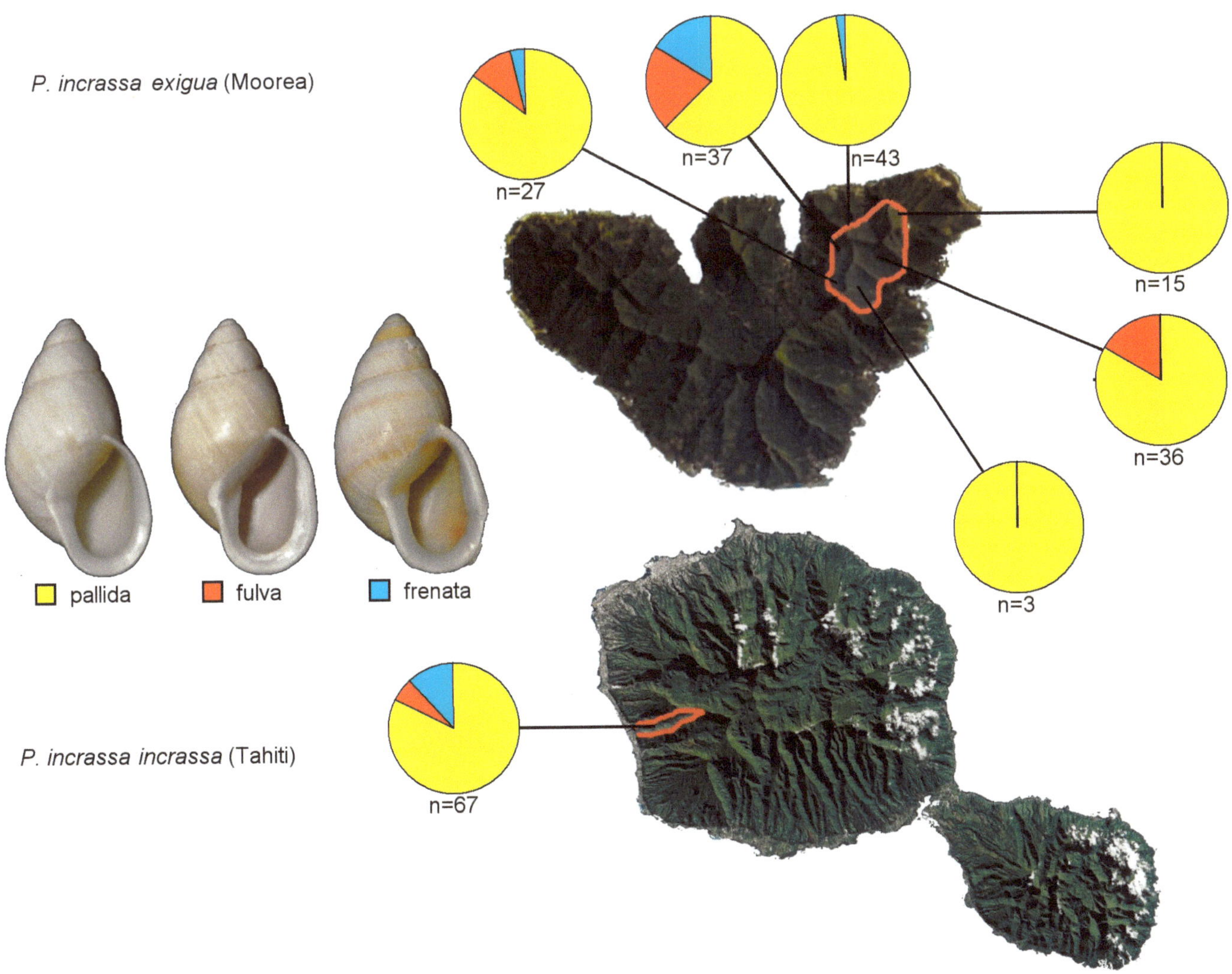

Note: This species is a close relative of the very similar to *P. taeniata*, but is narrower than sympatric populations of that species and is thinner shelled. It hybridised with *P. taeniata* naturally (Murray & Clarke 1980).
The nominate form on Tahiti was originally described as a variety of *P. clara* (Crampton 1916) but mitochondrial DNA phylogeny unambiguously shows it to be descended from the Moorean *P. taeniata*- '*exigua*' clade (Lee *et al.* 2009). Comparison of the shell and anatomy with Moorean specimens shows that '*P. clara incrassa*' cannot be separated from '*P. exigua*' and examination of the specimens upon which the molecular phylogeny is based shows that the Tahiti population arises from '*P. exigua*'. Accordingly, *incrassa* and *exigua* can only be separated at subspecies level.

Partula mooreana Hartman, 1880

Partula mooreana Hartman, 1880: 229

Partula mooreana Hartman, Hartman 1881a: 175, 1881b: 10, 1885: 10, 1892: 75. Paetel 1883: 153, 1889: 267. Garrett 1884: 59. Smith 1903: 448. Pilsbry 1909: 200. Crampton 1925: 22, 1932: 77. Kondo 1955: 149, 1968: 76. Clarke & Murray 1969: 31. Murray & Clarke 1980: 89. Murray *et al.* 1982: 316. Richardson 1990: 42

Sinistral (0.1% dextral in Crampton's specimens, n=6,048). Umbilicus narrow. Surface shiny with fine, dense wavy spiral striae clear on all whorls. 5 almost straight sided whorls, body whorl compressed laterally. Suture impressed. Aperture oblique. Lip reflexed, strongly thickened internally, narrowed in its upper third. Columella dilated at the insertion. A thick but transparent parietal callus and a well-developed parietal tooth are usually present.

Fig. 9.4.57 *Partula mooreana*: a – type (ANSP 59536); b-d – Mouaputa shells in Crampton's original figure (ANSP 279068)

Fig. 9.4.58 Crampton's (1932) *P. mooreana* figures (edited for position)

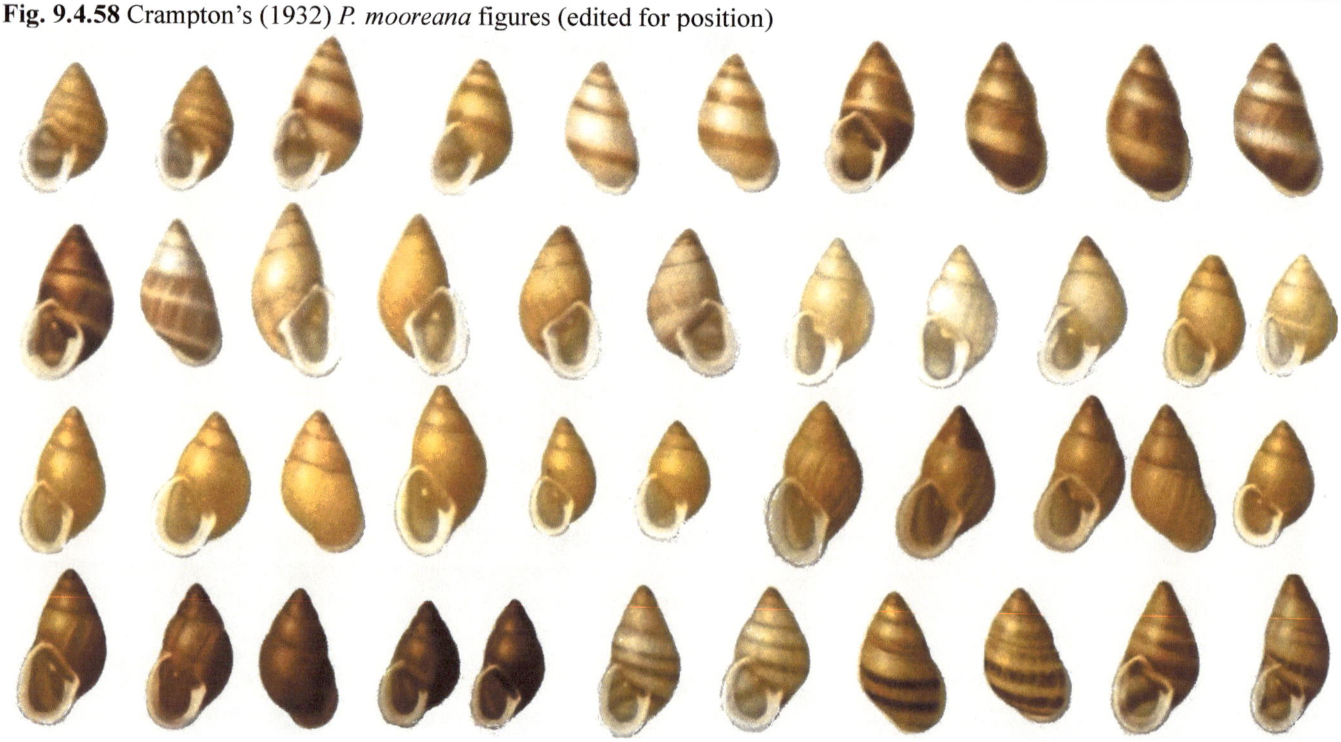

Colour: Crampton recorded the colour form fusca as the most frequent (63%), followed by pallida (15%), fulva (11%) and rare individuals of phaea, cestata, frenata, zonata and bisecta (n=5,959). There is also a rare variant of lyra.
Anatomy: Penis short (¼ oviduct length), subcylindrical and slender. Caecum short (⅓ penis length). Vas deferens subapical.
Type specimen: Syntypes ANSP 59536, CMNH 62.4221, MCZ 24818 & 25372.
Distribution: Upper parts of the valleys from Haapiti to Haumi and from Roroie to Tefeo. Isolated specimens have been recorded from the eastern valleys of Putuu and Faamaariri (Clarke & Murray specimens, NHMUK). Arboreal; associated with *Freycinetia* sp. (Murray & Clarke 1980).
Note: The western-most populations are more elongate, and closer to *P. olympia* in form, anatomically they are distinct as well, with a longer penis and a constriction between the caecum and penis.

Fig. 9.4.59 *Partula mooreana* geographical variation. Colour variation (main pie charts) and tooth development (bars) based on data from Crampton (1932), anatomical variation (small pie charts and inset drawings) based on specimens collected by Clarke & Murray in 1962 (NHMUK). 10 specimens dissected from each point, except 1-2 isolated eastern specimens (associated shells shown)

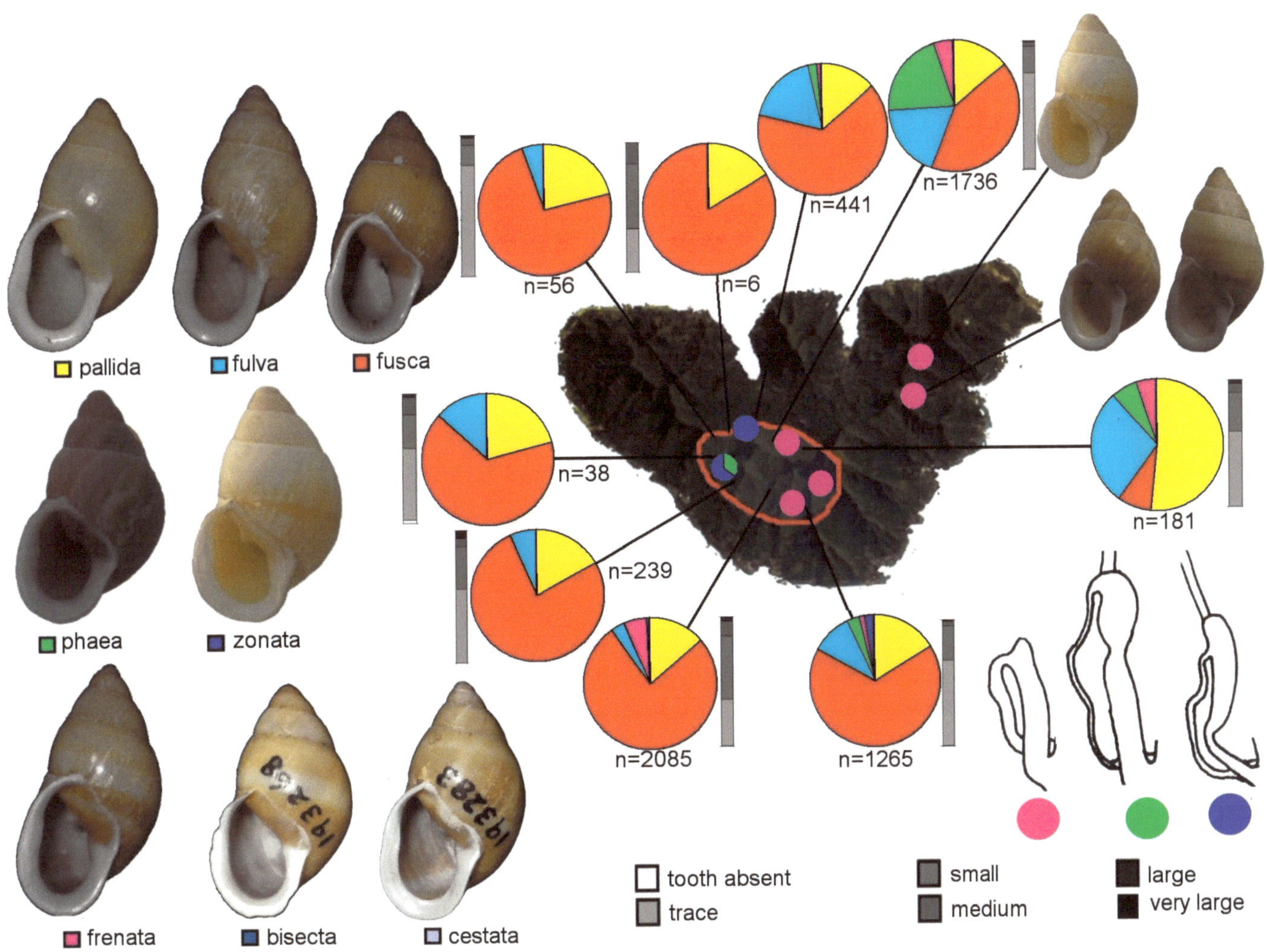

Partula tohiveana Crampton, 1924

Partula tohiveana Crampton, 1924a: 110, 1932: 34

Partula olympia Crampton, 1924a: 112, 1932: 42

Partula tohiveana Crampton, Crampton 1925: 12, 1932: 34. Kondo 1955: 43, 1968: 76. Clarke & Murray 1969: 31. Murray & Clarke 1980: 94. Murray *et al.* 1982: 316. Clarke *et al.* 1984: 101. Richardson 1990: 43

Partula olympia Crampton, Crampton 1925: 12, 1932: 42. Kondo 1955: 145, 1968: 76. Clarke & Murray 1969: 31. Murray & Clarke 1980: 96. Richardson 1990: 43

Sinistral. Elongate-conic. Umbilicus open. 5-5½ moderately convex whorls, flattened at apex. Spire with fine spiral striae, these are usually indistinct, fading on body whorl, or absent. Suture usually slightly impressed. Middle area of body whorl flattened or indented by lip. Columella straight. Aperture elongate, long axis inclined away from columellar axis. Lip rounded or slightly bevelled, thickened inwardly, sharply angled at columellar insertion, dentated near outer insertion. Lip thickened within. Callus thin, shining, with a weak palatal projection developing into a distinct tooth, variable in development but always present. Crampton (1932) noted that "The average degree of its growth diminishes consistently as we pass from the lower to the higher levels… In the absence of a single shell without this feature, and in the abundance of shells with its high development, *Partula tohiveana* is a unique species of Moorea."

Yellow-brown to dark brown, sometimes with a darker apex and dark striations, often banded. Lip white. Crampton (1932) recorded 47% as apex, 46% pallida and also fusca, fulva, cestata and frenata (n=731). Crampton's cestata should be regarded as zonata, there are also rare lyra shells. Around half of specimens (49%) have a dark apex.

Anatomy: Penis vermiform, elongate, slender, subcylindrical with slight medial constriction; ½ oviduct length. Vas deferens subapical, large, strongly convoluted. Penis with ridges and tubercles internally. Caecum twice as long as penis, with one large diaphanous ridge, with few folds becoming papillae. 8-9 subordinate ridges. *P. tohiveana olympia* had a smaller caecum (½ the size of the penis) with less developed papillae.

Subspecies: Crampton (1924a) described two species which Murray & Clarke (1980) considered conspecific. They are parapatric taxa forming a cline in all conchological features and hybridising freely in the laboratory (Murray & Clarke 1980). Anatomically they are recognisably different but very close. Genetically they are sister taxa (see p. 245, *olympia* misidentified as '*mooreana* 1' and 2). The two taxa are retained at subspecific level here:

Fig. 9.4.60 *Partula tohiveana*: a-d – *tohiveana*; a-b – *tohiveana* types (NHMUK); c – Clarke & Murray specimen 551 (NHMUK); d – two of Crampton's (1932) illustrated specimens (ANSP 272754); e-i – *olympia*; e-f – *olympia* types (NHMUK); g – *olympia* Clarke & Murray specimen 339 (NHMUK); h-i – Mouaputa valley specimens figured in original description (ANSP 279068)

P. tohiveana tohiveana Crampton, 1924 – The larger form recorded from the north-western end of the range: Tefeo to Irua.

P. tohiveana olympia (Crampton, 1924) **comb. nov.** – Differs in being smaller (averaging 2.74 mm shorter), more ovate, usually with more distinct wavy spiral striae. The columella is twisted slightly, more evidently in adolescent than in adult shells. Aperture elliptical, axis parallel to columellar axis. Lip less narrowed toward outer insertion. Parietal callus finely pitted microscopically, parietal tooth is less developed. Colour more variable. South and eastern part of the range: Maatea to Putuu and Mouaputa to Paparoa.

Type specimens:

Partula tohiveana Crampton, 1924 – Syntypes ANSP 2789067, 297068, A893; BPBM 92316; NHMUK 1935.10.15.14-5.

Fig. 9.4.61 Crampton's (1932) *P. tohiveana* figures; rows 1-2 – *P. t. tohiveana*, row 3 – *P. t. olympia* (edited for position)

Fig. 9.4.62 Geographical variation in *Partula tohiveana*: bars – tooth development; large pie charts – colours; small pie charts – anatomical variation (red – typical *tohiveana*, blue – typical *olympia*)

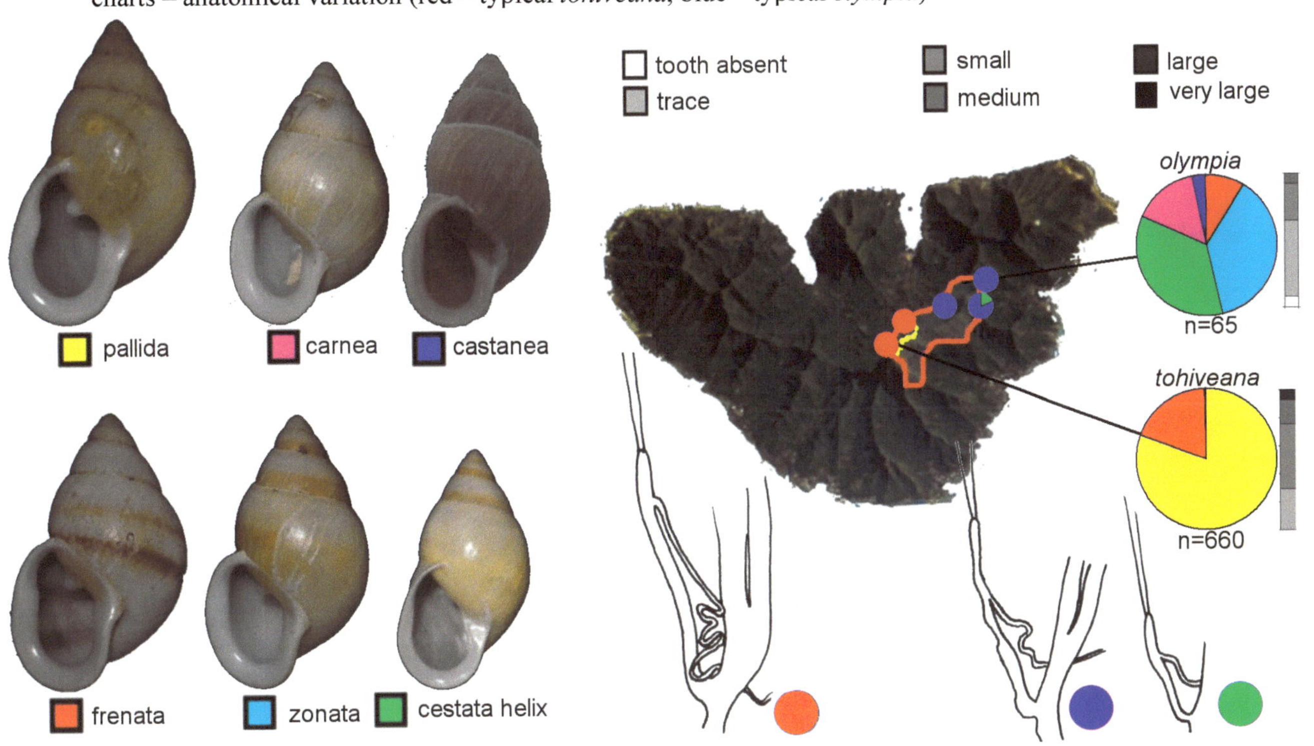

Only BPBM 92316 are labelled types but Crampton did not designate specific types but recorded that "Less than a hundred examples" were collected in 1919; all of these should be syntypes.

Partula olympia Crampton, 1924 – Syntypes ANSP 279064, 279065; BPBM 92312-5; NHMUK 1935.10.15.17-20. Only BPBM 92312-5 are labelled types ('paratypes') but Crampton did not designate specific types and all his 1923 material should be considered syntypes.

Distribution: Eastern mountain ridges of Moorea. *P. t. tohiveana* from inner slopes of Mt. Mouaputa at high levels: Tefeo, Fareaito and Irua. *P. tohiveana olympia* from Hotutea, Atiraa, Putuu, Mouaputa and Paparoa. The identity of intervening populations is not clear (Poua) and some mixing was apparent in Maatea, Haumi, Hotutea and Atiraa. Arboreal; associated with *Freycinetia* sp. (Murray & Clarke 1980).

Note: Crampton (1924a) noted that *P. tohiveana olympia* size and proportions varied across the island. Analysis of his data (Lundman 1947) showed that shell length is greatest in the middle, decreasing strongly to the north and weakly to the south. The width shows the opposite pattern. As a result there is a cline from the short, broad shells of the north to the long thin ones of the south.

Fig. 9.4.63 Living Moorean *Partula*: a – *P. mooreana*; b – *P. mirabilis*; c – *P. tohiveana*; d – *P. suturalis*; e-f – *P. taeniata*; g – *P. taeniata/mirabilis propinqua* (photos: a-c – D. Clarke, e-f – J.-Y. Meyer, g – T. Coote)

Huahine

Only three species of *Partula* are known from Huahine and there have been no taxonomic issues for this island's partulids. The island's species are notable in that, as noted by Pilsbry (1909), there were no dentate species. Pilsbry (1909) also commented incorrectly that "The structure of the kidney [short and triangular] raises a doubt as to the systematic position of *P. rosea* and *varia*. They may belong in or near the section *Samoana* rather than to *Partula* s. str."

Partula varia Broderip, 1832

Partula varia Broderip, 1832: 125

Bulimus varius Broderip, Pfeiffer 1841: 86, 1842: 124

Partula varia Broderip, Müller 1836: 33. Reeve 1842: pl75 f5, 1851: pl3 f17. Catlow & Reeve 1845: 163. Pfeiffer 1848: 70, 1853b: 448, 1856: 162, 1856: 273, 1859: 509, 1868: 157, 1877: 197. Gould 1852: 85. Adams 1855: 145. Frauenfeld 1869: 874. Paetel & Schaufuss 1869: 83. Schmeltz 1869: 71, 1874: 92. Pease 1871b: 473. Paetel 1873: 104, 1883: 153, 1889: 268. Gloyne 1878: 338. Hartman 1881a: 189 (excl. *strigata*), 1881b: 14, 1885: 218, 1892: 98. Pfeiffer & Clessin 1881: 301. Garrett 1884: 78. Smith 1903: 462. Pilsbry 1909: 156. Kondo 1955: 170, 1968: 76. Richardson 1990: 64

Bulimus roseus β. Pfeiffer, 1848: 70

Partula rosea var. Pfeiffer 1853b: 448, 1856: 273

Partula simplaria Morelet, 1853: 370

Partula glutinosa Pfeiffer, 1853b: 448

Partula glutinosa Pfeiffer, Pfeiffer 1854b: 85, 1955b: 265, 1865: 162, 1859: 509, 1868: 157, 1877: 198. Adams 1855: 145. Paetel & Schaufuss 1869: 83. Paetel 1873: 104, 1883: 153, 1889: 267. Kobelt 1880: 16. Hartman 1881a: 182, 1881b: 14, 1885: 218. Smith 1903: 463

Partula mucida Pfeiffer, 1855a: 98

Partula mucida Pfeiffer, Pfeiffer 1855b: 513, 1856: 162, 1868: 160, 1877: 206. Paetel 1889: 267

Table 9.4.3 Huahine *Partula* shell measurements

	N	L			W			L aperture			W/L		
		range	mean	sd	range	mean	sd	range	mean	sd	range	mean	sd
varia	4177	15.6-21.1	17.57	0.752	9.6-11.9	10.65	0.512	6.5-8.6	9.73	0.441	55.5-65.5	59.76	2.144
rosea	226	16.7-24.8	21.31	1.295	9.2-15.1	12.98	0.799	7.0-12.5	9.49	0.790	52.2-64.0	57.12	2.413
arguta	67	10.3-13.2	11.44	0.697	6.5-10.1	7.83	0.599	6.1-8.1	6.87	0.600	58.7-77.9	68.37	3.710

Fig. 9.4.64 Map, shells and anatomy of Huahine *Partula*, recorded localities of *P. arguta* marked: a-b – *P. varia*; c-d – *P. rosea*; e-f – *P. arguta*; a – Fare (Kondo 1955); b – Tefari (ANSP A1425); c – Fare? (Kondo 1955); d – Fareaito (ZSL 156); e – ZSL (mislabelled 1161); f – Mt. Turi (Kondo 1955); g – Tefari (ANSP 1425). Scale bar 5 mm.

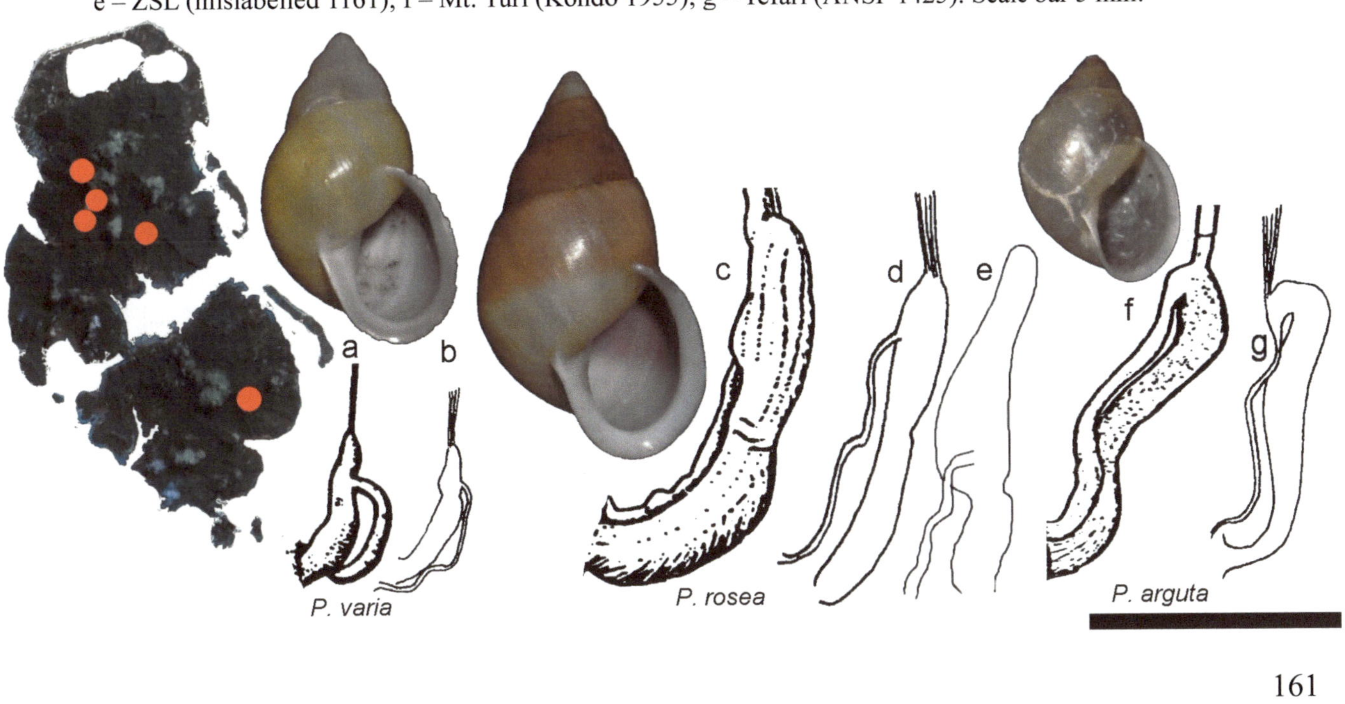

Partula varia var. *glutinosa* Pfeiffer, Pease 1871b: 473
Partula varia var. *pulchra* Pease, 1871b: 473 (nomen nudum)
Partula varia var. *simplex* Pease, 1871b: 473 (nomen nudum)
Partula pulchra 'Pease', Schmeltz 1874: 92. Gloyne 1878: 338. Paetel 1883: 153. Kilias 1969: 748 (all nomen nudum)
Partula simplaria Morelet, Schemltz 1874: 92 (misidentification, *P. rosea* fide Hartman 1884)
Partula lugubris Gloyne 1878: 338 (non Pease)
Partula adusta 'Garrett', Hartman 1881a: 179 (nomen nudum). Garrett 1884: 79 (as syn. of *varia*)
Partula perplexa 'Pease', Hartman 1881a: 185 (as syn. of *varia*, nomen nudum)
Partula bicolor 'Garrett' (non Pease), Hartman 1881a: 189 (as syn. of *varia*, nomen nudum)
Partula huaheinensis Garrett, 1884: 79 (as syn. of *varia*). Gloyne 1878: 338. Paetel 1883: 153, 1889: 267 (all nomen nudum).
Partula huahinensis Garrett, Smith 1903: 479 (misspelling, as syn. of *varia*) [N.B. Schmeltz 1874 '*Partula huaheinensis'* cited by Hartman 1881 is not a *Partula*]
Partula varia huaheinensis Garrett, Pilsbry 1909: 158 (description)

Pyramidally ovate; thin but strong. Umbilicus open. 5 convex whorls, body whorl rounded, or slightly subangular in front; very convex or saccate at base. Apex acute. Glossy, with sculpture of fine spiral striae, fading on body whorl but persistent around umbilicus. Suture not impressed. Aperture vertical; lip reflexed, evenly thickened within. Variable in size and shape; *pulchra* smallest form, *bicolor* larger and more robust than typical form, *simplaria* smaller than typical.

Crampton (MS) distinguished several colour forms: fusca pallida, fusca, fusca cestata, bicolor, fulva, phaea, bisecta, cestata and pulchra (Fig. 9.4.66). Typical form variable: light yellow (also Garrett's *adusta, bicolor, huaheinensis*) or dark chestnut. Light specimens sometimes with darker strigations, spire usually pale, rarely chestnut (Garrett's form *bicolor*), lip and sometimes base tinted dark brown or violet-brown, sometimes with two narrow bands on the body whorl or an ill-defined medial fulvous band. Dark specimens (Pfeiffer's form *mucida*) sometimes nearly black with a pale apex, lip dark or pale, often with a medial pale band of variable width. Aperture and lip white or purplish-brown.

Fig. 9.4.65 *Partula varia* syntypes: a – two *varia*; b – *huaheinensis* supposed type; c – *bicolor*; d – *simplaria*; e – *pulchra* supposed type; f – *glutinosa*; g – *mucida*. All NHMUK except b (ANSP 59519 – photo F. Borrero)

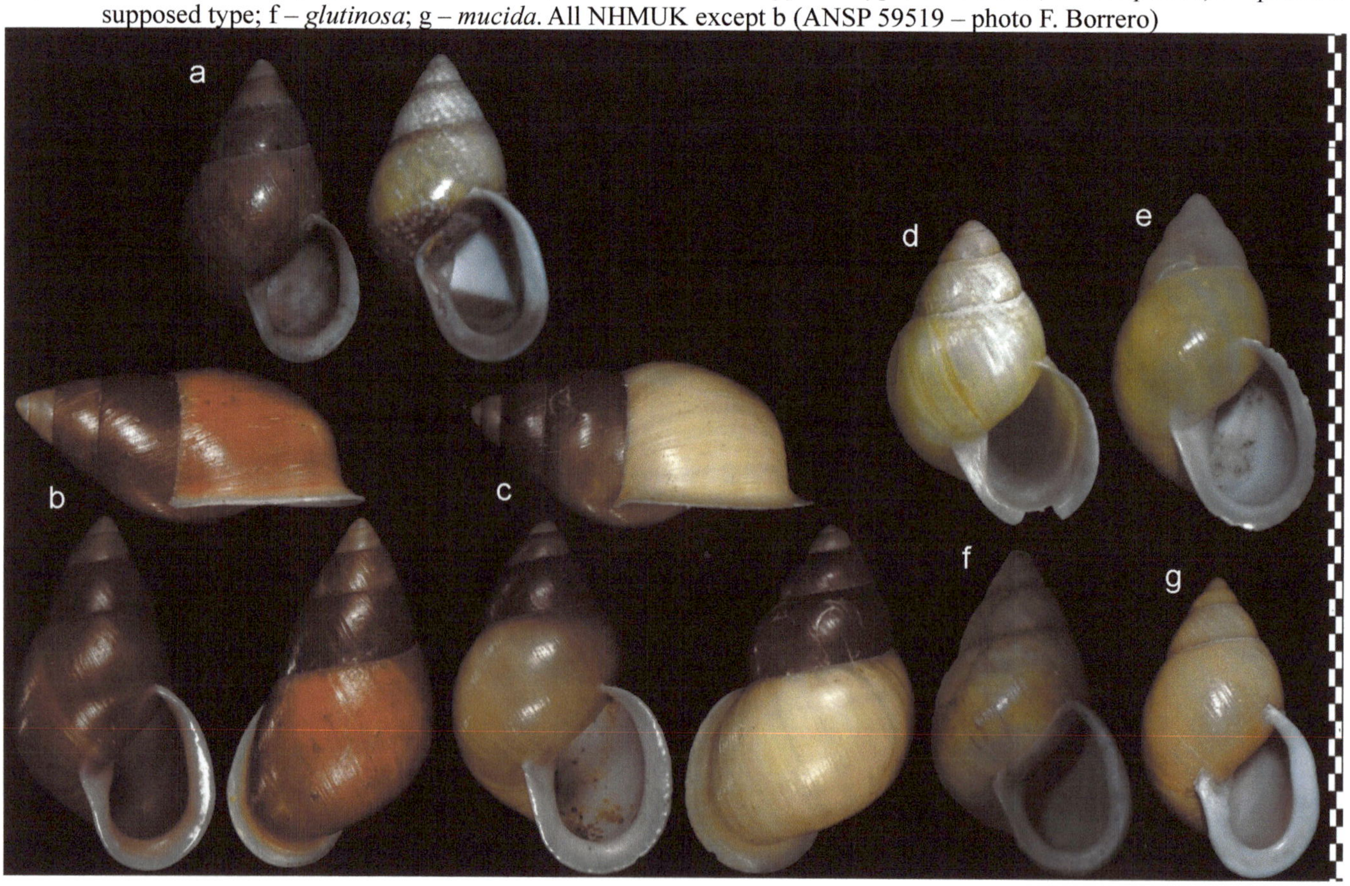

Anatomy: Penis moderately short, $^{1}/_{5}$-$^{1}/_{4}$ oviduct length; caecum well developed, about ½ penis length. Retractor muscle terminal on caecum. Vas deferens near base of caecum. Radula with 7-9 laterals, fewer than in *P. rosea.*

Type specimens:

Partula varia Broderip, 1832 – Syntypes AMNH 42258; FLMNH 112163, 76-8 ('paratypes'); MCZ 297854; NHMUK un-numbered; SMF 164979-80; ZMB 75718.

Partula simplaria Morelet, 1853 – Syntypes NHMUK un-numbered.

Partula glutinosa Pfeiffer, 1853 – Syntypes NHMUK un-numbered. Cuming 'P. glutinosa var.' may be the same series.

Partula mucida Pfeiffer, 1855 – Syntypes NHMUK un-numbered.

Partula bicolor Garrett, 1884 – Syntypes ANSP 59517, 74297.

Partula varia huaheinensis Pilsbry, 1909 – Syntypes ANSP 59519.

Partula varia var. *pulchra* Pease, 1871 – Supposed types of nomen nudum: FMNH 75473, 308260; MCZ 24878, 24920; NHMUK un-numbered; ZMB/Moll-74874.

Partula perplexa Pease MS., Garrett 1884 – Supposed syntypes of nomen nudum: MCZ 25358.

Distribution: All areas of Huahine. "The metropolis of the typical *P. varia* is in two valleys on the west coast of Huaheine, where they are very abundant on foliage, but it is generally distributed throughout all parts of the island." (Garrett 1884). Commonest in north (Faahiti). Arboreal, although some individuals were found on the ground (Operation Partula 1991).

Note: Some *P. varia* and *P. rosea* may be very similar, *P. rosea* has straighter sided whorls and is never as convex as *P. varia.* The most ovate forms of both were found in Faahiti in the north (form *simplaria* in *P. varia*). There may have been some hybridisation with *P. rosea*, resulting in similarity. Garrett's *huaheinensis* specimens resemble *P. rosea* in shape, but are anatomical *P. varia.* Other ambiguous populations include Haamene, Tepare and Tepepe. Atita and Tefari populations with high lateral tooth counts (similar to *P. rosea*'s 10-12) have a high proportion of abnormal teeth, suggestive of hybridisation.

Garrett (1884) noted some geographical variation in colour patterns with two forms in Hamene valley: *pulchra* and *bicolor*, the latter restricted to the upper part of the valley. There is some geographical variation in the maximum number of radula teeth in a row, increasing from south-eastern valleys to northern populations. This may have arisen through allopatric drift during times of heightened sea levels separating Huahine-Nui and the Huahine-Iti peninsula.

Fig. 9.4.66 Pilsbry's (1909) *P. varia* and *P. rosea* forms: top row – *P. varia*; other rows – *P. rosea* (1-2 rosea, 3 – purpurascens, 4-5- bipartita, 6-7 – bicolorata, 8-9 – straminea, 10-11 zonata, 12,17 – cognata, 13-16 estalliana)

Partula rosea Broderip, 1832

Partula rosea Broderip, 1832: 125

Partulus roseus Broderip, Beck 1837: 57

Partula rosea Broderip, Müller 1836: 32. Jay 1839: 57. Pfeiffer 1841: 85, 1842:121, 1848: 272; 1853b: 448; 1855c: 509; 1856: 162, 1856: 272, 1859: 509, 1868: 157, 1877: 157. Reeve 1842: pl175 f9; 1850: pl1 fl. Catlow & Reeve 1845: 163. Albers 1850: 186. Pease 1864: 675, 1871b: 473. Paetel & Schaufuss 1869: 83. Schmeltz 1869: 72, 1874: 92. Paetel 1873: 104, 1883: 153, 1889: 268. Gloyne 1878: 338. Hartman 1881a: 186 (excl. *simplaria*), 1881b: 14, 1885: 14, 1892: 98. Pfeiffer & Clessin 1881: 301. Garrett 1884: 67. Pilsbry 1901: 561, 1909: 156. Smith 1902: 461. Thiele 1931: 658. Pilsbry & Cooke 1934: 12. Kondo 1955: 169, 1968: 76. Richardson 1990: 53

Bulimus roseus Broderip, Pfeiffer, 1854a: 70 (exclusive of var. 5)

Partula purpurascens Pfeiffer, 1857b: 333

Partula purpurascens Pfeiffer, Pfeiffer 1859a: 511, 1868: 159, 1877: 201. Paetel 1883: 153, 1889: 268. Pilsbry 1909: 253 (var.)

Partula simplaria Morelet, 1853: 370

Partula simplaria Morelet, Pfeiffer 1859: 512, 1868: 160, 1877: 203. Pease 1871b: 473. Schmeltz 1874: 92.

Fig. 9.4.67 *Partula rosea* syntypes: a-e – *rosea* NHMUK; f – *rosea* MNHN 2000.27660; g – *straminea* ANSP 59510; h – *bipartita* ANSP 59509; i – *purpurascens* NHMUK; j – *zonata* ANSP 72012; k – *cognata* ANSP 59991; l – *bicolorata* ANSP 59507; m – sinistral specimen (H.G. Lee collection) (photos: f – M. Caballer, m – H.G. Lee)

Gloyne 1878: 338. Paetel 1883: 153, 1889: 268

Partula cognata 'Pease', Pease 1864: 675. Schmeltz 1874: 92. Pfeiffer 1877: 209. Gloyne 1878: 338. Hartman 1881a: 181, 1881b: 14, 1885: 218. Paetel 1883: 152, 1889: 266 (all nomen nudum). Pilsbry 1909: 253

Partula estalliana 'Garrett', Smith 1903: 461 (nomen nudum). Pilsbry 1909: 253 (var.)

Partula bipartita Pilsbry, 1909: 253 (var.)

Partula bicolorata Pilsbry, 1909: 253 (var.)

Partula straminea Pilsbry, 1909: 253 (var.)

Partula zonata Pilsbry, 1909: 254 (var.)

Dextral (a single sinistral shell known - Fig. 9.4.67). Pyramidally ovate. Umbilicus open. Thin. Moderately glossy; with fine growth-lines and distinct, close, wavy spiral striae, fading on body whorl. 5 very slightly convex whorls, body whorl angular at periphery near aperture. Apex obtuse. Aperture ovate, slightly oblique. Lip expanded, thickened internally, gradually tapering to the upper end. Columella thin, dilated above. Parietal callus transparent. Edentulous. The smallest variety was named 'cognata', the largest is 'zonata'.

Pink to violet, or yellow, colour varying geographically. Typically uniformly rose pink. Many have paler embryonic whorls or are yellowish with a pink sutural line. Colour variants were named by Pilsbry (1909) but not as taxonomic entities):

estalliana – Crampton's 'bisecta vera'. Faahiti valley only.

cognata – Crampton's 'bisecta alternata' with indistinct banding. "Very common" in Faahiti valley (Garrett 1884).

straminea – Crampton's 'pallida'. Very abundant according to Garrett (1884).

zonata – Crampton's 'bisecta lata' (NB, Pilsbry's zonata and Crampton's zonata are different patterns). Common.

bicolorata – Crampton's 'bicolor'. Common.

bipartita – A variant of bicolorata. "Frequent in the metropolis of the species but very rare elsewhere" (Garrett).

purpurascens – Common.

Typical *rosea* are Crampton's 'phaea' class. In addition a small number of cestata specimens have been recorded.

In juveniles straminea, estalliana, rosea and purpurascens all integrade, suggesting ontogenetic changes in some forms.

Anatomy: Penis large, muscular, ½ oviduct length. Caecum well developed, almost as long as penis. Vas deferens median on caecum. Radula with 10-12 lateral teeth, more than in other Huahine *Partula*.

Type specimens:

Partula rosea Broderip, 1832 – Syntypes: AMNH 42253; MCZ 297853 & 297858, MNHN-2000-27660, NHMUK un-numbered; QM MO35145(=958949); SMF164978; ZMB/Moll-20020064.00; ZSM 110102.

Partula purpurascens Pfeiffer, 1857 – Syntypes NHMUK un-numbered.

Fig. 9.4.68 Colour variation in *Partula rosea* and *P. varia*: pallida, pulchra/bisecta lata & cestata – NHMUK; bicolor – ANSP 59517; fulva, phaea fusca cestata & bisecta vera – UMZC 75347; fusca – ANSP 59519; purpurascens – ANSP 411388; bisecta alternata – ANSP 59991. Populations marked in red contain *varia*-like *P. rosea*

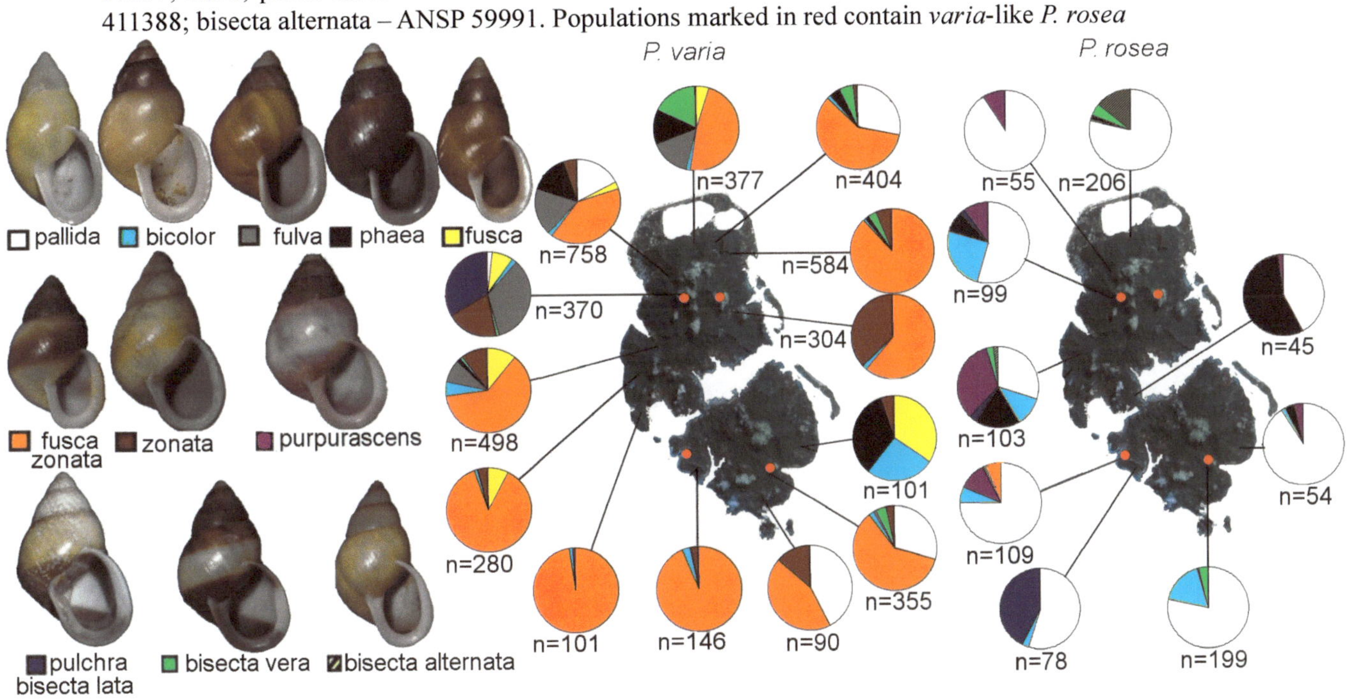

Supposed types of nomina nuda: *cognata* – ANSP 59991 ('lectotype' – Baker 1963), 2641, 59845, 59987-8, 463719, MCZ 24817, 25117 ('paralectotypes'); *estalliana* – MCZ 297859.
Supposed types of Pilsbry's (1909) colour varieties: bicolorata – ANSP 59507, 457190; bipartita – ANSP 59509, 59990; straminea – ANSP 59510; zonata – ANSP 72012, A1770, A1762B.
Distribution: All of Huahine, most abundant in north-west. Always rarer in samples than *P. varia* and largely restricted to *Pandanus* axils (D. Clarke pers. comm.). The most ovate *P. rosea* and *P. varia* were in Faahiti in the north (*P. rosea* form *cognata*). There may have been some hybridisation with *P. varia*, resulting in *P. varia*-like specimens in Hamene, Hopu, Tepare and Tepepe. Tefari *P. rosea* had low tooth counts (9 laterals, as *P. varia*) and was sympatric with *P. varia* with high tooth counts and frequent abnormalities. Hopu *P. rosea* had *P. varia* colours (fusca cestata). Radular lateral tooth numbers increase northwards, possibly due to separation of Huahine-Nui and Iti in periods of high sea levels.

Partula arguta (Pease, 1866)

Bulimus argutus Pease, 1866: 670
Bulimus argutus Pease, Pfeiffer 1868: 46, 1877: 46
Partula arguta Martens & Langkavel 1871: 55. Pease 1871: 473. Schmeltz 1874: 92. Pfeiffer 1877: 207. Hartman 1881a: 179, 1881b: 11, 1885: 218, 1892: 98. Paetel 1883: 152, 1889: 266. Garrett 1884: 62. Smith 1903: 457. Thiele 1931: 658. Kondo 1955: 171, 1968: 76. Zilch 1959: f460. Richardson 1990: 10
Partula (*Leptopartula*) *arguta* (Pease): Pilsbry 1909: 176

Ovate, extremely thin and fragile, tansluscent. Umbilicus minute. Surface shiny with weak growth lines and microscopic spiral striae, fading on body whorl. 4 convex whorls, spire very short; body whorl globose. Aperture somewhat oblique, more than half the total length, broadly ovate. Lip thin, evenly but narrowly expanded, basal margin more strongly arcuate. Columellar margin reflexed and dilated above; concave. Parietal callus a very thin film. Edentulous.
Corneous usually with a faint buff tint. Lip paler.
Mantle visible through shell: "beautifully maculated with black and white spots on grayish yellow ground" (Garrett 1884).
Anatomy: Penis long, ½ oviduct length. Caecum well developed, not clearly separated from penis. Interior of penis with several large pilasters, caecum with 5-6 papillate ridges, 3 rows of tubercles. Vas deferens subterminal. 7-9 radula laterals.
Type specimens: Syntypes ANSP 59912; MCZ 26353. ANSP 295292 are Pease duplicates, not types.
Distribution: Known only from isolated localities on Huahine. Garret (1884) recorded it from Hamene and Tepua: "in the upper portion of a mountain ravine… where it is rather common on the leaves of low shrubs and ferns". Collected at Faie in 1925, Fareaito in 1934 and 1991, Fitii in 1980 and 1991, and Pae Eo in 1994.

Fig. 9.4.69 *Partula arguta* types: a – ANSP 59912; b – MCZ 26353 (photo: a – F. Borrero)

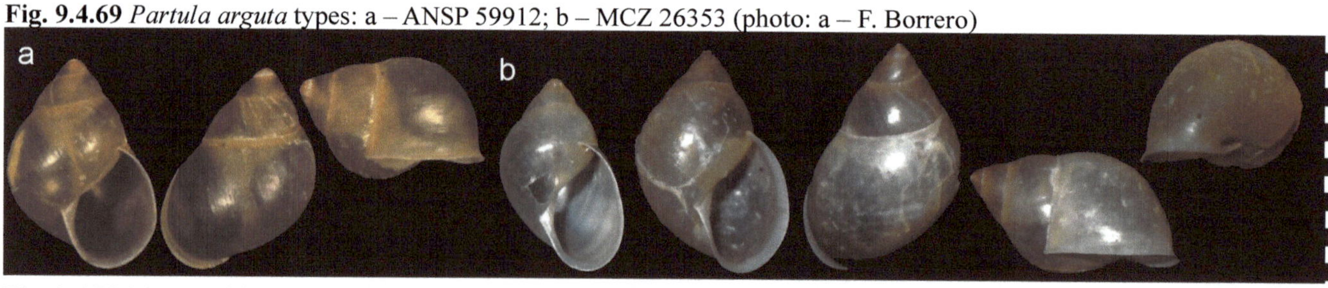

Fig. 9.4.70 Live Huahine *Partula*. a – *P. arguta*, b – *P. varia*, c-d – *P. rosea* (photos a, b, d – D. Clarke)

Raiatea

The Raiatean *Partula* are notable in including the first partulid to be described: *Partula faba* (Gmelin, 1791).

43 species have been described on Raiatea; half are not valid and, with the addition of three new species, a total of 23 Raiatean species are recognised here. In addition to these Pease (1871b) listed *P. stolida* from the island and Hartman (1881a) recorded *P. planilabrum* from 'Haamene' valley. The first appears to be a mislabelling and Garrett (cited in Smith 1903) noted that the second was an error for the valley of the same name on Tahaa island.

In order to manage the large number of Raiatean taxa there have been several attempts to group the species, initially based on shells and more recently on anatomy. Pilsbry (1909) removed *P. turgida* from the main *Partula* group (placing it in the subgenus *Leptopartula*) and divided the rest into three groups: the '*faba* group', '*dentifera* group' and '*hebe* group'. Crampton (MS) divided them further, separating out from the *faba*-group the *lugubris*-, *robusta*- and *fusca*-groups. Kondo (1955) added the *remota*- and *garretti*-groups and brought *P. turgida* back into the main *Partula* group, placing it in its own group, and accepting a total of 8 groups. Although these may have no phylogenetic meaning Kondo's arrangement is followed here for convenience with the combination of his *remota*- and *garretti*-groups:

faba-group – *P. faba* only. Large, robust shell. Penis long, typical *Partula*.

dentifera-group – Historically containing several species, here reduced to one (*P. dentifera*). Elongate shell, aperture usually auriculate. Penis similar to *P. faba* but varying geographically.

hebe-group – Three species. Globose shells. Penis much reduced.

lugubris-group – Four species. Elongately conical shells, thin in most species. Penis elongate, often clubbed.

robusta-group – Four species. Conical shells, aperture often auriculate. Penis robust.

fusca-group – Historically containing several species, here reduced to one (*P. navigatoria*). Elongately conical, aperture variable. Penis long.

garrettii-group – Six species (including Kondo's *remota*-group). Shell shortly conical, aperture often denticulate and columella twisted. Penis short. Note spelling corrected from Kondo (1955).

turgida-group – Three species. Shell very thin, globular, aperture simple. Penis large, robust.

faba-group

Partula faba (Gmelin, 1791)

Limax faba Martyn, 1784: 67 (unavailable work)
Auris Midae fasciata Chemnitz, 1786: 44 (unavailable work)
Voluta Auris Malachi var. γ Gmelin, 1791: 3437 (partim)
Helix Faba Gmelin, 1791: 3623
Bulimus australis Bruguière, 1792: 347
Voluta fasciata Chemnitz, Dillwyn 1817: 502
Helix faba Gmelin, Dillwyn 1817: 906. Wood 1818: pl33 f47
Bulimus faba Martyn, Férussac & Deshayes 1819: 40. Deshayes & Milne-Edwards 1839: 371. Pfeiffer 1848: 73
Partula Australis Férussac, 1821: 66
Partulus australis Férussac, Beck 1837: 57
Bulimus faba Deshayes (in Lamarck), 1838: 284
Bulimus inconstans Muhlfeldt, Anton 1839: 40
Bulimus tricolor Muhlfeldt, Anton 1839: 40 (nomen nudum)
Partula faba Gray, Sowerby in Gray 1839: 144
Partula faba Martyn, Reeve 1842: 175. Pfeiffer 1848: 73, 1853b: 44, 1877: 446. Pease 1870: 400. Hartman 1881: 182, 1885: 206. Garrett 1887: 57. Pilsbry 1909: 235. Kondo 1955: 112, 1968: 76. Richardson 1990: 24
Limax faba Martyn, Chenu 1845: 24
Bulimus faba Lamarck, Kuster, 1845: 20
Partula faba var. *subangulata* Pease, 1870: 401
Partula faba var. *subangulata* Pease, Pease 1871b: 458. Garrett 1884: 58
Partulus bulimoides (non Lesson) Pfeiffer 1877: 302
Partula amanda 'Garrett', Hartman 1881a: 180 (as var. of *faba*), 1881b: 6, 1885: 206. Paetel 1883: 152 (all nomen nudum). Garrett, 1884: 58. Smith 1903: 427
Partula bella 'Pease', Hartman 1881a: 180 (syn. of *amanda*, nomen nudum)
Partula biangulata 'Pease', Hartman 1881a: 180 (syn. of *faba*, nomen nudum)

Partula brunnea 'Pease', Hartman 1881a: 180 (syn. of *faba*, nomen nudum)
Partula dubia 'Garrett', Hartman 1881a: 181 (as var of *faba,* nomen nudum). Garrett 1884: 58
Partula marginata 'Garrett', Hartman 1881a: 184 (syn. of *faba*, nomen nudum)
Partula propinqua Hartman, 1881a: 195 (syn. of *faba*, nomen nudum), 1885: 206

Table 9.4.4 Raiatean *Partula* shell measurements. D – dextral, S – sinistral, species totals in bold

	N	L			W			L aperture			W/L		
		range	mean	sd	range	mean	sd	range	mean	sd	range	mean	sd
***faba* D**	8402	19.5-28.5	24.03	0.793	11.8-18.2	14.59	0.621	10.5-18.3	13.09	0.632	51.0-70.0	60.59	2.459
f. faba D	4782	19.5-28.5	23.87	0.665	11.8-17.9	14.58	0.611	10.5-15.9	13.08	0.667	-	60.50	2.608
f. faba S	1	-	25.80	-	-	15.70	-	-	11.09	-	-	61.00	-
***navigatoria* D**	1141	17.8-25.4	21.28	0.894	11.1-13.9	10.72	0.580	9.7-12.3	11.38	0.548	50.6-62.3	60.53	2.245
***navigatoria* S**	1	-	20.70	-	-	11.00	-	-	10.10	-	-	53.10	-
n. navigatoria	127	20.4-23.7	22.23	1.004	11.1-13.1	12.47	0.570	9.7-12.3	11.53	0.170	50.6-61.5	58.62	0.854
n. radiata	226	17.7-23.8	19.47	0.330	11.8-13.9	11.37	0.162	11.1-12.1	10.26	0.441	41.7-59.4	53.10	1.744
n. levilineata	2	17.8-18.8	18.30	-	11.1	11.10	-	9.7-9.9	9.80	-	56.6-61.5	59.05	-
dentifera	2413	17.3-25.6	20.31	1.147	9.4-15.1	11.95	0.604	6.8-14.3	11.39	0.578	48.5-68.5	59.40	2.180
d. callifera	78	17.7-20.8	18.95	0.726	11.3-13.1	12.30	0.415	9.5-11.7	10.69	0.434	55.5-63.5	59.81	1.658
d. candida	73	18.0-21.4	19.78	0.676	10.7-12.7	11.87	0.460	10.5-12.3	11.35	0.405	58.3-60.5	59.40	1.100
d. cedista	2	20.4-20.8	20.60	0.200	11.9-12.5	12.20	0.300	7.9-12.0	9.95	0.526	50.5-60.5	56.35	2.159
d. citrina	104	20.4-25.6	23.03	0.921	11.9-14.3	12.97	0.447	11.1-14.1	12.40	0.357	56.5-60.5	58.5	2.236
d. dentifera	260	16.5-22.4	19.12	0.860	9.4-12.7	11.07	0.502	9.5-12.1	10.80	0.575	54.5-65.5	58.98	2.224
d. formosa	654	18.0-25.0	21.39	1.203	10.7-15.1	12.38	0.694	9.9-14.3	11.99	0.658	54.5-65.5	57.50	1.309
d. imperforata	1315	16.5-23.5	19.87	1.220	9.9-13.3	11.80	0.604	9.3-13.1	11.18	0.578	48.5-68.5	59.40	2.180
auriculata	2282	15.3-22.1	18.53	0.867	10.3-13.9	12.22	0.489	9.1-12.7	10.38	0.260	59.5-74.5	65.89	2.190
protracta	1	-	19.80	-	-	11.80	-	-	10.30	-	-	59.60	-
dolichostoma	1	-	19.50	-	-	12.80	-	-	11.00	-	-	65.64	-
magistri	1	-	22.40	-	-	13.30	-	-	9.20	-	-	59.40	-
lugubris	550	15.0-22.0	17.75	0.870	8.9-13.1	10.46	0.462	8.1-12.3	9.61	0.507	52.4-67.5	58.98	2.282
l. lugubris	516	15.0-20.2	17.57	0.869	8.9-11.7	10.35	0.456	8.1-11.1	9.51	0.509	52.4-67.5	58.83	2.303
l. ovalis	34	18.3-22.0	19.90	0.857	10.7-13.1	11.84	0.541	10.1-12.3	10.87	0.468	55.5-63.5	59.50	1.894
leptochila	40	16.8-21.1	19.47	0.997	11.3-13.5	12.50	0.528	9.1-11.7	10.44	0.605	59.5-68.5	64.10	2.154
labrusca	10	17.1-22.4	18.66	0.603	12.7-14.3	11.71	0.603	7.2-9.8	7.93	0.402	56.3-64.0	61.19	3.470
dolorosa	5	16.5-18.1	17.33	0.487	9.5-11.1	10.26	0.542	9.1-9.5	9.38	0.160	56.6-61.5	58.90	1.743
garrettii	1638	13.5-20.5	16.27	0.764	8.3-12.9	10.30	0.449	7.1-11.1	8.96	0.407	57.5-85.8	76.46	2.512
g. garrettii	895	13.5-17.8	15.60	0.723	8.3-11.3	9.54	0.402	7.1-9.7	8.58	0.392	57.5-70.5	64.83	1.325
g. rustica	131	13.5-18.0	16.04	0.804	9.7-12.1	10.75	0.544	7.3-9.9	8.70	0.407	57.5-85.8	76.46	2.512
g. thalia	613	14.4-20.5	17.31	0.815	9.9-12.9	11.32	0.497	8.3-11.1	9.57	0.456	58.5-73.5	65.36	2.336
cuneata	1	-	17.30	-	-	10.20	-	-	8.70	-	-	58.96	-
levestriata	1	-	17.20	-	-	12.10	-	-	9.90	-	-	59.04	-
remota	93	16.4-19.0	17.49	0.808	11.1-13.7	12.35	0.496	8.9-11.1	10.01	0.480	65.5-76.5	70.64	2.303
tristis	5	16.6-17.0	16.62	0.041	10.5-11.1	10.82	0.073	8.9-9.5	9.30	0.413	63.5-66.5	64.50	1.323
atilis	15	16.5-20.2	18.17	0.900	11.1-13.1	11.99	0.541	9.3-10.9	10.11	0.332	65.5-76.5	71.60	2.166
hebe	3402	12.3-17.8	15.00	0.801	8.5-12.9	10.73	0.662	6.9-10.3	8.54	0.486	61.5-82.5	71.50	2.528
crassilabris	57	13.5-14.6	14.90	0.686	9.1-11.5	10.50	0.440	7.5-9.3	8.25	0.425	65.5-75.5	70.27	2.240
turgida	22	13.5-17.6	15.80	0.912	9.5-12.7	11.62	0.767	9.3-11.5	10.71	0.589	68.5-78.5	73.50	2.730
clarkei	5	14.0-15.5	14.63	0.455	11.5-12.0	11.78	0.199	8.6-9.8	9.12	0.465	77.4-82.1	80.11	1.671
meyeri	1	-	c16.5	-	-	10.50	-	-	7.24	-	-	c65	-
pearcekellyi	1	-	16.30	-	-	11.50	-	-	9.40	-	-	69.23	-

Fig. 9.4.72 Raiatean *Partula*. Top row: *P. auriculata, P. dolichostoma, P. protracta, P. magistri, P. crassilabris, P. hebe.* Second row: *P. lugubris, P. dolorosa, P. leptochila, P. labrusca, P. dentifera, P. pearcekellyi.* Third row: *P. garrettii, P. atilis, P. levistriata, P. cuneata, P. remota, P. tristis.* Fourth row: *P. turgida, P. clarkei, P. meyeri, P. navigatoria, P. faba*

Dextral (a single sinistral specimen known: BPBM 87363). Globose-elongate, very little variation in form. Umbilicus open. Thick, glossy with weak oblique growth lines and fine spiral striae on spire. Smoother and more glossy in subspecies *subangulata*. 5-6 convex whorls. Aperture oblong, slightly more than half as long as shell. Lip widely reflected, thick, upper third narrow. Columella vertical, simple on Raiatea (subspecies *faba*) but on Tahaa (subspecies *subangulata*) the internal thickening terminates in an elongate callous nodule near the insertion. Parietal callus thin, transparent, usually edentulous but a small denticle may be present in some populations. Garrett (1884) recorded a parietal tooth in only one of "about 6000 examples"; present in 2% on Raiatea (n=4,879), 14% on Tahaa (n=3,620).

Colour variable. Several varieties have been described:

bisecta – light brown with a narrow dark brown band at suture and a broad one covering lower part of body whorl. Pale areas may be partially obscured by oblique darker striations. Spire fading to whitish. Aperture and lip white or flesh-coloured. This is the typical form on Raiatea island.

fusca pallida – pale yellowish, a common variant on Raiatea.

strigosa – yellow to light brown with darker stigations (='var. *ventricosa*').

brunnea – mid-brown, without bands. Garrett (1884) recorded this as "not infrequent", presumably on Tahaa island.

castanea – dark chestnut brown.

frenata – brown with two dark narrow spiral bands.

zonata – brown with a paler subsutural band.

Fig. 9.4.73 *Partula faba* types: a – *faba* iconotype (Martyn 1786); b – *subangulata* syntype (NHMUK); c-d – *dubia* syntype (c – NHMUK; d – ANSP 59967); e-f – *amanda* syntype (e – ANSP 59415; f – NHMUK)

Fig. 9.4.74 *P. faba* variation: a-f – *P. f. faba*, g-n – *P. f. subangulata*. a, c-e – Vaiapu (JG); b – sinistral (BPBM 87363); f – Hamoa (JG); g – abnormal (NHMUK); h-i – Tapuamu (ANSP 291651); j-n – Pease specimens (ANSP 295243)

Anatomy: Penis long, at least ¾ oviduct length, curved, divided into three portions: clavate penis, constricted neck and curved, partially twisted caecum. Penial rectractor terminal. Interiorly these are papillate, with 5-6 ridges inside penis, neck smooth, caecum heavily papillose. Subspecies *subangulata* as *faba* but caecum slightly smaller, neck slender, stalk robust, cylindrical. The Vainae population of *subangulata* has an extreme constriction between the penis and caecum.

Type specimens:

Helix Faba Gmelin, 1791 – Specimen lost, holotype figure 67 Martyn (1784) is now the iconotype.

Auris Midae fasciata Chemnitz, 1786 – Holotype ZMC GAS 431.

Bulimus australis Bruguière, 1792 – Lost.

Partula Australis Férussac, 1821 – Lost.

Partula bella Hartman, 1881 – Nomen nudum, supposed syntypes MCZ 3639.

Partula propinqua Hartman, 1881 – Nomen nudum, supposed syntypes MCZ 25124 and 25308.

Partula faba var. *subangulata* Pease, 1870 – Syntypes AMNH 120662, MCZ 24948, NHMUK un-numbered, SMF 4953.

Partula dubia Hartman, 1881 – Syntypes ANSP 59965, 59967, NHMUK un-numbered. MCZ 24995 should probably also be regarded as types.

Partula amanda Garrett, 1884 – Syntypes ANSP 59414-5, NHMUK un-numbered.

Subspecies:

Partula faba faba (Gmelin, 1791) – Raiatea, all valleys, recorded from sea-level to the highest peak.

Partula faba subangulata Pease, 1870 **comb. nov.** – Tahaa, all valleys. Differs from nominate form in its more varied colour patterns and the presence of a vertical ridge on the columella. A parietal denticle is occasionally present, especially in northern populations.

Distribution: Raiatea and Tahaa, found in most valleys. Arboreal.

Note: Garrett (1884) thought that hybrids with *P. radiata, P. fusca* and *P. navigatoria* were "not uncommon." Pilsbry (1909) thought that *P. subangulata, P. amanda, P. navigatoria* and *P. fusca* might be subspecies of *P. faba*. Smith (1903) recorded two possible hybrids between *P. faba* and *P. radiata*.

Pilsbry (1909) considered the Tahaa *P. faba* a distinct species: "While the Tahaan forms of the *faba* type differ but little from Raiatean *faba*, yet their variation-forms are somewhat diverse, and the minor differences noted above seem fairly constant. I have therefore thought it best, in view of the minute specific subdivision of Raiatean *Partula*, to give *subangulata* specific rank." The form *amanda* does not appear distinguishable from *subangulata*.

Fig. 9.4.75 *P. faba* anatomical forms: *P. f. faba* after Kondo (1955), *P. faba subangulata* Hamene, Tahaa (ANSP A1386) and Vaimai, Tahaa (ANSP A993)

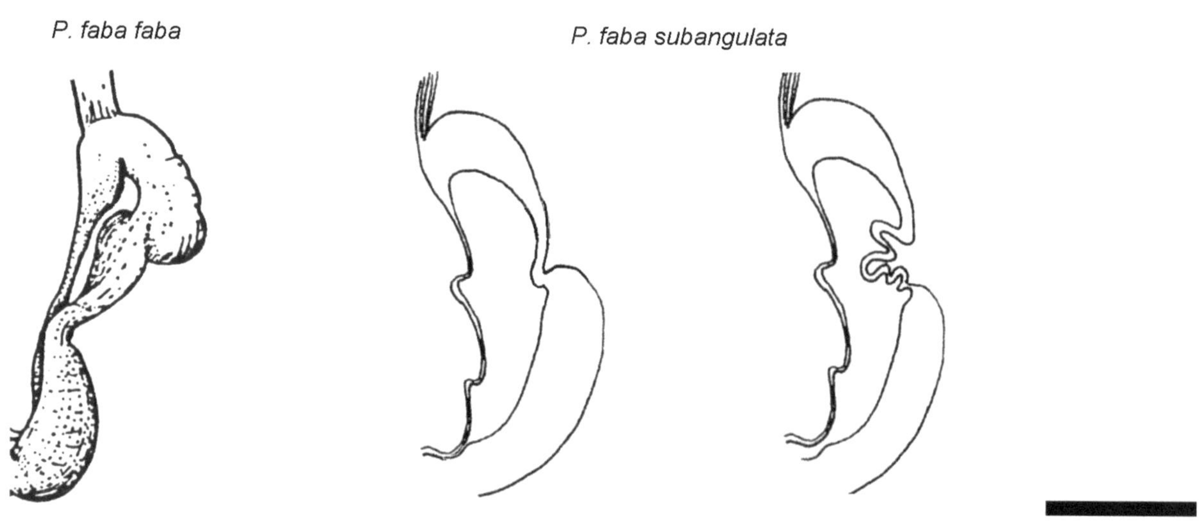

Fig. 9.4.76 *Partula faba* shell variation. Pie charts show shell colours with spire colour superimposed (Tahaa samples). Bar charts show tooth development. On Raiatea bar charts are shown only where some individuals showed tooth development. Raiatean populations with no variation are shown as red points with sample sizes superimposed.

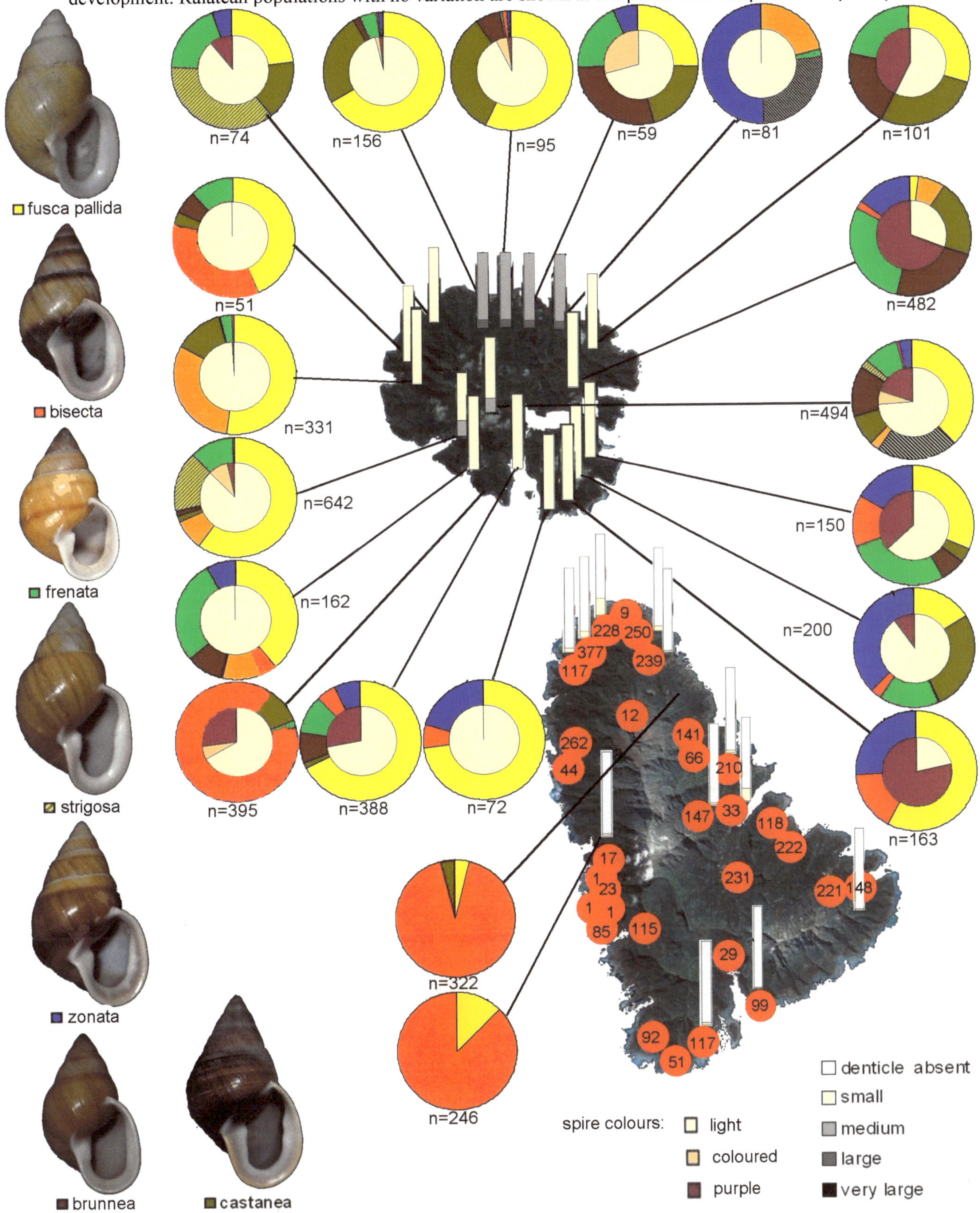

fusca-group

Partula navigatoria (Pfeiffer, 1849)

Bulimus (*Partula*) *navigatorius* Pfeiffer, 1849: 131

Partula radiata Pease 1864: 74

Partula nagivatoria Pfeiffer, Reeve 1850: pl 4 f21. Pfeiffer 1853b: 449, 1859: 512, 1868: 160, 1877: 205. Adams 1855: 145. Pease 1864: 675. Graeffe 1879: 85. Hartman 1881a: 184, 1881b: 7, 1885: 208. Garrett 1884: 76. Paetel 1889: 267. Smith 1903: 434. Pilsbry 1909: 161. Kondo 1955: 116, 1968: 76. Richardson 1990: 42

Partula fusca 'Pease' Carpenter in Pease 1865: 675 (as syn. of *navigatoria*, nomen nudum)

Partula vittata 'Pease' Carpenter in Pease 1865: 675 (as var. of *faba*, nomen nudum)

Partula protea 'Pease' Carpenter in Pease 1865: 675 (as var. of *faba*, nomen nudum). Schmeltz 1874: 92 (nomen nudum). Pfeiffer 1877: 209. Gloyne 1878: 338. Hartman 1881a: 182 (as var. of *fusca*), 1881b: 6, 1885: 207. Paetel 1883: 153, 1889: 268. Smith 1903: 430. Kondo 1955: 118, 1968: 76

Partula radiata Pease, Carpenter in Pease 1865: 675 (as syn. of *compressa*). Hartman 1881a: 185, 19881b: 7, 1885: 208, 1892: 74. Garrett 1884: 74. Ancey 1884: 344, 1889: pl1 f17. Smith 1903: 429. Pilsbry 1909: 232. Kondo 1955: 117, 1968: 76. Richardson 1990: 51

Partula faba, var., Carpenter in Pease 1865: 675 (misidentification)

Partula compressa 'Pfeiffer', Carpenter in Pease 1865: 675 (misidentification). Pease 1871: 453. Schmeltz 1874: 207. Paetel 1883: 152

Partula fusca Pease, 1866: 193

Partula vittata Pease, 1866: 194

Partula variabilis Pease, 1866: 203

Partula variabilis Pease, Pease 1866: 203, 1867: 81, 1871b: 473. Schmeltz 1869: 71, 1874: 207. Paetel 1873: 105, 1889: 268. Pfeiffer 1877: 201. Gloyne 1878: 338. Hartman 1881a: 189, 1881b: 7, 1885: 208. Smith 1903: 434

Partula fusca Pease, Pease 1871b: 473. Binney 1876: 245. Pfeiffer 1877: 205. Hartman 1881: 182, 1885: 207. Garrett 1884: 71. Pilsbry 1909: 240. Kondo 1955: 115, 1968: 76. Richardson 1990: 27

Partula vittata Pease, Pease 1871b: 473. Paetel 1873: 104. Pfeiffer 1877: 302. Garrett 1884: 75. Hartman 1881: 169, 1885: 207. Pfeiffer 1877: 200. Hartman, 1881: 189 Kondo 1955: 116, 1968: 76. Richardson 1990: 67

Partula terrestris 'Pease', Paetel 1873: 104 (nomen nudum), Gloyne 1878: 388 (nomen nudum). Garrett 1884: 75

Partula microstoma 'Pease', Paetel 1873: 104, 1883: 153, 1889: 267. Hartman 1881a: 184 (syn. of *vittata*), 1881b: 7, 1885: 207. Garrett, 1884: 74 (as form of *radiata*). Smith 1903: 429 (all nomen nudum)

Partula approximata 'Pease', Schmeltz 1874: 207, 1874: 80. Pfeiffer 1877: 208. Hartman 1881a: 179, 1881b: 6, 1885: 207. Paetel 1883: 152, 1889: 266 (all nomen nudum). Garrett 1884: 75 (as var. of *vittata*).

Partula terrestris 'Pease', Pfeiffer 1877: 208. Gloyne 1878: 337. Paetel 1883: 104, 1889: 268 (all nomen nudum). Garrett, 1884: 75 (as form of *vittata*)

Partula castanea 'Garrett', Hartman 1881a: 181, 1881b: 7 (as syn. of *terrestris*, nomen nudum)

Partula solidula Pease, Hartman 1881a: 187, 1881b: 7 (misidentification, partim)

Partula castanea Pease, Garrett, 1884: 76 (as syn. of *vittata*), nomen nudum)

Partula dentifera Pease, Hartman 1885: 207 (partim, misidentification)

Partula fusca vittata Pease, Pilsbry 1909: 242

Partula fusca approximata 'Pease' Garrett, Pilsbry 1909: 243

Partula fusca terrestris 'Pease' Garrett, Pilsbry 1909: 243

Partula levilineata Crampton MS, Kondo 1955: 129 (unpublished)

Partula levilineata Crampton, 1956: 6

Partula levilineata Crampton, Kondo 1968: 76. Richardson 1990: 36

Normally dextral (one aberrant sinistral shell known – MCZ 297856). Narrowly to moderately ovate. Umbilicus open or closed, varying geographically; narrow or closed in southern populations, wider in the north-west. Coarse growth lines and spiral striae on spire, may extend faintly onto body whorl. Spire conical, usually less than half of shell height. Suture impressed. 5-6 flattish whorls, last rounded, turgid. Aperture oblong, sides nearly parallel. Columella flattened, receding, nodular in old examples. Some populations acutely labiated on inner side or concave. Lip slanting, broadly expanded, moderately thick, thickened internally and sinuous above or semi-auriculate (7% of specimens). Parietal callus shining, a moderately distinct deeply recessed denticle sometimes present; well developed in some populations, absent or only present in older animals in others.

Fig. 9.4.77 *Partula navigatoria*: a – *navigatoria* syntype (NHMUK); b – *radiata* paralectotype (ANSP 59957); c – *fusca* syntype (ANSP 59430); d – *protea* syntype (NHMUK 1934-6-22.66-70); e – *microstoma* 'type' (MCZ 297857); f – *vittata* syntype (MCZ 24879); g – *variabilis* syntype (NHMUK 1934-6-22.31); h – *approximata* paralectotype (ANSP 59975); i – *approximata* lectotype (ANSP 59451); j – *levilineata* holotype (ANSP 272345); k-n – *castanea* paralectotypes (MCZ 297856); o – *terrestris* paralectotype (ANSP 59450) (photos: i-j – F. Borrero)

Whitish yellow to dark brown or black. Yellow forms may have chestnut base and sutural bands like *P. faba*. Dark forms may have an indistinct paler broad band at periphery. Pale shells may have darker streaks; strigation is most frequent in north-eastern populations. Apex pale, sometimes purple-black. Suture white margined. Lip usually white, often margined with purple-brown in northern populations, and tinged with brownish purple in the south. Parietal tooth in subspecies *radiata* often margined with pale purplish brown.

Body uniformly brown to black.

Anatomy: Penis long, ½-¾ oviduct length, inflated basally and terminally, constricted in middle. Caecum short, moderately inflated, laterally compressed. Vas deferens terminal. Penial retractor muscle apical. Everted penis spatulate.

Type specimens:

Bulimus (*Partula*) *navigatorius* Pfeiffer, 1849 – Syntypes NHMUK un-numbered.

Partula radiata Pease 1864 – Lectotype ANSP 59409, paralectotypes AMNH 42194; ANSP 2600, 59955, 59957; MCZ 24957. Lectotype designation by Baker (1963).

Fig. 9.4.78 *Partula navigatoria* variation: a – Tetoroa *navigatoria* (ZSL 180, wild collected); b-c – *radiata* × *navigatoria* (ZSL 974); d-g – Tepua *protea* (Garrett specimens ANSP 59972); h-i – Voaara *navigatoria* (Crampton specimens ANSP 294274); j-r – Pease duplicate *vittata* (MCZ 25011)

Partula fusca Pease, 1866 – Syntypes AMNH 118935, 42108; ANSP 59430; CAS IZ: 85759; MCZ 25263; FLMNH 111942 ('paratypes'); RMNH.MOL.56765 ('paratype').

Partula vittata Pease, 1866 – Syntypes MCZ 24879; SMF 157955. Other specimens ex Pease collection are not labelled as types (e.g. MCZ 25011) and are of uncertain status.

Partula variabilis Pease 1866 – Lectotype ANSP 59428; paralectotypes AMNH 42216, 118941, 42216; ANSP 463935; LACM 1877; MCZ 24994; NHMUK 1934-6-22.31-35; QM DM13130000(=1004346); RMNH.MOL.56764; SMF 4948. Lectotype designation by Baker (1963).

Partula protea Pfeiffer, 1877 – Syntypes MCZ 24996, 25349; NHMUK 1934-6-22.66-70.

Partula approximata Garrett, 1884 – Lectotype ANSP 59451 (ex Mus. Godeffroy); paralectotypes ANSP 456180, 59975 (ex Hartman); FLMNH 111939-40; MCZ 50851, 24927, 25197; NHMUK 1934-6-22.41-45; RMNH MOL 56765. Lectotype designation by Baker (1963).

Partula terrestris Garrett, 1884 – Lectotype ANSP 59450; paralectotypes ANSP 463720; MCZ 25296, 297859. Lectotype designation by Baker (1963).

Partula levilineata Crampton, 1956 – Holotype ANSP 272345; paratype ANSP 456713.

Partula castanea 'Garrett' – Supposed lectotype (nomen nudum) ANSP 59977a; paralectotypes MCZ 297856 (ex Pease, labelled Faaloa – the erroneous spelling for Faaroa bay used by Garrett in the brief description). Lectotype designation by Baker (1963).

Partula microstoma 'Pease' – Supposed types (nomen nudum) MCZ 25341, 297857.

Subspecies: *P. navigatoria* and its subspecies have been referred to as the *P. fusca* species group. Crampton (ANSP) kept *P. navigatoria* and *P. fusca* separate but placed the Tepua *P. protea* in *P. fusca*. There are two clear anatomical groups, one containing the western populations of *P. fusca* and *P. navigatoria*, with a modified version in *P. levilineata*, and another group containing all other populations. These two groups are interfertile as demonstrated by captive crosses between Tetoroa valley *navigatoria* and Hamoa *radiata* (Fig. 9.4.78b-c). The hybrid form is indistinguishable from *P. vittata*. This can be interpreted as a single species with three subspecies and hybrid forms:

P. n. navigatoria (Pfeiffer, 1848) – Voaara, Tetoroa, Huaru valleys. Tooroa '*vittata*' are probably this form. Includes *fusca* and *navigatoria*. Umbilicus open, wide. Columella simple.

P. navigatoria radiata (Pease, 1864) **comb. nov.** – East coast (Haamoa, Toahiva, Averarahi, Faaharoto, Vairua, Vairihi). Includes *radiata* and *microstoma*. Differs from nominate form in the umbilicus being narrower, lip acutely labiated on inner side, palatal projection more developed, parietal tooth present and columellar lip subnodose. Anatomically the caecum is more elongate, less inflated. Umbilicus rimate. Columella acutely labiated on inner side, lip subnodose. Parietal denticle usually present.

P. navigatoria levilineata Crampton, 1956 **comb. nov** – Marahi valley only, known from just two specimens. Smaller than nominate subspecies, umbilicus narrowly rimate, a weak labial ridge present near the outer margin of lip.

Fig. 9.4.79 *P. navigatoria* anatomical variation: a-d – *P. n. navigatoria*; e – *P. navigatoria radiata*; f – hybrid forms. a – Uparu (Kondo 1955), b – Uparu (ANSP A944), c – Voaara (Kondo 1955), d – Uturoto (ANSP A1413), e – Vairahi (ANSP A952), Vairua (ANSP A964) and Faaharota (ANSP A2747), f – Tetoroa × Haamoa (ZSL 974), g – Tuatua (ANSP 294465), h – Tepua (Kondo 1955); i – Marahi (Kondo 1955). Scale bar 10 mm.

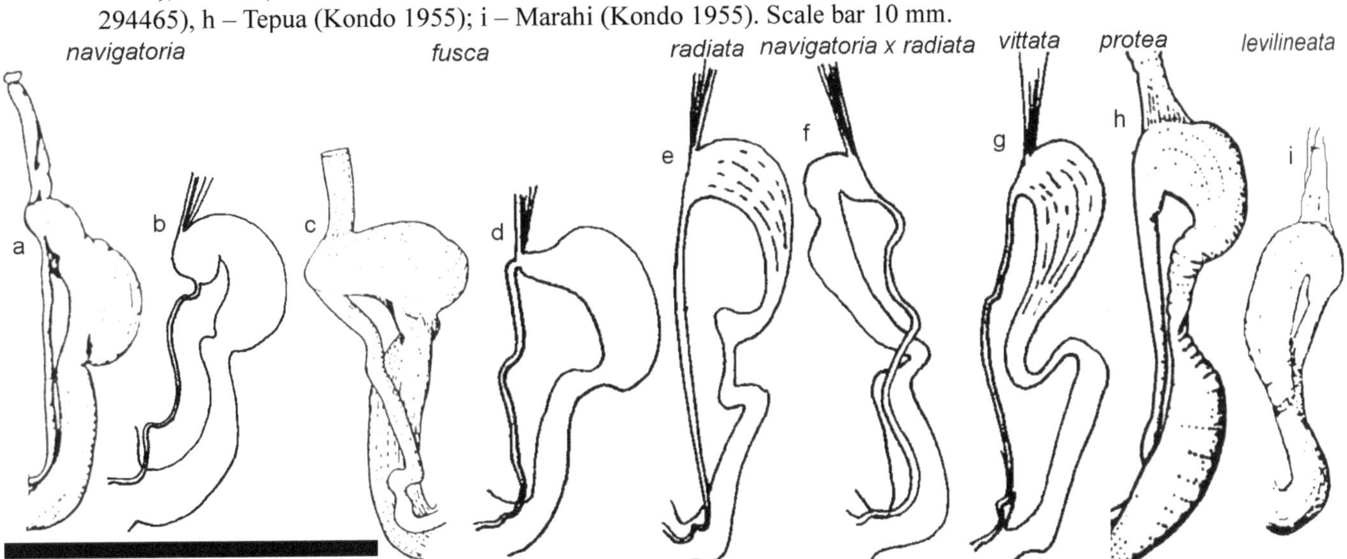

Fig. 9.4.80 *Partula navigatoria* geographical variation. Range boundaries of named taxa marked in yellow: c – *castanea*; f – *fusca*; l – *levilineata*; m – *microstoma*; n – *navigatoria*; p – *protea*; r – *radiata*; t – *terrestris*; v – *vittata*. Coded by currently recognised taxa: green – *P. n. navigatoria*; blue – *P. navigatoria radiata*; orange – *P. navigatoria levilineata*; white – *P. navigatoria* × *radiata*

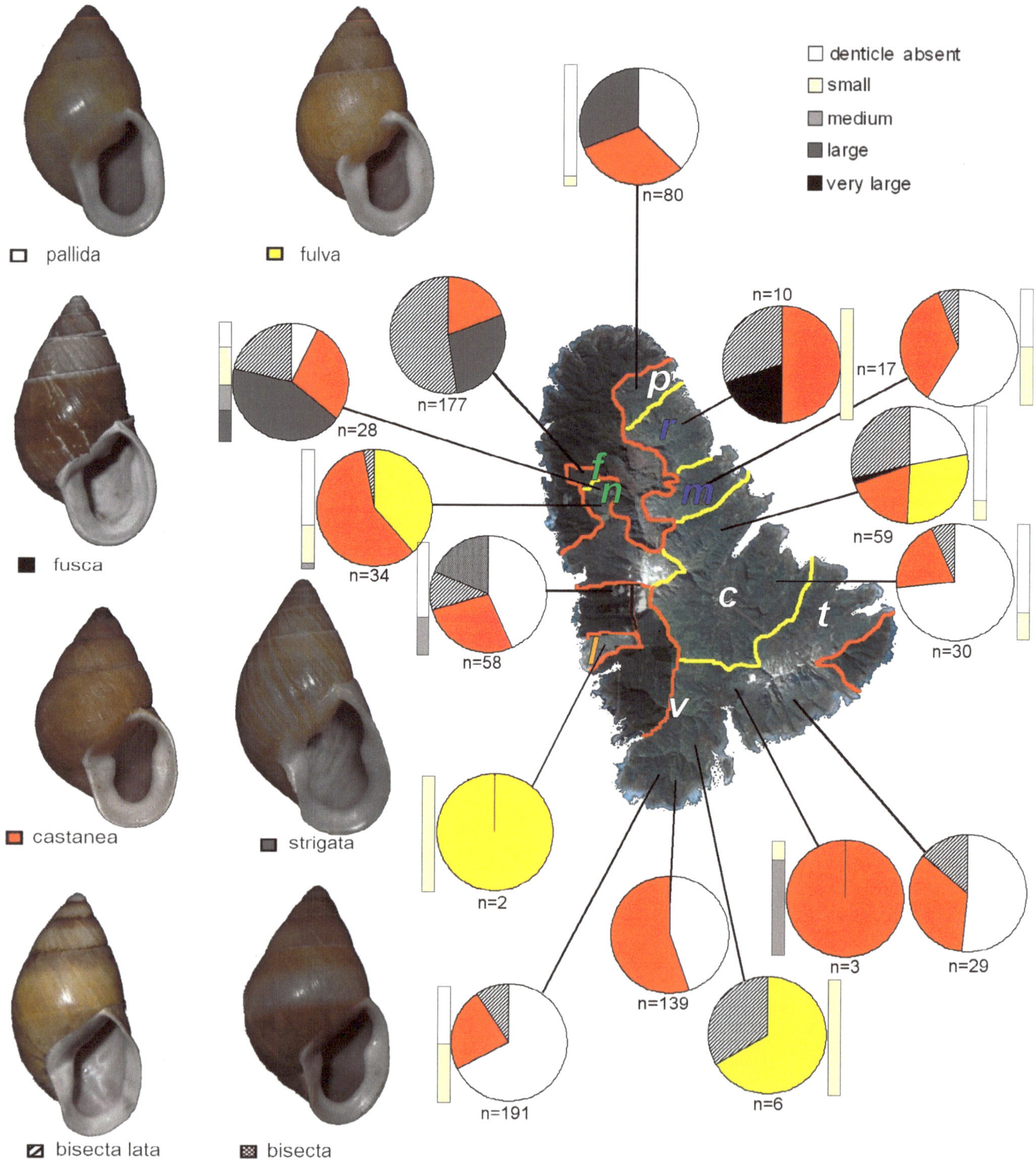

Parietal tooth very small or only a trace. Anatomically only a slight constriction between penis and caecum.

Hybrid populations – East coast south from Faaroa, west coast only in the southern valleys of Tuatau, Vaetuna, Uturoto and Vaihuti (*vittata, castanea* and *terrestris*), and a separate population in the north, in Tepua valley (*protea*). Anatomically as *P. navigatoria radiata* except constriction between penis and caecum more pronounced. Shells variable, a moderately developed parietal denticle may be present (*vittata*). Umbilicus narrow or closed in the south. Columella concave. Lip thinner in *vittata*, thinnest at upper end, forming a distinct sinus.

Distribution: Raiatea, present in most valleys. Terrestrial, described by Garrett (1884) as being "not uncommon under decaying vegetation" (*protea*), "quite common beneath decaying vegetation and among piles of loose stones" (*radiata*) or "frequently found adhering to the lower portion of the trunks of trees and shrubs" (*microstoma*). Crampton (1956) recorded *P. levilineata* as arboreal, however he did not collect the specimens himself and this may be an error.

Note: This species has long been a source of confusion. Garrett (in Smith 1903) observed that "*P. protea, fusca* and *navigatoria* are the most difficult species to study of any inhabiting Raiatea. They are all connected with each other by hybrids, and all are truly protean forms." Garrett (1884) also noted that "Hybrids between *protea* and the arboreal *P. faba* are not uncommon, and are usually found adhering to the lower parts of the trunks of trees." On what basis he ascribed them to hybrids was unexplained. "In Vaioara, hybrids between *fusca* and *navigatoria*, and between the two former and *faba*, are so frequent as to be very embarrassing in the separation of the three species collected in that valley. Like the Tepua hybrids, all those between the two ground species and the arboreal *faba* live on the lower parts of the trunks of trees." There has even been confusion with the *P. lugubris*-group; Hartman (1885) observed "The type examples [meaning 'typical', not the type series] of *fusca* (Coll. Pse.) are immature shells of the uniform dark fuscous *ovalis*, two quarts of which were included in the Pease duplicates, labelled *P. ovalis* by Pease." Hartman (1881) considered *P. lugubris*, *P. ovalis*, *P. protea* and *P. fusca* to be a growth series. Garrett (1884) simplified matters by correctly placing three manuscript names in *vittata – terrestris, castanea* and *approximata*, and *microstoma* in *P. radiata*. Contrarily, Hartman (1881) considered *P. microstoma* to be *P. vittata.* Pilsbry (1909) considered *P. vittata* "Doubtfully distinguishable from *P. fusca.*" This confusion arises from the hybrid origins of these populations.

dentifera-group

Partula dentifera Pfeiffer, 1853

Partula dentifera Pfeiffer, 1853b: 447

Partula dentifera Pfeiffer, Pfeiffer 1854b: 85, 1855c: 264, 1856: 162, 1856: 264, 1868: 159, 1877: 201. Adams 1855: 145. Carpenter in Pease 1865: 675. Pease 1871b: 473. Schmeltz 1874: 207, 1874: 80. Hartman 1881a: 181, 1881b: 7, 1885: 210, 1892: 74. Paetel 1883: 153, 1889: 266. Garrett 1884: 84. Smith 1903: 439. Pilsbry 1909: 216. Kondo 1955: 123, 1968: 76. Richardson 1990: 21

Partula callifera Pfeiffer, 1857b: 333

Partula callifera Pfeiffer, Pfeiffer 1859a: 511, 1868: 159, 1877: 201. Carpenter in Pease 1865: 675. Pease 1871b: 473. Gloyne 1878: 33. Hartman 1881a: 180, 1881b: 8, 1885: 210, 1892: 74. Paetel 1883: 152, 1889: 266. Garrett 1884: 60. Smith 1903: 440. Pilsbry 1909: 214. Kondo 1955: 123, 1968: 76. Richardson 1990: 14

Partula citrina 'Pease' Carpenter in Pease 1865: 675 (as syn. of *faba*, nomen nudum)

Partula faba var. Carpenter in Pease 1865: 675 (misidentification). Hartman 1881a: 180, 1881b: 6, 1885: 207

Partula imperforata 'Pease' Carpenter in Pease 1865: 675 (as syn. of *dentifera*). Schmeltz 1869: 71, 1874: 92. Paetel 1873: 104, 1883: 153. Pfeiffer 1877: 209. Gloyne 1878: 338. Hartman 1881a: 183, 1881b: 8, 1885: 210, 1892: 74 (all nomen nudum). Garrett 1884: 54. Smith 1903: 440. Pilsbry 1909: 160. Kondo 1955: 120, 1968: 76. Kilias 1869: 247. Richardson 1990: 35

Partula citrina Pease, 1866: 195

Partula labiata 'Pease', Schmeltz 1869: 71, 1874: 92, 1874: 80. Carpenter 1865: 675. Pease 1868: 227. Paetel 1873: 104, 1883: 153, 1889: 267. Pfeiffer 1876: 33, 1877: 209 (all nomen nudum)

Partula citrina Pease, Pease 1871b: 473. Schmeltz 1874: 81. Pfeiffer 1877: 200. Hartman 1881a: 180, 1881b: 6, 1885: 207, 1892: 74. Paetel 1883: 152, 1889: 266. Garrett 1884: 64. Smith 1903: 430. Pilsbry 1909: 218. Kondo 1955: 119, 1968: 76. Richardson 1990: 18

Partula megastoma 'Pease', Schmeltz 1874: 71, 1874: 92. Carpenter in Pease 1865: 675. Pfeiffer 1877: 209 (all nomen nudum)

Partula callistoma 'Pease', Schmeltz 1874: 207, 1874: 81 (all nomen nudum)

Partula formosa 'Pease', Graeffe 1879: 85. Hartman 1881a: 182, 1881b: 8, 1885: 210, 1892: 74 (all nomen nudum). Garrett 1884: 60. Smith 1903: 440. Kondo 1955: 121, 1968: 76. Richardson 1990: 27

Partula decorticata 'Pease', Hartman 1881a: 181 (as syn. of *dentifera*, nomen nudum)

Partula Peaseana 'Garrett', Hartman 1881a: 186 (as syn. of *dentifera*, nomen nudum)

Partula raiatensis 'Garrett', Hartman 1881a: 186 (as var.of *dentifera*), 1881b: 8, 1885:210 (all nomen nudum). Garrett 1884: 54 (as var. of *imperforata*). Smith 1903: 440
Partula recta Pease, Hartman 1881b: 8 (misidentification), 1885: 210
Partula dentifera formosa Garrett, Pilsbry 1909: 218
Partula cedista Crampton MS, Kondo 1955: 122 (unpublished)
Partula cedista Crampton, 1956: 2
Partula candida Crampton, 1956: 4
Partula cedista Crampton, Kondo 1968: 76. Richardson 1990: 17
Partula candida Crampton, Kondo 1968: 76. Richardson 1990: 16

Fig. 9.4.81 *P. dentifera* specimens: a – *dentifera* syntype (NHMUK); b – *citrina* syntype (ANSP 59564); c – *formosa* paralectotype (NHMUK 1934-6-22.126); d – *megastoma* 'paralectotype' (ANSP 59456); e – *callifera* Crampton, Haamoa (ANSP 294220); f – *imperforata* syntype (NHMUK 1934-6-22.173-6); g – *imperforata* colour variety (BPBM 87365); h – *raiatensis* lectotype (ANSP 59488); i – *candida* syntype (ANSP 272348); j – *cedista* holotype (ANSP 272254); k-l – *formosa* colour variants, Vaihuti (ANSP 294480) (photo: b – P. Callomon)

Ovate-conical to elongate-conical. Shiny. Very narrowly umbilicate ('chink-like' – Garrett 1884). Conical spire with acute apex. 4¾-5½ largely flat whorls, penultimate whorl more convex, last slightly shorter than spire, convex and impressed anteriorly in the middle. Suture margined. Smooth, slightly shining; ornamented with spiral striae, fading on the body whorl. These vary in development and are usually indistinct. Columella almost vertical, slightly folded. Aperture oblique, narrowly auricular. Lip thickened, margins subparallel, right margin deeply curved above, with a strong tooth-like tubercle in the middle. Its face angularly ridged, strongly labiated within. A parietal tooth may be present, varying in development geographically (most frequently developed in Vairahi valley: 84%, n=47). The aperture changes from with age: from a vertical columella, becoming more inclined and plicate; older subadults with a partial lip, with a less developed labial tooth and more openly auriculate aperture than full adults (Fig. 9.4.82).

Pale yellow to dull corneous brown. Sutural line sometimes white. Apex in most is fawn coloured, similar to the spire but slightly darker (e.g. in Vairahi 4% [n=47] had white apices that were not due to decortications, and in Averahi the apex was distinctly reddish in 3% [n=214]). Darkest on protoconch, sometimes extending almost to penultimate whorl. Aperture and lip white. Garrett (1884) recorded a very rare variety of ruddy brown colour with a purple-black apex and flesh-coloured lip, not located by Crampton. Rarely banded.

Anatomy: Penis long, usually ½ oviduct length. Constriction between penis and caecum slight. Penis with 5-branched pilasters, caecum with papillate ridges. Vas deferens subterminal.

Type specimens:

Partula dentifera Pfeiffer, 1853 – Syntypes NHMUK un-numbered.

Partula callifera Pfeiffer, 1857 – Apparently lost. Three supposed types have been located, but none are types. ANSP 59946 and 59455 were from Hartman and Garrett respectively and as both were collected by Garrett they cannot be the types. The NHMUK specimens are annotated 'types ? Clench' 'no Crampton ?35', suggesting that Crampton did not consider them types. They do not match Pfeiffer's description and are misidentified *P. auriculata* (see p. 186: Fig. 9.4.85f). The types were collected by Cuming but cannot be located in NHMUK.

Partula citrina Pease, 1866 – Syntypes AMNH 42142; ANSP 295264; MCZ 24885, 25027.

Partula formosa Garrett, 1884 – Lectotype ANSP 59453; paralectotypes AMNH 118946; ANSP 59454; MCZ 25193; CAS: IZ: 87967, 87969; FLMNH 111935; NHMUK 1934-6-22.126-130; SMF 4959. Lectotype designation by Baker (1963).

Partula imperforata Garrett, 1884 – Lectotype ANSP 59502a, paralectotypes AMNH 118945, 42224; CAS IZ: 87970; FLMNH 112004-5, 112012; MCZ 25262; NHMUK 1934-6-22.173-6; ZMB Moll-74332, 75634. Lectotype designation by Baker (1963).

Partula raiatensis Garrett, 1884 – Lectotype ANSP 59488; paralectotypes ANSP 458797, 59951; FLMNH 112076; NHMUK 1934-6-22.160-3; YPM IZGP 015862; ZMB/Moll-74331, 75719. Lectotype designation by Baker (1963).

Partula cedista Crampton, 1956 – Holotype ANSP 272254.

Partula candida Crampton, 1956 – Syntypes ANSP 272348, 294226-8; USNM 5490.418872. Although USNM 5490.418872 and ANSP 272348 are labelled holotypes Crampton (1956) did not designate any specific types, and all should be regarded as syntypes.

Partula labiata 'Pease' – Supposed holotype (nomen nudum) ANSP 59460; 'paratype' MCZ 24882.

Partula megastoma 'Pease' – Supposed lectotype (nomen nudum) ANSP 5945a; 'paralectotypes' ANSP 59456, MCZ 25019 (Baker 1963).

Subspecies: Several supposed species are best regarded as subspecies of *P. dentifera*.

P. d. dentifera Pfeiffer, 1853 **comb. nov.** – Recorded from Avera, Averarahi, Averaiti, Faaiti, and the northern part of Faaroa valley. Garrett (1884) described *P. dentifera* (referring to the nominate form) as being mainly distributed in Vairahi valley, an error for Averarahi (Crampton MS). Distinguished by its elongate-conical form: straw-yellow colour: rather shining surface, reduced, rimate umbilicus, small oblong to obauriform aperture, resulting from the prominent labial denticle, heavily calloused, carinate and internally labiated lip.

Fig. 9.4.82 *Partula dentifera* aperture development, all shells from Faaroa (JG)

Fig. 9.4.83 Geographical variation in colour and tooth development in *Partula dentifera* populations, subspecies ranges labelled

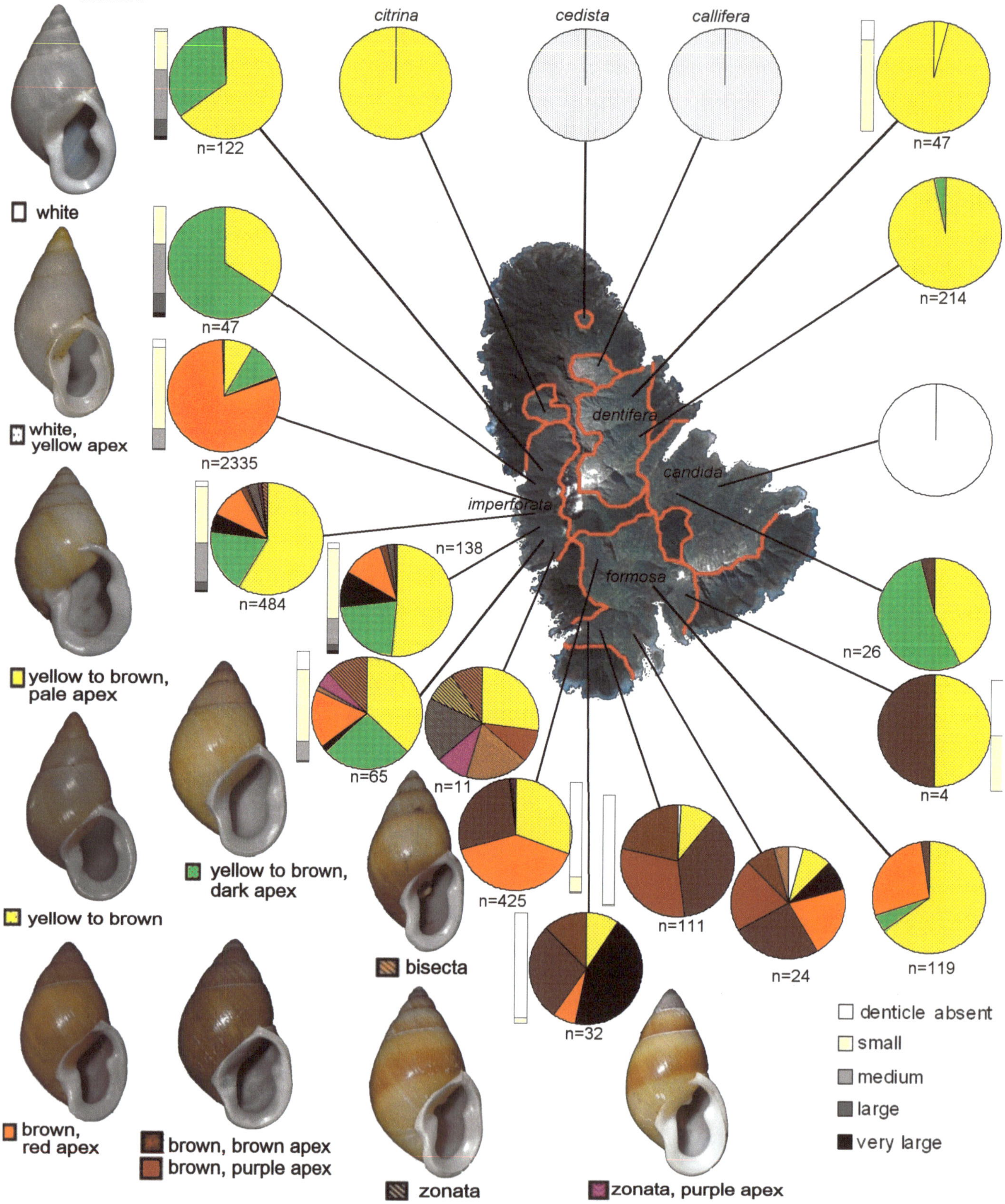

Partula dentifera citrina (Pease, 1866) **comb. nov**. – Uparu Valley, Tepua and Voaara valleys. Garrett (1884) noted that he originally labelled this "a somewhat rare species" and that many specimens labelled *P. citrina* were pale, slender forms of *P. faba*. It was not in fact rare as Garrett (1884) explained: "When I first discovered it in 1861, I took but few examples, in consequence of not penetrating far enough into the valley to find its headquarters. A more extended research in 1873 revealed its specific centre, and I took about eight hundred specimens in various stages of growth, and many of the adults were in a gravid condition." Only known from three collections: Garrett 1861 from upper Uparu valley (type series), Garrett 1873 from the same locality (about 800 specimens), and Crampton 1908-9 in Uparu (73 adults, 9 subadults) and upper Voara valley (66 adults, 82 subadults). Crampton (MS) noted that Garrett did not collect the species from Voaara even though he collected specimens of other species from the valley. Differs from nominate form in whorls being slightly more convex, aperture more strongly auriform, labial tubercle more developed, columella slightly dilated and slightly guttered, an oblong callous nodule below the gutter, labial projection normally absent (very small in one specimen) and labial carina rarely present. It is always pale yellow with faintly darker or greenish streaks, apex yellow or with a red tint. Anatomically the penial caecum is shorter than in other subspecies.

Fig. 9.4.84 *P. dentifera* anatomical variation. a & b – Averarahi (Kondo 1955), c – Fareparahi (Kondo 1955), d & e – Uparu (Kondo 1955), f – Uparu (ANSP A946), g & h – Voaara (ANSP A951), i – Utuara (Kondo 1955), j – Tooroa (ANSP A960), k – Tooroa (ANSP A960) and Utuara (ANSP A1391), l – Vaiaau (Kondo 1955), m – Tutira (ANSP 294480). Scale bar 5 mm.

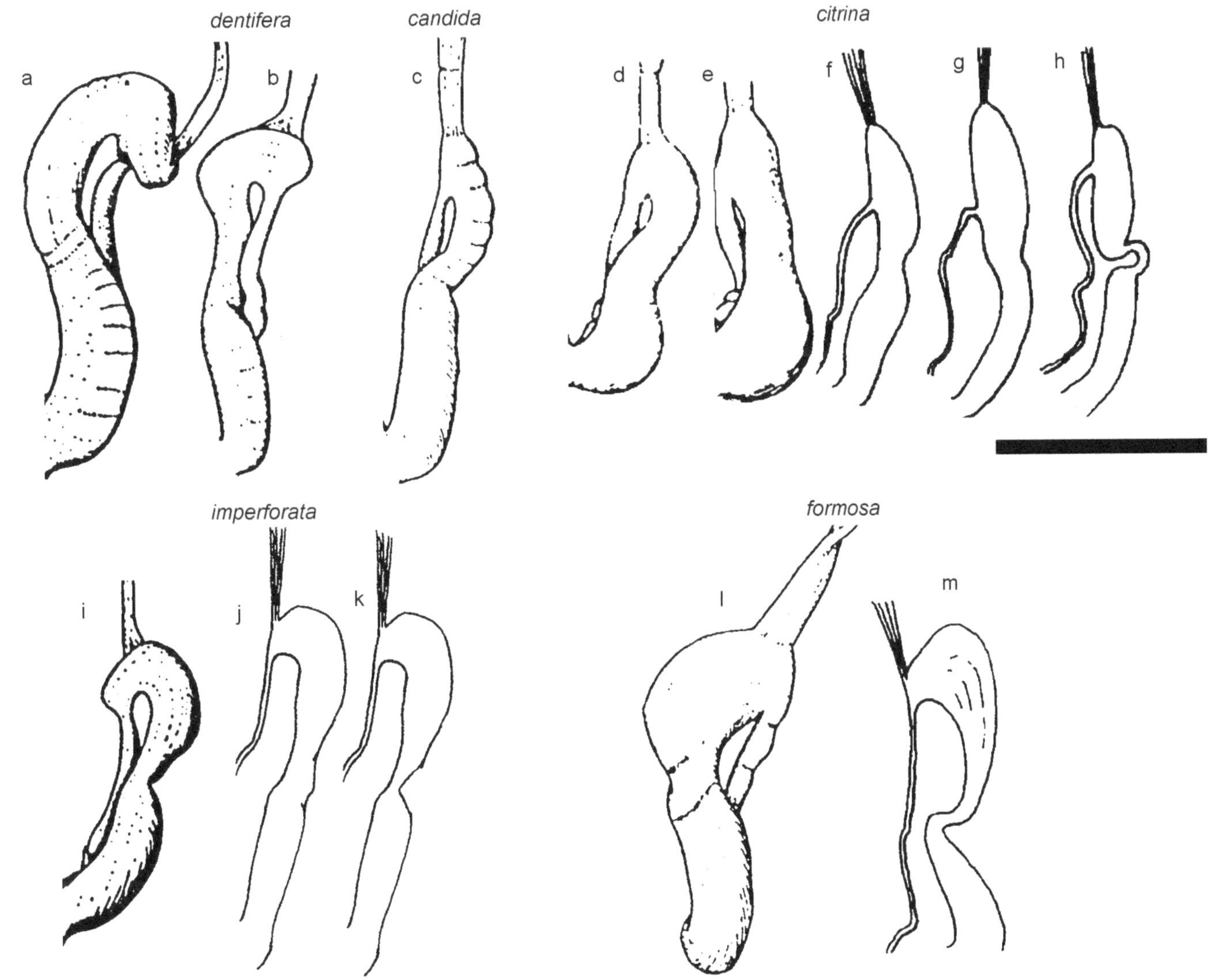

Partula dentifera callifera (Pfeiffer, 1857) **comb. nov**. – Restricted to the higher parts of Haamoa valley. Differs from nominate form in being more ovate, with an inflated body-whorl, more open umbilicus open, more developed sculpture, columella tubercular at top, folded within. White, lacking periostracum in adults. The spire is usually yellow. Shows only a small range of variation in shade. Anatomy not known.

Partula dentifera formosa (Garrett, 1884) – Vaiaau, Faaroa, Tivae, Tuatau, Vaeanae and Faatemu valleys. The Faaroa record is probably from the head of the valley, spreading over from Vaiaau valley: 25 specimens were collected, fewer than other sites. Garrett recorded it only from Faatemu and Vaiaau, recording its great abundance in the coastward and low-level parts of Faatemu, declining in abundance to the northwards. Crampton found it more widespread: "This account differs from what Garrett recorded for his time, and the disparities call for explanation. The author's personal collections of Formosa in Faaroa valley were found in the very centre of that wide and open area, in company with P. hebe, P. faba, and P. vittata. Garrett's correct assignment of these associated species to this valley is acceptable evidence that he actually surveyed the area in question. [His failure] to record the presence of formosa may have been inadvertent it seems more probable that the existing colony has been established since his time by migration either from Faatemu Valley or from Vaiaau Valley, or from both.

"The conclusion is otherwise when the additional valleys of present occupation are considered. Intrinsic evidence makes it virtually certain that Garrett did not explore them, as his did the major valleys of Faatemu, Vaiaau and Faaroa. The colonies now existing in Vaianae on the east, and in Tuatau, Vaihuti and Tivae on the west differ distinctly from those of Faatemu and Vaiaau; they comprise well-represented classes of dark colored shells that are absent elsewhere, while the shells are much smaller in all of their direct dimensions. These distinctions must have required considerable time for their productions; they are reasonably interpreted as the results of genetic variation subsequent to the invasion of the smaller coastal valleys in question by migrants from Faatemu or Vaiaau. Additional circumstantial proof that Garrett did not survey the valleys from Tuatau to Vaihuti is the author's discovery in these localities of a novel and distinctive species belonging to the hebe group, namely Partula remoata [sic]." (Crampton MS). Differs from nominate form in the umbilicus being completely closed, more developed sculpture, slightly impressed suture margined with a rugose white line. Smith (1903) recorded a parietal tooth present in around half of specimens but Garrett (1884) described the parietal region as "edentate", however Crampton collected dentate individuals in Vaianae, Vaihuti and Tivae (small or medium in 15%). A prominent palatal projection is present. Pale yellowish white to dark brown, rarely completely white. Spire frequently tinged rose-red, sometimes purple brown. Some shells are medium brown with a submedial light band, light brown with a darker band or with two narrow bands. Anatomically the penial caecum is shorter than in *P. d. dentifera*.

Partula dentifera imperforata (Garrett, 1884) **comb. nov**. – Small western valleys (Utuara, Toloa, Hapai, Murifanna, Ereeo, Hanua, Marahi, Tehurui, Tooroa, Utuara, Vaiaau, Paaoio, Tetoroa) and Mt. Tefatua, from sea-level to 700 m. Differs from nominate from in umbilicus being narrower, sometimes closed, oblong aperture, subnodose columella, subnodose and frequent parietal tooth. Plain light yellowish brown, brown or deep brown, or banded, usually uniform and diffused by a few with darker transverse streaks on a lighter background. Almost always imperforate in Hapai valley (typical form), Toloa is occupied by the lighter, rarely imperforate, rose apexed *raiatensis* form. Garrett's *Partula raiatensis* differed from *P. dentifera* in having a strong parietal tooth and a less prominent palatal projection. The apex was generally rosy. Hartman (1881) placed this in *P. dentifera* whereas Garrett referred it to *P. imperforata*. Garrett (1884) united the two *imperforata* populations of Toloa and Hapai on the basis of an examination of around 2,000 specimens. Constriction between caecum and penis more pronounced than in *P. d. dentifera*.

Partula dentifera cedista (Crampton, 1956) **comb. nov**. – Known only from the holotype from Vairua valley, 450 ft. Differs from nominate form in shape: fusiform-conic with an elongately conic spire, subsutural area distinctly impressed on the last half of the body whorl, umbilicus reduced by a reflexed fold of the columella, columella vertical, with a weak tubercle, parietal tooth well-developed. Dull grey-white, earlier whorls tinged with light fawn, periostracum present only as microscopic pale fawn fragments.

Partula dentifera candida (Crampton, 1956) **comb. nov**. – Small south-eastern valleys (Farepaiti, Fareparahi, Faaroa and Maufenua. Crampton (1956) recorded that this form spread into the edges of Faaroa valley but was scarce and not found at lower altitudes there. Differs from nominate form in being imperforate, distinctly auriform and having a more inclined aperture. Shell completely devoid of epidermis, even in embryos. White, sometimes with a fawn tinted apex. Penial caecum shorter and narrower than in *P. d. dentifera*.

Distribution: Recorded from most valleys on Raiatea. Several subspecies can be recognised in different valleys. Arboreal (*P. d. dentifera* was recorded "in vast numbers on foliage": Pilsbry 1909).

Note: Pilsbry (1909) included the Tahaa species *P. virginea* in this group (and listed as a synonym of *P. imperforata* by

Richardson 1990) and defined them as arboreal species with "the usually light yellow color of the shell, the summit white, yellow, pink, or purple-tipped; lip white, thickened within, more or less strongly toothed within the outer lip and excised above the tooth. Columella and parietal wall often toothed. The species are closely related to some of the *faba* group." Although these taxa are closely related, they are kept as separate species here.

robusta-group

Contains five species (four as defined by Kondo [1955] plus one new species), only one of which (*P. auriculata*, including *robusta*) is known from more than shell material.

Partula auriculata Broderip, 1832

Partula auriculata Broderip, 1832: 125

Partula maura 'Muhl.' (non Grateloup), Anton 1839: 40 (nomen nudum)

Partula auriculatus (Broderip) Beck 1837: 58

Bulimus auriculatus Broderip, Jay 1839: 57, 1841: 204. Pfeiffer 1841: 80, 1842: 111

Partula auriculata Broderip, Müller 1836: 33. Reeve 1842: 175, 1850: no 11. Catlow & Reeve 1845: 162. Pfeiffer 1848: 72, 1856: 264. Gould 1852: 84. Pease 1864: 675, 1866: 201, 1871b: 473. Schmeltz 1869: 71, 1874: 97. Paetel 1873: 104, 1883: 153, 1889: 266. Hartman 1881a: 180, 1881b: 7, 1885: 209, 1892: 74. Garrett 1884: 58. Smith 1903: 436. Pilsbry 1909: 247. Kondo 1955: 124, 1968: 76. Kilias 1969: 246. Richardson 1990: 11

Bulimus Otaheitanus Pfeiffer 1848: 71 (misidentification)

Partula Otaheitana Reeve 1851: pl2 (misidentification)

Partula compacta 'Pease' Carpenter in Pease 1865: 675 (as syn. of *auriculata*, nomen nudum)

Partula robusta 'Pease' Carpenter in Pease 1865: 675 (as syn. of *auriculata*, nomen nudum). Garrett 1884: 58 (syn. of *auriculata*). Hartman 1881a: 186, 1881b: 1 (var. of *auriculata*, nomen nudum). Smith 1903: 436. Kondo 1955: 125

Partula compacta Pease, 1866: 200

Partula compacta Pease, Pease 1867: 81; 1871b: 473. Schmeltz 1869: 71, 1874: 92. Paetel 1873: 104, 1883: 152, 1889: 266. Pfeiffer 1877: 207. Hartman 1881a: 181, 1881b: 7, 1885: 209, 1892: 74. Garrett 1884: 55. Smith 1903: 437. Pilsbry 1909: 248

Partula tahitana Bruguière, Pease 1871b: 473 (nomen nudum)

Partula tahitiensis Pease, Schmeltz 1874: 92 (nomen nudum)

Partula solidula 'Pease', Smith 1903: 436 (misidentification)

Partula tahulana Anton, 1893: 40 (nomen nudum)

Partula auriculata robusta Pease, Pilsbry 1909: 248

Obese to ovate-pyramidal. Thick. Umbilicus open, narrow. Surface dull with rough growth lines and dense spiral striae on spire. 5-6 convex whorls. Lip moderately expanded, thickened internally. Aperture slightly (*robusta*) or strongly auriculate (*auriculata*). Columella thick, oblique and broad, notched at the upper insertion. Palatal projection indistinct but usually present (in 85% of *robusta* and well developed in all *auriculata*). Parietal callus thick, raised. Small, low parietal tubercle, sometimes indistinct in *robusta*, more developed in *auriculata*.

Typically light brown with oblique chestnut strigations, to dark brown. Darker on spire, becoming purple-black towards apex. Lighter variety yellowish with faint brown streaks, apex pale. Darker varieties are either uniformly dark chestnut with pale steaks or dark with a supramedial yellow band of variable width. Banded forms are most common in *auriculata*, with more variation in *robusta*. Lip usually faintly brown-edged appearing to have been lightly washed, this is more intense on the inner face of the carina. This is most apparent in brown and banded shells; present in 50% of striated shells. Apex and lip colours are not correlated.

Anatomy: Penis and caecum short, separation indistinct, ¼-½ oviduct length. Caecum semi-cordate. Vas deferens medial on caecum (*robusta*) or submedial (*auriculata*).

Type specimens:

Partula auriculata Broderip, 1832 – Syntypes ANSP 59480; NHMUK un-numbered ('Huaheine'); SMF 164982; ZMB/Moll-88365 (ex Krammann collection).

Partula robusta Smith, 1903 – Lectotype ANSP 5944a; paralectotypes ANSP 59440, 59986, 463798; MCZ 3651. Lectotype designation by Baker (1963).

Partula compacta Pease, 1866 – Lectotype ANSP 59983; paralectotypes AMNH 42179; FLMNH 111891, 1934-6-22.71-5; MCZ 24993; NHMUK (un-numbered); SMF 4952, 164983. Lectotype designation by Baker (1963).

Fig. 9.4.85 *Partula auriculata*: a-h – *auriculata*; a-b – syntypes (NHMUK); c – Hamoa valley specimen (ANSP 294449); d – specimen (UMZC 75294); e, g & h – specimens (ANSP 295317); f – supposed *P. callifera* type (NHMUK); i-n *robusta*; i – paralectotype (ANSP 59444); j & m –Vaitaporo valley (ANSP 294443); k-l – Vaipao valley (ANSP 294445); n – Pease specimen (mixed with *P. bilineata* ANSP 59440); o-q – *compacta*; o – syntype (NHMUK); p-q – specimens (UMZC 75292)

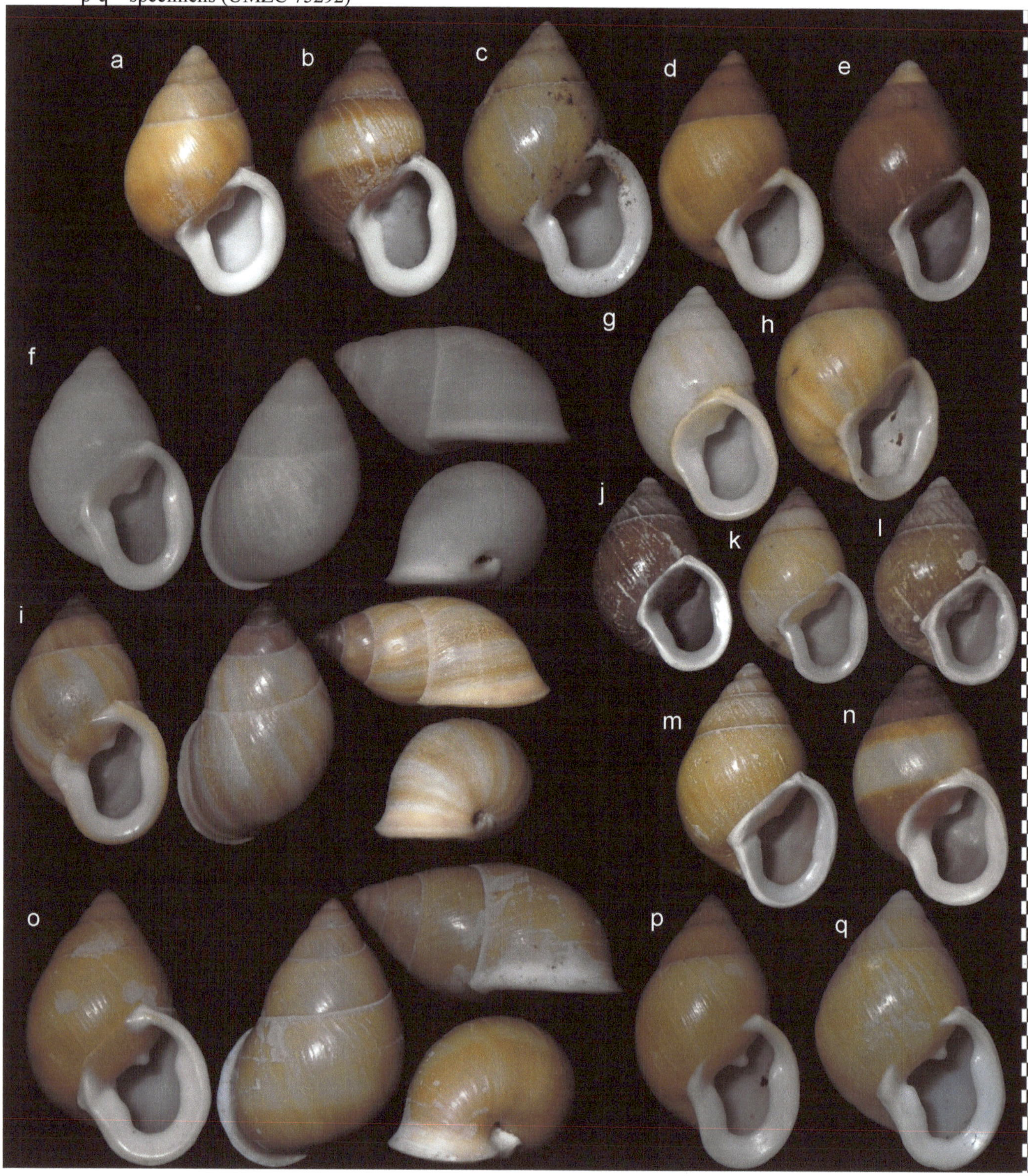

Distribution: Northern Raiatea – *robusta* recorded from Vaetahe, Faafau, Pufau and Vaitaporo, *auriculata* recorded from Hamoa, Uturoa, Vaitaporo, Tepua and Vairaua valleys. Both forms have been recorded from Vaitaporo, although Kondo's (1955) record of *P. auriculata* may be a misidentification. Arboreal.
Note: Crampton (MS) considered *compacta* to be synonymous with *auriculata* and noted that Garrett had misidentified *robusta* as *compacta* due to the plaque in NHMUK bearing Broderip's four specimens having two *auriculata* and two *robusta*.

Partula protracta Crampton, 1956

Partula protracta Crampton, 1956: 12

Partula protracta Crampton, Richardson 1990: 51

Ovately conical, thick. Umbilicus compressed. Apex somewhat obtuse; last whorl evenly rounded. 5½ flatly convex whorls. Suture impressed only on the terminal third of body whorl. Sculpture indistinct; growth lines most prominent near the lip. Columella vertical, broad at base, and shouldered; slightly curved, with a shallow notch at base. Parietal tooth small. Aperture slightly oblique. Lip well developed and thickened, subauriculate, with a low but definite carina near the outer border. Parietal callus thin and shining; arched border thickened at ends.
Uniformly pale fawn.
Anatomy: Unknown; no spirit material.
Distribution: Huaru valley. Known from a single immature specimen in 1908 and an adult in 1909. Arboreal.
Type specimens: Holotype ANSP 272346; paratype ANSP 456714.
Note: Noted by Crampton (1956) to be "sufficiently like that of the unquestionable members of the "*robusta* group" to justify its assignment to that section".

Partula dolichostoma Crampton, 1956

Partula dolichostoma Crampton, 1956: 11

Partula dolichostoma Crampton, Richardson 1990: 22

Ovate-conic, thick. 5 flatly convex whorls. Spire stoutly conic; suture slightly impressed. Surface dull, coarsely marked with irregular growth lines. Foveae absent on the protoconch and faint spiral striae only present on last whorl. Aperture narrow, almost perfectly elliptical, almost vertical. Columella virtually uniform in breadth and devoid of a notch. Lip moderately reflexed, inwardly incrassate, and with simple insertions. A low, narrow carina present along lip, extending to just within border of parietal callus. Callus thin, shiny, with a nearly straight border, thickened only near ends. No parietal tooth.
Deep purple-brown, which grades into deep brown on the second whorl and to less intense brown on the third whorl. Penultimate and fifth whorls transversely marked with irregular chestnut-brown and lighter fuscous strigations.
Anatomy: Unknown; no spirit material.

Fig. 9.4.86 Anatomical variation in *P. auriculata*, a & b – Hamoa (Kondo 1955), c – Tepua (ANSP A959), d & e – Vaetahe (Kondo 1955), f – Vaitaporo (ANSP A1415). Scale bar 5 mm.

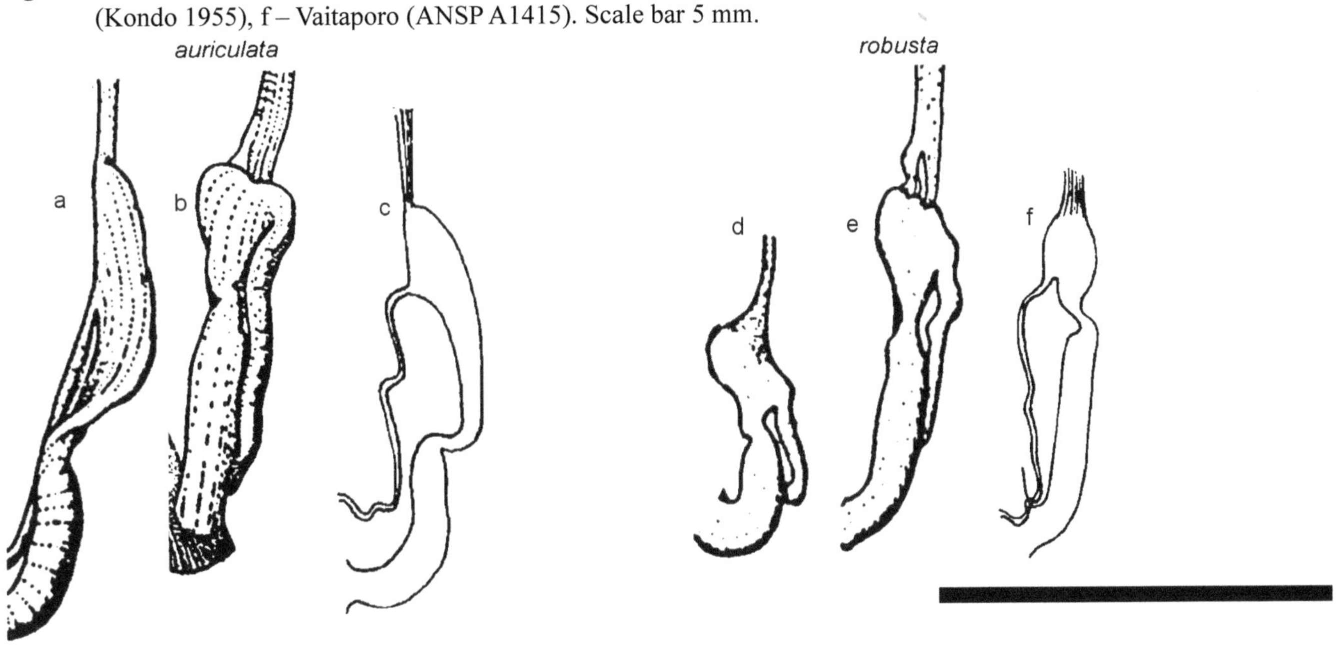

Distribution: Upper portion of Huaru valley. Known from a single specimen. Arboreal.
Type specimens: Holotype ANSP 272342.

***Partula magistri* sp. nov.**

Ovately conical, shell thick. Apex acute. Umbilicus open. Surface highly glossy with fine spiral striae. Spire half of shell height, suture raised. 5 moderately convex whorls. Body whorl flattened dorsally, giving a distinct humped outline to the shell in profile, and compressed near the aperture. Aperture elongate, distinctly auriculated by internal palatal thickening. Lip expanded and strongly thickened internally. Columellar margin evenly expanding above, flattened. A thin, transparent callus between columella and insertion of lip. No parietal denticle.

Pale yellow with irregular dark strigations, body whorl slightly darker yellow than other whorls. Apex rosy, fading but colour distinct on second whorl and detectable on third. Suture clearly white margined. Lip and inside of aperture white.

Type specimens: Holotype NHMUK 20140799, Vaiapu valley 3.ix.1992, J. Gerlach.

Distribution: Known from a single dead specimen from Vaiapu valley.

Etymology: *magistri* – Latin for teachers or professors, named after Professors Bryan Clarke, Jim Murray and Michael Johnson, the geneticists who studied *Partula* from 1962.

Similar species: Closest to *P. auriculata* but less globular, more conical, body whorl distinctly flattened. The aperture is thinner in *P. auriculata*, the columella curved and there is usually a parietal denticle present. The colouration is similar but no rose apexed *P. auriculata* have been recorded. It is geographically isolated from any members of the *P. robusta* group to which *P. auriculata* belongs.

Measurements of type specimen: 22.4 × 13.3 mm, aperture length 9.2 mm. 5 whorls.

Fig. 9.4.87 Rarer *robusta*-group types: a – *P. dolichostoma* holotype (ANSP 272342); b – *P. protracta* holotype (ANSP 272346); c – *Partula magistri* holotype (NHMUK 20140799) (photos: a-b – F. Borrero)

lugubris-group

This comprises a series of largely thin-shelled species extending to the higher parts of Raiatea: *P. lugubris* (including *ovalis*), *P. leptochila, P. dolorosa* and *P. labrusca* according to Crampton and Kondo (1955). Pilsbry (1909) placed *P. ovalis* and *P. lugubris* in the *faba*-group.

Partula lugubris Pease, 1865

Partula (?var.) *lugubris* Pease, 1865: 672

Partula ovalis Pease, 1866: 194

Partula lugubris Pease, Pfeiffer 1868: 158. Pease 1871b: 473. Schemltz 1874: 207. Garrett 1884: 77. Hartman 1885: 208. Pilsbry 1909: 234. Kondo 1955: 128

Partula ovalis Pease, Pease 1871b: 473. Pfeiffer 1877: 205. Kondo 1955: 129

Oblong-ovate. Umbilicus narrow. Moderately thin, dull, with fine growth-lines crossed by spiral striae. Spire conical. 5

Fig. 9.4.88 *robusta*-group geographical variation, *robusta* and *auriculata* forms separated by dashed yellow line

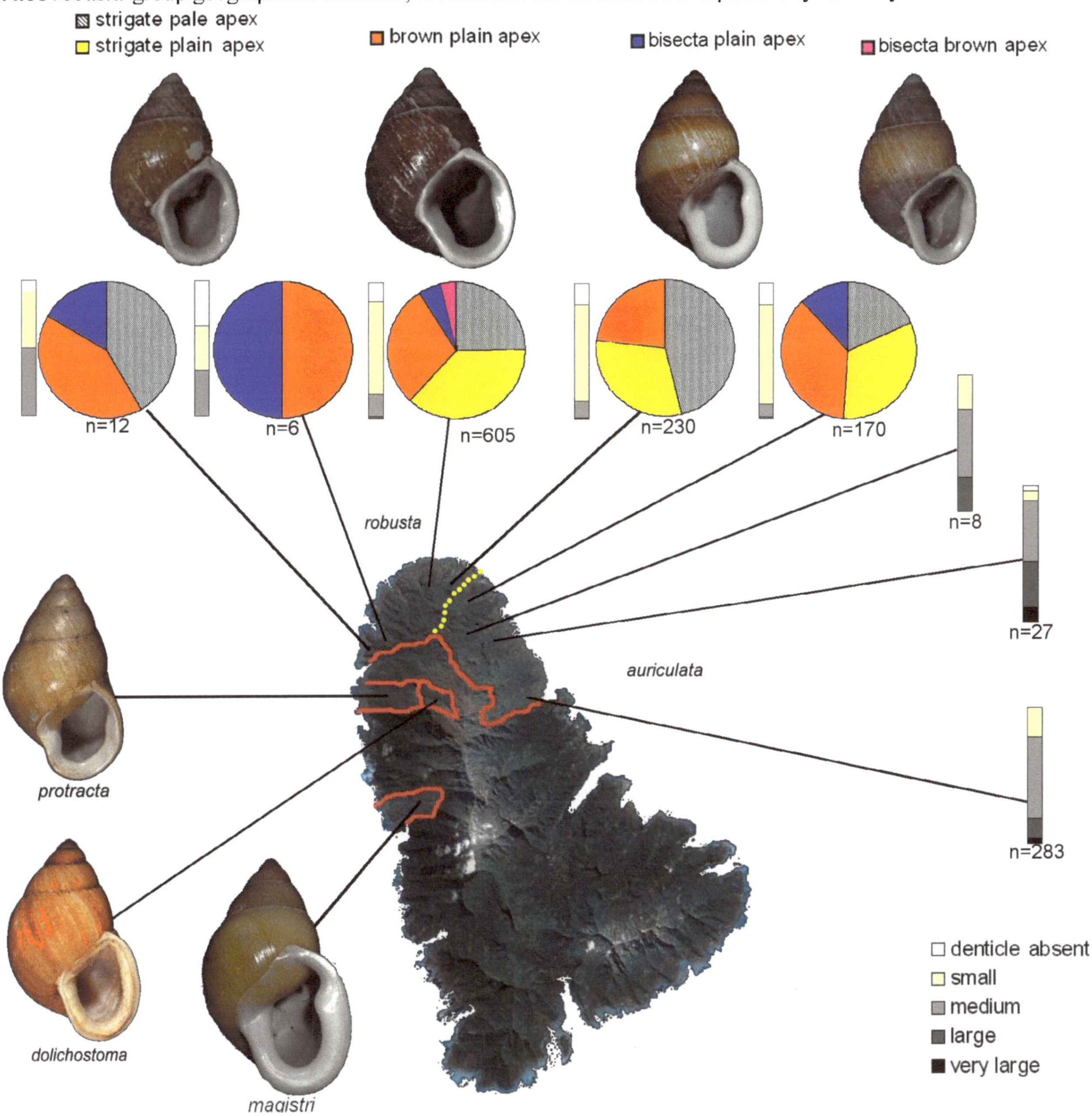

moderately convex whorls. Aperture ovate. Lip narrowly expanded, slightly thickened ('*lugubris*') or distinctly thickened with a slight indication of a palatal projection ('*ovalis*'), thinner near the upper insertion. Columella dilated above, simple. Thin, shiny parietal callus, edentulous or with a slight nodule in some '*ovalis*' (19%; n=37).

Pale yellow to dark chestnut-brown. Pale forms have a white lip and a darker spire, with a brown apex, sometimes with a narrow brown peripheral band. Dark forms paler above and usually with a narrow pale peripheral band. Lip tinted purple-brown, flesh-colour inside. Crampton (MS) identified five colour patterns: plain light brown (a very rare form – 2% of shells from Utuara valley, n=516), plain brown (61%), brown with a narrow light medial band (24%), brown with a wide pale median band (3%) and brown with a narrow darker band (11%).

Anatomy: Penis long, ½ oviduct length. Constriction between penis and caecum indistinct. *P. l. lugubris* has a distinctive form, with the base of the caecum inflated.

Fig. 9.4.89 *Partula lugubris*: a-c – *lugubris* syntypes (NHMUK 1934.6.22.61-65); d – *ovalis* paralectotype (ANSP 59448); e-g – *lugubris* ex Pease (ANSP 59447); h – *ovalis* Pease duplicate (ANSP 294276); i-j – *ovalis* UMZC 75300; k-p – *lugubris* ex Garrett (k-n – ANSP 59446, o-p – ANSP 59960)

Type specimens:

Partula lugubris Pease, 1865 – Syntypes AMNH 42123; ANSP 2623, 59447; MCZ 3641, 25288; NHMUK 1934.6.22.61-5; SMF 4956.

Partula ovalis Pease, 1866 – Syntypes AMNH 42132; ANSP 294276; MCZ 5007. Baker (1963) listed ANSP 59448a.

Subspecies: The two confused forms can be combined into a single species, with very close, but distinct subspecies.

Partula lugubris lugubris Pease, 1865 **comb. nov.** – Northern (possibly Tehurui, Uturara, Erreo, west Faaroa. Lip not thickened.

Partula lugubris ovalis (Pease, 1866) **comb. nov.** – Southern (Vaiaau, Tehurui). Lip internally thickened.

Distribution: Restricted to south-west Raiatea, from Vaiaau and the western edge of Faaroa, north to Ereeo: Vaiaau, Faaroa, Tehurui, Utuara (='Hapai'), Erreo (='Exaao'). Terrestrial.

Notes: Hartman (1881) considered *P. ovalis* to be a synonym of *P. protea,* which he considered a juvenile *P. fusca*, but they are clearly distinct. *P. lugubris* was considered by Pilsbry (1909) to be a very thin member of the *faba*-group, lacking lip callus and teeth, smaller and more lightly built than *P. fusca*. Garrett (1884) writes: "I have found hybrids between *lugubris* and *imperforata*, the latter a strictly arboreal species." This seems a highly questionable statement and it is not known which specimens he was referring to here. The only site with both *P. lugubris* and *P. ovalis* was Tehurui but it is not known if the species were actually sympatric. Crampton did not collect in the valley personally and his notes record only: "Tehuri Valley. July 21. Aft. [afternoon] Joseph and Maniha. Poor lot." It is possible that the *P. lugubris* specimens were collected on the way to Tehuri rather than in the valley itself. Garrett (1884) recognised that *lugubris* was "smaller, thinner, more attenuated and more variable in color and fasciation" than *ovalis* but united them in some parts of his text; he seems to have considered *ovalis* as the primary form with *lugubris* as a derivative. Pilsbry (1909) considered *P. ovalis* synonymous with *P. lugubris*, differing

Fig. 9.4.90 *Partula lugubris* of known origin: a-i – *lugubris*, j-p – *ovalis*; a-c – Tehurui (ANSP 294280); d – Faaroa (ANSP 294283); e – Exaao (ANSP 294282); f – Hapai (ANSP 295324); g – Utuara (ANSP 294281); h-i – Mt. Tefatua (BPBM 87366); j-k – Tehurui (ANSP 294278); l-p – Vaiaau (ANSP 294277)

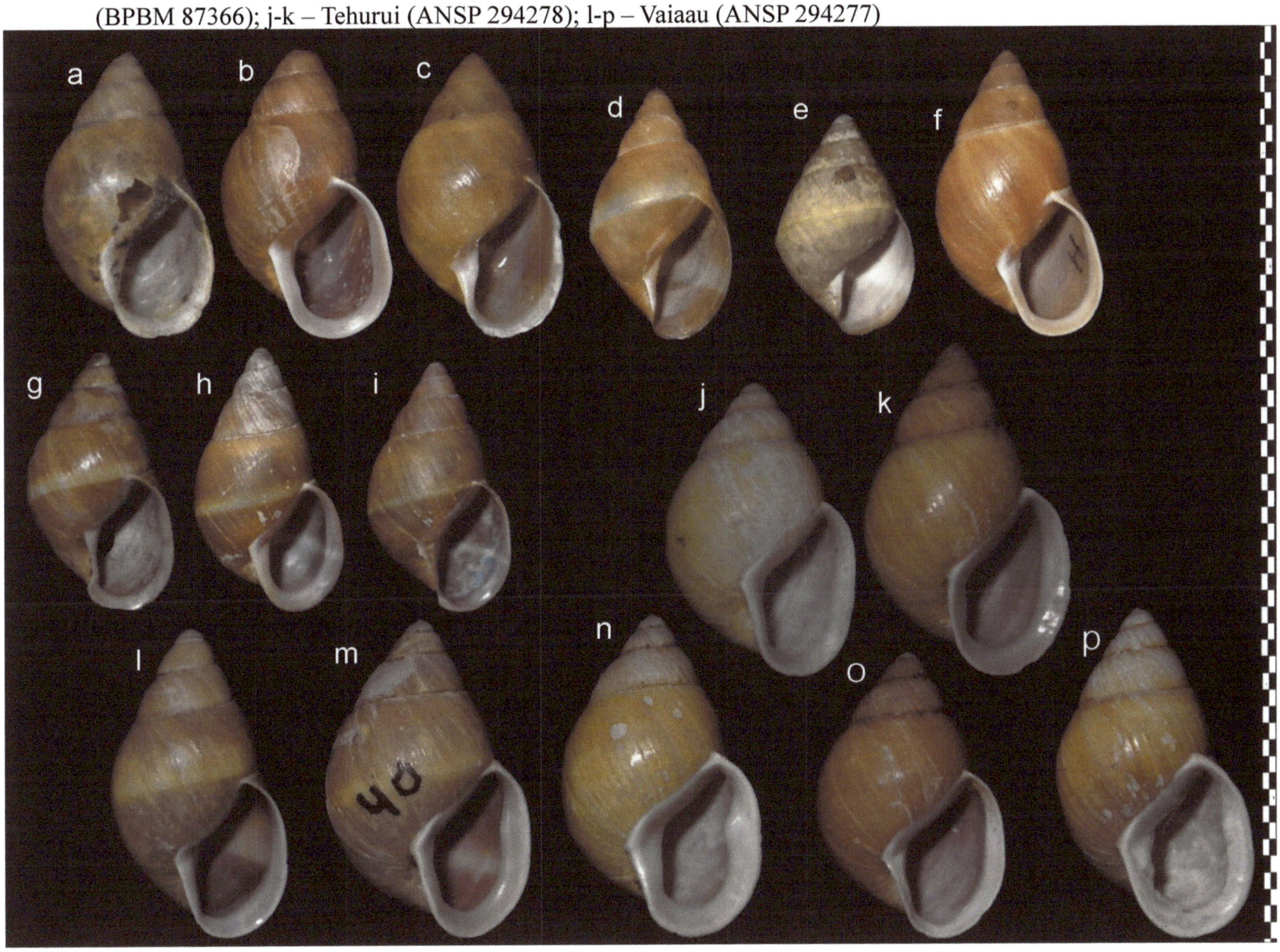

only by being a little larger, more solid, with a stronger lip-callus. Crampton regarded both as typical and unmistakable. Two of the samples dissected by Kondo (1955) appear to have become mislabelled. BPBM 215312 is labelled 'P. lugubrious [sic], Utuara' but matches the figure of *P. ovalis* and contains two specimens. BPBM 215313 is labelled 'P. ovalis, Vaiau' and contains a single specimen which is clearly the basis for Kondo's (1955) figure of the single specimen of *P. lugubris* that he had available.

Partula leptochila Crampton, 1956

Partula leptochila Crampton, 1956: 10

Partula leptochila Crampton, Kondo 1955:126, 1968: 76. Richardson 1990: 36

Ovate-conic, moderately thick. 5-5½ whorls. Umbilicus open. Spire stoutly conic, with moderately convex whorls. Spiral striae fading on body whorl, persistent on subsutural and basal areas. Aperture usually oblong-ovate, but variable. Lip reflexed, somewhat flaring, calloused within, more or less bevelled, and narrowed towards the outer insertion. Columella widens towards junction with body whorl; sulcus at line of contact is horizontal or only slightly inclined. Parietal callus very thin and shining; median portion and its slightly arched border usually more tenuous; no parietal tooth.

Light to dark brown with three colour variants: light brown with irregular darker strigations; mid- to dark-brown with a purple apex; brown with a median paler band, indistinctly bordered purple-brown on spire.

Anatomy: Penis short, ¼ oviduct length. Otherwise as *P. lugubris*.

Distribution: Upper portion of Huaru and Voaara valleys at 200-400 ft, and parts of the Temehani plateau. Largely restricted to Huaru (41 live adults, 3 live subadults, 30 dead shells), only one shell found in the highest part of adjacent Voaara. 6 live collected (3 adults, 1 subadult, 2 juveniles) collected on Temehani plateau in 1991. Largely terrestrial.

Type specimens: Syntypes ANSP 272343, 294284 (& A978), 294285.

Partula labrusca Crampton & Cooke, 1953

Partula labrusca Crampton & Cooke, 1953: 151

Partula labrusca Crampton, Kondo 1955: 126, 1968: 76. Richardson 1990: 36

Ovate-conic. Shell moderately thick. Umbilicus open. 5-5½ convex whorls. Very glossy, weak growth lines, fine, distinct spiral sculpture on spire only and sometimes faintly persistent around umbilicus. Suture moderately impressed, not marginated. Aperture nearly vertical, slightly inclined. Columella straight, stoutly triangular. Lip simple, thin, thickened slightly inwards, contracting aperture. Parietal callus very thin, dull; edentulous.

Dark brown with purple tint, an indistinct 1.5 mm wide median paler band. Lip pale or purple to purple brown.

Fig. 9.4.91 Anatomical variation in *P. lugubris* group: a – Utuara (Kondo 1955); b – Mt. Tefatua (BPBM 87366); c-d – Vaiaau (c – Kondo 1955; d – ANSP A970); e-g, i-j – Temehani (e – Kondo 1955 and BPBM 139357; f – Kondo 1955 and BPBM 139341; g – ZSL 245 10/12/92; i – ZSL 245 13/2/93; j – ZSL 245 14/12/92); h – Huaru (Kondo 1955). Scale bar 5 mm.

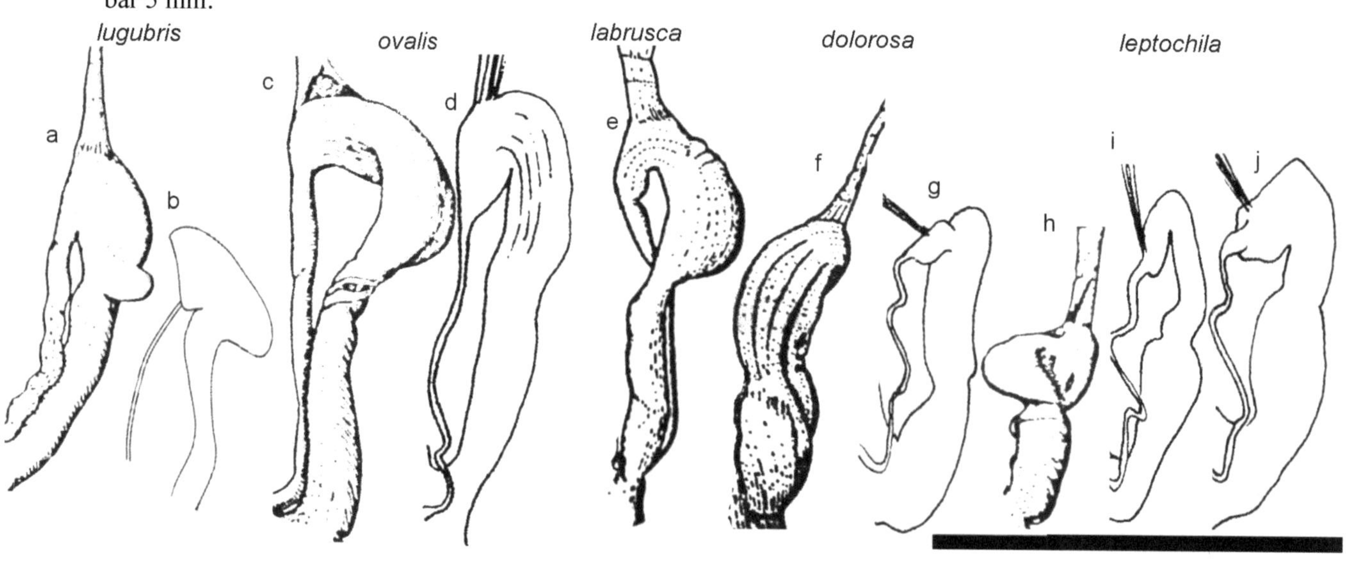

Fig. 9.4.92 Rarer species of the *lugubris*-group. a-b – *dolorosa* (a – syntype ANSP 294474; b – ZSL 245 10/12/92); c-e – *labrusca* (c – syntype ANSP 294475; d – ZSL; e – BPBM 139357); f-k – *leptochila* (f – Temehani ZSL 245 14/12/92; g – syntype ANSP 272343; h-i – ANSP 294284; j – ZSL 184; k – ZSL 245 13/2/93 (photos: a, c, g – F. Borrero)

Anatomy: Penis long, ½ oviduct length.
Type specimens: Syntypes ANSP 294475; BPBM 9976.
Distribution: Temehani plateau 1400-2000 ft. Found in localised pockets of vegetation (3 adults, 2 immatures and one dead shell in 1934; 12 live in 1992).

Partula dolorosa Crampton & Cooke, 1953

Partula dolorosa Crampton & Cooke, 1953: 153

Partula dolorosa Crampton & Cooke, Kondo 1955: 127, 1968: 76. Richardson 1990: 22

Narrowly ovate-conic, very thin. Umbilicus elongate, rimate. 4¾-5 whorls. Suture narrow, deeply impressed. Surface dull, strong growth lines, strong wavy spiral striae on all whorls. Aperture elliptical, vertical. Lip very thin, slightly reflexed, weakly thickened internally. Columellar margin only very slightly curved, columella triangular. Parietal callus indistinct; tooth absent. Columellar margin of aperture less bevelled than rest. Lip exceptionally thin.
Pale brown, uniform or with darker transverse strigations. A pale median or sub-median band, only slightly paler than the ground colour but clear due to darkening of adjacent areas. One specimen has more marked colour contrast. Apex often tinted purple. Lip tinged pale brown.
Body dark slate-grey, mottled.

Fig. 9.4.93 Distribution of *lugubris*-group

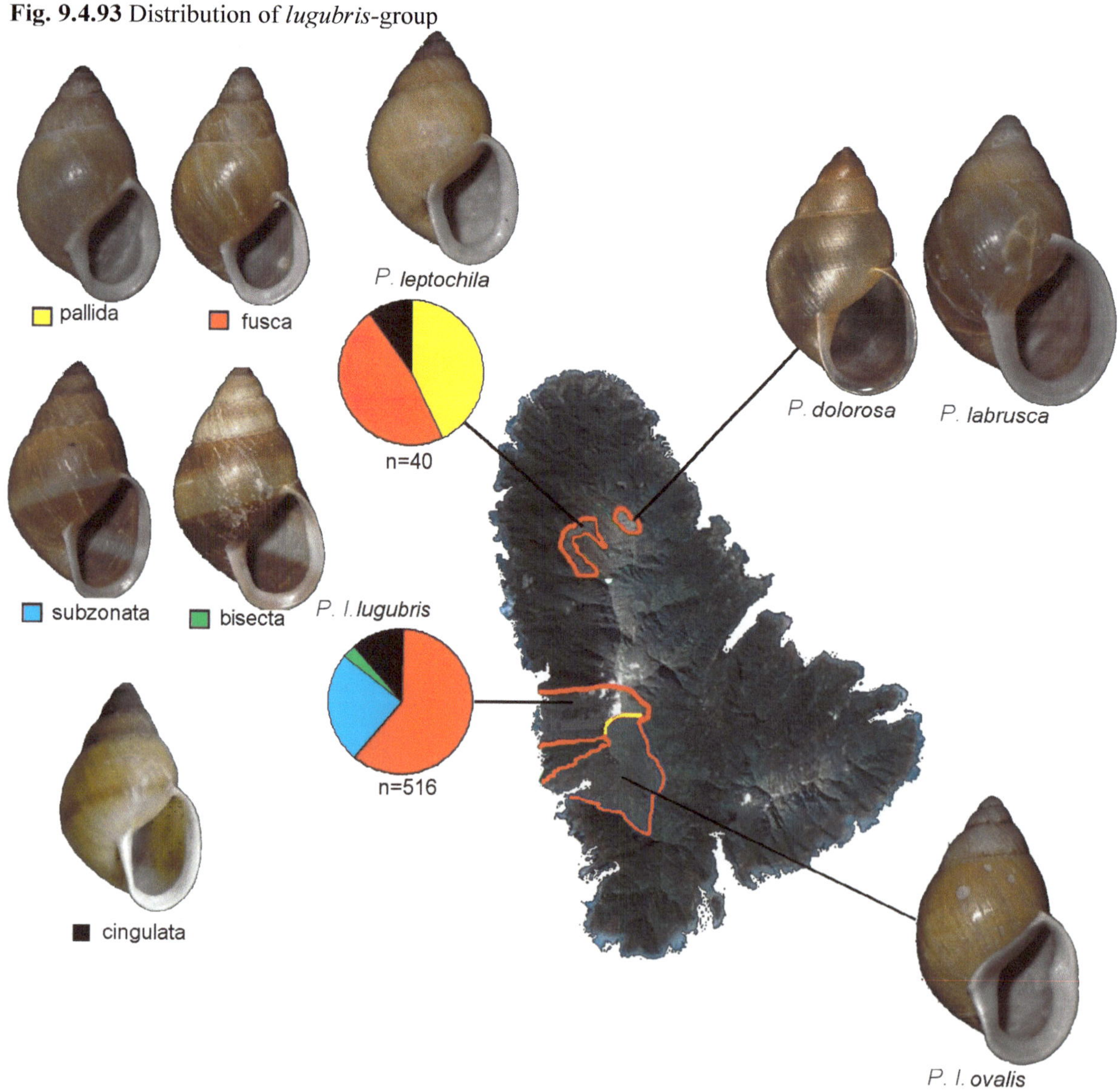

Anatomy: Reproductive anatomy as *P. lugubris* but stouter.
Type specimens: Syntypes ANSP 294474; BPBM 9975.
Distribution: Temehani plateau 2000 ft on *Pandanus* (Crampton & Cooke 1953). Five adult and five immatures collected in 1934, 9 collected in 1992 (ZSL).

garrettii-group

Contains *P. rustica, P. garrettii, P. thalia, P. cuneata, P. levistriata, P. remota, P. tristis* and *P. atilis* according to Crampton and Kondo (1955) and included in the *hebe*-group by Pilsbry (1909). Crampton (1956) placed several species here on the basis of the shape of the aperture, columella, lip and weak labial carina. Kondo (1955) separated out the *remota*-group but this is included here, with a total of four species.

Partula garrettii Pease, 1865

Partula auriculata var., Carpenter 1864: 675
Partula Garrettii Pease, 1865: 672
Partula abbreviata 'Pease' (non Mousson), Carpenter in Pease 1865: 675 (as syn. of *auriculata*, nomen nudum)
Partula auriculata 'Pease' (non Broderip), Carpenter in Pease 1865: 675 (as syn. of *auriculata*, nomen nudum)
Partula rustica 'Pease' Carpenter in Pease 1865: 675 (as syn. of *auriculata*, nomen nudum)
Partula rustica Pease, 1866: 199
Partula rustica Pease, Pease 1867: 81, 1871b: 473. Schmeltz 1868: 71, 1874: 207. Pfeiffer 1877: 204. Gloyne 1878: 338. Hartman 1881a: 187, 1881b: 9, 1885: 211. Paetel 1883: 153, 1889: 268. Garrett 1884: 77. Smith 1903: 438. Pilsbry 1909: 227. Kondo 1955: 135, 1968: 76. Richardson 1990: 55
Partula garrettii Pease, Pfeiffer 1868: 158, 1877: 198. Schmeltz 1874: 207. Graeffe 1874: 80. Hartman 1881a: 182, 1881b: 10, 1885: 211, 1892: 75. Garrett 1884: 56. Smith 1903: 441
Partula Garrettii Pease, Pease 1871b: 473. Garrett 1884: 56. Hartman 1885: 211
Partula gonocheila (non Pfeiffer) Schmeltz 1874: 92 (nomen nudum)
Partula crassilabris Gloyne (non Pease), Gloyne 1878: 228 (syn. of *garrettii*, nomen nudum)
Partula Peasii 'Garrett' (non Cox), Hartman 1881: 188 (nomen nudum)
Partula Thalia 'Garrett', Hartman 1881a: 188, 1881b: 7. Paetel 1883: 153 (all nomen nudum)
Partula thalia 'Garrett', Garrett 1884: 69 (description). Hartman 1885: 209, 1892: 74. Smith 1903: 437. Pilsbry 1909: 231. Kondo 1955:133, 1968: 76. Kilias 1969: 248. Richardson 1990: 61
Partula Paeseiana 'Garrett', Hartman 1885: 209 (ex errore)
Partula garretti Pease, Pilsbry 1909: 228 (unjustified emendation). Kondo 1955: 134, 1968: 76. Kilias 1869: 247. Richardson 1990: 28

Fig. 9.4.94 *garrettii*-group anatomical variation. a – Voaara (Kondo 1955), b & c – Uparu (ANSP A746), d – Tevaitoa (ZSL 597 & 810), e – Tooroa (Kondo 1955), f – Ereeo (ANSP A791), g – Tahara (Kondo 1955), h – Hauru (ANSP A1431B), i – Temehani (ZSL 185); j – Vaihuti (Kondo 1955), k – Paaoio (Kondo 1955, and BPBM 139608), l – Voaara (Kondo 1955)

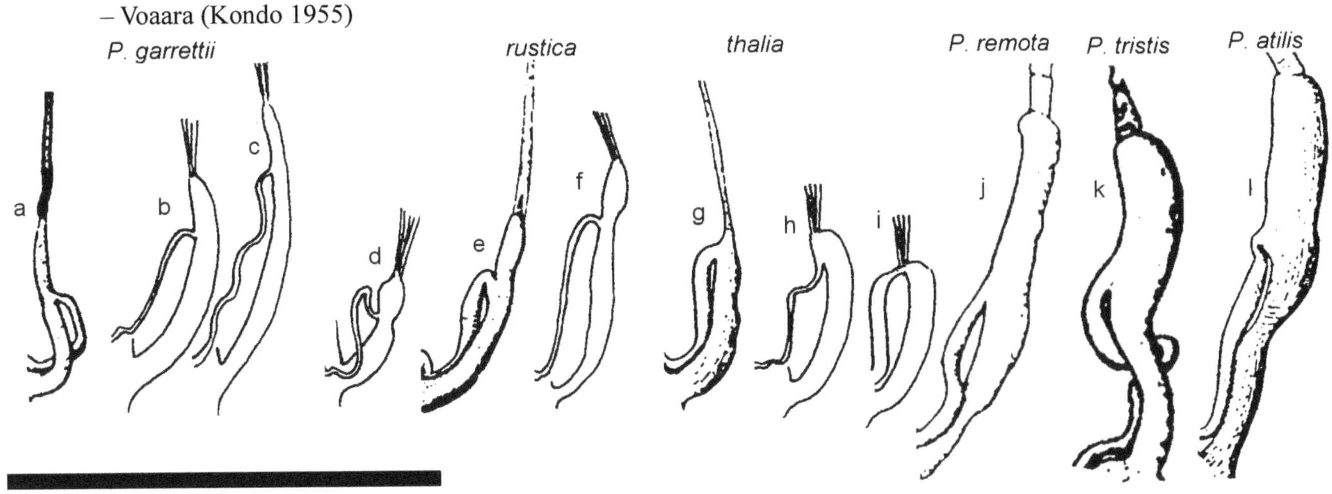

Oblong-ovate to ovately conic, thick. Umbilicus narrow, rimate or closed. Surface slightly shining, with growth lines and minute spiral striae on the spire, extending to the body whorl. 4½-5 whorls, only slightly convex. Suture lightly impressed, edged with a white line. Aperture ovate to subauriform. Lip narrowly expanded, thickened within, face convex or keeled, with a strong internal rib. Columella vertical, sinuous, with a strong tubercle or nodule above middle in the more northern populations, angulated at its junction with basal margin making it appear to "push in towards the aperture" (Garrett 1884). Parietal wall with a smooth transparent callus, sometimes a small, deeply recessed denticle, most developed in the north.
Whitish or yellowish to light brown. Spire paler with white apex or pink-tined with purplish apex. Aperture white within, lip white. A form with a brown base and sutural band is common. Subspecies *rustica* darker, chestnut brown, sometimes with a pale band or pale brown on the body whorl (5% of specimens, n=38, from Paaoio valley – BPBM 139600) and the spire banded in 21%. Aperture dark flesh-coloured within.
Anatomy: Penis short, ⅓ oviduct length, slender, a distinct caecum usually present, very reduced or absent in northern populations (*thalia*) and may be inflated in southern populations (*rustica*). Vas deferens submedian to subapical.
Type specimens:
Partula garrettii Pease, 1865 – Syntypes AMNH 118940, 42030; ANSP 2649, 59501; FLMNH 111963-4; MCZ 24894, 25306; IRSNB 317683; ZMB/Moll-74325, 74330, 75631-2.
Partula rustica Pease, 1866 – Lectotype ANSP 59480; paralectotypes AMNH 42128; ANSP 2507, 464473; MCZ 25140, 25302; SMF164976. Lectotype designation by Baker (1963).
Partula thalia Garrett, 1884 – Lectotype ANSP 59476; paralectotypes ANSP 459107; ZMB/Moll. 75721, 74335. Lectotype designation by Baker (1963).
Partula abbreviata Pease, 1865 – Supposed type of MCZ 2505 (nomen nudum).
Subspecies: Three species are best regarded as subspecies of a single species; distinct conchological forms can be recognised but integrade and are all anatomically very similar. Hybrids between them have long been reported (Hartman 1881; Garrett 1884; Smith 1903). Such intermediate forms may include BPBM 139600.
Partula g. garrettii Pease, 1865 – central range (Vaioara, Uparu, Vaiapu, Voaara). Columella typically strongly sinuous.
Partula g. rustica (Pease, 1866) **comb. nov.** – south range (Toloa, Tooroa, Tevaitoa, Vaetahi, Hanua, Ereeo). Umbilicus more open than nominate, containing a spiral groove. Lip narrower near upper insertion.

Fig. 9.4.95 *Partula garrettii* types: a-b – *garrettii* (a – ANSP 59501; b – ZMB 74325); c – *thalia* (ANSP 59476); d – *rustica* (ANSP 59480a) (photos: a, c – F. Borrero)

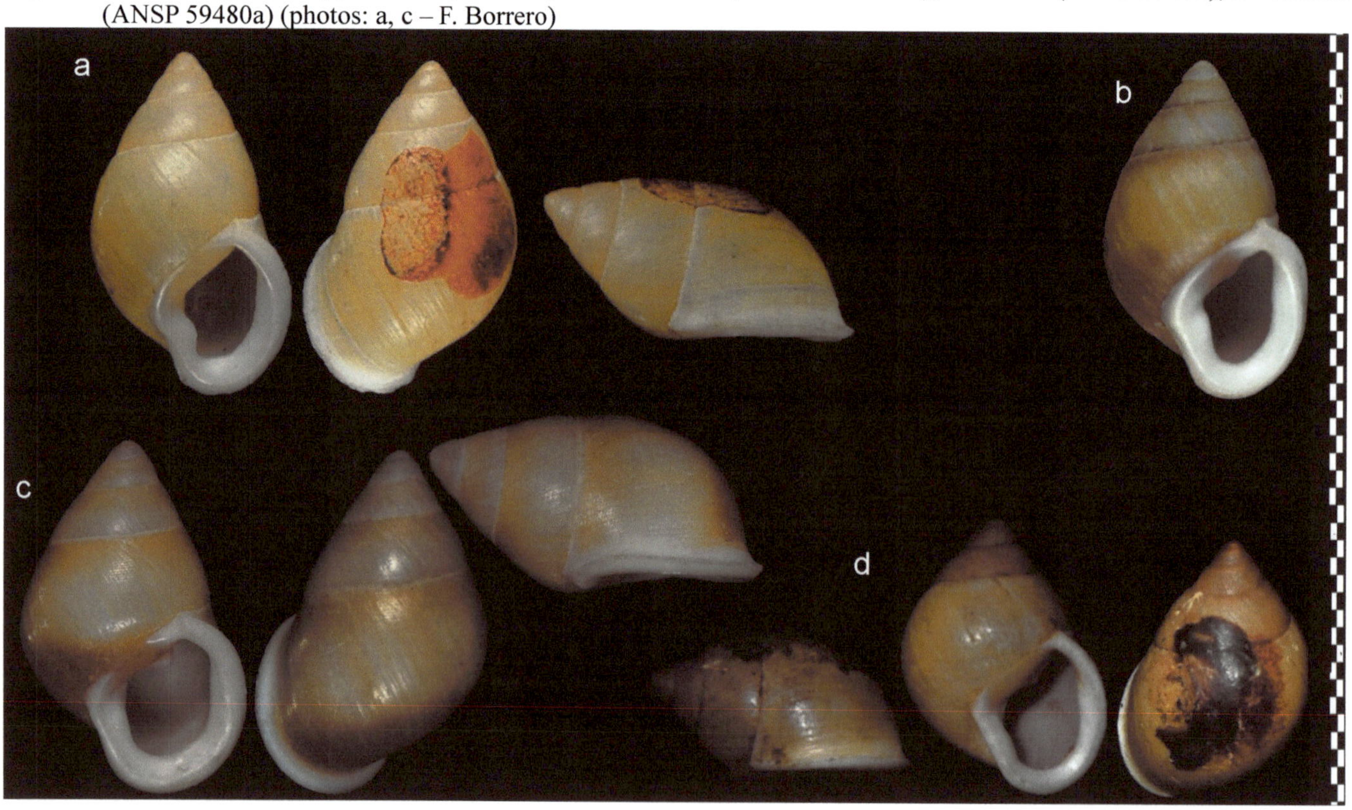

Fig. 9.4.96 *Partula garrettii*: a-l – *garrettii*; m-r – *rustica*; s-w – *thalia*; a-d – Voaara (ANSP 294436); e-h – Uparau (ANSP 294435); i – Tooroa (JG); j-l – Paaoio (BPBM 139600); m-p – Ereeo (ANSP 294432); q-r – Tooroa (ANSP 294434); s-v – *thalia* variation (UMZC 75308); w – Temehani (ZSL 185)

Partula g. thalia (Garrett, 1884) **comb. nov.** – north range (Tahara, Huaru, extending onto Temehani plateau). Aperture more auriform than other subspecies, a tubercular tooth usually present on the parietal callus. Almost always pale with a rose to purplish apex. Crampton (MS) recognised uniform and banded varieties. Plain shells are light yellow brown or deeper brown. Brown shells are more fulvous than in the corresponding form of the nominate subspecies. Bandeds have deep brown subsutural and basal bands. All the lower half of the shell is deeply tinted.

Distribution: Vaioara, Uparu, Vaiapu, Voaara, Toloa, Tooroa, Vaetahi, Hanua, Ereeo (upper part only), Tahara and Huaru valleys, extending onto Temehani plateau. Arboreal: "in prodigious numbers on bushes" (Pilsbry 1909) but *rustica* recorded as terrestrial: "in great abundance beneath decaying vegetation" (Pilsbry 1909). From 50 ft upwards.

Fig. 9.4.97 *garrettii*-group geographical variation. *P. tristis* locality marked by a blue point

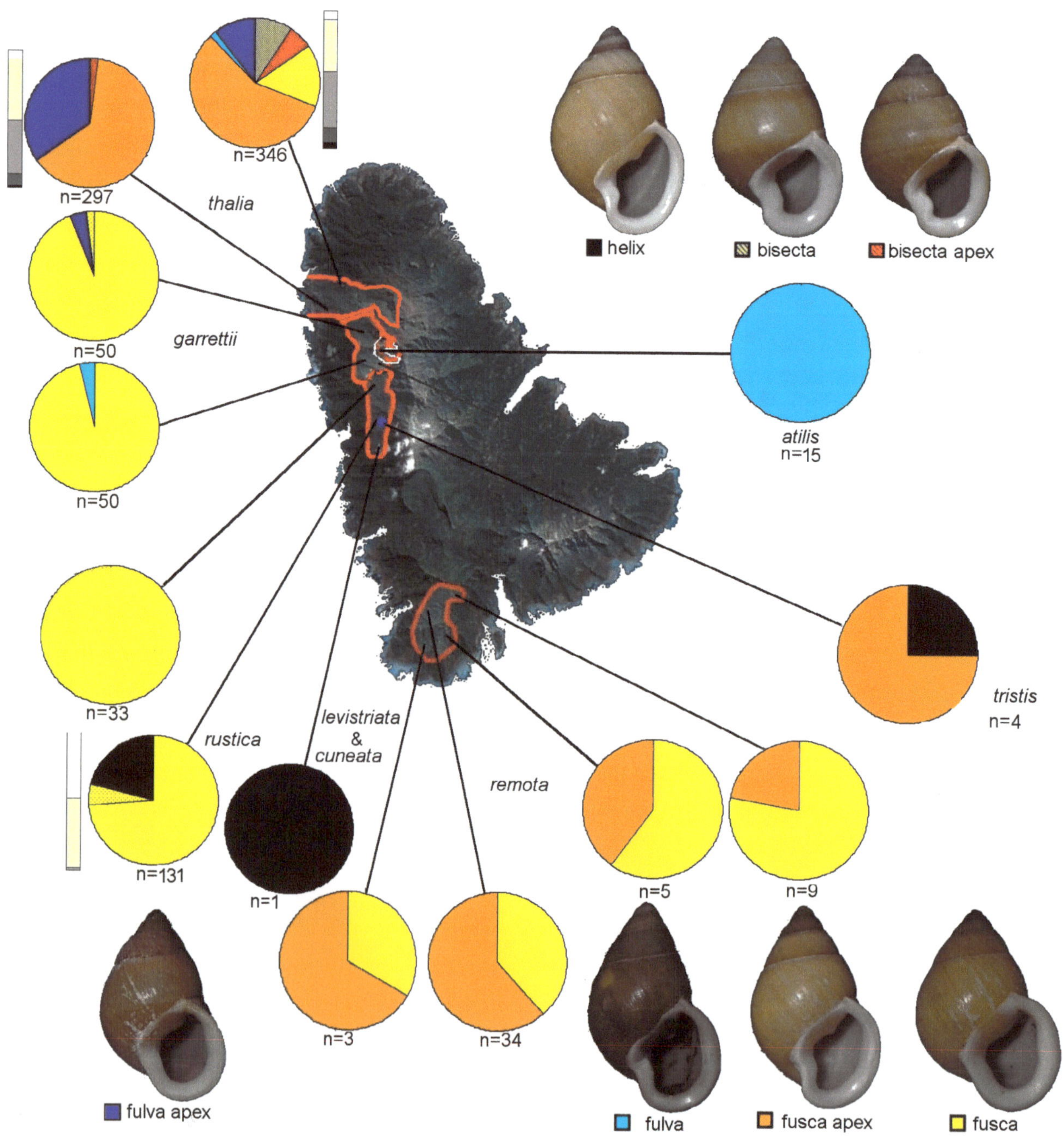

Note: *P. rustica* was the last section of his unfinished monograph that Crampton wrote – October and November 1953. He continued to work on *Partula* for at least a month after this as two Tooroa *P. rustica* specimens (ANSP 294434 – Fig. 9.4.96q-r) are labelled by him '?tristis Dec 53'. He found these specimens particularly confusing and originally labelled them 'probably lugubris' and then corrected this to 'An ally, certainly but colony may differ from lugubris – Aug 24 / 53'.

Partula cuneata Crampton, 1956

Partula cuneata Crampton, 1956: 9

Partula cuneata Crampton, Kondo 1968: 76. Richardson 1990: 20

Elongate ovate-conic, moderately solid. Umbilicus openly rimate. 5 moderately convex whorls. Suture impressed. Spire conic, with somewhat pointed apex. Surface smooth and shining, with weak growth lines. Spiral sculpture absent on apex, weak on spire and upper third of last two whorls. Aperture unusually small, roundly elliptical, with vertical axis. Lip well developed, calloused internally, and uniform in width. Lip distinctly bevelled and labial carina present as an uninterrupted ridge. Columella only slightly widened; columellar lip not thickened, and bearing a shallow notch close to inner insertion. Parietal callus exceedingly thin, with a very weak and slightly arched border. Parietal tooth little more than a trace.

First two whorls uniform purple-brown with a pale creamy-corneous spire, third whorl bicoloured owing to disappearance of deep colour below suture. The deeply coloured spiral diminishes in width as it continues onto fourth whorl, until it vanishes at beginning of last whorl. Lip white and shining.

Anatomy: Unknown; no spirit material.

Distribution: Ereeo valley. Arboreal.

Type specimens: Holotype ANSP 272256.

Note: Crampton (1956) stated that the revolving bicoloured pattern was unique among Raiatean *Partula*, however a faded version of this is apparent in *P. levistriata,* some *P. garrettii* and one *P. tristis*. Crampton (1956) placed it in the *garrettii*-group. This species is sympatric with *P. levistriata.*

Partula levistriata Crampton, 1956

Partula levistriata Crampton, 1956: 8

Partula levistriata Crampton, Kondo 1968: 76. Richardson 1990: 36

Broadly ovate-conic, and moderately solid. Umbilicate. 4¾ convex whorls. Suture slightly impressed. Spire stoutly conic, with an obtuse apex. Surface rough, with well-marked growth lines. Spiral sculpture fine and continuous throughout. Aperture roundly oval, with vertical axis. Lip well developed, not distinctly bevelled, moderately incrassated, and curved inwardly towards its upper insertion. Labial carina low and continuous. Columellar lip with a slight but definite inward thickening. Parietal callus very thin, only well marked at ends. Parietal tooth indicated by a very weak trace.

Pale yellowish corneous with irregularly spaced transverse strigations of warm brown. Protoconch dull dark brown.

Suture margined white. Lip white and shining.

Anatomy: Unknown; no spirit material.

Distribution: Ereeo valley. Arboreal.

Type specimens: Holotype ANSP 272344.

Note: A single specimen collected by Crampton who noted (1956) that it resembled *P. navigatoria levilineata* closely. Placed by Crampton (1956) in the *garrettii*-group with uncertainty, on the basis of the aperture, columella, lip and weak labial carina. It differs from *P. navigatoria levilineata* in being broader, having a more tumose last whorl, more open umbilicus, broader, oval aperture which is only vaguely subauriculate. In addition the lip curves inward at its outer insertion, the labial ridge is continuous and the columellar lip is uniform in width, not triangular at its terminus. This species is sympatric with *P. cuneata*.

Partula remota Crampton, 1956

Partula remota Crampton MS, Kondo 1955: 131 (unpublished)

Partula remota Crampton, 1956: 5

Partula cuneata Crampton, Richardson 1990: 53

Globose-ovate, solid, and umbilicate. 5 whorls. Spire stoutly conical, with convex whorls and obtuse apex; last whorl tumose. Suture impressed, margined with a narrow white line excepting on the apex. Columella vertical, broad, and simple. Surface smooth; weak growth lines and fine spiral striae fading on body whorl. Aperture vertical. Lip wide and thick, inwardly heavily thickened, and somewhat bevelled, with a well-developed carina throughout. Parietal callus thin, shiny, with a weakly accentuated border. Parietal tooth always present.

Colour uniformly corneous fawn throughout, or with pale rose-brown suffusion on spire. Apex never deeply coloured, and spiral bands are absent. Lip pure white and shining.
Crampton (MS) noted the limited colour range of this species: the light corneous-luteous epidermis is always devoid of the yellowish tinges of *P. hebe* colonies that retain the epidermis (especially in Vaiaau). *P. remota* is rarely decorticated completely but most have scattered spots of narrow transverse lines of decortication, some are more extensively decorticated on the apex, around the umbilicus or immediately outside of the lip. The only colour variation was 57% (n=106) having a reddish tinge on the spire, 43% uniform. The reddish colour varies in extent but never reaches the body whorl. All embryonic shells pure white (Crampton MS).
Anatomy: Penis moderately long, ½ oviduct length. Penis and caecum cylindrical, not constricted between them. Caecum longer than penis, terminally inflated, containing seven main rows of papillae and seven less distinct rows, penis with six inconspicuous pilasters. Vas deferens medial.
Type specimens: Syntypes ANSP 272347, 294777-80 (& A790).
Distribution: Tuatau, Vaetuna, Uturoto and Vaihuti valleys. Almost all collected from Vaihuti valley. Arboreal.
Note: Crampton (1956) considered the lack of any earlier specimens to indicate that Garrett had not surveyed the extreme southern part of Raiatea. In shape this species is very similar to *P. hebe.*

Partula atilis Crampton, 1956

Partula perpinguis Hartman, 1881: map (nomen nudum)
Partula pinguis Garrett Hartman, 1881: 185 (nomen nudum)
Partula pinguis Garrett, 1884: 77 (nomen nudum)
Partula atilis Crampton MS, Kondo 1955: 132 (unpublished)
Partula atilis Crampton, 1956: 13
Partula atilis Crampton, Kondo 1968: 76. Richardson 1990: 10

Stoutly ovate-conic, and solid. Umbilicus openly pyriform. 5 whorls. Spire short and stout, with slightly convex whorls; the last whorl is somewhat tumose, and the apex is obtuse. Surface dull and rough with coarse growth lines. Spiral striae not wavy, well separated; more or less obsolete on last whorl. Suture vaguely impressed on body whorl. Aperture broadly oval and auriculate, with a vertical axis. Lip wide, variably keeled, bevelled and strongly thickened inwardly, indented near its

Fig. 9.4.98 Types of rarer *garrettii*-group species: a – *levistriata* (ANSP 272344); b – *cuneata* (ANSP 272256); c – *remota* (ANSP 272347); d – *atilis* (ANSP 272255) (photos: all F. Borrero)

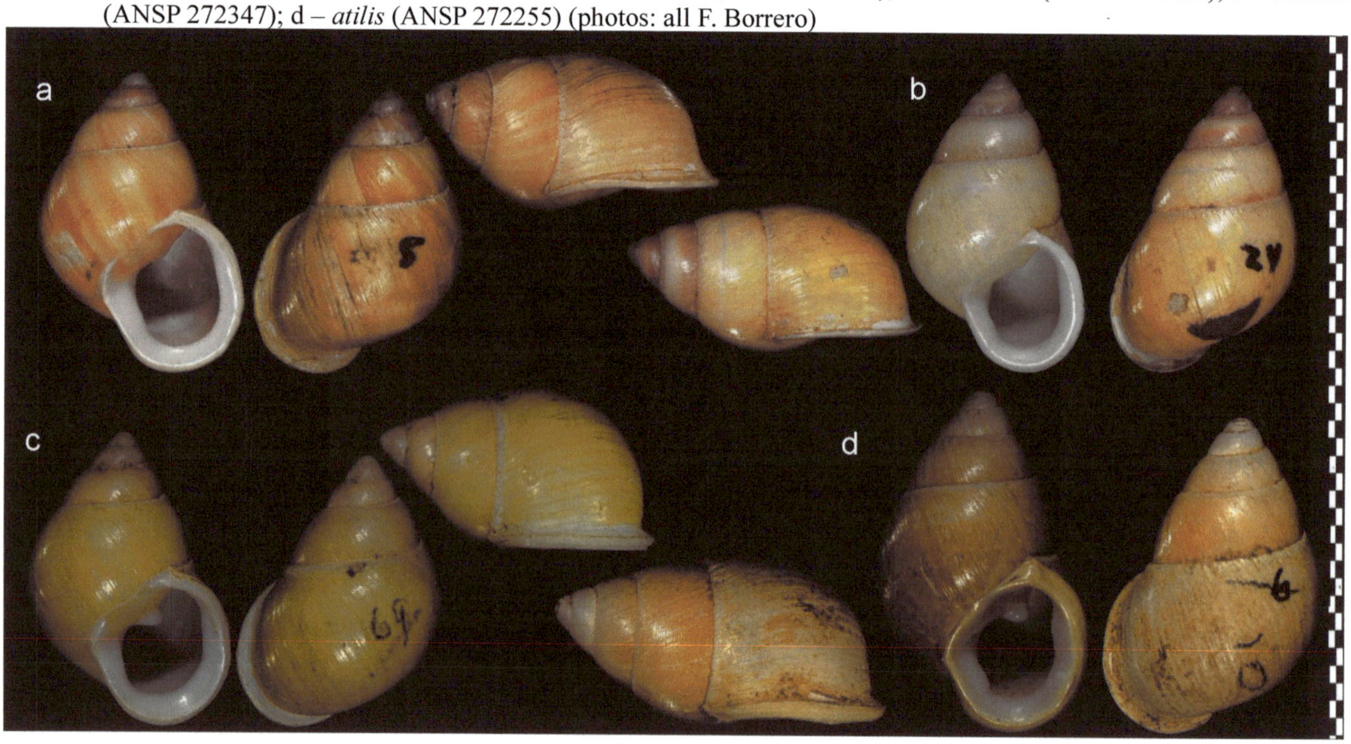

lowermost curvature. Base of columella widely triangular. Parietal callus thick, opaque; thickened and slightly arcuate border continuous with labial carina. Parietal tooth low and broad, present in all shells but varying in development: small in 80% (n=25), moderate in 16% and large in 4%.
Uniform dull ochre-brown, or (one specimen) deeper brown. Suture margined with a thread-like white line. Lip and callus white.
Anatomy: As *P. remota*. Egg capsules unusually small and stout for a *Partula* of this size: 2.9 × 2.3 mm (Crampton MS).
Type specimens: Syntypes ANSP 272255, 294437-8.
Distribution: Uppermost portions of Voaara and Uparu valleys. Terrestrial. Garrett (1884) recorded '*P. pinguis*' from Vaioara (=Voaara), considering it a synonym of *P. rustica*. Known from 12 specimens collected by Garrett and 19 collected by Crampton in 1908-9 (12 collected alive).
Note: This species differs from *P. rustica* in being larger, broader, the spire more stoutly conical, with more convex whorls, the body whorl more inflated, aperture broad, auriculate and with a small upper sinus, the lip more incrassate, the labial carina more distinct and a parietal tooth always present, whereas it is small or absent in *P. rustica*. Crampton (MS) noted that the differences were marginal, however they are anatomically distinct.

Partula tristis Crampton & Cooke, 1953

Partula tristis Crampton & Cooke, 1953: 150

Partula tristis Crampton, Kondo 1955: 132, 1968: 76. Richardson 1990: 62

Ovate-conic. Thin. Umbilicus narrow, compressed. 5 slightly convex whorls, suture impressed. Surface dull. Faint, irregular growth lines, spiral sculpture well developed on all whorls. Aperture elliptical, vertical. Lip roundly reflexed, slightly expanded, with little or no internal thickening; inner end appressed over umbilicus. Columella twisted and wide above, with a slightly notched appearance. Parietal callus thin and dull, arched; slightly thickened along outer margin in older animals. No parietal tooth.
Light grey to dark brown, with darker brown transverse streaks (inconspicuous in darker shells). Apex may be purple-brown. The shell may be uniform or banded with a narrow median light band on the body whorl, a broad darker suprasutural band (brown or purplish) or a medial dark band on the spire. Lip white, faintly tinted with background colour.
Anatomy: Penis moderately long, ½ oviduct length. Caecum as long as penis, no clear constriction between them.
Type specimens: Syntypes ANSP 294473 (ex BPBM 139608, body preserved under that number); BPBM 9977 (ex 139607), 139553. The original description only referred to BPBM 9977 a and b as 'cotypes', however, as the description was explicitly based on all four adults they should all be regarded as syntypes. Although ANSP 294434 are labelled as *P. tristis* this is an error for *P. garrettii rustica*.
Distribution: Tooroa ('Paaoio') valley, 75-150 ft (collected in 1934). Arboreal: 'found on trees and shrubs in the central portion of Tooroa Valley' (Crampton & Cooke 1953). BPBM 139553 is from the same valley but 600-1500 ft, 'on tree trunks, shrubs, & dead leaves' (Gessler, Kondo & Anderson).

Fig. 9.4.99 *Partula tristis* syntypes: a – typical colour form (ANSP 294473); b – colour variety (BPBM 139553) (photos: a – F. Borrero)

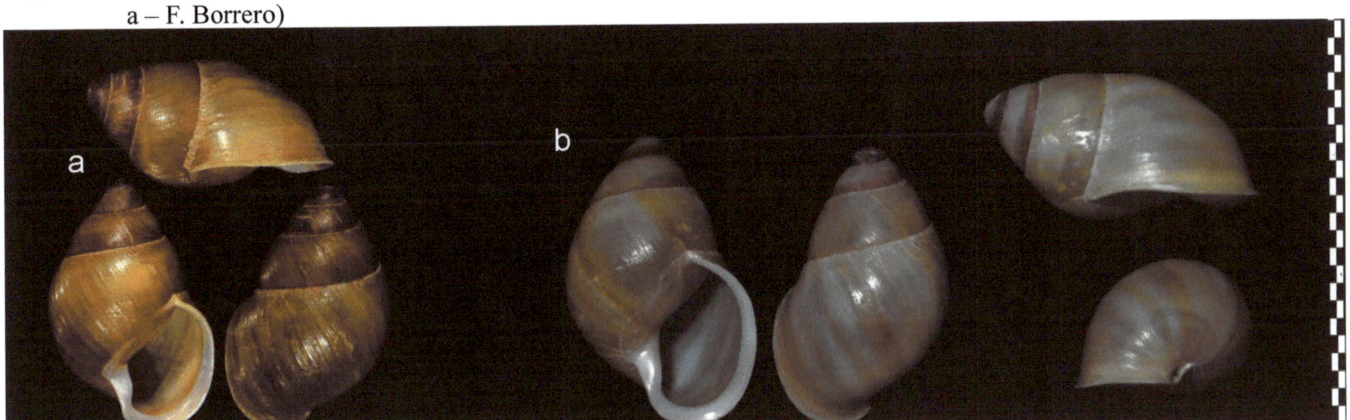

hebe-group

Contains *P. hebe* and *P. crassilabris* according to Crampton (MS) and Kondo (1955), with a third species added here. Pilsbry (1909) also included the *garrettii*-group within the *hebe*-group as all are "rather small forms the spiral striae are usually somewhat better developed than in most other Raiatean species."

Partula hebe (Pfeiffer, 1846)

Bulimus Hebe Pfeiffer, 1846: 39
Bulimus hebe Pfeiffer, Pfeiffer 1847: 268
Bulimus Hebe Pfeiffer, Pfeiffer 1848: 68, 1853b: 453
Partula Hebe Pfeiffer, Reeve 1851: pl4. Pfeiffer 1853b: 453, Pfeiffer 1856: 162, 1856: 268, 1859: 514, 1868: 161, 1877: 208. Adams 1855: 145. Frauenfeld 1869: 874. Paetel & Schaufuss 1869: 83. Schmeltz 1869: 72. Pease 1871b: 401, 473. Paetel 1873: 104, 1883: 153. Gloyne 1878: 338. Kobelt 1880: 275. Hartman 1881a: 183, 1881b: 9, 1885: 211, 1893: 75. Garrett 1884: 53. Smith 1903: 438. Pilsbry 1909: 223. Kondo 1955: 137, 1968: 76. Richardson 1990: 33
Partula Hebe, var. *bella* Pease, 1871b: 473. Paetel 1873: 104 (all nomen nudum)
Partula globosa 'Pease', Carpenter in Pease 1865: 675 (as var. of *hebe*). Schmeltz 1874: 207, 1874: 80. Pfeiffer 1877: 209. Gloyne 1878: 338. Hartman 1881a: 182, 1881b: 9, 1885: 211 (as var. of *hebe*), Paetel 1883: 33, 1889: 267 (all nomen nudum)
Partula hebe var. *bella* Pease, Hartman 1881a: 193, 1881b: 9, 1885: 211 (as var. of *hebe*). Garrett 1884: 54. Smith 1903: 438
Partula hebe bella 'Pease', Pilsbry 1909: 224. Kilias 1969: 246
Partula ventrosa 'Garrett', Hartman 1881a: 189 (as var. of *hebe*), 1881b: 9, 1885: 211 (all nomen nudum)
Partula ventricosa Garrett, 1884: 54 (as var. of *hebe*)
Partula ventricosa Garrett, Hartman 1885: 211

Globose-concial. Umbilicus open. Moderately thick, surface shiny, with fine spiral periostracal striae. Low conical spire with 4¼-4½ flat whorls. Columella short, wide in oblique view. Aperture subsemicircular. Lip usually narrowly expanded, calloused within, distinctly carinate. A well developed parietal tooth is almost always present (99.6%, n=3,402).
White, typically lacking epidermis. Lip white. Var. *bella* has a well-preserved epidermis and the apex is darker, frequently rosy and is sometimes banded. Var. *ventricosa* retains a white to fulvous epidermis, apex often purple-brown, or rarely reddish, never rosy. Var. *globosa* differs from *bella* in being slightly smaller and more globose, rosy apex is less well defined.
Anatomy: Penis short, ⅓ oviduct length, slender. Caecum shorter than penis. Vas deferens subapical.
Type specimens:
Bulimus Hebe Pfeiffer, 1846 – Syntypes AMNH 42273 (3 specimens collected by Mallet); MCZ 297855 (2 specimens, Mallet on label); SMF 164969. These are all labelled as types from the Cuming collections, only the AMNH and MCZ specimens specify 'Mr. Mallet' as Cuming's collector, as in the original description.
Partula hebe var. *bella* Hartman, 1881 – Syntypes MCZ 3639. Supposed syntypes, but actually 'types' of Pease's nomen nudum: FLMNH 111999; NHMUK 1934-6-22.156-9; ZMB/Moll-74873.
Partula ventricosa Garrett, 1884 – Syntypes ANSP 59497.
Partula hebe var. *globosa* Hartman, 1881 – Nomen nudum, supposed types MCZ 25275.
Distribution: Most valleys in southern Raiatea – Vairahi, Averarahi, Faaiti, Maufenua, Faaroa, Farepaitia, Fareparahi, Opoa,

Fig. 9.4.100 *Partula hebe* syntypes: a – *hebe* (MCZ 297855); b-c – *bella* (NHMUK); d – *ventricosa* (ANSP 59497)

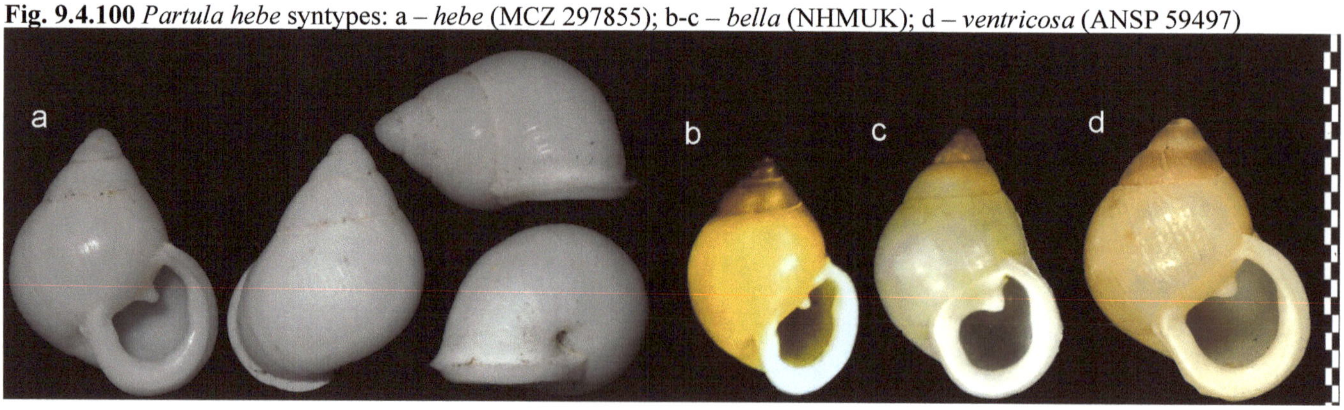

Hotopuu, Toahiva, Vaianae, Faatemu and Vaiaau valleys, and on Mt. Tefatua.
There was significant geographical variation in the frequency of colour forms (considered to be varieties by Garrett and Hartman) but these are not accompanied by and anatomical or conchological difference and so are not considered subspecies. Var. *bella* was present in Opoa "in large numbers" (Garrett 1884) and Hotopuu (the source of the captive population). Var. *ventricosa* was recorded from the west coast (Garrett 1884) and Mt. Tefatua at 700 m (Moore BPBM). Garrett (1884) restricted the typical form to "two contiguous valleys", apparently Faaroa and Averarahi. Arboreal, found "in great profusion on the foliage of bushes" in Faaroa (Garrett 1884), less abundant further north.
Note: Smith (1903) reported three possible hybrids between *P. crassilabris* and *P. hebe bella*, and a possible hybrid between *P. rustica* and *P. hebe*. There are no clearly overlapping populations but their ranges are adjacent.

Partula crassilabris Pease, 1866

Partula crassilabris 'Pease' Carpenter in Pease 1865: 675 (as syn. of *hebe*, nomen nudum)
Partula crassilabris Pease, 1866: 199
Partula crassilabris Pease, Pease 1871a: 81. Schmeltz 1874: 207. Pfeiffer 1877: 208. Gloyne 1878: 338. Hartman 1881a: 181 (excl. *rustica*), 1881b: 9, 1885: 211, 1892: 75. Garrett 1884: 66. Smith 1902: 438. Pilsbry 1909: 225. Kondo 1955: 136, 1968: 76. Kilias 1969: 247. Richardson 1990: 20
Partula Otaheitana Bruguière, Reeve 1851: pl2 f11 (misidentification)
Partula Hebe var., Carpenter 1864: 675

Semi-globose. Umbilicus open. Dull, with faint growth-lines and fine spiral striae on all whorls. Spire straight sided. 4½ slightly convex whorls, body whorl round. Aperture small, rounded. Lip expanded, thick, thickened within, carinae, narrowed above. Columellar margin broad but lacking tubercles or nodes. Parietal wall with a thin transparent callus with a very small tooth deeply recessed, absent in 9% of shells (n=57).
Uniformly brown to pale brown. Base darker, spire usually purplish-brown. Periphery usually with a yellow line. Aperture

Fig. 9.4.101 *Partula hebe* variation (ANSP 295316)

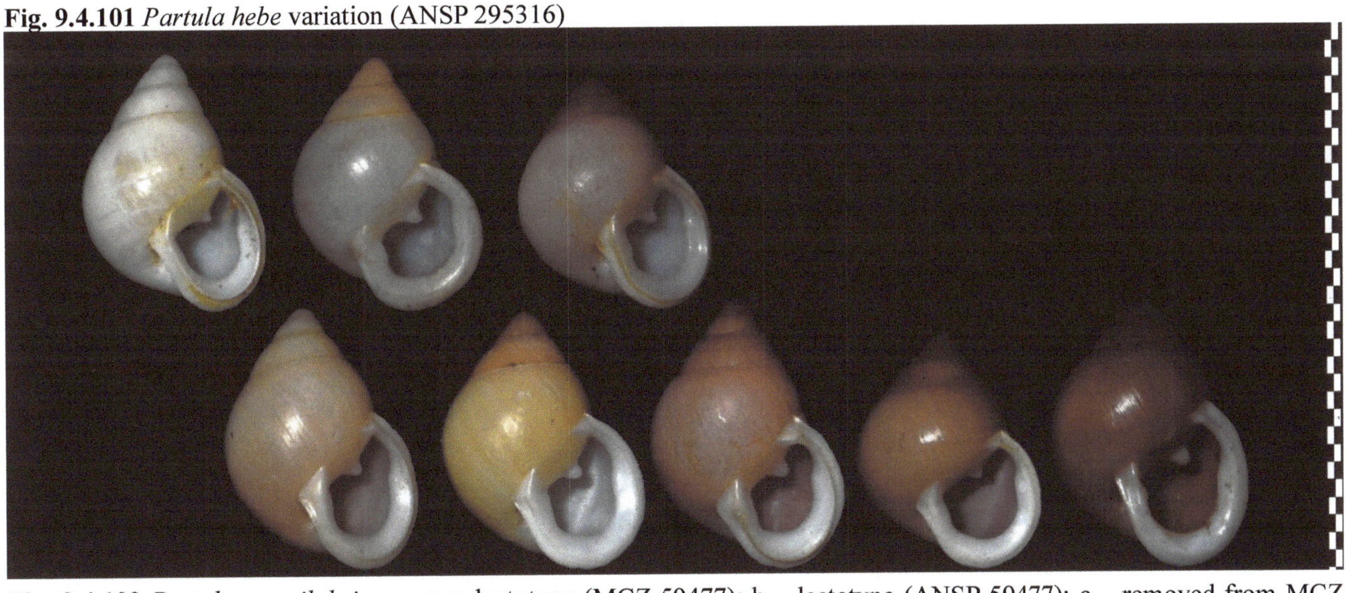

Fig. 9.4.102 *Partula crassilabris*: a – paralectotype (MCZ 59477); b – lectotype (ANSP 59477); c – removed from MCZ 25011; d-e – specimens UMZC 75172

dark flesh-coloured inside. Lip white or brownish to outer edge. Crampton (MS) defined 8 colour forms varying from uniform to a rarer form with a pale median band. Plain ones vary from whitish corneous tinged pale brown, medium brown, darker chestnut brown. Apex variable, usually dark but including light shells with a light apex (Pilsbry's 'albino').

Anatomy: As *P. hebe*; penis 1/3 length of oviduct.

Type specimens: Lectotype ANSP 59477 (erroneously listed as holotype by Pilsbry [1909]), paralectotypes AMNH 118937, 42093; ANSP 59478, 464354; CAS: IZ: 85758; FLMNH 111897-8; MCZ 25131, 25139, 25292; RMNH MOL.56766; SMF 164968; ZMB/Moll-74328-9, 75628-30. Lectotype designation by Baker (1963).

Fig. 9.4.103 *hebe*-group geographical variation showing colour pattern (apex colour superimposed) and tooth size variation. *P. pearcekellyi* locality marked in red

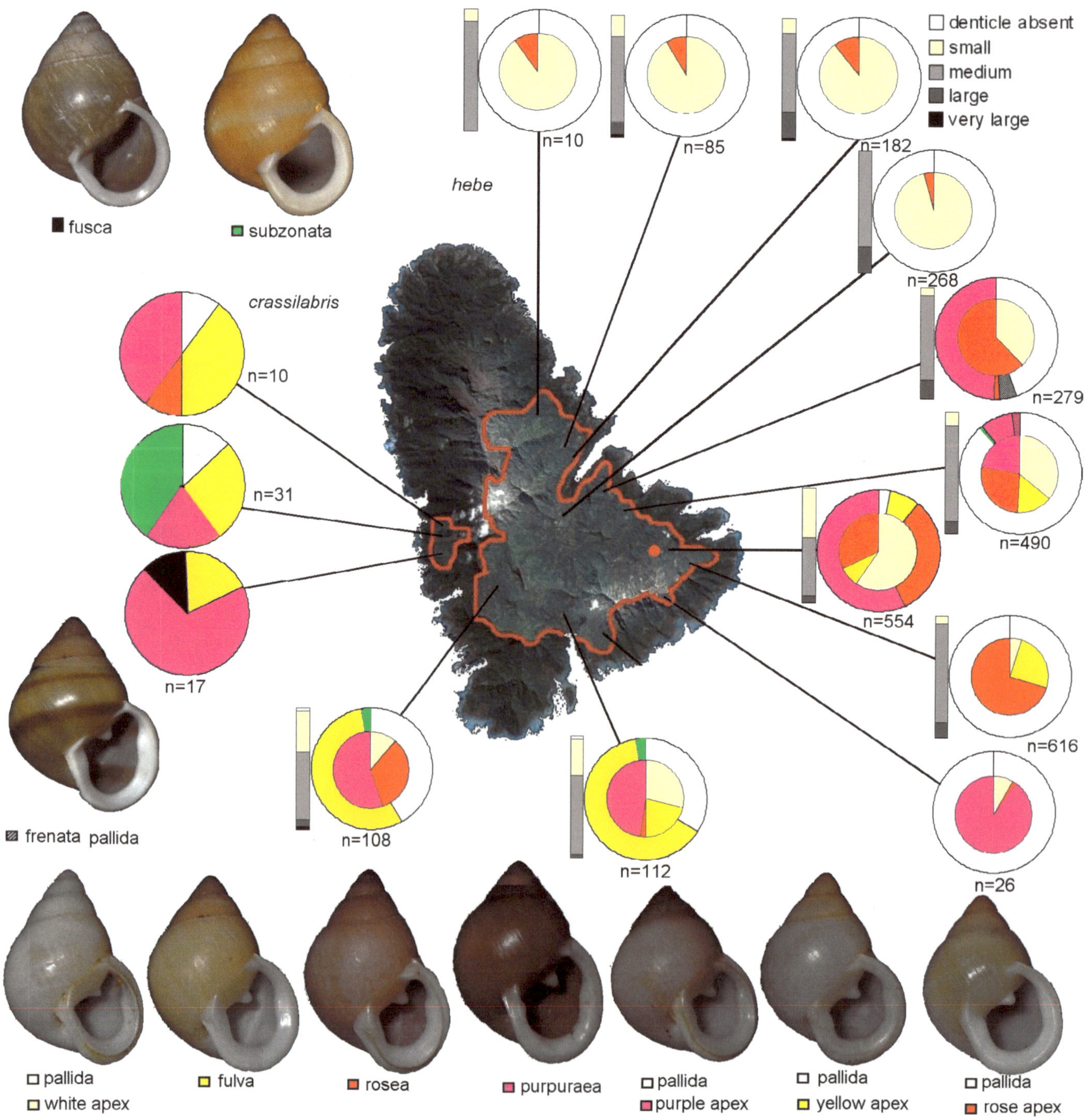

Distribution: Restricted to the west side of Raiatea, in Utuara, Hapai, Ereeo, Marahi and Tehuri valleys. Crampton collected relatively few specimens as he spent little time in the rugged and remote area occupied by this species: Utuara (21 live and 15 dead 1908, 10 live and 18 dead 1909), Tehurui (19 live 1908), Marahi (1 fresh dead 1908). Pease had 30 specimens and Garrett (1884) found it "very abundant, lurking beneath decaying vegetation" in Hapai valley. Terrestrial.
Note: Considered to by synonymous with *P. rustica* by Hartman (1881) as he considered them to differ only in shape: *P. crassilabris* being smaller, rounder and lacking an indented columella.
Smith (1902) reported a possible hybrid between *P. rustica* and *P. crassilabris*, and several possible hybrids between *P. crassilabris* and *P. thalia*. These are probably just variants at the ranges of these species are well separated.

Partula pearcekellyi **sp. nov.**
Ovately conic. Shell moderately thin but strong. Apex acute. Umbilicus open. Original surface eroded, spiral striae visible on fragments of periosracum. Spire about 1/3 shell length, suture impressed. 4½-5 slightly convex whorls. Aperture rounded, half length of shell. Lip thin, slightly and evenly expanded but not thickened. Columellar margin dilated above but not flattened. No distinct parietal callus.
Light brown; periostracum eroded in all specimens, even live collected juveniles retain only fragments.
Anatomy: Not know; the only spirit material is of juveniles, which appear to be developing *P. hebe*-type anatomy.

Fig. 9.4.102 *Partula pearcekellyi*: a – holotype (NHMUK 20140800); b – subadult (BPBM 136081); c – comparative specimen of *P. hebe*; d – comparative specimen of *P. crassilabris*

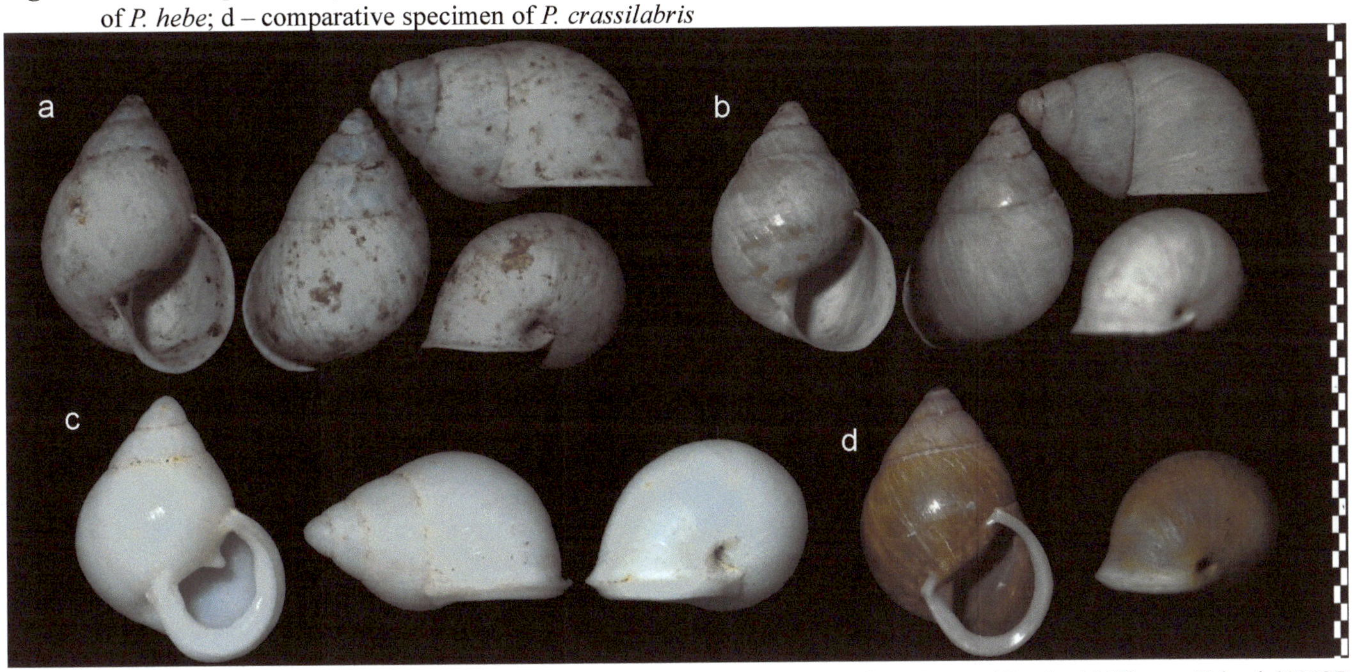

Fig. 9.4.105 Anatomy of *hebe*- and *turgida*-groups: a – Faaroa (Kondo 1955); b – Utuara (Kondo 1955); c – Tehuri (ANSP A789); d – Paaoio (BPBM 136085); e – Hotopuu (ZSL 117 & 316). Scale bar 5 mm.

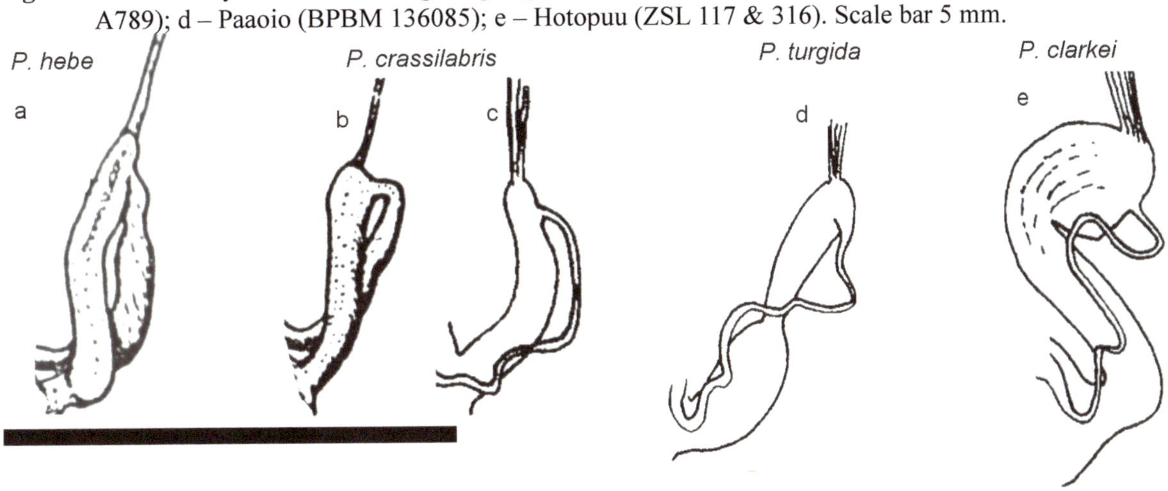

Type specimen: Holotype NHMUK 20140800, Vaitaratoa valley, 2.ix.1992, J. Gerlach. Other specimens: BPBM 136081 dead subadult; BPBM 136079-80, 136082 four juveniles, all Toahiva valley, collected by Y. Kondo & D. Anderson, 1934.
Distribution: Known from six specimens from Toahiva valley, of which 4 juveniles were live collected in 1934 'on ferns, shrubs, tree trunks, & dead leaves' (BPBM 136079-82). At the same time 1 juvenile and 26 adult *P. hebe* were collected, all live. This may suggest that *P. pearcekellyi* was a rare canopy species.
Etymology: Named after Paul Pearce-Kelly, of the Zoological Society of London.
Similar species: This species is closest to juvenile *P. hebe* in general form. It can be distinguished by the lack of any trace of a parietal tooth, which develops in *P. hebe* as the lip starts to expand, thus there are few *P. hebe* with expanded lips but without a parietal tooth. Crampton (MS) recorded only four edentulous specimens (0.05%; n=3,402). In addition the umbilicus is deeper in *P. hebe*, extending as a narrow perforation for a full whorl's depth. No periostracum is present in *P. hebe*.
Measurements of type specimen: 16.3 × 11.9 mm., aperture length 8.0 mm., 5 whorls.

turgida-group

Contains only *P. turgida* according to Kondo (1955). Two further thin-shelled species are added to the group here. This group is probably polyphyletic and is used for convenience only.

Partula turgida (Pease, 1865)

Bulimus turgidus Pease, 1865: 670

Bulimus turgidus Pease, Pfeiffer 1868: 12, 1877: 17. Pease 1871b: 473.

Partula turgida Pease, Hartman 1881a: 188, 1881b: 12, 1885: 217, 1892: 457. Garrett 1884: 56. Smith 1902: 457. Kondo 1955: 130, 1968: 76. Richardson 1990: 62

Partula (*Leptopartula*) *turgida* Pease, Pilsbry 1909: 178

Ovate, body whorl inflated. Extremely thin, fragile. Umbilicus not perforate; shortly, openly rimate. Glossy, with growth-lines and minute, close, spiral striae. Spire very short, conic. 4½, convex whorls, the last very large, evenly swollen. Aperture oblique, broadly ovate, ⅔ shell length. Lip thin, narrowly expanded, columellar margin dilated and flattened above, not thickened. Parietal callus thin, shining, margin distinctively sinuous, becoming straighter with age.
Pale horn-coloured slightly transparent, suture margined white. Crampton (MS) noted that some had a "very narrow revolving band of deeper color, vague in outline, just above the suture. When present, it is evident only on the ventral aspect of the last whorl." This was present in Crampton's Vaipao shell, figured, but not described, by Garrett (1884) and Pilsbry (1909).
Anatomy: Penis long, ⅓ oviduct length, twisted at constriction between penis and caecum. Caecum thin walled, papillate.
Type specimens: Probable syntypes ANSP 59552; MCZ 25367, 26354.
Distribution: Uturoa (='Vaipao'), Voaara and Tooroa (='Paaoio') valleys, 300-400 ft (Kondo 1955). Widely distributed but rare. Arboreal; a canopy species which was probably more widely distributed than the confirmed localities indicate.
Note: Pilsbry (1909) placed *P. turgida* in *Leptopartula*, along with *P. arguta* of Huahine, these being small, ovate, thin-shelled species; this grouping is artificial (Kondo 1955). Garrett found this species to be very rare, collecting only 9 (unknown locality). Crampton collected two from Uturoa and Voaara; he thought that Garrett's specimens probably also came from Uturoa as older Raiateans who remembered Garrett said he had apparently lived there for longest when on the island (Crampton MS). Kondo (1955) described *P. turgida* as 'very rare' and although the Mangareva Expedition found 34 animals, they were all from a single locality (Tooroa). Of Kondo's specimens 5 subadults 14.3-15.5 mm. long were gravid, whereas juveniles 14.1 mm. and under contained no embryos. This indicates precocious maturity at a size of 14.3 mm. and above (BPBM 136089).

Partula clarkei sp. nov.

Partula turgida Cunningham & Daszak 1998: 1139 (misidentification)

Globularly ovate, shell thin but strong. Apex acute. Umbilicus open, narrow and only partially covered by lip. Surface glossy with fine spiral striae. Spire short, suture impressed, not margined. 4½ whorls, first two straight sided, remainder strongly convex, penultimate whorl appearing to bulge above lip in dorsal view. Aperture rounded, half length of shell. Lip thin, slightly expanding but not thickened. Columellar margin dilated above but not flattened. No distinct callus between columella and insertion of lip.
Opaque white, with a fleshy brown tint to apex and body whorl, in some specimens this may be a rose colour.
Body grey, mantle pale with dark edge and mottling visible through shell.
Anatomy: Penis elongate, ½ oviduct length, tubular. Caecum club-shaped with a subterminal vas deferens.
Type specimens: Holotype NHMUK 20140801, paratype NHMUK 20140802, collected by Dave Clarke and Edwin Blake during the 'Operation *Partula* 1991' expedition, 1.ix.1991.

Distribution and status: Collected from Tevaitoa valley, Raiatea, ix.1991, 5 adults, 4 subadults, 9 juveniles collected. Maintained in captivity until 1st January 1996, now extinct.
Etymology: Named after Dave Clarke, of Zoological Society of London, member of the team that collected the original founder population.
Similar species: When collected this species was identified as *P. turgida* (Pease, 1864) to which it bears a superficial resemblance. *P. turgida* has a more ovate shell, lacking the inflated body whorl and a uniformly thin shell. It was thought that the thicker, opaque shell of *P. clarkei* might have resulted from an abnormally high calcium content in the captive diet, but wild collected individuals that had been in captivity for only one year show the same shell characteristics. Furthermore, the extinct species *P. arguta* also had a thin, *P. turgida*-like shell, but captive populations retained the diaphanous, fragile shell. *P. clarkei* has a longer spire and more convex whorls, giving a higher shouldered appearance to the shell. The apex of *P. turgida* is less acute, the umbilicus obscured by the columella. It also has a distinctly white margined suture and a uniform pale brown colour. Anatomically the two species could be distinguished by the obviously inflated caecum in *P. clarkei* (Fig. 9.4.105).
Measurements of types: Holotype – 15.5 × 12.0 mm, aperture length 9.8 mm; paratype 14.0 × 11.5 mm, aperture length 8.6 mm, 4½ whorls.

Partula meyeri Burch, 2007

Partula meyeri Burch, 2007: 2

Conical. Thin, translucent. Umbilicus narrow. Approximately 4½ convex whorls, ornamented with distinct spiral striae, fading on body whorl but remaining distinct on base. Sutures moderately indented. Aperture moderately expanded and slightly thickened, simple. Lip narrowly expanded, not thickened, incomplete, not continuing on the parietal side.
Brown, lip paler.

Fig. 9.4.106 *turgida*-group: a – *P. turgida* syntype (ANSP 59552); b-c – *P. turgida* Tooroa specimens (b – BPBM 136089, c – BPBM 136085); d – *P. meyeri* holotype (BPBM 255947, from Burch 2007); e – two *P. clarkei* specimens (ZSL); f – *P. clarkei* holotype (NHMUK 20140801)

Body dark grey in preserved animal. Mantle mottled with dark and white patches and stripes.
Anatomy: Not known.
Type specimens: Not located; holotype listed as BPBM 255947 (Burch 2007) but not received or accessioned by BPBM.
Distribution: 950 m on the highest summit of the island of Raiatea, Mount Tefatua (Toomaru), French Polynesia, by Jean-Yves Meyer in 2005. A rare species; three animals were seen in 2005 and the only other collection from the site (J. Moore 1927) does not include this species.

Fig. 9.4.107 Recorded localities of thin-shelled Raiatean *Partula*

Fig. 9.4.108 Live Raiatean *Partula*: a – *P. faba*; b – *P. navigatoria radiata*; c – *P. meyeri*; d – *P. labrusca*; e – *P. hebe*; f – *P. garrettii*; g – *P. clarkei* (photos: c – J.-Y. Meyer; e-g – D. Clarke)

Tahaa

Tahaa supported six species of *Partula* most of which were clearly closely related to Raiatean species. One species, *P. faba*, was present on both islands (see p. 167).

Table 9.4.5 Tahaa and Bora Bora *Partula* shell measurements

	N	L			W			L aperture			W/L		
		range	mean	sd	range	mean	sd	range	mean	sd	range	mean	sd
faba subangulata	3620	21.8-27.7	25.26	0.993	13.6-18.2	15.22	0.655	12.6-18.3	13.65	0.629	56.7-66.7	62.63	2.341
umbilicata	2572	15.0-18.3	16.68	0.201	8.7-10.3	12.21	0.135	8.7-10.3	9.07	0.120	72.7-87.8	72.94	0.705
bilineata	285	11.7-19.5	18.33	0.798	11.7-19.5	12.21	0.501	8.9-12.4	10.11	0.495	60.8-77.1	66.50	2.263
planilabrum	1296	17.8-23.7	20.41	0.834	10.2-16.7	11.91	0.430	9.2-15.7	10.89	0.371	56.1-70.8	59.01	2.295
eremita	1	-	16.90	-	-	11.00	-	-	9.10	-	-	65.10	-
sagitta	19	16.2-19.8	17.77	0.757	9.7-10.9	10.03	0.332	8.1-9.3	8.59	0.333	54.5-60.5	56.24	1.712
lutea	778	15.3-20.5	17.49	0.810	8.9-11.7	10.99	0.423	8.1-11.1	9.63	0.454	52.4-67.5	62.79	2.222

Fig. 9.4.109 *Partula* of Tahaa: a – *P. eremita*; b – *P. bilineata*; c – *P. umbilicata*; d – *P. sagitta*; e – *P. planilabrum*; f – *P. faba subangulata*

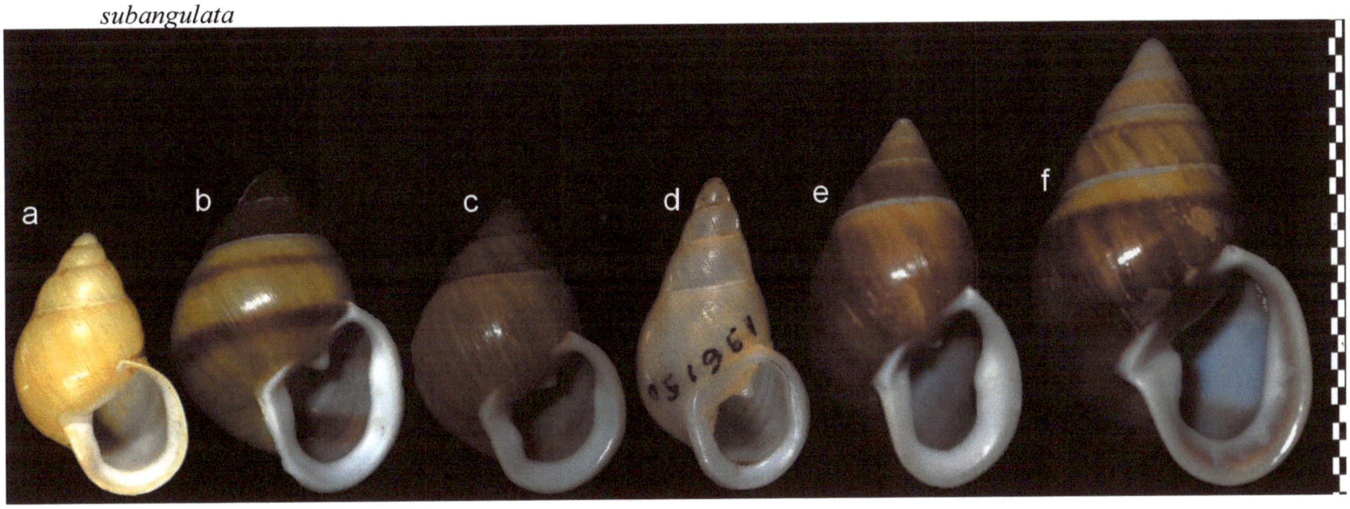

Fig. 9.4.110 Anatomy of the *Partula* of Tahaa (a-h) and Bora Bora (i-j): a – Tapuaamu '*virginea*' (ANSP A806); b- Patio (Kondo 1955); c – Vaiare (ANSP A808); d – Hurepiti (Kondo 1955); e – Hamene (ANSP A814); f – Faaha (Kondo 1955); g – Afatatea (ANSP A798); h – Hamene (Kondo 1955 and BPBM 136150); i – Mt. Pahio (Kondo 1955); j – Pahia (ANSP A1414) and unknown (NHMUK Cheeseman 1925). Scale bar 5 mm.

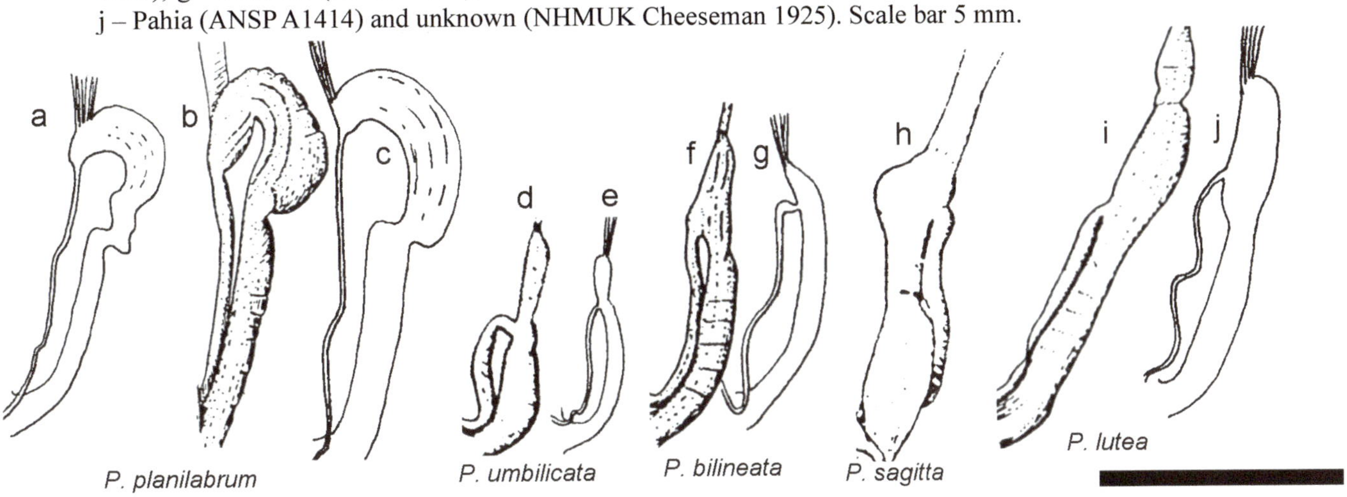

Partula planilabrum Pease, 1864

Partula planilabrum Pease, 1865: 672

Partula virginea 'Pease', Carpenter in Pease 1965: 675 (as var. of *solidula*, nomen nudum). Schmeltz 1874: 81 (nomen nudum). Binney 1876: 245 (unavailable). Hartman 1881a: 189, 1881b: 8, 1885: 8, 1889: 210 (nomen nudum). Garrett 1884: 61. Smith 1902: 440. Pilsbry & Cooke 1934: 14

Partula planilabrum Pease, Binney 1865: 245, 1876: 245. Pfeiffer 1868: 156, 1877: 195. Pease 1871b: 473.

Fig. 9. 4. 111 *Partula planilabrum*: a & b – types (NHMUK); c – *P. virginea* lectotype (ANSP 59474); d-g – Pahura *planilabrum* (ANSP 291634); h-n Tapuamu N *virginea* (ANSP 291630); o-p – Faahiu *virginea* (ANSP 291623); q-r – Apu-Apu *virginea* (ANSP 291639) (photo: c – P. Callomon).

Schmeltz 1874: 81. Graeffe 1879: 85. Hartman 1881a: 185, 1881b: 7, 1885: 156, 1892: 74. Paetel 1883: 153, 1889: 74. Garrett 1884: 3. Smith 1903: 436. Pilsbry 1909: 244. Kondo 1955: 138, 1968: 75. Richardson 1990: 50

Partula solidula Pease, 1871b: 473 (misidentification)

Partula imperforata virginea 'Pse.' Garrett, Pilsbry 1909: 222

Elongately ovate, umbilicus open. 5-5½ almost flat sided to slightly convex whorls, slightly glossy. Aperture oblong-oval, somewhat auriculate, toothed. Lip oblique and broadly flattened, thickened within, projecting outwardly. Parietal tooth present but variable in development. Columella straight, angular at base, with variably developed oblong nodule on face.

Yellowish corneous to chestnut-brown, almost always pale below white margined suture. Body whorl sometimes encircled with broad yellowish-brown band. Rarely light brown with indistinct darker streaks but no bands. Lip whitish, frequently tinged violet or carnelian. Colour varies geographically with five variants recognised by Crampton (MS): pallida, fusca, pallida strigosa, fusca strigosa, frenata (Crampton MS labelled northern populations as zonata, but are indistinguishable from the 'frenata' of the southern populations) and bisecta. All the frenata and zonata recognised by Crampton appear to be bisecta.

Anatomy: Penis large, robust, ½ oviduct length; caecum curved and inflated, separated from penis by distinct constriction, slender and elongate at base. Specimens identified as *P. planilabrum* and *P. virginea* are anatomically indistinguishable.

Fig. 9.4.112 *Partula planilabrum* colours. *planilabrum* populations marked in red, *virginea* marked in yellow. Mean shell length shown on sampling points (in mm).

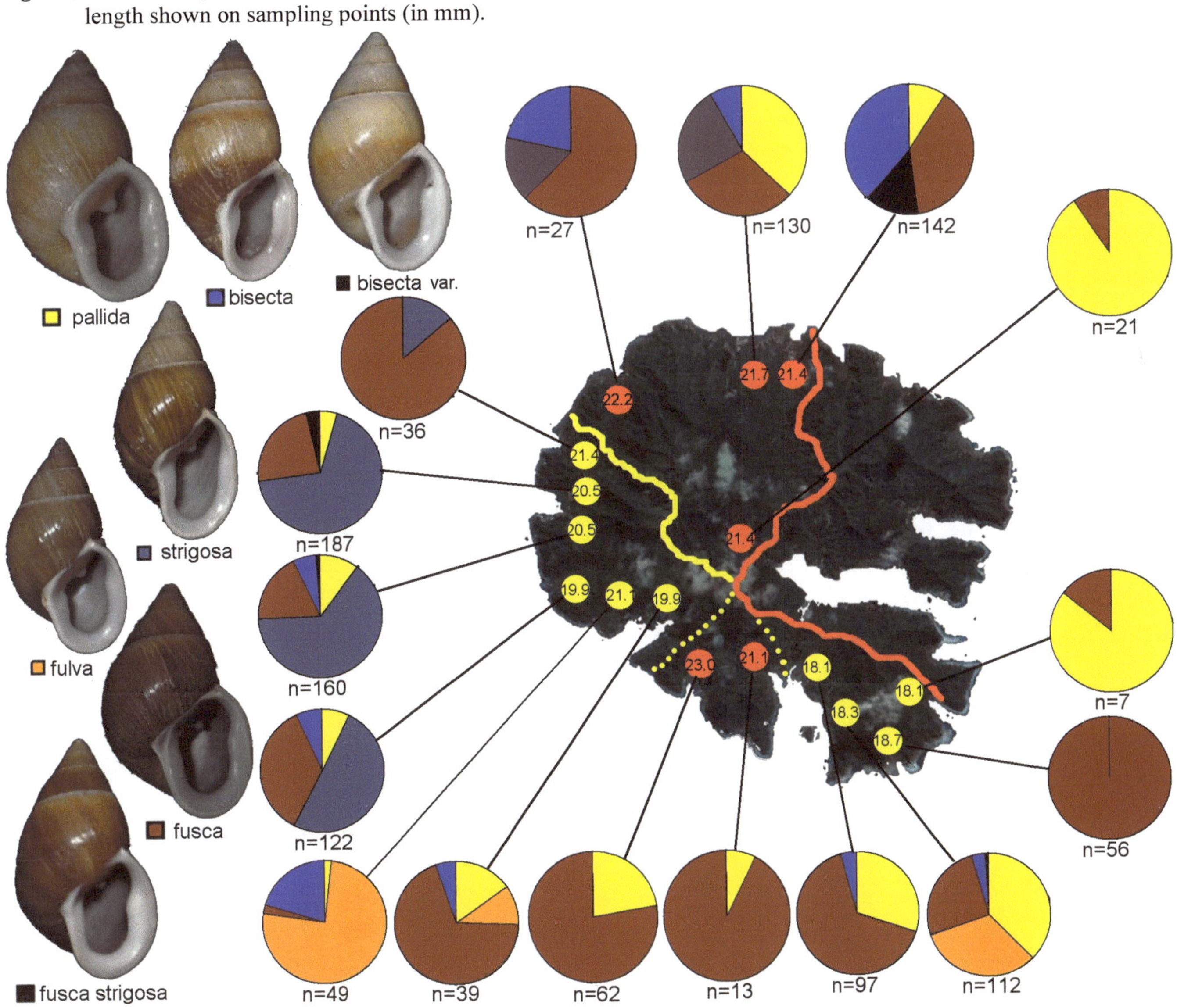

Type specimens:

Partula planilabrum Pease, 1865 – Syntypes AMNH 42082; ANSP 2638, 59978; FLMNH 112066; MCZ 25004, 25298; NHMUK un-numbered; SMF4957; YPM IZGP 015857.

Partula virginea Garrett, 1884 – Lectotype ANSP 59474; paralectotypes AMNH 42264; ANSP 59474, 59952; MCZ 25243. Lectotype designation by Baker (1963).

Distribution: South and west Tahaa – Garrett (1884) recorded *P. planilabrum* only from the eastern side of the island: Haamene valley "where it is common, and, though usually lurking beneath decaying vegetation, is sometimes found adhering to the trunks of the wild banana. It is found, though less abundant, in a valley north of its specific centre, but does not occur in the intermediate valley Faa-apa, the home of *bilineata*" (Garrett 1884). Crampton collected it much more widely. According to Garrett (1884) *P. virginea* was restricted to Vaipiti valley, north to Murifanna valley, on the west coast of Tahaa. He described it as occurring in Vaipiti "in the greatest profusion on the foliage of shrubs". Crampton collected this form in broadly the same range: Hurepiti, Otuna, Tapuamu, Vaiora, Faataoto, Faatia, Apu Apu and Apu Mao.

Note: *P. virginea* refers to the smaller shells of the southern populations. These were less elongate, the mouth less auriculate. The differences were significant (length t=44.63, width t=54.55, $P_{2,1294}<0.0001$ for both), but as these are the only differences the species are synonymised here. Crampton (MS) recorded both forms sympatrically only in Apu-Mao and Apu-Apu in the south of the island, and in these valleys almost all shells were of the *virginea* type (Apu-Mao 104 *virginea*, 'some' *planilabrum*; Apu-Apu 129 *virginea*, 'some' *planilabrum*). There are no preserved *planilabrum* specimens from those valleys.

Pilsbry (1909) considered *P. virginea* this to be synonymous with *P. dentifera imperforata* (var. *raiatensis*) of Raiatea: "The distinction is based chiefly on geographic distribution but also to some extent on the diverse variations of the two forms. Pale specimens of *virginea* usually have the apex purple-tipped (not pink, as in *raiatensis*); but the darker shells have the embryonic whorls white. The only closely related or similar shell on Tahaa is *P. planilabrum*, which belongs to the *faba* series." Although *dentifera* and *planilabrum* are similar they are maintained here as separate species.

Fig. 9.4.113 *Partula umbilicata*. a – lectotype (ANSP 59452); b, c & i – UMZC 75316; d, e, g & j – Pease specimens (ANSP 295237); f – UMZC 75170; h – NHMUK

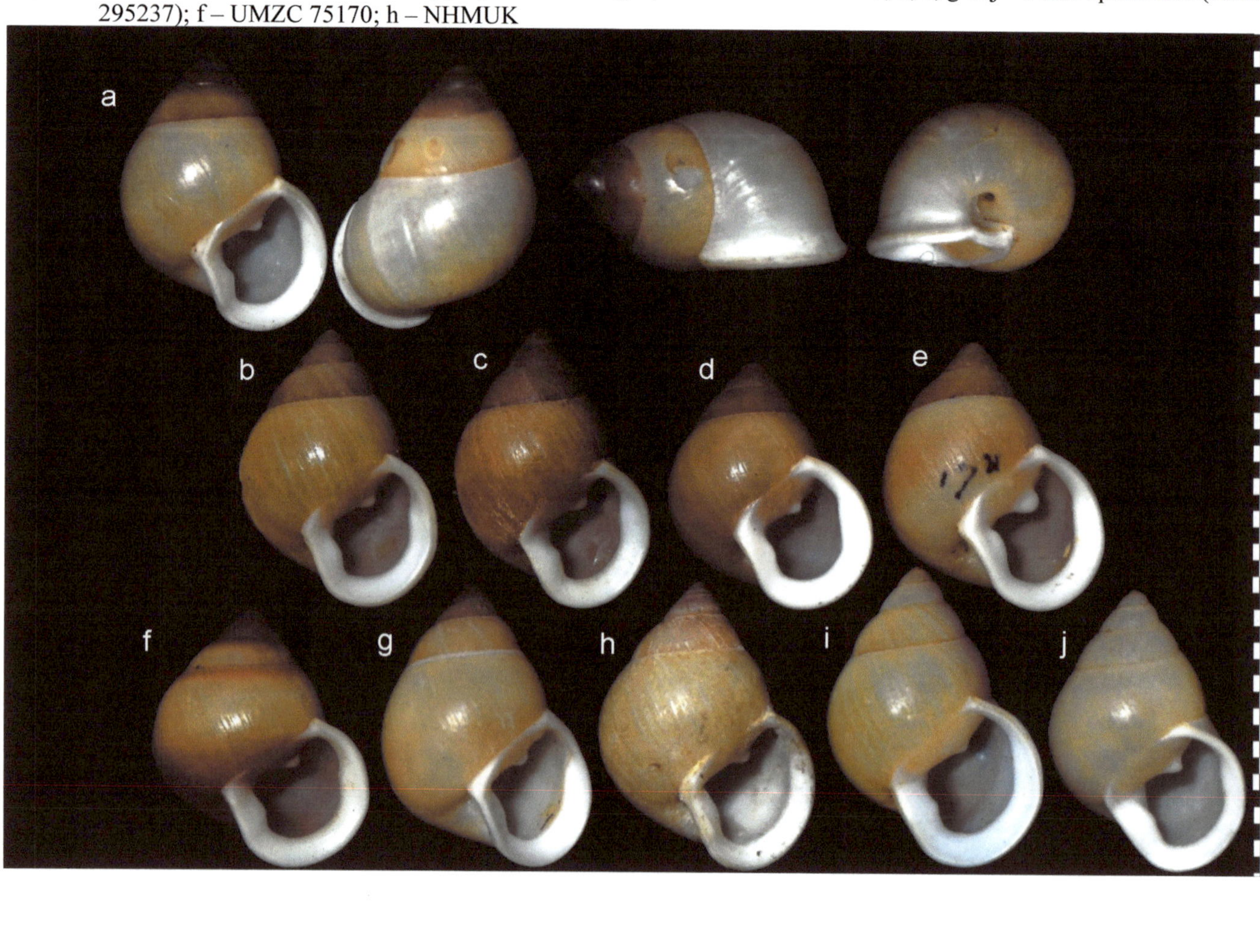

Partula umbilicata Pease, 1866

Partula umbilicata 'Pease', Carpenter in Pease 1865: 675 (as syn. of *auriculata*)

Partula umbilicata Pease, 1866: 200

Partula umbilicata Pease, Pease 1867: 81; 1871b: 474. Schmeltz 1869: 71. Paetel 1873: 104, 1883: 153, 1889: 268. Binney 1876: 245. Pfeiffer 1877: 207. Graeffe 1879: 85. Hartman 1881a: 189, 1881b: 8, 1885: 8, 1892: 75. Garrett 1884: 61. Smith 1903: 437. Pilsbry 1909: 229. Kondo 1955: 141, 1968: 75. Richardson 1990: 63

Partula auriculata var. Pease, 1864: 675. Hartman 1885: 211

Globosely ovate, umbilicus open and margins of umbilicus roundly angulate. Surface dull, with fine spiral striae. 4½ convex whorls, the body whorl inflated. Aperture rounded-oval, toothed. Lip internally thickened, obliquely flat, moderately expanded. Columella angular at base, with a tooth-like callus. Parietal tooth varies from small to well developed but is always present. Yellowish or brownish horn-colour, faintly streaked darker. Apex reddish-brown or dark purple, occasionally wholly dark chestnut-brown. In some specimens a broad yellowish horn-coloured band is present. Rarely they may be yellowish horn-coloured with a chestnut band. Outer lip cream-white behind, white on the face. Colour varies geographically.

Anatomy: Penis very short, cylindrical, ¹/₅ oviduct length; caecum elongate with a narrow, subcylindrical stalk and a slight submedian bulge. Vas deferens submedian.

Type specimens: Lectotype ANSP 59452; paralectotypes ANSP 59452, 464355; MCZ 25300, 25310; SMF164975. Lectotype

Fig. 9.4.114 *Partula umbilicata* variation in colour (pie charts) and tooth development (bars)

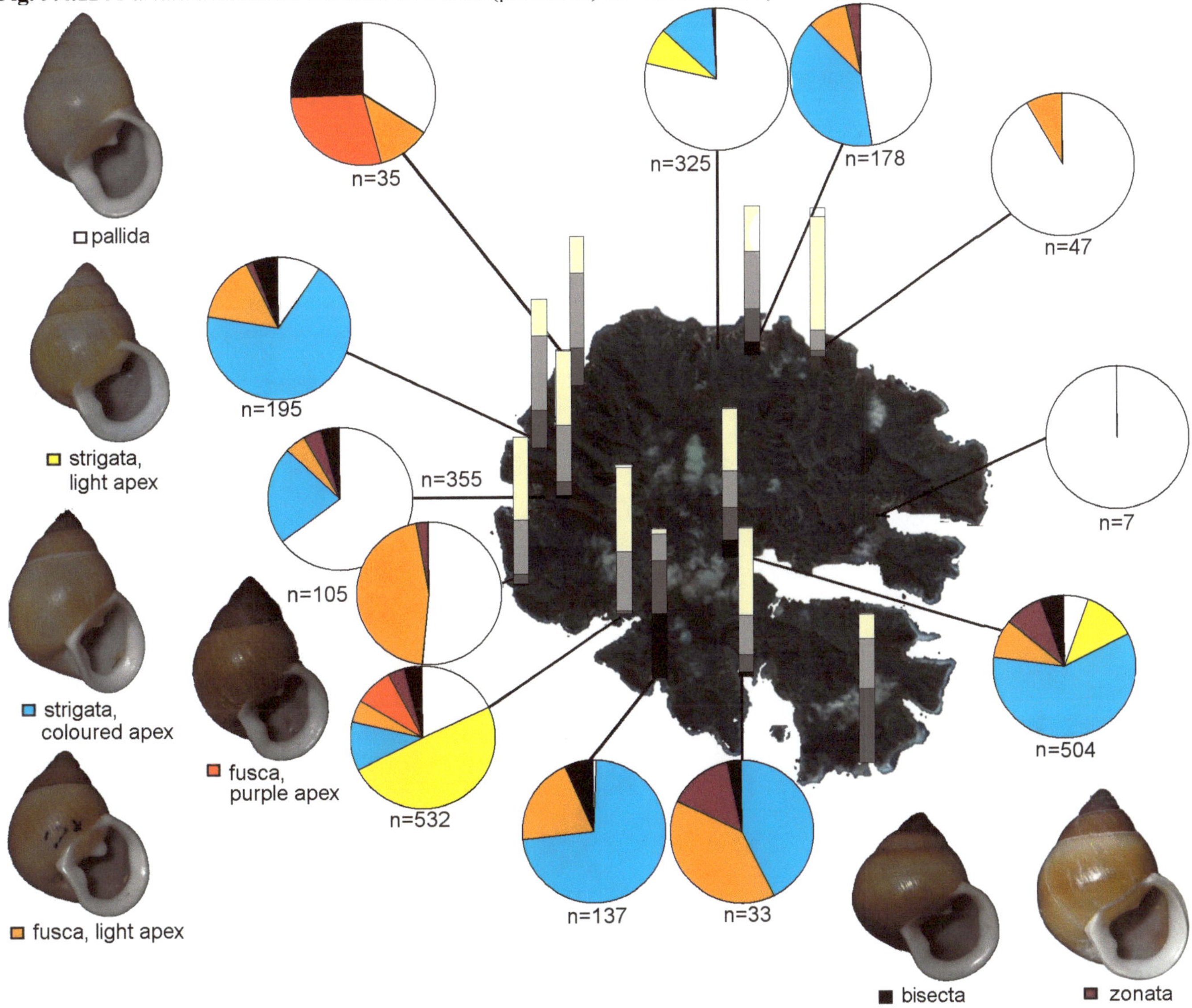

designation by Baker (1963).
Distribution: All areas of Tahaa. Garrett (1884) recorded it from Haamene in the east to Murifanna on the north-west (except Faa-apa), but not from the south. However, Crampton recorded it in all regions. Garret (1884) noted that in Haamene: "they are found in prodigious numbers on the foliage of low bushes".

Partula bilineata Pease, 1866

Partula bilineata 'Pease' Carpenter in Pease 1865: 675 (as syn. of *auriculata*)

Partula bilineata Pease, 1866: 201

Partula bilineata Pease, Pease 1867: 81, 1871b: 473. Schmeltz 1874: 81. Binney 1876: 245. Pfeiffer 1877: 195. Gloyne 1878: 338. Graeffe 1879: 85. Hartman 1881a: 180, 1881b: 8, 1885: 8, 1892: 74. Paetel 1883: 152, 1889: 266. Garrett 1884: 62. Smith 1903: 437. Pilsbry 1909: 245. Kondo 1955: 139, 1968: 75

Partula auriculata Carpenter (non Broderip) 1864: 675

Ovate-conic. Umbilicus open. Solid, smooth and glossy, with light growth-lines only. Spire almost straight sided. Lip broad, expanded and strongly thickened within. The columellar lip has a low nodule on its inner face, sometimes hardly noticeable. Parietal tooth deeply placed, well developed.

Whitish-corneous, pale yellow or brownish yellow. There is usually a narrow brown subsutural band and a wider subperipheral one, sometimes the whole base is dark chestnut, or the shell may be dark with a light equatorial band. Aperture almost white inside, lip white. Apex dark. Suture margined white. Crampton (MS) recognised four intergrading colour varieties: pallida, frenata, 'zonata' and zonata suffusa, all of which are variants of frenata. Almost all specimens are frenata, pallida is very rare.

Anatomy: Penis moderately long, slender, 1/3 oviduct length; caecum well developed, 1/3 penis length, no distinct constriction between penis and caecum.

Type specimens: Lectotype ANSP 59438, paralectotypes AMNH 42081; ANSP 4639333; MCZ 24903; SMF 194973. Lectotype designation by Baker (1963).

Distribution: Afatetia and Faahaiti valleys. Garrett (1884) recorded it only from 'Faa-apa valley' on the east coast. It was abundant on banana trees and at the bases of ferns (Garrett 1884). Kondo (1955) cited Crampton specimens from Faaha valley, which is presumably what Crampton labelled Faahaiti.

Fig. 9.4.115 *Partula bilineata, P. eremita* and *P. sagitta*: a-c – *P. bilineata*; a – type (ANSP 59438); b – zonata suffusa form (ANSP 291623); c – frenata form (NHMUK); d – *P. sagitta* type (ANSP 291602); e – *P. eremita* type (ANSP 291603) (photos: d-e – F. Borrero)

Partula eremita Crampton & Cooke, 1953

Partula eremita Crampton & Cooke, 1953: 157

Partula eremita Crampton & Cooke, Richardson 1990: 22

Ovate, moderately thin, umbilicus compressly pyriform. 5 convex whorls, suture somewhat impressed, margined only on last whorl. Dull with weak growth lines and distinct wavy spiral striae on all whorls, extending onto margin of lip. Aperture vertical, broadly oval. Lip only slightly thickened, bevelled and expanded to thin edge; columellar insertion simple. Columella narrow with a low swelling on inner face. Parietal callus thin, irregularly opaque and incomplete. No parietal tooth.
Yellowish corneous, with irregular faint darker streaks. Lip white within, tinged light brownish outwardly.
Anatomy: Unknown; known from a single shell.
Type specimens: Holotype ANSP 29160.
Distribution: Hurepiti valley near the foot of Mount Tete. Known from a single specimen collected by Crampton in 1908.

Partula sagitta Crampton & Cooke, 1953

Partula sagitta Crampton & Cooke, 1953: 155

Partula sagitta Crampton & Cooke, Kondo 1955: 140, 1968: 75. Richardson 1990: 56

Lanceolate-conic. Umbilicus open, compressed. 5-5½ straight sided whorls. Suture not impressed, marginated increasingly distinctly towards lip. Surface smooth, with fine growth lines and indistinct straight spiral lines. Aperture small and nearly circular. Lip wide, inwardly thickened, rounded and shelf-like on outer edge. Outer insertion of lip drawn upwards on body whorl, inner insertion slightly and roundly notched. Columella simple and nearly straight. Parietal callus thin, transparent and shining, with a well developed parietal tooth.
Faint yellowish brown, usually faded to white.
Body yellowish, contrasting with the silvery grey mantle.
Anatomy: Penis large, robust, 1/3 oviduct length. Caecum knob-shaped, neck well defined, stalk broad. Penis strongly rugose internally, with 7-9 vertical branching and anastomosing pilasters with one pilaster the largest and longest. Pilasters end abruptly in neck of caecum to be succeeded by profuse growth of papillae. Vas deferens subapical.
Type specimens: Syntypes ANSP 291602; BPBM 9974; SMF 16496.
Distribution: Hamene valley, on the slopes of Mt. Purauti, 750 ft. On tall tree 40 ft from ground. Only collected by the Mangareva Expedition in 1934 (19 adult and 17 immatures).

Fig. 9.4.116 Range of *Partula bilineata, P. eremita* and *P. sagitta*, with colour variation in *P. bilineata*

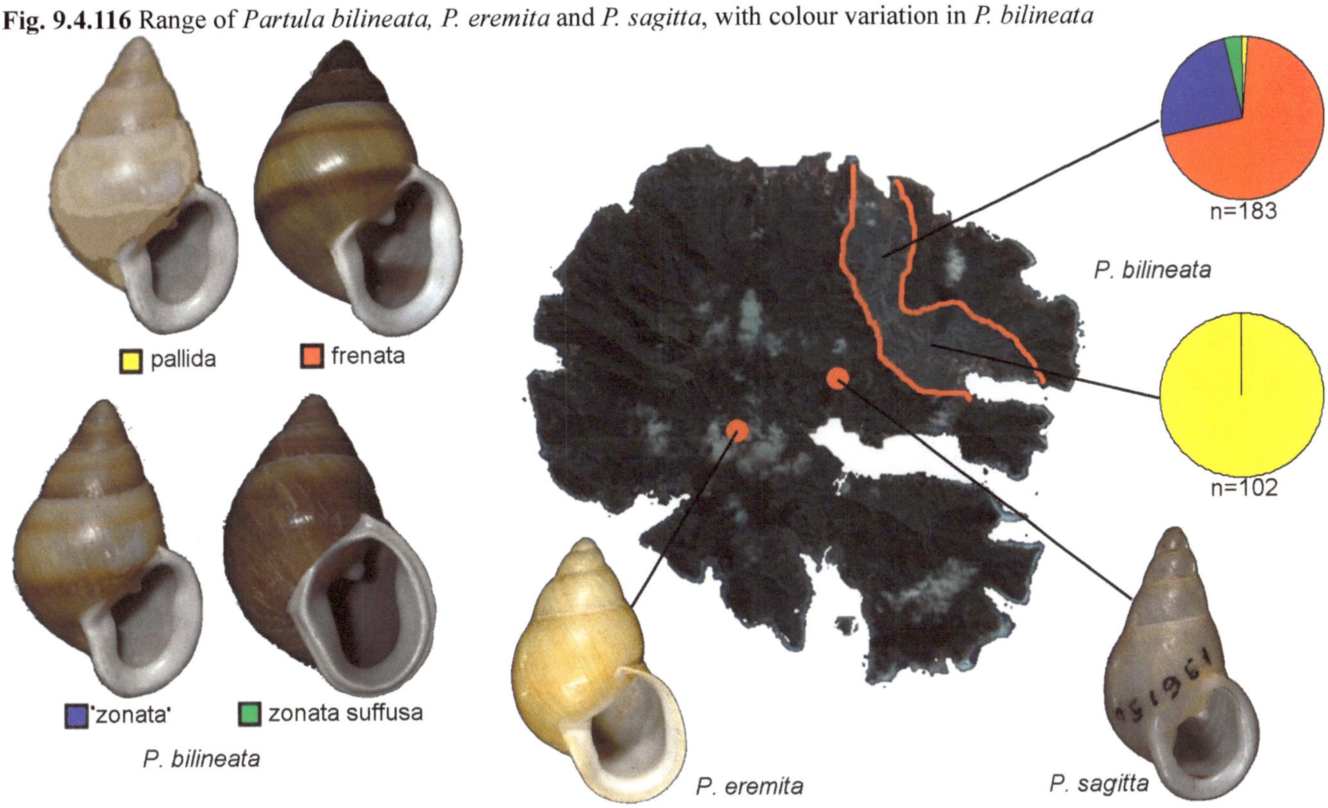

Bora Bora

Only a single, well-defined species has been recorded on Bora Bora. Measurements given on p. 209.

Partula lutea Lesson, 1831

Partula lutea Lesson 1831: 325
Bulimus (*Partula*) *luteus* Lesson, Pfeiffer 1842: 117, 1848: 229
Bulimus luteus Lesson, Férussac & Deshayes 1850: 123
Partula solidula Reeve, 1850: pl4
Partula solidula Reeve, Pfeiffer 1853b: 452; 1855b: pl64 f15-16. Schmeltz 1877: 92 (non Reeve)
Partula lutea Lesson, Pfeiffer 1853b: 453. Adams 1855: 145. Pease 1871b: 473. Hartman 1881a: 184 (partim), 1881b: 8, 1885: 8, 1892: 75. Garrett 1884: 53. Paetel 1889: 267. Smith 1903: 441. Pilsbry 1909: 255. Kondo 1955: 173, 1968: 75
Partula lilacina Pfeiffer, 1855c: 334
Partula lilacina Pfeiffer, Pfeiffer 1859: 514, 1868: 160, 1877: 207. Pease 1871b: 473

Obesely ovate-conic. Solid. Umbilicus narrow. 5 whorls, convex on spire, last two straight sided. Glossy with spiral striae on spire, fading on body whorl, but present near suture and umbilicus. The first two whorls slope steeply, making the apex distinctively conical. Aperture oblique. Lip reflexed, thickened within. Columellar margin only slightly dilated at insertion, with a callus on inner margin near insertion. Parietal callus transparent, thickened near lip. Edentulous.

Shell colour uniformly pale yellow or brown, apex concolorous or dark. *P. lilacina* Pfeiffer, 1855 is the normal yellow form worn thin on the spire to reveal the purple-brown shell. Crampton (MS) identified three colour varieties: pallida, fusca and purpuraea-helix (yellow with a revolving purple band on the spire). Of these pallida predominated (60%, n=838), followed by fusca (31%) and purpuraea-helix (9%). These showed notable geographical variation, but no clear pattern. Lip and thickened part of parietal callus white.

Anatomy: Penis moderately long, ½ oviduct length; caecum large. Penis with several pilasters, caecum densely papillose with papillae forming irregular ridges, 5 of which meet to form one main ridge.

Type specimens:
Partula lutea Lesson 1831 – Syntypes MNHN-IM-2000-27659.
Partula solidula Reeve, 1850 – Syntypes MCZ 297852; NHMUK un-numbered; SMF164970.
Partula lilacina Pfeiffer, 1855 – Holotype NHMUK un-numbered.

Fig. 9.4.117 *Partula lutea* colour variation

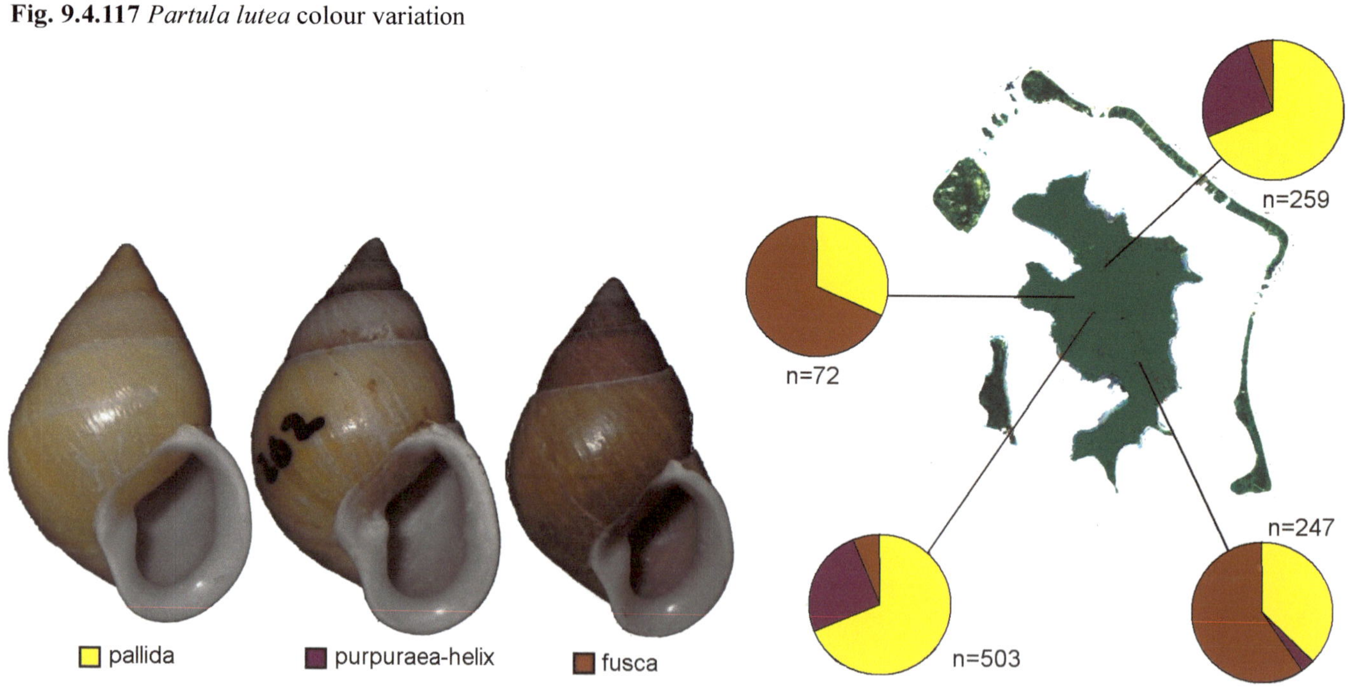

Distribution: Bora Bora, "widely diffused throughout the island, on the trunks and foliage of trees and bushes" (Garrett 1884). Note: Garrett (1884) considered the variation to be minimal: "notwithstanding its wide range over the island, it has not developed a single local variety. In fact it exhibits less variation than some of the species at the other islands which are restricted to single valleys."

Maupiti

On Maupiti Partulidae are known only from two subfossil shells photographed in 2010 (J.-F. Butaud pers. comm.; Fig. 9.4.118) and recent shells seen in 2012 at 360 m (J.-Y. Meyer pers. comm.). The former appear to be close to *P. lutea.* In addition there is a shell labelled *Partula thalia* from Maupiti (ANSP 3921521, ex C. Heller collection), this appears to be a mislabelling. No *Partula* were found on Maupiti by Garrett or by K.P. Emory in 1925 (BPBM material). These data suggest that *P. lutea* were introduced to Maupiti after 1925.

Fig. 9.4.118 *Partula lutea*: a – *lutea* type (MNHN 2000-27659); b – *lilacina* type (NHMUK un-numbered); c – *solidula* type (NHMUK un-numbered); d-h – Pahia valley variation (ANSP 290763); bottom row – *Partula* cf. *lutea* from Maupiti (photos: a – M. Caballer, Maupiti shells J.-F. Butaud)

9.5 Austral Islands

Introduced populations of *P. hyalina* have been recorded on Raivavae, Rimatara, Rurutu and Tubuai (see p. 135). A new species has been found on Rurutu and Rimatara.

Table 9.4.6 Morphometrics of Austral and Tuamotu *Partula*

	N	L			W			Aperture L			W/L		
		range	**mean**	**sd.**	**range**	**mean**	**sd.**	**range**	**mean**	**sd.**	**range**	**mean**	**sd.**
Australs													
cootei	58	15.7-21.6	18.58	1.218	9.2-12.5	10.47	0.516	7.1-10.9	9.55	0.625	47.7-75.4	56.49	2.197
hyalina	41	14.5-17.2	16.80	0.740	8.6-9.7	9.72	0.447	8.2-9.7	9.20	0.452	56.1-61.2	56.61	2.946
cootei × *hyalina*	61	15.8-20.0	17.54	1.042	9.2-11.1	10.21	0.446	8.8-10.3	9.33	0.393	52.4-64.0	58.31	2.681
Tuamotus													
makatea *	18	19.4-22.1	20.98	1.042	11.2-13.9	12.54	0.734	9.2-11.6	10.36	0.940	41.6-52.5	49.06	4.631

* n=8 for length

Partula cootei **sp. nov.**

Dextral. Elongately conic. Umbilicus narrow, rimate. 4½-5 slightly convex whorls, glossy with fine spiral striae on all whorls. Suture impressed. Aperture rounded oblong, nearly vertical. Lip thickened and expanded, thickened internally but minimally at parietal insertion. Parietal callus thick. Columella angled, expanding at insertion, impressed and plicate.

Uniformly white or with a poorly defined broad corneous greenish medial band on all whorls.

Anatomy: Penis short, stout, slightly inflated terminally with a small, indistinct caecum. Vas deferens inserts subterminally. Some specimens appear to be hybrids with *P. hyalina*, with a narrower penis but a subterminal vas deferens insertion. Some individuals are anatomically identical to *P. hyalina.* 67% of conchological *P. cootei* are anatomically typical and 33% apparent hybrids (n=6).

Type specimens: Holotype BPBM 278892, paratypes BPBM 278893, all ex BPBM 136858. Collected by D. Anderson on the Mangareva Expedition, 29.viii.1934. Mato Naa, 'both sides and top', Rurutu. '10-90 feet; medium; on ferns, shrubs, tree trunks, & cliffs.'

Distribution: Rurutu and Rimatara. On Rimatara known only from subfossils from the north and south-west of the island (MNHN Rm25b – Gargominy & Fontaine, 30.x.2004, marsh sediment in cave 10 m. a.s.l., north-west of Amaru, near the runway; MNHN Rm30 – Gargominy & Fontaine, 02.xi.2004; cave sediment 5 m a.s.l., south-west of the island). On Rurutu recent shells were found in woodland near the Peva Iti River (T. Coote, 17.iv.2005, above Peavaiti River 100 m a.s.l.). The Mangareva Expedition (1934) collected *P. cootei* mainly from patches of native forest in the higher areas (950-1150 ft) around Mt. Manureva, down to 200 ft with a few records from 50-200 ft around Moerai, on Mato Arei and at the foot of the bluff north of Moerai on ferns and shrubs. These were mixed with apparent *P. cootei* × *hyalina* hybrids on the western slopes of Mt. Manureva at 400-800 ft, extending up to 1150 ft in the north. Further hybrids were found with *P. hyalina* on the south side of the large mountain north of Avera 1000-1200 ft on ferns and shrubs and on Mato Naa. *P. hyalina* dominated in the lowland areas: Mato Arei (alongside *P. cootei*), around Moerai, Moto Arapi, Anapoto, Oromana, Amaru, the foot of Mato Naa and southwest of Hauti. Some were recorded from the middle slopes to the summit of Mt. Manureva (1200 ft).

Notes: This species was confused with *P. hyalina*. It is longer, has a more ovate aperture and lacks the angled columella of that species. The shell is also coloured, unlike most *P. hyalina*. The distinctions are obscured by apparent hybridisation.

The anatomical and conchological data suggest that the original *P. cootei* hybridised with *P. hyalina* giving a range of morphotypes. In contrast, in the Cook Islands only *P. hyalina* is present and is typical in all features (three dissected, all identical). The first records of *Partula* from Rurutu are from Cuming (Broderip 1928), which appear to be *P. hyalina*. Garrett (1879) examined shells collected on Rurutu by C. de Gage which included a large number of *P. hyalina* (Garrett 1879). Garrett could find no differences between these and the Tahiti specimens, he noted some variation in length but found the shape, colour and aperture constant in all. These are likely to have been collected in coastal areas.

Conchological white *P. hyalina* in the Austral islands are anatomically 89% *P. hyalina* and 7% aphallic (n=27, Kondo MS and new dissections). Coloured shells resembling *P. hyalina* are 60% *P. cootei*, 20% apparent hybrids and 20% *P. hyalina* (n=10).

Specimens examined:

P. cootei – BPBM 136795-6, 136829, 136831-3, 136836, 136846-8, 136850, 136856, 136858 (part), 136861-3, 136865 (part), 136872, 136875-6, 136879-881, 144856, 144878-9; MNHN Rm25b, Rm30; T. Coote specimens.

P. cootei × *hyalina* – BPBM 136791, 136833, 136847-8, 136865 (part), 136858, 136875
P. hyalina – BPBM 136766-7, 136773-5, 136777-8, 136783-88, 136790-1, 136792-4, 136797-804, 136815-22, 136828, 136830, 136833-5, 136837, 136840-5, 136849, 136851-5, 136857, 136858 Part), 136860, 136867-71, 136873-8, 136882-3, 136887-91, 136894-911, 136933-6, 136938-47; NHMUK; T. Coote specimens.

Fig. 9.5.1 *Partula cootei* and hybrids with *P. hyalina*. a – large coloured form (BPBM 136850); b – holotype (BPBM 278892 ex 136858); c – large white form (BPBM 136861); d – hybrid (BPBM 136865); e – Rurutu (T. Coote specimen); f – Rimatara subfossil (MNHN Rm 25b); g – anatomical *cootei* (BPBM 136856); h – coloured *hyalina* (BPBM 136887)

Fig. 9.5.2 Anatomy of Austral islands *Partula* and comparison with *P. hyalina*: a-d Rurutu (a – BPBM 136856; b & c – BPBM 136865; d – Kondo MS); e-f – Cook Islands (e – Kondo MS; f – BPBM 215795); g-h – Tahiti (g – Kondo 1955; h – various samples)

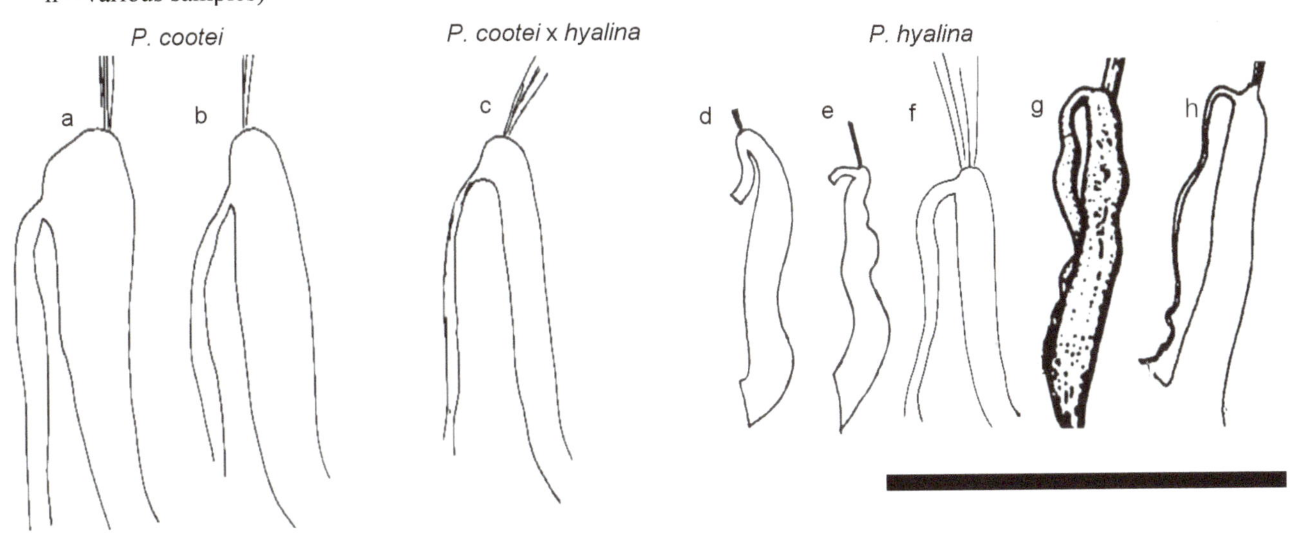

9.6 Tuamotu islands

Anaa atoll in the Tuamotus was recorded as the type locality of *Partula otaheitana amabilis* by Reeve (1840), but this can be discounted as an error in labelling. Garrett (1884) stated "Both Dr. Pfeiffer and Reeve described the latter species [*amabilis*] from specimens in the Cumingian collection, and both quote Anaa, a low coral island, as its habitat. Having resided five months on that island, and searched all parts for shells, I did not find a single *Partula* there, or on any other low coral island." However, an unidentifiable fragment of a subfossil shell was found by J.-F. Butaud (pers. comm.) (Fig. 9.6.1). Well preserved subfossils have been located on Makatea.

Makatea is a limestone plateau, reaching a height of 113 m. and covering 28 km^2 (Sartori *et al.* 2014). The island was inhabited in prehistoric times, and since 1908 has been heavily exploited by guano mining. The first partulid material was collected by Edgar Aubert de la Rüe in 1955 (Soyer collection 1969 in MNHN). These were incorrectly labelled *Partula nodosa*.

The age of these subfossil deposits is not known; Sartori *et al.* (2014) note that the island was heavily degraded by mining and this may have caused the extinction of many of the endodontid species. However, collections made in the 19th century would probably have included any partulids present, so it seems that extinction predates the mining.

Fig. 9.5.3 Location of *Partula* records from Austral islands (Tuamotu islands marked with a white point). Islands occupied by *P. hyalina* coloured blue. Rimatara: red dots – *P. cootei*. Rurutu: red dot – 2005 *P. cootei*; yellow – 1934 *P. cootei*; green – 1934 *P. cootei* × *hyalina*; blue – 1934 *P. hyalina*

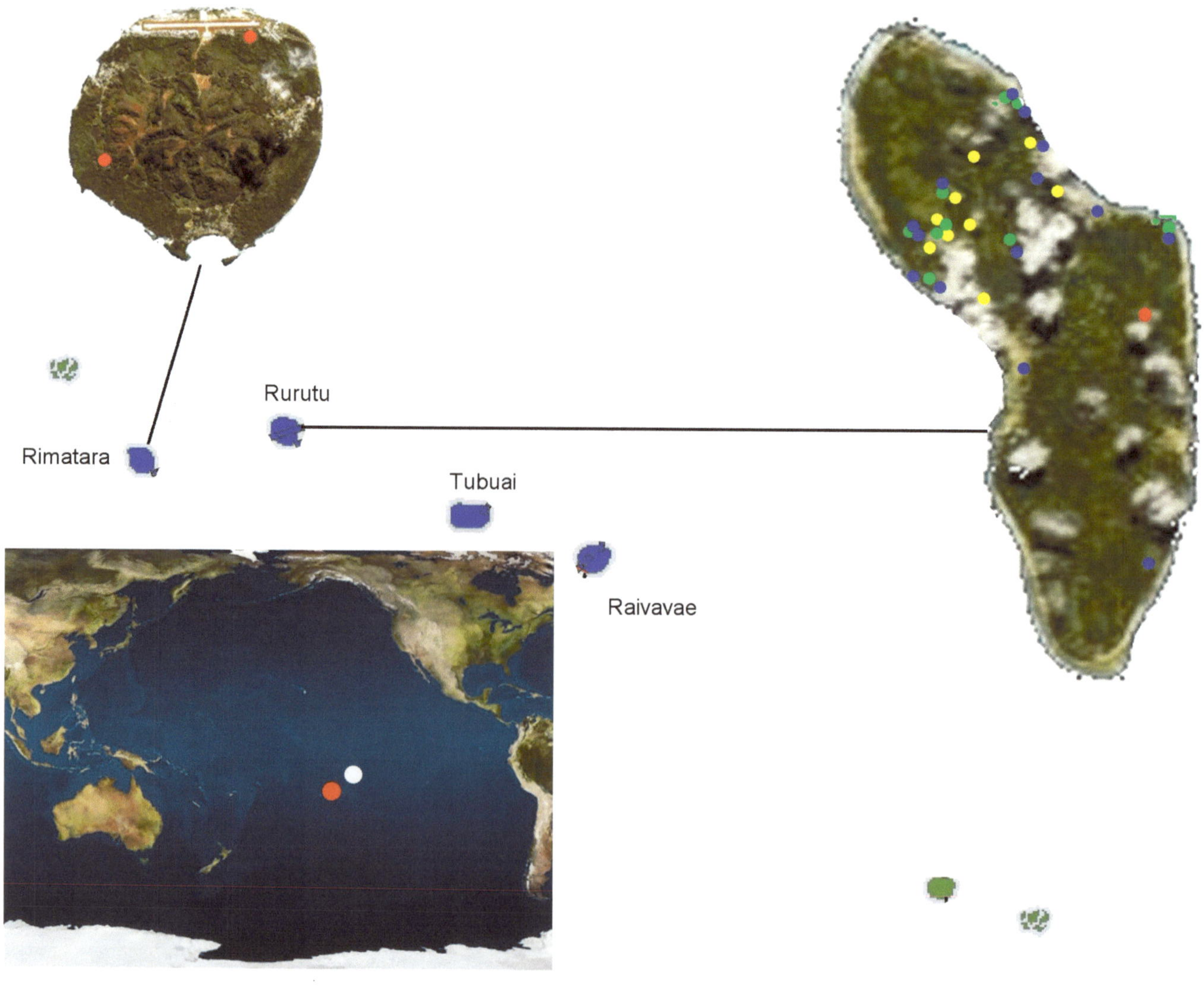

***Partula makatea* sp. nov.**

Dextral. Elongately conical. Umbilicus narrowly perforate. Surface shiny in some specimens. 5-5½ straight sided whorls, ornamented with distinct wavy spiral striae. Apical whorls with very strong, regular striae, becoming more wavy towards base, fading in some specimens but usually distinct on basal ½ of body whorl and sometimes distinct on all whorls. Apex abruptly obtuse. Body whorl slightly compressed and aperture sometimes moderately auriculated. Lip expanded and moderately thickened except at parietal insertion. A minute, deeply recessed parietal ridge is present, this runs inwards towards the columella and is distinct only when the aperture is tilted back. This is variably developed: absent in 65%, usually small, rarely distinct. It starts to develop in subadults with 4½ whorls. Parietal callus not distinct, apparently usually absent or thin and dehiscent.

Original colour not known; some fragments are reddish-brown, many have a white sutural band.

Anatomy: Unknown – only known from subfossil material.

Type specimens:

Holotype – MNHN IM-2000-30801; cave, west coast 3 km. south of Temao, 10 m, 17.xi.2005, O. Gargominy & B. Fontaine (Mk13).

Paratypes – MNHN IM-2000-30801; 3 specimens; cliff fall 1 km. north of Temao, 10 m., 16.xi.2005, O. Gargominy & B. Fontaine (Mk12).

Distribution: Makatea island, known from seven samples in MNHN collected by E.A. de la Rüe in 1955, and O. Gargominy and B. Fontaine in 2005 (Soyer colln. 1969, de la Rüe, 1955; 1 shell and 3 fragments; Mk 04 – cave, Moumu, 30 m, 14.xi.2005, 1 shell and 2 fragments; Mk10 – soil in coconut plantation south-east of Momu, 5 m, 15.xi.2005, 10 fragments; Mk12 – cliff fall 1 km. north of Temao, 10 m., 16.xi.2005, 4 shells and 85 fragments [>80 individuals, 3 with some colour]; Mk13 – cave, west coast 3 km. south of Temao, 10 m, 17.xi.2005, many fragments, 2 with some colour; Mk15 – west coast, 1.5 km south of Temao, 20 m., 17.xi.2005, 1 fragment; Mk21 – in rocks on plateau west of Anapoto, 5 fragments).

Fig. 9.6.1 Distribution of records of *Partula* in the Tuamotus. Inset – fragment from Anaa atoll, scale bar 1 cm (photo: J.-F. Butaud)

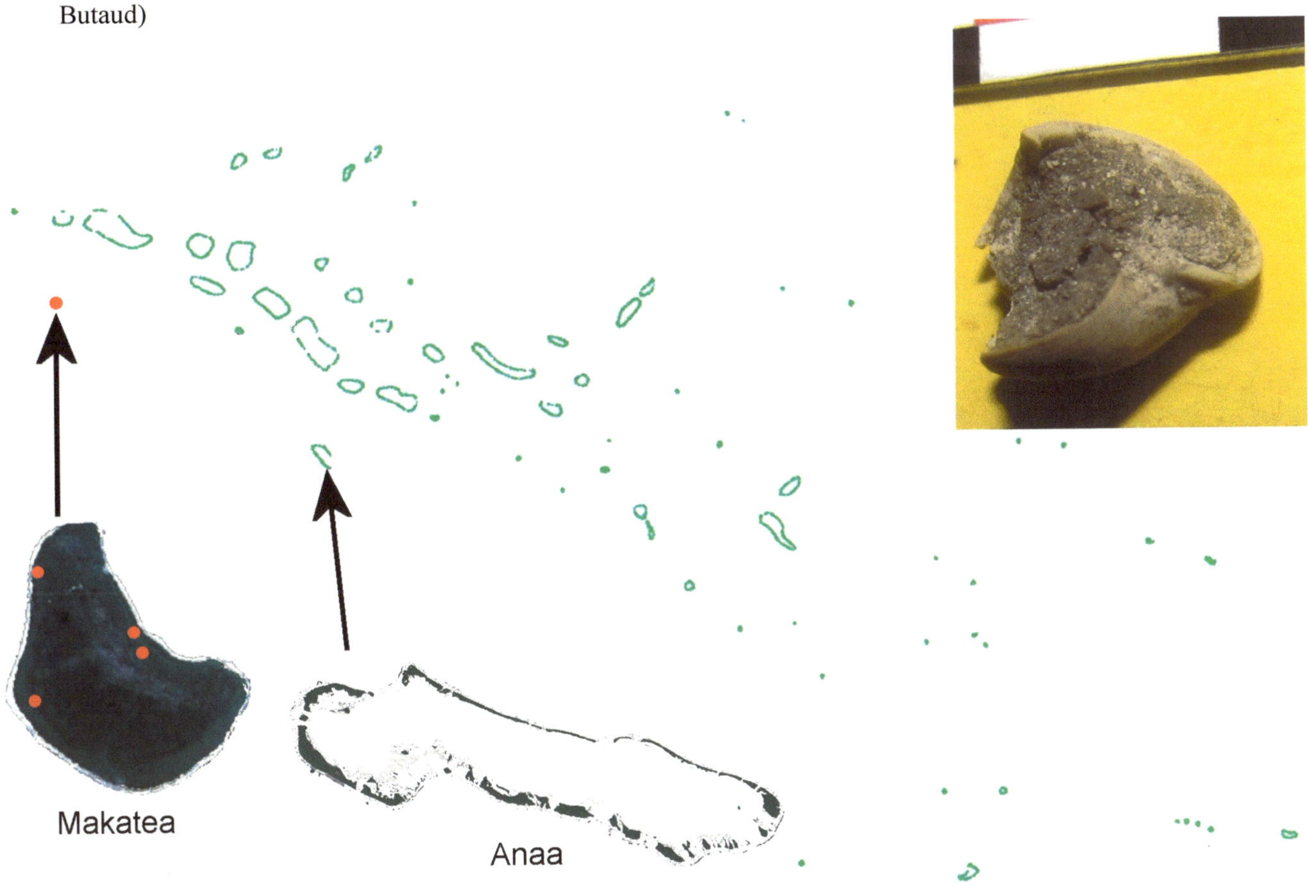

Fig. 9.6.2 *Partula makatea.* a – holotype (MNHN IM-2000-40801); b – Mk12; c – paratype (MNHN IM-2000-40802); d – Soyer; e-h – fragments with colour traces (e-g – Mk12; h – Mk 13); i-j – Mk13

10. Ecology of Partulidae

Habitat and diet

There are limited data on the general habitat preferences of partulids; most species have been categorised as 'arboreal' or 'terrestrial' (e.g. in Garrett 1884) but their exact microhabitat preferences have been studied only in a few cases (mostly the Moorean species covered in Murray *et al* 1982). In addition some host plant preferences were recorded on the labels of specimens collected by the 1934 Mangareva Expedition and in the field notes of Clarke, Murray and Johnson. On Tahiti and Moorea T. Coote has recorded the height above ground of the few surviving species.

For the majority of species the only quantified ecological data are dietary. Partulids are generally detritivores or herbivores, but there are notable exceptions. *Eua zebrina* is herbivorous but regularly consumes other snails: 25% contained shells in the stomach (51% in juveniles, only 17% in adults) (Cooke 1928). 97% of these had been live when swallowed but only the shells digested. Carnivory was also reported for *Partula incrassa exigua* (Johnson *et al.* 1993); this was confirmed in the present study where 38% of snails dissected contained shells of other snails. Kondo (MS) reported carnivory in *P. clara* and *P. hyalina*, but in these only 3% had fed on snails. All prey consumed by partulids are arboreal, except in *E. zebrina* in which 92% of prey items were of terrestrial species. There are no specialisations for shell-swallowing in any partulid. The data on carnivory are summarised in Table 10.1. These data indicate that snail-eating is significant only in *E. zebrina* and *P. incrassa*. It is not accidental in any of the species as the prey items are all too large to be consumed without considerable effort, unlike the ants recorded in the guts of other species. It is likely that the primary reason for swallowing these shells is access to calcium, rather than protein, but why these three species should have a greater demand for calcium than other partulids is not apparent.

In the present study gut contents of wild-collected preserved animals were examined to identify dietary components. Light microscopy enabled a range of different items to be categorised and identified, in the case of plant tissues this was achieved through comparison with plant material collected by T. Coote on Tahiti (Fig. 10.1). No attempt was made to quantify volumes of food items as this would be confounded by the digestibility of different items and the unknown length of time between snail collection and fixation. Instead the number of different items was recorded in each snail using gut content samples of approximately 4 mm^3. This gives an estimate of the range and relative diversity of food types.

These dietary data have been combined with the sparse habitat data to provide an ecological categorisation of almost every species. Diets were categorised by the most frequent item as: detritivore, fungivore or herbivore. Species were divided into specialists where the most frequent item formed >90% of identified items or generalist where it formed 50-90%. If no item formed more than 50% the species was categorised as an overall generalist.

Most partulid species are recorded from low- to mid-vegetation (although there may be under-recording of the less accessible higher canopy species). As would be expected, diet ranges correspond to the diversity of potential food sources,

Table 10.1 Carnivory in Partulidae. Data from published sources (Cooke 1928), the present study (JG) and unpublished manuscripts (Cooke, Kondo). In addition Johnson *et al.* (1993) recorded 'tornatellids' in the diet of *P. incrassa exigua* on Moorea

Prey	**Partulid**						
	Eua zebrina	***P. incrassa***	***P. clara***	***P. hyalina***			
		Moorea	Tahiti	Tahiti	Cooks		Austral
	Cooke 1928	JG	Kondo	Kondo	Kondo	Cook	Kondo
Assiminaeidae – *Omphalotropis*	32	-	-	-	-	-	-
Achatinellidae – *Elasmias apertum*	3	-	-	-	1	2	-
Lamellidea oblonga	-	-	-	-	-	2	-
Tornatellides oblongus	-	-	-	-	-	3	-
Tornatellinops variabilis	-	-	-	1	1	15	-
Bradybaenidae – *Bradybaena similaris*	-	-	-	3	-	-	-
Helicarionidae – *Diastole conula*	-	5	1	1	-	-	-
Hydrocenidae – *Georissa parva*	-	3	3	2	-	-	-
Helicinidae – *Aphanaconia*	60	-	-	-	-	-	-
Sturanya	26	-	-	-	-	-	-
Vertiginidae – *Nesopupa*	4	-	-	-	-	-	-
Subulinidae – *Opeas*	23	-	-	-	-	-	-
Subulina	83	-	-	-	-	-	-
% with snails prey	25%	38%	3%	3%	20%	?	0
number sampled	691	16	61	129	10	?	21

Fig. 10.1 Gut contents of Partulidae: a – fungal spores; b – fungal conidia; c – hyphae, conidia and fern spore; d – sporangium; e – filamentous alga; f – foliose liverwort; g – moss leaf; h – *Etlingera cevuga* fibres; i – *Hibiscus tiliaceus* hairs; j – unidentified plant fibres; k – *Inocarpus fragifer* hairs; l – ant head; m – partially digested *Aphanoconia* shell; n – intact *Aphanoconia*; o – ant

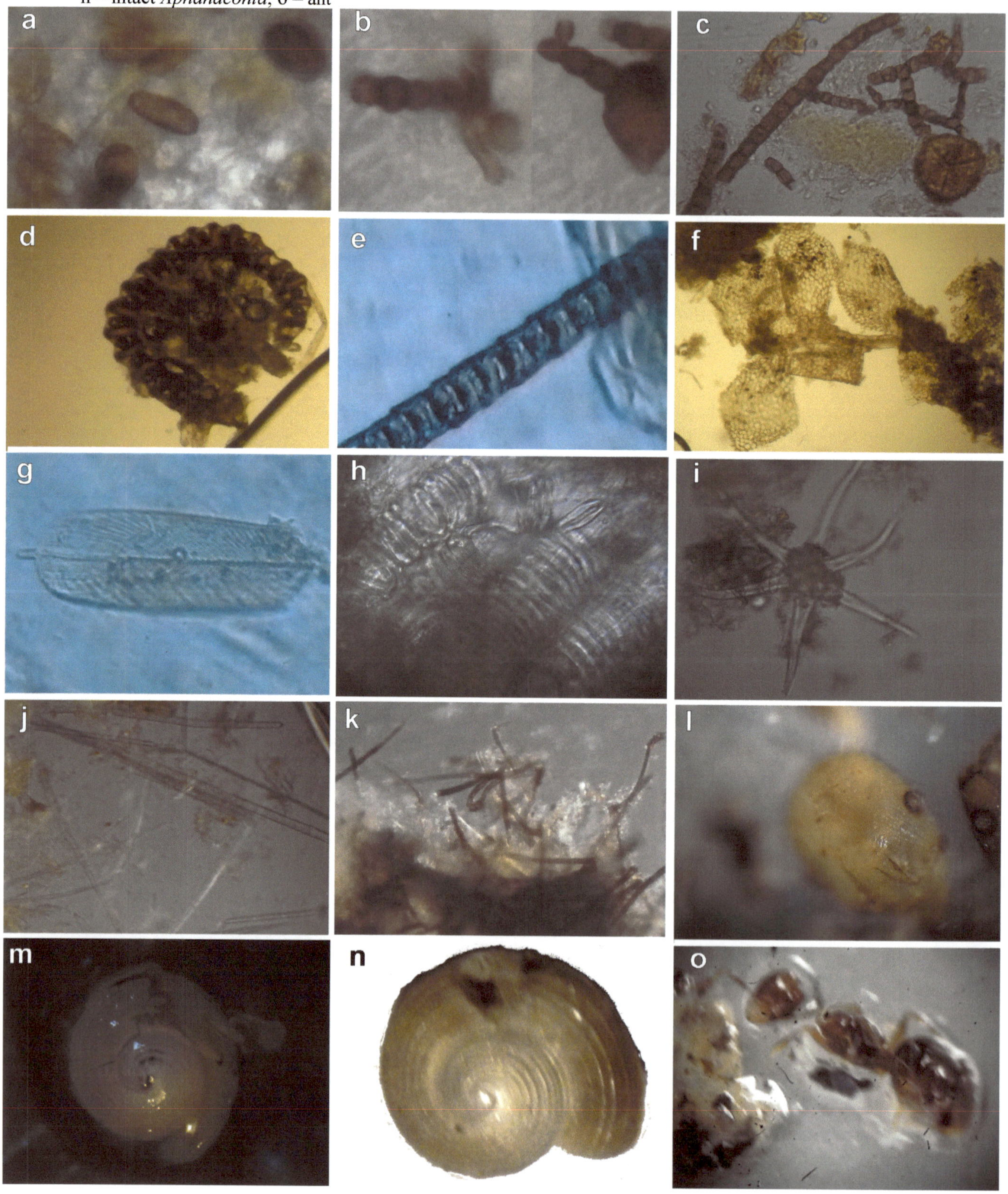

Table 10.2 Ecological categorisation of Partulidae. Specialists: F = *Freycinetia*, P = *Pandanus*.

Sources: 1 – Quoy & Gaimard 1832; 2 – Garrett 1884; 3 – Garrett 1887; 4 – Pilsbry 1909; 5 – Crampton 1916; 6 – Crampton 1925; 7 – Cooke 1928; 8 – Crampton 1932; 9 – Cooke & Crampton 1930; 10 – Crampton & Cooke 1953; 11 – Kondo 1955; 12 – Crampton 1956; 13- Murray *et al.* 1982; 14 – Tonge 1991; 15 – Hopper & Smith 1992; 16 – Cowie & Cook 1999; 17 – Bowden 2000; 18 – Cowie *et al.* 2002; 19 – Burch 2007; 20 – Rundell 2010; 21 – Pelep & Hadfield 2011; 22 – Smith 2013; 23 – Brodie *et al.* 2014; 24 – D. Sischo pers. comm.; 25 – Slapcinsky & Krauss 2016; 26 – Crampton field notes; 27 – BPBM notes; 28 – Murray *et al.* field notes; 29 – J. Gerlach field notes; 30 – Coote field notes; 31 – B. Holland pers. comm.; 32 – C. Bick pers. comm.; 33 – W. Sutherland pers. comm.; 34 – D. Clarke pers. comm. and ZSL field notes

	ground	arboreal				diet	source
		low <2m	mid 2-10m	top >10m	canopy		
Sphendone insolita	+	-	-	-	-		25
Eua expansa	-	+	+	-	-	plant specialist	3
Eua globosa	+	-	-	-	-	plant specialist	11
Eua montana	-	+	-	-	-	fungus generalist	26
Eua zebrina	+	+	-	-	-	plant generalist	7, 16
Eua sp.	-	+	-	-	-	?	27
Palaopartula calypso	-	-	+	-	-	generalist	20
P. leucothoe	-	-	+	-	-	plant specialist	27
P. thetis	-	-	+P	-	-	plant specialist	20, 25
Partula affinis	-	+	+	(+)	-	plant generalist	27, 34
P. arguta	-	(+)	(+)	+	+	plant generalist	2, 28
P. assimilis	-	-	+	-	-	fungus specialist	30
P. atilis	+	-	-	-	-	plant specialist	2, 12
P. auraniana	?	?	?	?	-	fungus specialist	
P. aurantia	-	+F	+F	-	-	plant generalist	8, 13
P. auriculata	+	+	+	-	-	plant generalist	2, 26, 28
P. bilineata	+	+	-	-	-	plant specialist	2, 5
P. clara	-	12%	88%	-	-	plant generalist	30
P. clarkei	-	+F	-	-	-	?	34
P. compressa	+	(+)	-	-	-	generalist	2, 27
P. cootei	-	+	+	-	-	detritus specialist	27
P. cramptoni	-	?	?	?	-	generalist	-
P. crassilabris	+	-	-	-	-	plant generalist	2
P. cuneata	-	+	+	-	-	?	12
P. cytherea	-	+	-	-	-	detritus generalist	27, 30
P. dentifera	-	+	-	-	-	generalist	2, 28
P. desolata	?	?	?	?	?	?	-
P. diminuta	+	+	-	-	-	generalist	2, 8, 27
P. dolichostoma	-	+	+	-	-	?	12
P. dolorosa	-	+	+	-	-	generalist	27
P. emersoni	-	-	-	+	?	plant generalist	6
P. eremita	?	?	?	?	?	?	-
P. faba	-	+	+	+	-	plant generalist	2,28, 29
P. flexuosa	-	+	+	-	-	plant generalist	27
P. garrettii	+	+	+	-	-	generalist	2, 26, 28
P. gibba	-	+	+	-	-	plant generalist	6, 15
P. sp. (Rota)	-	+	+	-	-	plant generalist ?	24
P. grisea	-	+	+	-	-	detritus generalist	27
P. guamensis	100%	-	-	-	-	detritus specialist	21
P. hebe	-	+	(+)	-	-	plant generalist	2, 27
P. hyalina	-	38%	62%	-	-	plant specialist	5, 30
P. incrassa	-		+	+	-	generalist	8, 13
P. jackieburchi	(+)	+	+	-	-	plant generalist	27
P. labrusca	-	100%	-	-	-	generalist	29
P. laevigata	-	+	-	-	-	?	30
P. lanceolata	-	+	+	-	-	?	27
P. langfordi	-	+	+	-	-	?	22, 27
P. leefei	-		+	-	-	generalist	23
P. leptochila	+	+	-	-	-	generalist	12, 34
P. levistriata	-	+	+	?	-	?	12
P. lirata	-	+	+	-	-	plant specialist	27
P. lugubris	+		+	-	-	plant generalist	2, 27
P. lutea	-		+	-	-	generalist	2
P. makatea	?	?	?	?	-	?	-
P. meyeri	-	100%	-	-	-	?	19
P. micans	-	+	+	-	-	plant generalist	32
P. mirabilis	-	68%	32%	-	-	fungus generalist	8, 13
P. mooreana	-	88% F	12% F	-	-	fungus generalist	8, 13
P. navigatoria	+	(+)	-	-	-	plant generalist	2, 28, 29

with an increasing reliance on living plant matter the further the snail lives from the ground. Terrestrial species are generalists or general herbivores (67%), with 13% being specialist detritivores or specialist herbivores. The greatest range of dietary types was found in low- and mid-level vegetation. In both, herbivory was the most frequent strategy (low- 57%, mid- 59%). Low-vegetation species tended towards generalism (only 25% of herbivores were specialists) compared to mid-vegetation (55%). At higher levels herbivory predominated: 66% in high-vegetation and 100% in the canopy (Fig.10.2).

Distribution and competition

The majority of partulid species are single island endemics (70%, n=103) and the majority of islands are occupied by a single species (72%, n=125). Excluding introduced populations, sympatric partulid species are known from Palau, Carolines (Ponape), Marianas, Samoa, Tonga, Fiji, Societies and the Marquesas. Of these, only two islands in the Marianas (Guam and Rota), three Samoan islands (Savi'i, 'Upolu and Tutuila), the Societies and the Marquesas have more than two sympatric species. Those from Samoa and the Marquesas contained no competing species; islands occupied by single species supported a generalist or generalist herbivore, whilst more diverse islands had a wider range of specialists.

Table 10.2 (continued)

	ground	arboreal				diet	source
		low <2m	mid 2-10m	top >10m	canopy		
P. nodosa	-	-	+	-	-	plant generalist	2, 5
P. obesa	-	+	+	-	-	generalist	276
P. otaheitana	-	77%	23%	-	-	plant generalist	2, 5, 30
P. pacifica	?	+	+	-	-	generalist	17, 33
P. pearcekellyi	-	-	+	-	?	?	27
P. planilabrum	+	+	-	-	-	plant generalist	2
P. producta	+	+	+	-	-	plant generalist	2, 5, 27
P. protracta	-	+	+	+		?	12
P. pyramis	?	?	?	?	?	?	-
P. radiolata	-	-	+	-	-	plant generalist	6, 15
P. radiosa	?	?	?	?	?	?	-
P. remota	-	+	+	?	-	plant specialist	12
P. rosea	-	-	+P	-	-	plant generalist	2, 28, 34
P. rufa	+	(+)	-	-	-	detritus specialist	27
P. sagitta	-	-	-	-	+	plant generalist	10
P. salifana	-	+	+	-	-	generalist	6
P. similaris	-	+	+	-	-	generalist	27
P. suturalis	-	20%	80%	-	-	plant generalist	8, 13
P. taeniata	-	52%	48%	-	-	detritus generalist	8, 13
P. tohiveana	-	3% F	97% F	-	-	detritus generalist	8, 13
P. tristis	-	-	+	-	-	plant specialist	27
P. turgida	-	-	-	+	+	plant specialist	27, 28
P. umbilicata	-	+	-	-	-	plant generalist	2
P. vanikorensis	+	-	+	-	-	plant specialist	1
P. varia	(+)	+	(+)	(+)	-	plant generalist	2, 14, 28, 34
Samoana abbreviata	-	-	+		-	plant specialist	7, 16
S. alabastrina	-	?	?	?	?	generalist	-
S. annectens	-	-	-	+	+	plant specialist	2, 27, 28
S. attenuata	-	32%	68%	+	+	detritus generalist	2, 30, 34
S. bellula	-	+	(+)	-	-	plant specialist	27, 34
S. burchi	-	43%	57%	-	-	fungus specialist	28
S. conica	-	-	+	-	-	fungus specialist	7, 16
S. cramptoni	-	?	?	?	?	detritus specialist	-
S. decussatula	-	-	+	-	-	detritus specialist	3
S. diaphana	-	-		+	+	plant specialist	27, 30
S. fragilis	-	-	+	-	-	generalist	6, 15
S. ganymedes	-	-	+	-	-	plant specialist	3
S. gonochila	<1%	2%	97%	-	-	?	4, 34
S. inflata	-	-	+	-	-	plant specialist	3
S. margaritae	-	-	+	-	-	plant specialist	27
S. medana	-	+	+	-	-	?	27
S. minuta	-	+	+	-	-	plant specialist	27, 34
S. dryas	-	-	+	-	-	generalist	27
S. pilsbryi	-	+	+	-	-	plant specialist	27
S. stevensoniana	-	-	+	-	-	plant generalist	27
S. thurstoni	-	-	+	?	-	detritus specialist	9, 16, 18

Guam has two pairs of competing species: two generalist herbivores and two generalists. Evidence for ecological separation between these potential competitors is weak; anecdotally *P. gibba* may tend to occur lower on plants than *P. radiolata*, and *P. salifana* and *S. fragilis* have not been recorded from the same localities. Historically (Crampton 1925) *P. gibba* was the most abundant Guam partulid, but since its recent decline *P. radiolata* has become the more abundant species and has also extended its range. This suggests that in the past competitive exclusion may have limited *P. radiolata*.

There is an extraordinary diversity of partulids in the Society islands, especially on Raiatea. Despite the number of species on each island (2-24) the number of species found in sympatry is relatively low; the maximum number of sympatric species in any valley is 7 (Tooroa valley, Raiatea). Considering the ecology of the species, it is apparent that there is very little geographical overlap between potentially competing species. 103 Society islands valleys (41%) are occupied by more than one species in any of the ecological categories, but even these are usually separated by more subtle ecological differences.

Moorean species have been demonstrated to have microhabitat preferences that create ecological separation (Murray *et al.* 1993). A particularly notable specialisation is the close association of *P. mooreana* and *P. tohiveana* with *Freycinetia* plants. The axils of these plants appear to provide ideal conditions for partulids, trapping decomposing leaf litter and retaining a damp microhabitat. *Freycinetia* plants are a favoured plant for many Society island partulids (Fig. 10.3). In some cases these preferences were plastic; in the absence of *Freycinetia* specialists the generalists *P. suturalis* and *P. taeniata* (note that Murray *et al.* 1993 categorised *P. taeniata* as a generalist based on its host plant diversity, whereas it is classed here as a generalist detritivore, based on diet) increased their use of *Freycinetia*, suggesting competitive exclusion by the specialist taxa. Further

Fig. 10.2 Dietary ranges in partulid species for which both diet and microhabitat are known (n=70) and (right) a canopy specialist, *Partula arguta* (photo: D. Clarke)

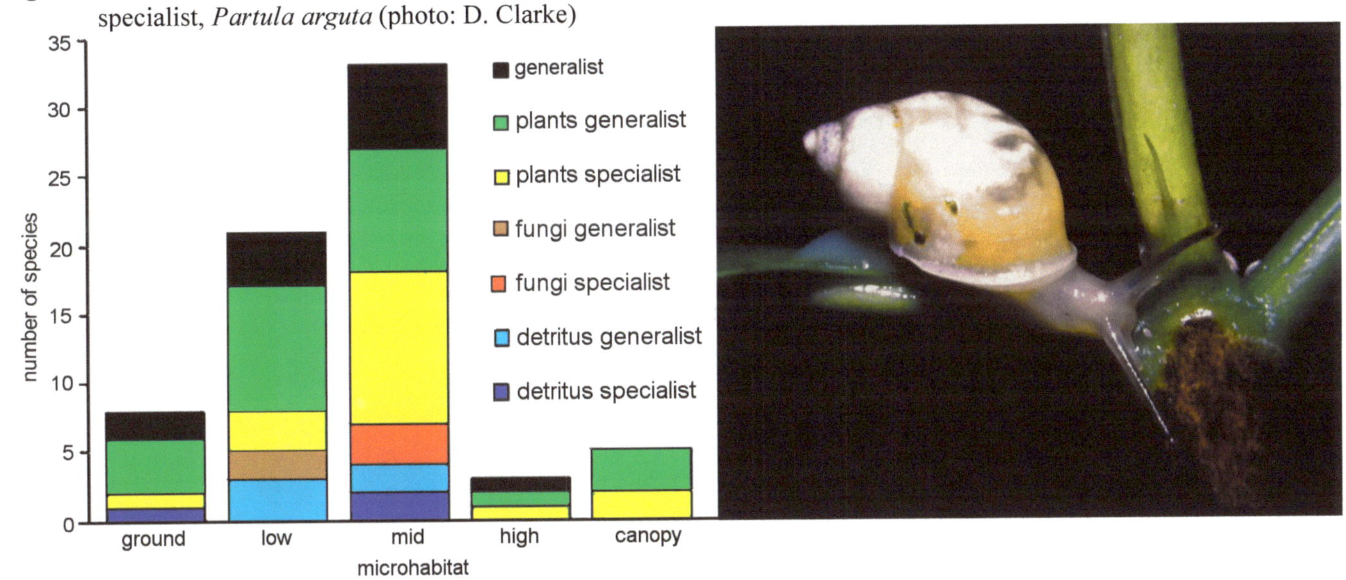

Fig. 10.3 *Freycinetia* on Raiatea (left) and *Partula varia* on *Freycinetia* on Huahine in 1991 (right, photo: D. Clarke)

indications of competition between partulids may be found in the increase in records of *Samoana* species on Tahiti and Moorea since the early 1990s, possibly as a result of competitive release, following the loss of most *Partula* species on these islands.

On Huahine there is no clear dietary separation between the *Partula* species, but they may have been separated by microhabitat preferences: *P. arguta* in the canopy, *P. rosea* high on trees (largely associated with *Pandanus*) and *P. varia* lower down (D. Clarke pers. comm.).

On Tahaa there were numerous valleys where the generalist herbivores *P. faba, P. planilabrum* and *P. umbilicata* coexisted. *P. faba* was found higher on vegetation than the other species, suggesting little direct competition, except between *P. planilabrum* and *P. umbilicata.* The ecological data for this island are very limited but Garrett's (1884) notes indicate geographical variation in *P. planilabrum* arboreality, being terrestrial in the typical *P. planilabrum* and arboreal in the '*virginea*' form. Both overlapped with the ecologically similar *P. umbilicata*, but '*virginea*' was only abundant where *P. umbilicata* was scarce, whereas both typical *P. planilabrum* and *P. umbilicata* occurred in abundance in the same sites (Crampton collection data). This may suggest a behavioural shift towards terrestriality in *P. planilabrum* where *P. umbilicata* was common.

With more species, Raiatea had more potential for overlap. There were several valleys with 2-3 generalist herbivores but these were separated by vertical stratification. Raiatea was the only island to have supported sympatric generalists, in the case of *P. dentifera* and *P. garrettii.* These species both appear to have occupied low vegetation, however where they were sympatric the arboreal *P. g. garrettii* was replaced by the terrestrial *P. garrettii rustica* (i.e. in Ereeo, Hanua and Tooroa).

The most extensive ecological overlap appears to have existed on Tahiti, with several valleys containing multiple generalist herbivores. Excluding overlap between the 'edge' species *P. clara/hyalina* and the deeper forest *P. otaheitana*, there were 20 valleys where *P. otaheitana* coexisted with *P. affinis*, 8 with *P. nodosa*, 3 with *P. jackieburchi* and 3 with *P. producta.* No evidence of ecological separation can be found for *P. nodosa* and *P. otaheitana*; however, the data on *P. nodosa* (Garrett 1884; Crampton 1916) are particularly thin. *P. affinis* and *P. otaheitana* have the same habitat preferences, but when sympatric populations are examined a dietary shift is apparent; *P. affinis* becoming a plant specialist and *P. otaheitana* a generalised detritivore (e.g. Vavii to Oofu samples).

Where *P. otaheitana* and *P. producta* overlapped a behavioural shift was apparent in the latter. In isolation both were arboreal (86% of isolated *P. producta* populations being at least partly arboreal, n=7) but in the one valley where they were fully sympatric *P. producta* was largely terrestrial (64% in Faarahi valley). Both species also occurred in Vaitunamea but there *P. producta* was largely arboreal, and no mixed colonies existed. Similarly, both species were present and arboreal in Taharua but *P. otaheitana* was very scarce (7:137 collected by Burch). A similar but less marked shift was also apparent in the fully sympatric Faarahi population where 94% of *P. otaheitana* were arboreal (n=34), compared to 84% to 59% when in isolation (Tereehia n=77 and Orofere n=61, respectively [BPBM 1970-99, 114, 116, 179]). Competition between the two species appears to have driven sympatric *P. producta* towards terrestriality and *P. otaheitana* to greater arboreality.

A similar behavioural shift may have occurred in the Mt. Aorai population of *P. jackieburchi*, which was collected on the ground whilst *P. otaheitana* was on shrubs, ferns and tree trunks (BPBM specimen data). Elsewhere the two species appear to have had complete ecological overlap and this lack of separation may be the cause of the apparent relictual status of *P. jackieburchi* populations.

In the Austral islands, *P. cootei* was the only native *Partula* species and so would be expected to be a generalist, using all available niches. However the available data indicate that it was a detritus specialist. It is possible that this species was originally a generalist but had been forced into a detritivore niche through competition by the introduced generalist herbivore *P. hyalina*.

Population sizes

Historically partulids were highly abundant (Garrett 1884; Crampton 1916, 1921, 1925, 1932; Fig. 10.4) but no quatitative data exist prior to the 1960s. Relative abundances may be obtained from Crampton's valley collections, but even these are confounded by detectability and the effort put into collecting specific forms. The only useful population data are for 12 species on Moorea, Huahine and Raiatea (Table 10.3) which confirm that population densities could be remarkably high, with an average of 1,039 per hectare recorded on Moorea. Locally densities may have reached 20 m^{-2} (Johnson *et al.* 1993). Similar maximum densities have been recorded on Guam (*P. gibba* to 17.2 m^{-2}; Smith *et al.* undated). *P. navigatoria* data are the only density estimates for terrestrial species.

No host plant preferences were apparent on Raiatea (JG data): *P. faba* being found at equal densities on *Ardisia elliptica, Gliricidia sepia, Pandanus* sp. (highest density at 0.3 per plant, compared to 0.04-0.1 for other species) and *Psychotria* sp. None were found on *Artocarpus atilis, Cocos nucifera, Hibiscus tiliaceus* or *Cecropia peltata. Partula dolorosa* was found at similar levels on *Pandanus* sp. (0.1 per tree), *Psychotria* sp. (<0.2), and *Ardisia elliptica* (0.13 per tree).

Reproduction

There is evidence of some degree of self-fertilisation in many partulids. This has been demonstrated to occur in some Moorean *Partula* (Murray & Clarke 1976a, b) and all captive *P. gibba* are descended from a single snail. Low genetic diversity in the wild in this species indicates that self-fertilisation may be normal in this species (Johnson *et al.* 1977), supported by the observation that captive breeding rates from self-fertilisation are higher than when snails are paired. Similar high rates of self-fertilisation have been recorded in *P. hyalina, P. clara* and *P. taeniata* with significant rates in *P. radiolata, P. clarkei* and *P. suturalis* (ZSL data: Fig. 10.5). In *P. taeniata* self-fertilisation accounts for 20% of offspring in early reproductive life, declining later, with an overall frequency of 2% (Murray & Clarke 1976a). This ontogenetic change is not found in *P. suturalis* which has a 1-2% overall rate but with wide individual variation (Murray & Clarke 1976b). The lack of allozyme variation in Society islands *Samoana* may suggest high selfing rates in that genus (Table 10.4) and high rates of self-fertilisation have

Table 10.3 Population densities (per hectare) of adult partulids. *P. hyalina* includes percentage of marmorata forms. Sources: 1 – Murray *et al.* 1993; 2 – Murray & Johnson field notes; 3 – ZSL field notes; 4 – JG field notes; Kondo MS

Moorea (1)	***taeniata***	***mirabilis***	***tohiveana***	***mooreana***	***suturalis***
Fareaito 551	4500	1500	2800	0	2400
Fareaito 610	8200	2800	2900	0	1600
Maramu 575	1800	4800	0	2400	3000
Maramu 618	3400	4300	0	6100	2000
Matapoopoo	9100	3000	0	0	300
Tefeo	4300	4300	0	0	3900
Faatoai	3300	0	0	0	800
Moruu	2700	0	0	0	3400
Atimaha	3000	0	0	3300	2600
Hotutea	2700	0	700	0	1900
Raiatea (4)	***faba***	***labrusca***	***dolorosa***	***navigatoria***	
Haamoa	3096	0	0	4000	
Temehani	192	100	94	0	

Tahiti (2)	***otaheitana***	***affinis***	***hyalia***	***S. attenuata***
Mt. Mara	763	0	0	0
Faroa	590	1483	1450	17
S of Pierae	5000	1000	7200 (69%)	200
Fareara	0	475	0	0
Taapeha	0	233	2134 (55%)	0
Fareara	0	0	950 (92%)	0
Mahaena	0	0	550	100
Huahine (2,3)	***varia***	***arguta***		
Fahiti (2)	2800	200		
Faie (3)	21000	0		
Maroe (3)	36667	0		
Marquesas (3,5)	***S. pilsbryi***	***S. bellula***	***S. minuta***	***S. gonochila***
	287-598	916	3.6	17210

Fig. 10.4 *Partula* abundance. Left – *P. gibba* on Saipan in 1920. Crampton remarked on the great abundance here, counting 21 snails on the single *Caladium* leaf held by his son in this photograph (Crampton 1921, 1925). Right – a rare recent observation of multiple *Partula* on a single leaf, *P. mirabilis* × *taeniata propinqua* on Moorea (photo: T. Coote)

been confirmed for *S. attenuata* and *S. gonochila* (Fig. 10.6). Neonates apparently produced by self-fertilisation have similar survival rates to the average for each species and generation (e.g. weekly mortality of *P. clara* produced by self-fertilisation or by cross-fertilisation is 0.01, for *P. taeniata* self-fertilisation gives 0.03 compared to 0.04). No data are available that would allow self-fertilisation to be distinguished from parthenogenesis.

In addition, sperm storage can be prolonged, reportedly up to two years in *P. suturalis* (Davidson *et al.* 2009). Given that self-fertilisation can occur in most species, the duration of sperm storage may be difficult to determine. In species with frequent self-fertilisation (from ZSL data) apparently prolonged sperm storage (to 1065 days in *P. clara*) may be confounded by selfing. In occasional self-fertilisers apparent sperm storage is more moderate, at a maximum of 541 days (*P. mooreana*). Rare self-fertilisers may have more meaningful estimates of sperm storage duration (56 days in *P. lirata*, 75 *S. ganymedes*, 657 *P. otaheitana*, 1155 *P. faba*). The only non-selfing species with data on sperm storage is *P. garrettii* with a duration of 161 days.

Different courtship behaviours were described in *P. suturalis* and *P. taeniata*, with one snail moving over the other's shell in straight lines or figure-of-eight patterns (Lipton & Murray 1979), but these have not been recorded subsequently and their significance remains unknown. Courtship in *P. rosea* × *varia* is simple, with no systematic patterns in movement (pers. obs.). Mating is often non-reciprocal, taking the form of two successive mating sessions with role-reversal, although reciprocal mating has been recorded in *P. taeniata* (Murray & Clarke 1966, 1976a).

Partulids are all ovoviviparous, giving birth to one (occasionally two) young at a time in most species (Fig. 10.6). *P. hyalina* is an exception, with frequent twinning and very rarely producing up to 8 young at a time (ZSL data). Crampton

Table 10.4 Recorded partulid heterozygosity values. Data from Johnson *et al.* (1977, 1986a, b, 2000)

group	species	heterozygosity	n	group	species	heterozygosity	n	
Moorea	*P. auranita*	0.148	38	Guam	*P. gibba*	0.000	15	
	P. incrassa	0.070	305	Samoa	*S. conica*	0.027	19	
	P. mirabilis	0.134	133	Societies	*S. attenuata*	0.001	48	
	P. mooreana	0.170	70		*S. burchi*	0.000	6	
	P. suturalis	0.160	501		*S. diaphana*	0.006	53	
	P. taeniata	0.139	380	Marquesas	*S. bellula*	0.003	42	
	P. tohiveana	0.103	86		*S. medana*	0.014	9	(misidentified as *S. decussatula*)
Tahiti	*P. affinis*	0.067	3		*S. ganymedes*[1]	0.018	29	(mixed *ganymedes* and *S. pilsbryi*)
	P. jackieburchi	0.135	33		*S. inflata*	0.009	29	
	P. otaheitana	0.130	31		*S. gonochila*	0.007	62	

Fig. 10.5 Maximum sustained captive fecundity (derived from 6 month means to minimise effect of pathological extremes; ZSL data). Blue – paired snails (crossing and self-fertilising); red – definite self-fertilisation; white – no data.

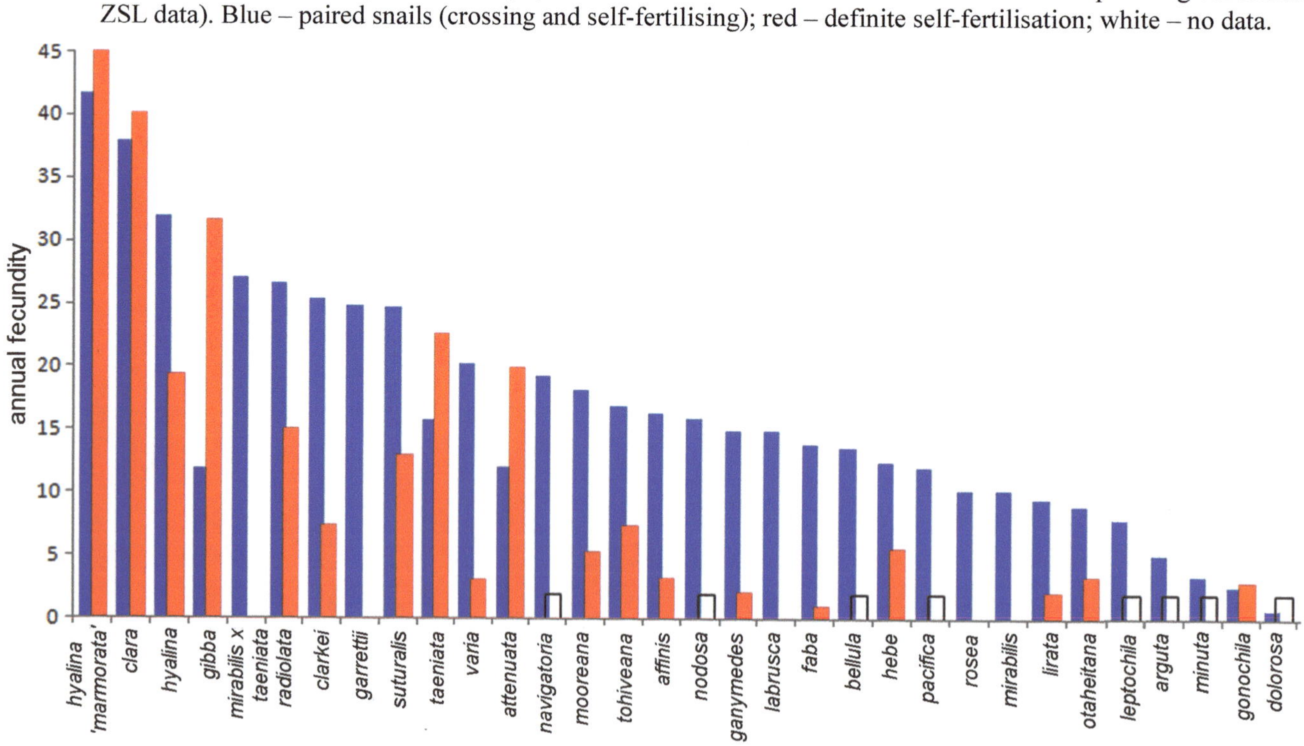

(1932, MS) recorded differences in the degree of egg membrane calcification, and his data were expanded by Kondo (1955). However, Kondo (MS) subsequently realised that the lack of calcification in some specimens was due to the acidic nature of some of formalin in which Crampton fixed his samples. Kondo later restricted his study of egg-shells to fresh material and concluded that all partulid eggs were calcified except in *Eua globosa* and Tahitian *Partula clara* and *P. hyalina*. The Rurutu population of the latter species were calcified, leading Kondo (MS) to consider that this might be the ancestral population of the species. It is more likely that the calcification arises from hybridisation with *P. cootei*, the original inhabitat of the island, which may naturally have been calcified. Samples of *P. hyalina* and *P. cootei* from Rurutu show varying levels of calcification. On Moorea Murray & Clarke (1968, 1980) showed that fresh egg capsules were calcified in all populations except for eastern *P. taeniata*. These were variable: 77% of individuals showing at least partial calcification at Hotutea and 97% at Atiraa, but in the northeast calcification fell to 19% in Paopao and 51% in Paparoa, 21% in Vaipua and 25% in Vaipohe.

Very rarely captive snails will prematurely produce individual eggs (0.01%, n=70,047; ZSL data, excluding uncalcified *P. clara* and *P. hyalina*); occasionally babies extracted from these have been viable. This has been recorded at very low levels in several species: *P. affinis* (0.09%, n=1,639), *P. faba* (0.04%, n=2,484), *P. garrettii* (0.05%, n=4,124), *P. mooreana* (0.03%, n=3,955) and *P. otaheitana* (0.23%, n=444). More frequent egg laying has been recorded in speies that may have developmental issues due to a hybrid origin (*P. rosea* × *varia*: 1.38%, n=145; Fig. 10.7), or failed to adapt to captivity: *P. leptochila* (3.33%, n=30), *S. bellula* (0.91%, n=110), and *S. minuta* (5.88%, n=17).

Crampton (and Kondo to a lesser extent) amassed an enormous volume of data on reproductive rates of partulids (Fig. 10.8) based on dissections of 82,714 field preserved specimens. This provides a very accurate estimate of what can be loosely termed clutch sizes. Despite the number of data points it is not possible to derive actual reproductive rates directly from these data, for although 'clutch' sizes were meticulously counted, no data on gestation times exist for any of the species. Accordingly, these can only be used as relative reproductive rates, assuming that gestation periods do not differ significantly between species. Whilst this is unlikely to be true for all taxa it does provide a useful relative measure. These data also cannot be regarded as true clutch sizes as only one young is produced at a time in most species, thus the number of eggs and embryos within a partulid at any one time (what Bick *et al.* [2014] termed the 'instantaneous clutch size') reflect the reproductive potential of that individual at the point of collection, not an actual 'clutch', nor the total reproductive output for that breeding season. Temporal and spatial issues further complicate comparison as samples were collected in different years and times of the year. These are recorded and can be taken into account to some extent, but again limit analysis to relative comparisons. In some cases Crampton sampled a species repeatedly in the same valley but may have collected in different localities or microhabitats within the valley. Thus caution needs to be exercised in analysis of these data or comparison to captive data (Bick *et al.* 2014), although there is a correlation between the maximum rate of captive reproduction sustained over a month (thus excluding short-term stress-induced multiple births) and the 'instantaneous clutch size' (Fig. 10.9a; R^2=0.854, t=11.84, $P_{2,24}<0.001$).

Fig. 10.6 Proportions of number of births per individual reproductive event and (right) *Partula rosea* × *varia* newborn and

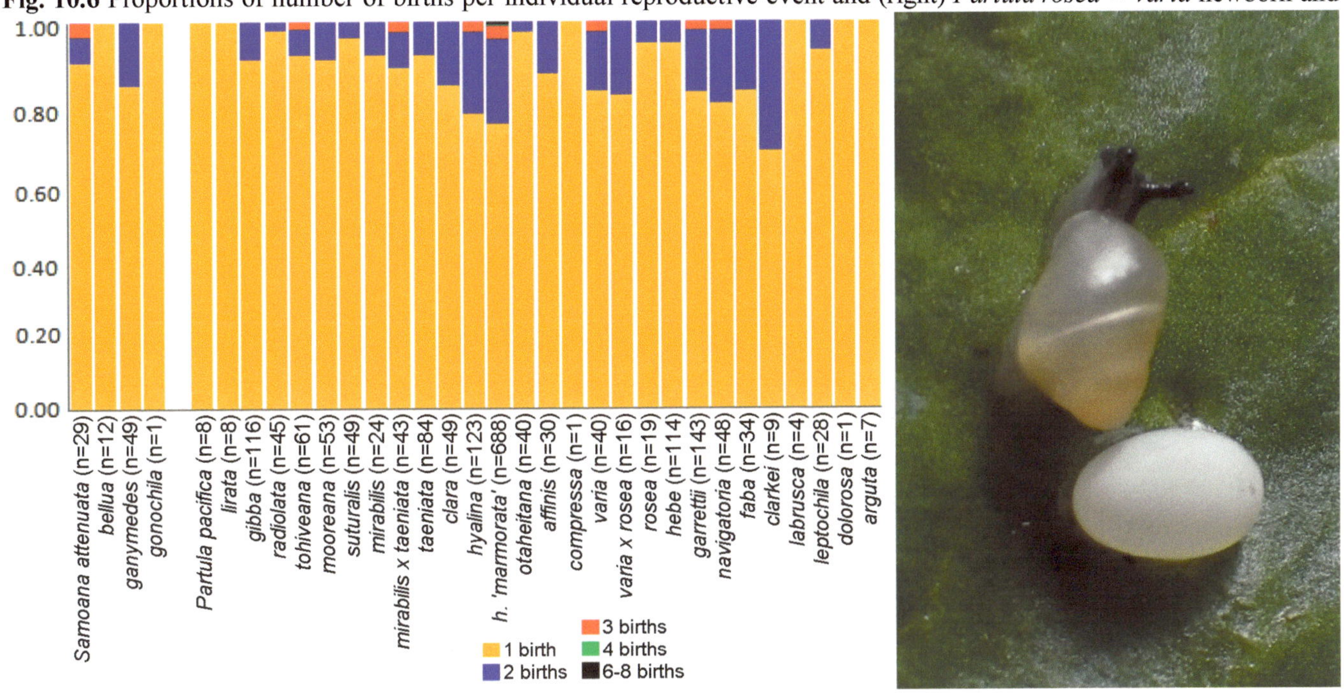

A comparison with longer-term (6 month) sustained reproduction gives a similar but weaker relationship (R^2=0.553).

Maturity (as indicated by the development of the lip) may be attained as early as 2 months, but is normally reached around 6 months of age (Table 10.5) and may be delayed for more than 2 years. Notably early maturing species are *P. navigatoria* (mean 134 days), *P. hyalina* (139), *P. gibba* (141), *P. garrettii* (149) and *P. radiolata* (149). Exceptionally slow maturing species are *S. ganymedes* (325), *P. faba* (328), *P. affinis* (307), *P. rosea* (349) and *P. pacifica* (665). Some species have appreciably delayed onset of first breeding (217 days after development of the adult shell lip in the case of *P. compressa*). Others may be fully fertile from this point (e.g. *P. gibba*), precocious breeding (before the shell lip formed) was identified in *S. attenuata* and *P. turgida* (Crampton MS). With its exceptionally large instantaneous clutch size *P. turgida* had the potential for very fast population growth, this is probably associated with strong population fluctuations in this erratically recorded species.

Based on survival rates of the most recent captive generations (Fig. 10.8) indicate that species can be divided into fast-breeding, intermediate and sustained breeders (Fig. 10.10). Maximum recorded longevity (Table 10.5) is similar to the longevity inferred from survival rates in most species. Excluding species that did not become estalished in captivity (*P. dolorosa, P. lirata, P. otaheitana, S. gonochila* and *S. minuta*) only two can be categorised as fast-breeding (*P. clara* and *P. hyalina*), and two as sustained breeders (*P. mooreana* and *S. attenuata*). The majority are intermediate (*P. affinis, P. clarkei, P. faba, P. garrettii, P. gibba, P. hebe, P. labrusca, P. leptochila, P. mirabilis, P. navigatoria, P. nodosa, P. pacifica, P. radiolata, P. taeniata P. rosea, P. suturalis, P. tohiveana, P. varia S. bellula* and *S. ganymedes*). Inevitably this classification will be partly a reflection of natural species' biology and differences in adaptation to captive conditions.

A mark-recapture study demonstrated that longevity of wild *P. taeniata* can exceed 5 years (Murray & Clarke 1984). Wild adult mortality in this species was estimated at 50% (Murray & Clarke 1984), but may be an overestimate as a 5 year lifespan would equate to an annual mortality rate of under 16%.

Fig. 10.7 Partulid 'instantaneous clutch sizes'. Grey – limited data (n<10), green – close to average clutch of all species (2.4), red – relatively high rates (>3.0), blue – relatively low rates (<2). Ordered by genus and island group (west to east)

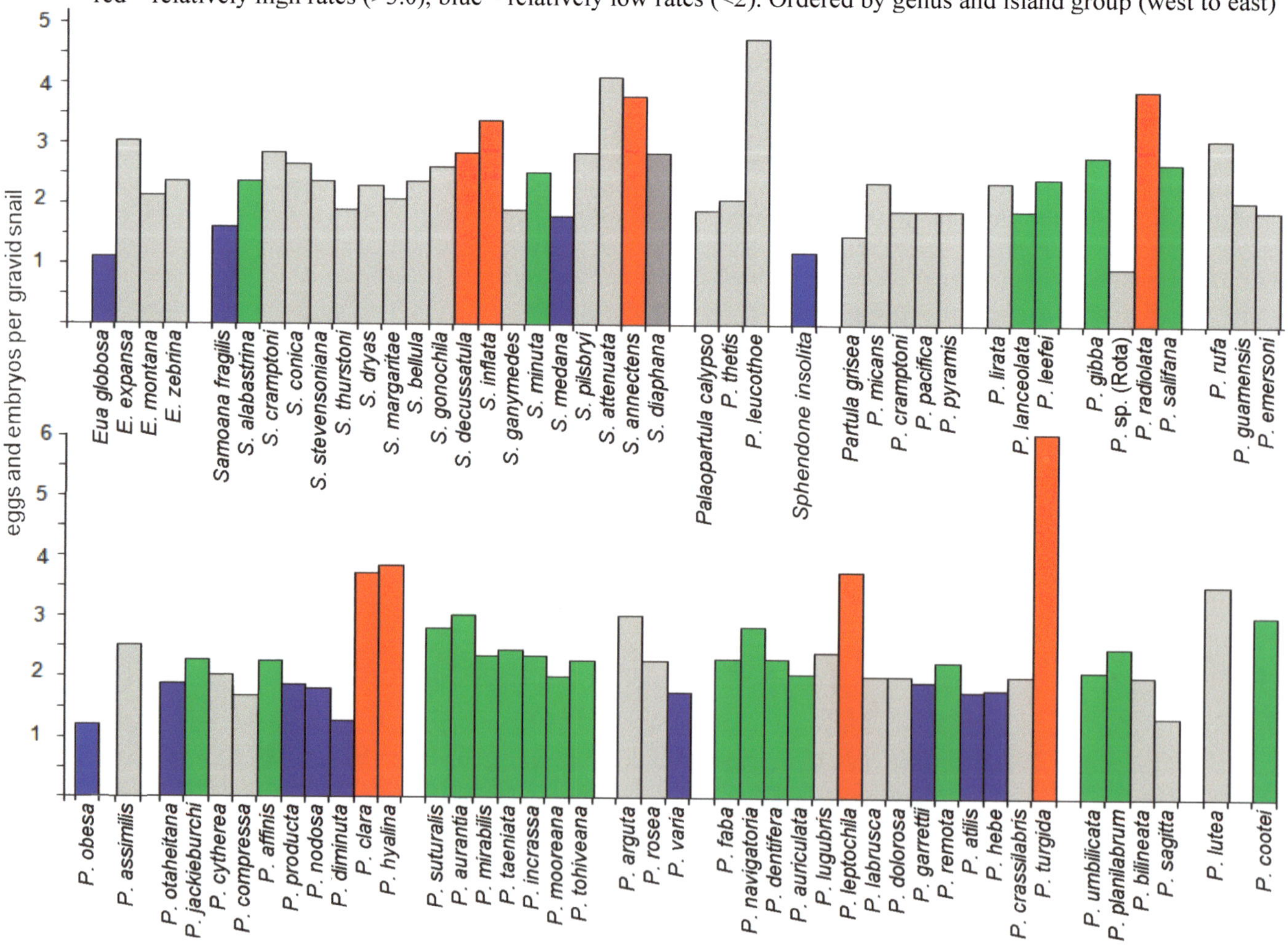

Fig. 10.8 Life history of captive partulids, data from ZSL. Lines – survival rate (left axis; black line – parental generation, red– most recent filial generation), shading – weekly reproductive rate (right axis).

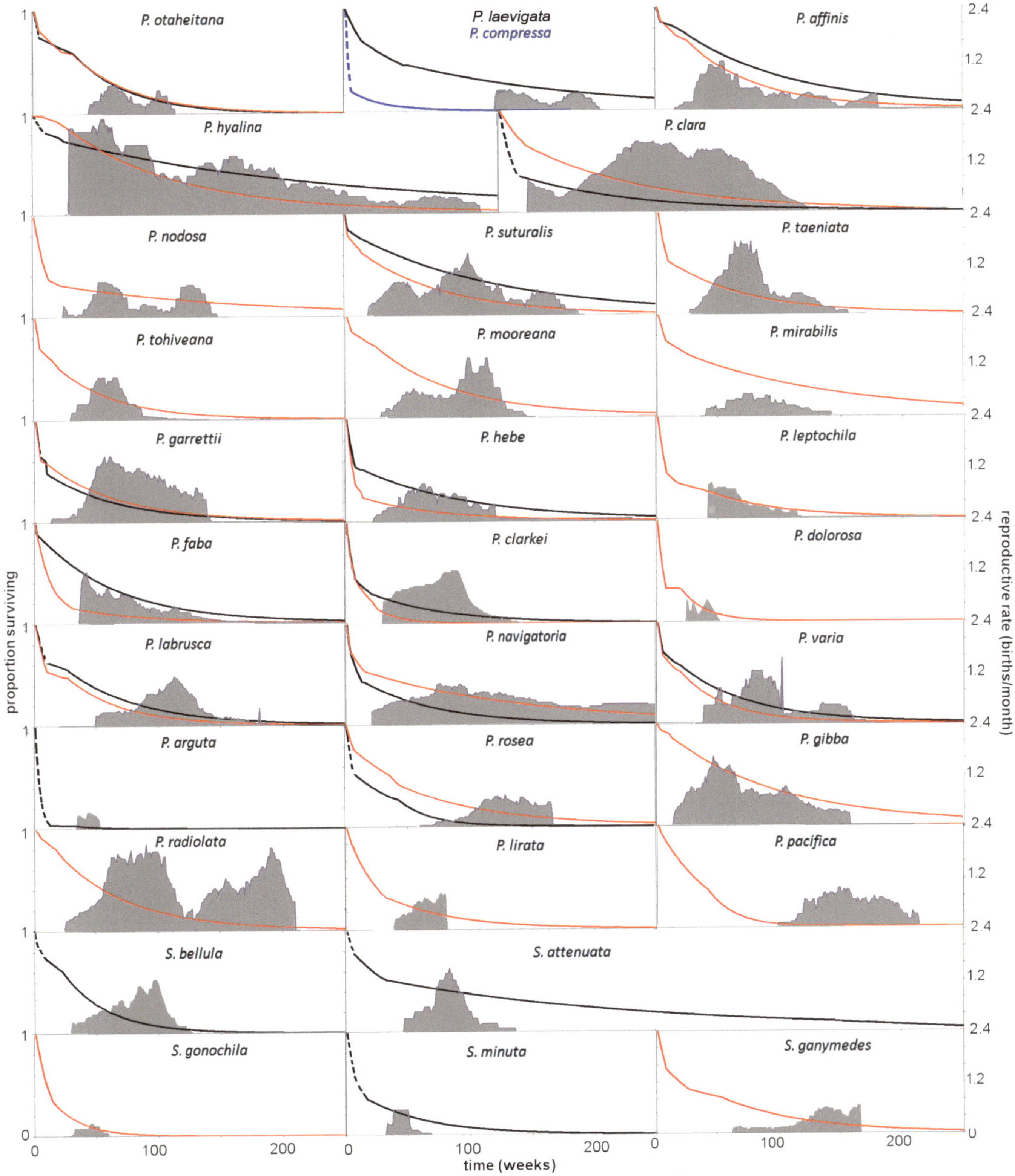

Longevity in captivity is highly variable (Table 10.5); to a large extent this will be due to differences in adaptation to captive conditions. Several species are known to live at least 6 years in captivity, with a maximum record of 7.7 years in *P. faba*. In some cases the rate of reproduction declines and may cease some weeks before death. The gap between the last date of reprodution and the date of death in captivity exacerbates this effect due to the difficulty of assigning parentage to any specific individual in the colonies. Reproductive senescence was studied in *P. rosea* × *varia* (n=20); these gave birth every 23.0±4.24 days, except for twinning events whiere the next birth followed 31.7±10.07 days later. Births ceased 1-2 months before death (38-68 days, mean 51.7±15.18). This suggests that the long gap between last recorded birth and death in some species (4.7 years in *P. hyalina* '*marmorata*') is not a real effect.

Predators, parasites and pathogens

Natural predators are speculated to have included extinct birds and the native kingfisher on Moorea (Clarke 1986). Crampton (1916) recorded some rat predation. Kondo (1973) attributed shell damage to *Samoana burchi* to rats, as did Solem (in Emberton 1982). Clarke (1986) proposed rats, feral chickens and pigs as predators on Moorea. Photographs of *Partula lanceolata* from Cicia, Fiji show clear signs of rat damage (photos provided by G. Brodie). Other recorded predators are

Table 10.5. Captive maturity and longevity data (ranges, means and standard deviations). 'maturity' – age at development of lip (days); 'birth' – time (days) between lip development and first birth; 'last' - last recorded birth (years); 'age' – maximum recorded longevity (years)

	maturity	birth	last	age		maturity	birth	last	age
S. attenuata	231-(247±22.5)-273	102	2.7	2.7	*P. labrusca*	200-(221±27.9)-259	154	3.6	3.6
S. bellula	c230	14	2.7	2.8	*P. laevigata*	358	496	-	4.3
S. gonochila	c230	70	1.3	2.2	*P. leptochila*	>434	?	2.8*	3.0*
S. ganymedes	231-(325±70.5)-413	7	3.0	3.4	*P. lirata*	238	133	1.9	2.1
S. minuta	?	14	?	2.3	*P. mirabilis*	189-(236±30.1)-287	105	4.8	5.7
P. affinis	194-(307±93.9)-476	42	4.7	4.7	*P. m.* × *taeniata*	77-(196±112.6)-385	?	3.0	3.0
P. arguta	>250	7	1.5*	1.5*	*P. mooreana*	168-(246±68.9)-408	28	4.6	4.8
P. clara	140-(207±41.5)-245	7	4.5	4.5	*P. navigatoria*	119-(134±33.8)-203	42	5.9	5.9
P. clarkei	168-(197±30.3)-245	21	2.9	3.1	*P. nodosa*	147-(189±29.2)-245	7	2.9	2.9
P. compressa	>364	217	1.6	2.0	*P. otaheitana*	125-(297±116.9)-462	91	6.9	6.9
P. dolorosa	?	?	0.1*	0.3*	*P. pacifica*	483-(665±168.6)-882	224	2.1	2.6
P. faba	222-(328±64.4)-462	37	5.2	7.7	*P. radiolata*	84-(149±40.7)-224	7	2.9	5.5
P. garrettii	70-(149±52.2)-250	21	4.6	4.6	*P. rosea*	307-(349±36.1)-371	126	2.5	3.1
P. gibba	84-(141±41.2)-189	3	2.8	2.9	*P. suturalis*	56-(192±59.5)-266**	35**	5.4	6.7
P. hebe	112-(168±42.5)-252	28	2.8	3.2	*P. taeniata*	140-(249±72.7)-433**	49**	4.3	5.0
P. hyalina	63-(139±50.9)-217	21	3.8	7.4	*P. tohiveana*	155-(181±39.1)-238	56	4.4	4.6
P. h. 'marmorata'	70-(176±73.1)-343	3	2.6	7.3	*P. varia*	140-(207±36.1)-252	133	5.2	7.1

*estimated

**Murray & Clarke (1966) recorded median age of lip development for *P. suturalis* as 353 (219-373) days and for *P. taeniata* 153 (104-289). First births were recorded 192 (121-373) and 145 (110-232) days later.

Fig. 10.9 *Partula* reproduction (data from Fig. 10.8 and ZSL). a) correlation between instantaneous clutch rate and maximum rate of reproduction in captivity sustained over a month. Species with little regular captive breeding shown in blue. b) relationship between age of maturity (mean, s.d.) and maximum sustained captive reproductive rate. Green – high reproducing (*P. hyalina, P. clara, P. gibba*), red – late reproducing (*P. pacifica*)

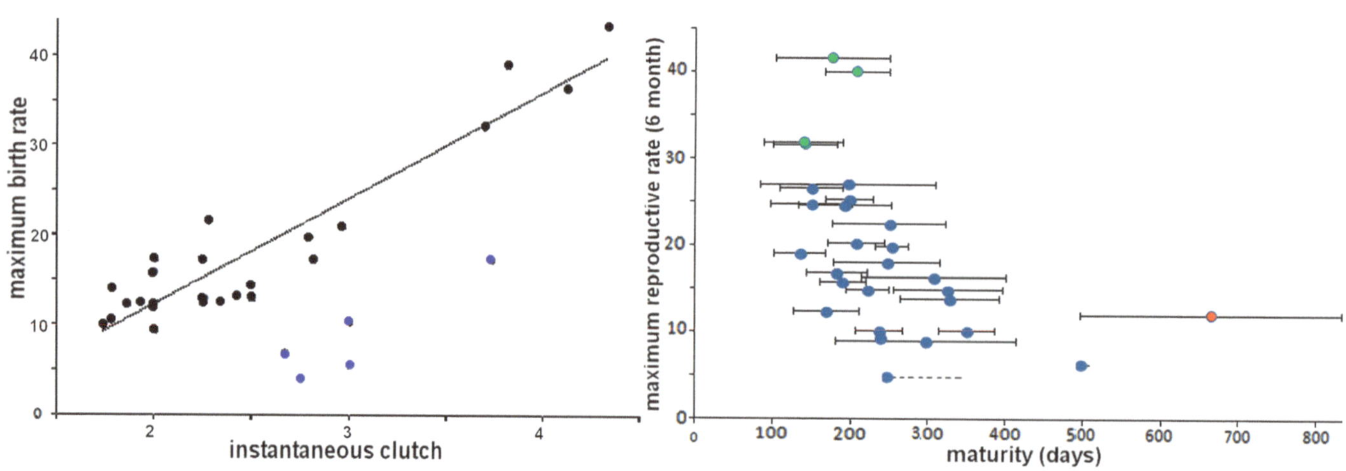

Euglandina rosea and *Platydemus manokwari* (see p. 271-272).

Pilsbry (1909) and Kondo (1955) speculated that partulids may have been excluded from the larger Fijian islands by *Placostylus* and Kondo thought that the absence of partulids from some other islands might be due to geochemical factors. The presence of *Placostylus* may explain the absence of partulids on the main Fijian islands, although they are sympatric in the Solomons and Vanuatu.

The parasitic mite *Riccardoella* sp. was recorded on *Partula* collected from Raiatea in 1991 (Pearce-Kelly & Clarke 1992); its origin and impacts are unknown. Bacterial and protist parasite infections have been recorded in captive snails (Cunningham *et al.* 1996; Cunningham & Daszak 1998; Selman *et al.* 2004; Cwiertna *et al.* 2010) and wild-collected samples (E. Flach pers. comm.).

Fig. 10.10 Classification of species according to longevity and reproduction in captivity, species extinct in captivity marked as red points, species which did not persist for more than one filial generation marked as grey points. Ellipse represents one standard deviation from the mean of all species and encloses the 'intermediate' species

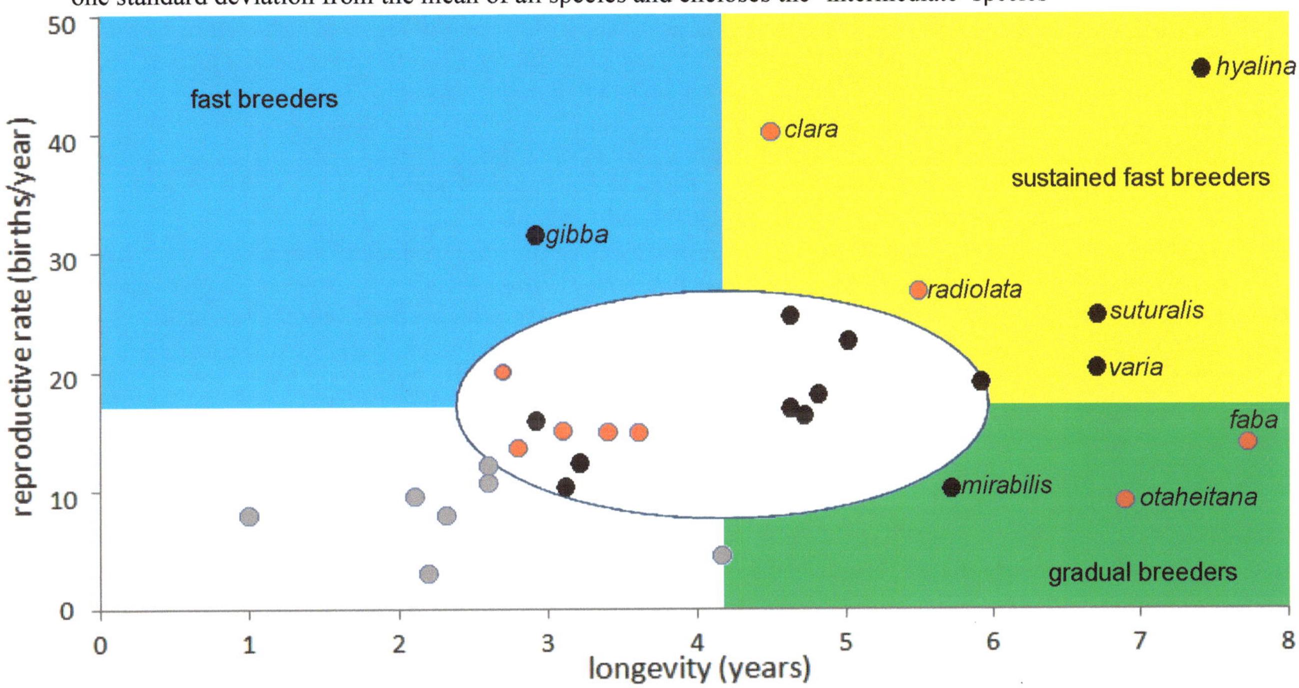

Fig. 10.11 Adult and new-born *Partula varia*

11. Evolution of Partulidae

Family origins

Martens (1860) placed *Partula* between *Bulimulus* and *Achatinella*; Fisher (1885) using the jaw and radula considered it a subgenus of a very broad *Bulimulus* ; Kobelt (1891) and Semper (1865) placed it with *Placostylus* in the Bulimulidae. Pilsbry (1900) grouped it with Achatinellidae, Amastridae, Pupillidae and the buliminoids in the Orthurethra, due to the elongate kidney and the lack of venation on the lung. This arrangement led him to consider Partulidae to be primitive (Pilsbry 1909). Thiele (1931) reverted to plcing it as a genus of Bulimulidae and Boss (1982) doubted they were orthurethran; this is contradicted by molecular studies supporting Pilsbry's arrangement (Wade *et al.* 2001, 2006). These analyses all considered the family to be primitive within the Stylommatophora, although it was subsequently noted that the chromosome count of 29 (Scvortzoff 1966) was higher than other supposedly primitive genera (Burch 1965; Murray 1975).

Tillier (1989) supported their orthurethran status due to the typical orthurethran pallial complex including the synapomorphic orthurethran long kidney, considering them closest to the Amastridae or Enidae. According to Wade & Mordan (2000) and Wade *et al.* (2001, 2006) the Partulidae are a basal family of the poorly resolved Orthurethra. Some relationship to *Draparnaudia* has been suggested and Bouchet & Rocroi (2005) placed both taxa in Partuloidea. Tillier & Mordan (1995) agreed that *Draparnaudia* was close to the Partulidae but concluded that its precise phylogenetic position was impossible to ascertain. A link between the taxa has not been supported by molecular phylogenies (Wade *et al.* 2006). In their neighbour joining tree Wade *et al.* (2006) found Partulidae to be basal to all other orthurethrans, but their Bayesian tree grouped it with *Buliminus* (Enidae) in a clade that also contained Achatinellidae, Cochlicopidae, Amastridae and Vallonidae. Herbert & Mitchell (2009) placed Partulidae basal to Enidae/Cerastidae and Vertiginidae.

Despite the lack of consensus on the relationships between Partulidae, Achatinellidae, Enidae and *Draparnaudia*, partulids probably have a continental Asian/Melanesian origin (e.g. *Draparnaudia* is restricted to New Caledonia). The divergence of the living taxa is relatively recent at around 5 million years ago (inferred from data in Lee *et al.* 2014) but, given its distinctiveness, the family is likely to have a much older origin. It is possible that early partulids or their close relatives are all extinct following displacement by another family in the western Pacific. In this regard it may be relevant to note that partulids and *Placostylus* are rarely found on the same island and almost never in the same locality.

Divergence of genera

The most basal genus appears to be *Eua*, indicating a Melanesian origin for the family. This may have spread eastwards, colonising Tonga-Samoa (*Eua*) and Palau (*Sphendone*). Basal species may have been terrestrial, a trait retained in both these genera. On Palau a shift to arboreality gave rise to the remaining genera, with a clade of *Samoana* and *Palaopartula*, and the true *Partula*. Both *Samoana* and *Partula* originated as a result of dispersal eastwards (Fig. 11.1-2). These genera appear to have diverged around 4.0 million years ago (this and all other dates inferred from Lee *et al.* [2014]).

Partulids have clearly colonised different islands on a great many occasions, with inter-island dispersal being a major component of their evolutionary history. Crampton (1916, 1925, 1932) was of the opinion that their distribution was shaped primarily by subsidence of ancient land bridges, a hypothesis that was refuted long ago (Cockerell 1925).

Fig. 11.1 Phylogeny of partulid genera and species of *Eua, Palaopartula* and *Sphendone* (modified from Lee *et al.* 2014) based on mtDNA. Genera coloured separately. Grey lines – hypothetical placement of species based on morphology. Possible duration of occupation of ranges shown in light green

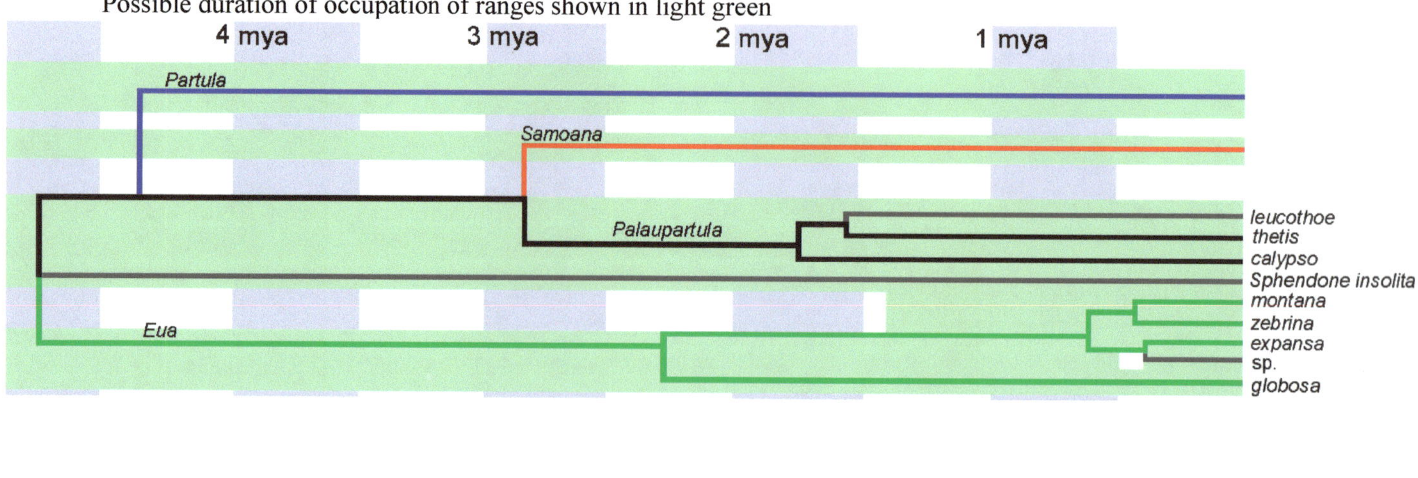

Eua

Eua has a basal split into two major groups: Tongan *E. globosa* and the Samoan species. The separation of the Tongan and Samoan lineages dates to around 1.9 million years ago, at which time the larger islands of the Samoan group would have been in existence (Price *et al.* 2014). In Samoa *E. expansa* is basal, *E. zebrina* and *E. montana* are sister species. This matches a pattern of step-wise colonisation of the islands in an easterly direction, with allopatric speciation on each island, or it may

Fig. 11.2 Partulid dispersal across the Pacific in the past 4.5 million years. White – existing but unoccupied islands; grey – submerged islands; green – *Eua*; purple - *Sphedone* and *Palaopartula*, red – *Partula*; yellow – *Samoana*

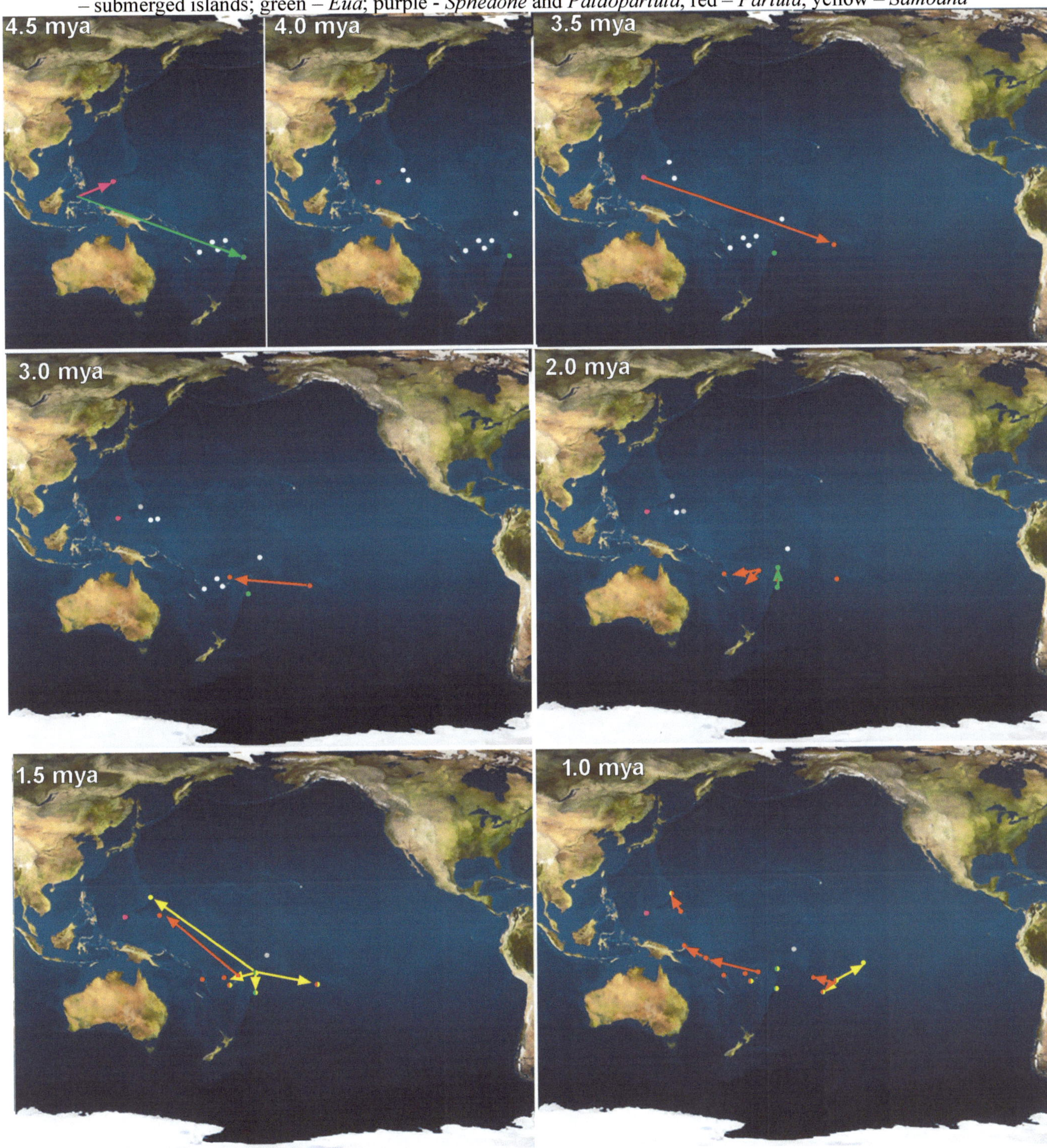

arise from repeated radiations, of which only the most recent survives. The covering of Savi'i by lava 0.4 million years ago (Konter & Jackson 2012) means that *E. expansa* and *E.* sp. represent two waves of recent colonists, probably from 'Upolu.

<u>*Palaopartula*</u>

Palaopartula is restricted to the Palau islands and the three living species diverged around 1.5 million years ago, although the genus formed around 0.8 million years earlier. There are genetic data in *Palaopartula* only for *P. thetis* and *P. calypso*. The species may have diverged on separate islands but during the last glacial maximum 120,000 years ago many of the Palau islands would have been united by lowered sea levels (Fig. 11.3). During this period the isolated species expanded their ranges, resulting in their present day sympatry. This may account for the low level of diversity in the genus and the very high level of haplotype similarity on different islands. The anatomical divergence between *P. calypso* and *P. thetis* suggests some degree of prezygotic isolation. *P. leucothoe* may have arisen from the *P. thetis* lineage allopatrically.

<u>*Sphendone*</u>

Sphendone appears to be of early origin. It is a geographically restricted monotypic genus; its cryptic habit may offer limited dispersal opportunities. In appearance the genus is closest to the terrestrial *Partula guamensis* and *P. rufa* of the nearby Caroline islands. In the absence of genetic data for these two species they are retained in *Partula*, but some affinity to *Sphendone* may be possible.

<u>*Samoana*</u>

Samoana originated in the west, around 2.3 million years ago with divergence of the living taxa starting 0.5 million years later. The genus has high levels of diversity (21 species) and takes two forms: thick-shelled, often generalist species; and thin-shelled, arboreal species. Where *Partula* species are present *Samoana* are constrained to the thin-shelled form (e.g. Marianas and Societies), suggesting competitive exclusion of *Samoana* by *Partula*. In the Marquesas islands there are no *Partula* but not all *Samoana* are thick-shelled, it is notable though that only 25% of species (n=8) are thin-shelled. In the Austral islands there are only two species: *S. dryas* and *S. margaritae*. The latter is the only partulid species on Rapa, whereas *S. dryas* coexists with introduced *P. hyalina* on Raivavae. *P. margaritae* is a thick shelled generalist whereas *S. dryas* is a thin-shelled arboreal specialist.

The most basal living species are *S. fragilis* of the Marianas and *S. thurstoni* of Samoa which diverged from one another 1.4 million years ago. Other Samoan members of the genus are part of the main *Samoana* clade which suggests a early Samoan origin for the genus. These Samoan species appear to have evolved allopatrically on the different islands, with *thurstoni* isolated on Ofu, and *abbreviata, conica* and *stevensonia* on Tutuila and Savi'i. Within the Samoan radiation *S. conica* is of interest; it is the only species to occur on more than one island and the only sinistral species in the genus. If reversed chirality arises as a result of character displacement creating prezygotic isolation (as in *Partula*, see p. 244, 256) then

Fig. 11.3 Maps of the Palau and Solomon islands showing current islands and the probable extent of dry land during the last glacial maximum (blue). Left – Palau, *Palaopartula* islands in red; right – Solomons and eastern New Guinea with names placed next to probable original ranges: *Partula grisea* on Manus with recent expansion eastwards, *P. similaris* on New Britain, with recent expansion to New Guinea. Guadalcanal (shown with red and blue stripes) was probably originally occupied only by *P. micans*, with *P. flexuosa* being a recent introduction. Similarly, the New Britain population contains a mixture of *P. grisea* and *P. similaris* genotypes

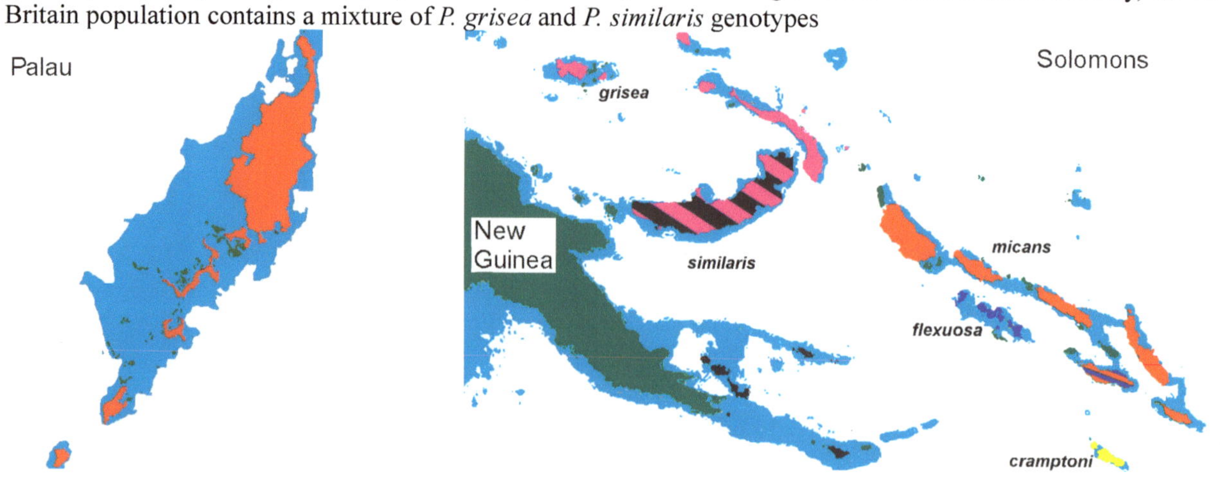

this may indicate that *S. conica* is of relatively recent origin, being ancestrally dextral and shifted to sinistrality in sympatry with *S. stevensoniana* on Savi'i and with *S. abbreviata* on Tutuila. It is the only *Samoana* on 'Upolu but is also sinistral there. This suggests that the 'Upolu population is the most recent, and has not had sufficient time to revert to dextrality (although the island is 3.2 million years old – Price *et al.* 2014). *S. conica* is very closely related to *S. stevensoniana* which may indicate chiral divergence between the two taxa on Savai'i followed by range expansion by *S. conica*. The Tutuila and Savai'i populations are moderately distinct genetically. These species have diverged within the last 1.3 million years, in accordance with the recent origin of Tutulia (1.5 million years ago) and the coverage of Savi'i by lava 400,000 years ago (Konter & Jackson 2012). The distinct lineage of *S. thurstoni* is a relatively old one but is restricted to Ofu, the youngest island occupied by *Samoana* at around 560,000 years old, with extensive volcanism until 240,000 years ago – Hart & Jackson 2014). This means that *S. thurstoni* must have colonised Ofu in the recent past from a much older island.

The Samoan radiation gave rise to two separate dispersal lineages: *S. fragilis* of the Marianas arising by long-distance dispersal from the *S. thurstoni* line, and the more recent dispersal from the main Samoan group to Polynesia around 900,000 years ago. The highly isolated nature of *S. fragilis* and its apparent failure to diversify over 1.8 million years may suggest that it is a relatively recent arrival in its present range, having originally occupied an intervening island group. Possible candidates may be the raised phosphatic islands in the Kiribati group and Nauru, and the high islands of the Caroline group. Although these do not appear to be suitable for *Samoana*, the similar island of Makatea in the Tuamotus was occupied by *Partula*. The Carolines appear to be suitable for *Samoana*, but are only occupied by *Partula* (Kosrae and Pohnpei, with no partulids on 200 m high Yap). Few *Samoana* species are present on islands occupied by *Partula* but it is not impossible for them to colonise, as is the case in the Marianas and Societies. *S. fragilis* may be constrained from diverging by the presence of *Partula*.

Fijian *S. alabastrina* and Tongan *S. cramptoni* have not been placed genetically, but appear to be related to Samoan species. Around 5 million years ago Fiji, Tonga and Vanuatu were in close proximity. Although the Tongan islands are ancient and could have been colonised by early partulids, they were all flooded by high sea levels during the last interglacial period, with the exception of 'Eua. Consequently only this island supports partulids today.

The most plausible scenario for the colonisation of Polynesia is the dispersal of a Samoan taxon to the Society islands and simultaneously, or shortly afterwards, to the Marquesas. This resulted in two related radiations in the different archipelagos. In the east the Marquesas species are notable for their highly varied shell morphologies. This group appears still to be in the process of adaptation and divergence within the archipelago. By the time that *Samoana* colonised the eastern Pacific the Marquesas had all been formed, the youngest island being 1.3 million years old Fatu Hiva (Duncan & McDougall 1976). The Marquesas species show a pattern of allopatric speciation and inter-island dispersal, giving no more than three species per island, and no sympatric sibling species. The only exceptions to this simple pattern are *S. gonochila* and *S. medana* which show a mixture of morphotypes within the same populations. *S. gonochila* shows local divergence within Nuku Hiva island; at present it i;s not know if this is incipient adaptation or an ecophenotypic effect. *S. medana* contains three distinct conchological forms typical *medana*, 'diaphana' and 'lebronneci', of which the latter is anatomically distinct. This partial separation may suggest divergence but with subsequent dispersal (possibly by human agency) mixing the forms.

11.4 Phylogeny of *Samoana* based on mtDNA (after Lee *et al.* 2014) and allozymes (Johnson *et al.* 2000). Key as Fig. 11.8

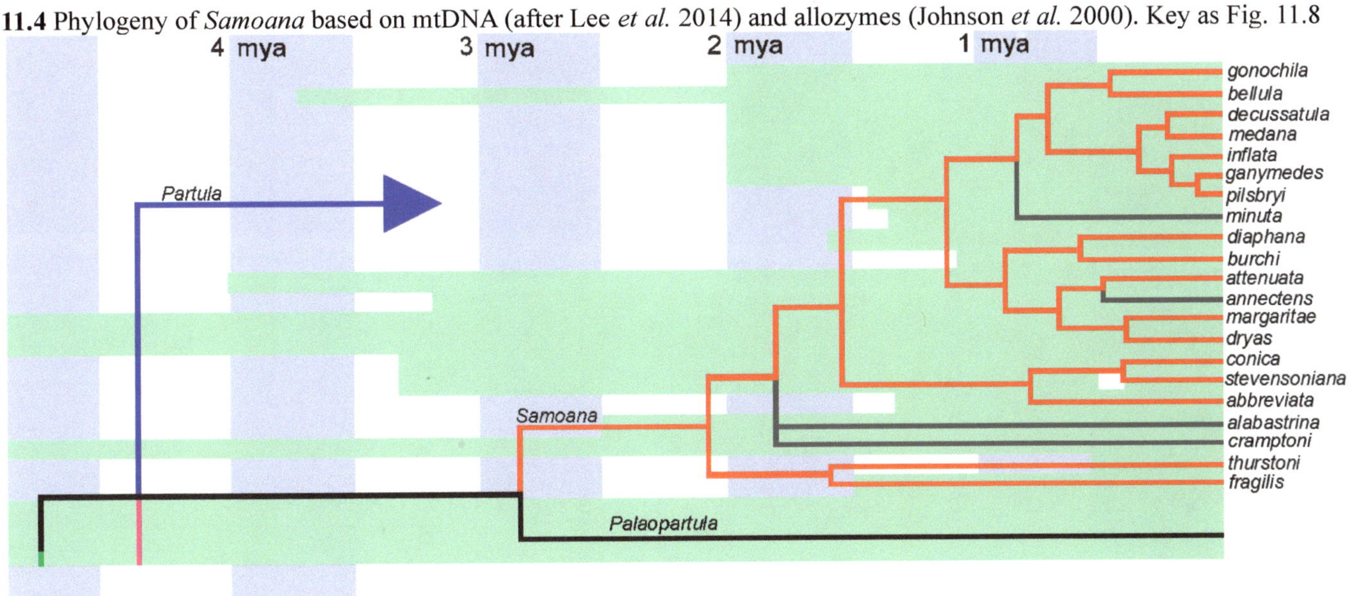

In the Society islands *Samoana* distribution appears to be the result of allopatric speciation and subsequent inter-island movement. *S. attenuata* is present on all of the Society islands occupied by partulids, except on Huahine which is occupied by its allospecies *S. annectens*. For the Society islands *Samoana* the phylogeny suggests an ancestral split between proto-*attenuata* on an unidentified island (with Huahine's *S. annectens* as the earliest offshoot) and proto-*burchi* on Tahiti. Proto-*attenuata* then colonises each island separately and randomly. Proto-*burchi* divides into two ecologically separated forms – *S. burchi* and *S. diaphana*, with the latter diverging anatomically as a result of selection for prezygotic isolation. It also colonises Moorea twice, giving a polyphyletic Moorean linage. Under this scenario the only ecological speciation occurred between *S. burchi* and *S. diaphana*, the thin shell of the latter indicating a high-altitude specialisation. The three species on Tahiti appear to be largely geographically separate, with *S. attenuata* at relatively low altitudes, *S. burchi* at high altitudes and *S. diaphana* as a canopy species at high altitude. The very low heterozygosities recorded in this radiation (p. 228) indicate a high frequency of self-fertilisation, which would make these species well adapted to inter-island dispersal. This may also explain why another branch of the *attenuata* ancestral lineage was able to colonise the Austral islands around 600,000 years ago, with divergence of the lineage 300,000 years later. This is long after the islands formed as a result of movement of the Cook-Austral hotspots (20 million years ago). Present-day islands occupied by *Samoana* date from 5.1 (Rapa) and 6.5 million years ago (Raivavae) (Duncan & McDougall 1976). Within the Australs *S. dryas* remains typical of the Society islands derived *Samoana*, being thin shelled and ecologically restricted. In contrast *S. margaritae* is found in a variety of habitats and has reverted to a more basic, thicker shelled *Samoana* form.

S. attenuata populations on different islands are morphologically distinctive despite close genetic similarity (Lee *et al.* 2014). This can be attributed to genetic drift arising from a series of founder effects. The Tahitian population is the most distinct in colouration (Fig. 11.5), being plain brown. Moorean and Raiatean snails are strongly mottled, differing in shade. ANOVA identifies significant population differences in length (F=27.59, P<0.0001) and width (F=126.30, P<0.0001) with Moorean populations being smaller than all others (Tukey range test: q>9.54, $P_{2,297}<0.001$ for all comparisons). Other taxa with statistically significant morphometrics are the Samoan *S. conica* (length – F=83.44, P<0.0001; width – F=6.610, P=0.0014) where Tutuila populations differ from those of 'Upolu and Savi'i (length F>7.47, $P_{2,888}<0.001$).

Lee *et al.* (2014) questioned why the genus is most diverse in the east when it originated in the west, and why there are no *Samoana* between the extremes. They thought it probable that there had been species on other islands but that these had died out for unknown reasons which had also prevented subsequent recolonisation. *Samoana* seem to be constrained by the presence of *Partula*; where the two genera coexist there are never more *Samoana* species than *Partula* (Guam 1 *Samoana*: 3 *Partula*, Rota 1:1, Bora-Bora 1:1, Tahaa 1:6, Raiatea 1:19, Huahine 1:3, Moorea 2:7, Tahiti 4:10) and the only radiations

Fig. 11.5 *Samoana attenuata* phenotypic variation: a – Tahiti (Haapaponi – T. Coote); b – Moorea (Tefeo – T. Coote); c – Raiatea (Averarahi – F. Jaq); d – Raiatea (Mt. Toomaru – J.-Y. Meyer); e-f – shell measurements showing means, standard deviations and ranges (n: Tahiti = 67, Moorea = 9, Raiatea = 240, Tahaa = 25, Bora-Bora = 2)

of *Samoana* are in the absence of *Partula* (Marquesas and Samoa), with a limited radiation on Tahiti. The Tahiti radiation is predominantly at high altitude, where *Partula* are relatively scarce. It may simply be the case that as *Partula* colonised the islands first *Samoana* was forced into this pattern of saltatory colonisation, rather than for any competitive reason.

<u>*Partula*</u>

There have been several speculations on the evolution of *Partula* species (summarised in Cowie 1992). Recent phylogenies (Lee *et al.* 2014) indicate an origin 3.6 million years ago and identify two main clades in *Partula*: a western one subdivided into south-western (Vanuatu/Fiji) and far-western (Micronesia/New Guinea/Solomons and Futuna), and the eastern Polynesian species.

The western group diverged 2 million years ago into the south-western and far-western groups, with species of Vanuatu and Fiji dating from 1.7 million years ago, present-day Fiji species may have diverged 500,000 years later. Both Vanuatu and Fiji have ancient rock but much of Vanuatu was submerged in the Quaternary and Fiji has been tectonically unstable. Most Fijian partulids are found in the Lau islands, and the Lau ridge is notable for its low volcanic activity since the Quaternary, although there were eruptions on Mago 280,000 years ago (Nunn 1990). The two Fijian *Partula* have been recorded from the same islands but are anatomically very similar. There are no clear physical reasons why the two species should remain separate. The records from different islands are scattered, and no collection has found both species in the same locality at the same time. It is possible that the different records may represent isolated and temporary colonisations, with very little real overlap between the two species.

In the far west, the Micronesian species originate as a clade 1.4 million years ago, and diverge into Marianas and Carolines lineages 200,000 years later. This fits with the relatively recent reduction in volcanic activity in these islands. Although the islands have been in existence for at least 11 million years (Chuck island) Pohnpei was reshaped volcanically as recently as 2 million years ago, with localised eruptions up to 900,000 years ago and on Kosrae to 1-2 million years ago (Hafiz Ur *et al.* 2013). The Marianas were submerged 2 million years ago, and must have been colonised by *Partula* since that date. Data for species within these groups are largely limited to the divergence of *P. gibba* and *P. radiolata* 700,000 years ago. The coexistence of six species without any significant anatomical divergence is exceptional. These species are not evenly distributed and only Guam is occupied by more than two species. It is probable that most originated allopatrically, with *P. radiolata* and *P. salifana* on Guam (separated by altitude), *P.* sp. (and subfossil *P. desolata*) on Rota and *P. langfordi* on Agiguan. *P. gibba* is a frequent self-fertiliser or parthenogenetic species which makes it an excellent colonist, and it is the only species to be found throughout the Marianas. Because it rarely, if ever, cross-fertilises it can coexist with any other partulid without hybridisation. Thus there are no potential hybridising sympatric partulids in the Marianas.

The species of New Guinea and the Solomons may have originated 2.0 million years ago although the present lineages are only 1.2 million years old. The Ontong Java plate on which their islands are situated is ancient but has been submerged and raised numerous times, and was most recently obducted 4 million years ago (Neall & Trewick 2008). This would mean that present-day partulids would have colonised since then, initially over a large area with ranges becoming fragmented as islands subsided or were covered by lava in frequent periods of extensive volcanism. The Pleistocene glacial maximum 120,000 years ago would have lowered sea levels to such an extent that the Solomons islands of Santa Isabel, Choiseil and Guadalcanal would have been united. The islands of the New Georgia group would have formed a separate single island (Neall & Trewick 2008). These groupings largely correspond to the present-day distribution of the *Partula* species in the islands and may have been the origins of the separate taxa (Fig.11.9). *P. grisea* and *P. similaris* diverged within the past 400,000 years, but their genotypes are currently mixed on New Britain (Ó Foighil *et al.* 2011; Lee *et al.* 2014).

Futuna seems to have been colonised from New Guinea 1.2 million years ago. There has been extensive recent volcanic activity, especially on Wallis to 80,000 years ago (Price *et al.* 2014). There is variation in the partulid populations but this does not appear to represent established species, there may be incipient speciation within this variation.

In discussing the origins and distributions of partulids Crampton (1916, 1932) took the view that the individual Society islands and, by inference, some other Pacific islands had been formed as a result of the submergence of larger land masses. Since then abundant geological evidence has accumulated to show that the vast majority of the island occupied by partulids are the remnants of emergent volcanoes and have never been connected. The only exceptions are the islands of Raiatea and Tahaa in the Societies, which lie within a single shallow lagoon and probably formed a single island at some stage in the past. Consequently all partulid populations (other than on Raiataea-Tahaa) have colonised by inter-island dispersal.

Partula of the Australs (excluding recently introduced *P. hyalina*) appear to be descended from a Societies islands ancestor, possibly the Moorea-Tahitian *taeniata* clade. They occupy Rurutu (12 million years old with some activity 1.1 million years ago – Duncan & McDougall 1976) and Rimatara (possibly 27 million years ago). None is naturally present on Rapa, Raivavae, Tubuai or Mangaia, all of which are at least 5.1 million years old (Duncan & McDougall 1976; Maury *et al.* 2013). Thus, colonisation of the Australs was not limited by geology but by chance dispersal.

11.6 Phylogeny of *Partula* species (modified from Lee *et al.* 2014, Brodie *et al.* 2016) based on mtDNA. Key as Fig. 11.1

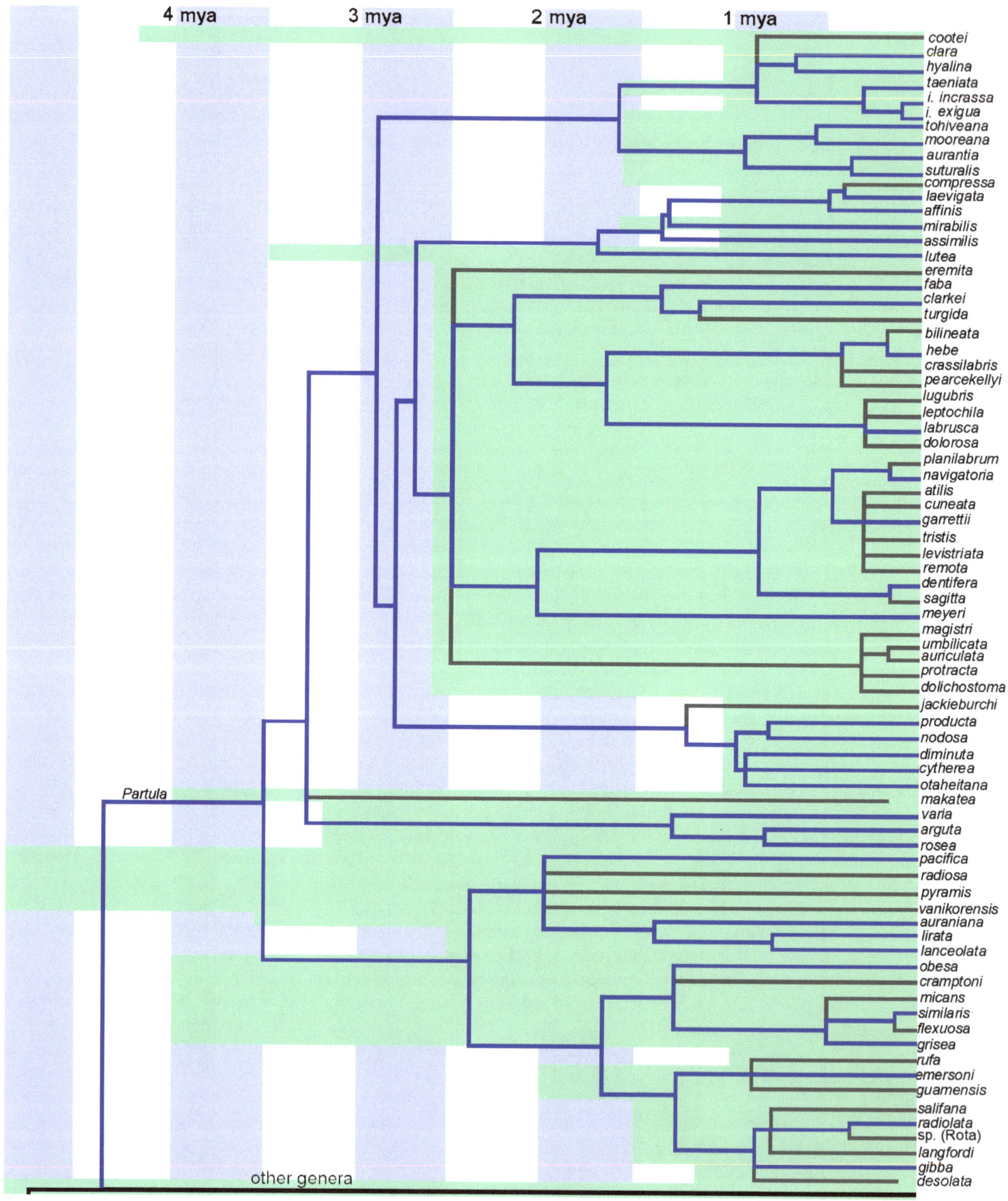

In the east the Society Islands were colonised by a single founding *Partula*, which Lee *et al.* 2014 estimated at around 3.27 million years ago, based on the age of Bora Bora, the oldest Society island occupied by partulids. However, Bora Bora may be 4.0 million years old, Maupiti 4.6 million years old (Uto *et al.* 2007) and other submerged islands may have been available for colonisation even earlier. Within the archipelago the colonisation pattern is complicated, with four lineages, forming monophyletic clades in the western Leeward islands but with several inter-island dispersal events in the eastern Windward islands (Fig. 11.7). It is notable that there is a significant mismatch between the age of island lineages and their radiations, with radiations on Tahiti, Moorea and Huahine occurring 1.2-1.3 million years after the start of the lineages. The exception is Raiatea where the discrepancy is only 300,000 years. This suggests either that the extant radiations are only the most recent radiations on the first three islands, or that the lineages originated on another island prior to the colonisations. This original divergence would have occurred on Maupiti or Bora Bora.

The Cook Islands *P. assimilis* is descended from a Moorean form (the original *P. mirabilis* clade) along with the Bora Bora *P. lutea* (Lee *et al.* 2014). Anatomically *P. assimilis* is what would be expected from a *P. mirabilis* clade descendant evolving in isolation, with a reduction in sperm competition leading to the minimisation of the caecum. *P. lutea* in contrast has the long penis and large caecum of a species facing significant reproductive competition, despite the lack of any sympatric *Partula*. This suggests that a second *Partula* was probably present on the island until relatively recently, but became extinct as a result of island erosion causing range contraction.

At 1,122 km the colonisation of the Cook islands is the furthest natural dispersal event in *Partula* (although less than the original 880,000 km for the original proto-*Partula*, or the 5,804 km for the ancestor of *Samoana fragilis*). The origin of *P. assimilis* is estimated at about 1.2 million years ago, and the island it inhabits, Rarotonga, is 1.1 million years old. The closeness of these dates may indicate that *P. assimilis* colonised the island immediately after its formation, or that it originated elsewhere earlier. Rarotonga formed as part of one of three 20 million years old Cook-Austral hotspot chains (Duncan & McDougall 1976) and it is possible that the ancestor of *P. assimilis* colonised another hotspot island before reaching Rarotonga. The other isolated species apparently of Society islands origin are *P. cootei* of the Australs and *P. makatea* of the Tuamotus. These show some resemblance to Huahine's *P. varia* in general shape and in the presence of a low parietal ridge.

The molecular phylogeny indicates an early division between the Huahine species and the rest of the Society Islands group. Within Huahine *P. arguta* and *P. rosea* are sister species; as high vegetation specialists their separation from the generalist *P. varia* is probably due to ecological specialisations. Further specialisation into canopy (*P. arguta*) and *Pandanus* associated (*P. rosea*) forms would have followed. Recent variation patterns may have arisen from some degree of population isolation and allopatric divergence may be hypothesised. There would have been isolation between the two parts of the island, Huahine-Nui and Huahine-Iti during periods of elevated sea levels.

Species of the other islands fall into two clades. The first of these comprises the Tahitian *P. otaheitana* group (including *P. producta, P. diminuta* and *P. nodosa*) and the sister groups of Raiatea and Moorean *P. mirabilis*. The second group comprises the main Moorean radiation. Of the four clades on Tahiti, the *P. otaheitana* clade is the only one not to have originated on Moorea. Its origin is not known and it may have started on an ancient volcanic island, such as Maupiti or Tupai. Within this clade there were several dextral geographically isolated species (*P. diminuta, P. nodosa* and *P. producta*), with a more widespread *P. otaheitana*, represented by several anatomical forms, and the enigmatic *P. jackieburchi*. It can be hypothesised that two waves of colonisation occurred, the first with *P. jackieburchi* which has subsequently been displaced almost everywhere by the second colonist, *P. otaheitana*. The penial caecum of *P. jackieburchi* is unique in *Partula*; it could

Fig. 11.7 Pattern of colonisation and radiation in the Society islands, from Bora Bora or Maupiti

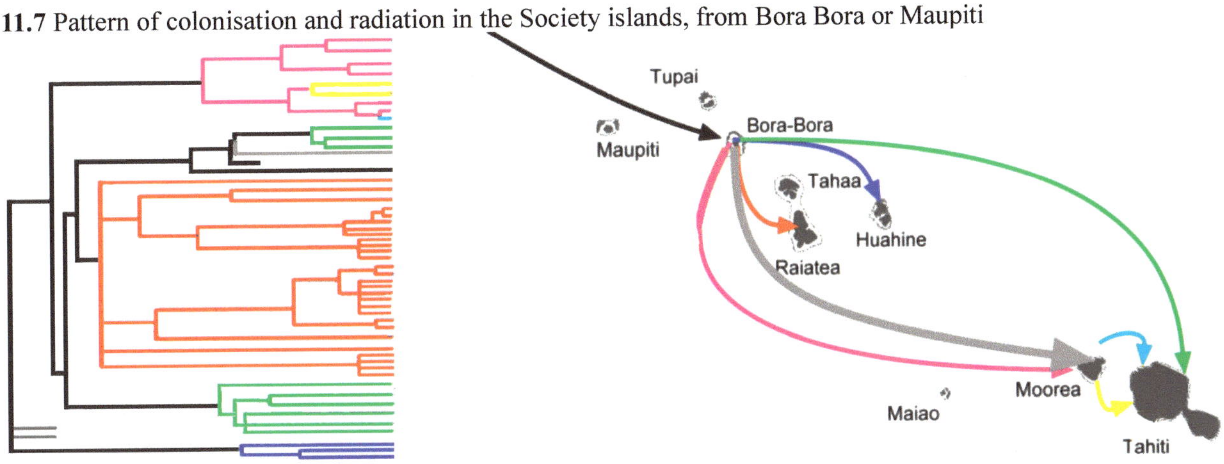

be interpreted either as incipient development of a distinct caecum or degeneration. Functionally this long, thin walled caecum would be likely to expand greatly, but would be easily deformed, giving poor attachment. It is likely that *P. jackieburchi* would be less successful in sperm competition with *P. otaheitana* but more competitive than *P. affinis*. It could be hypothesised that the ancestral *P. jackieburchi* evolved sympatrically with a species like *P. affinis*. This was able to colonise Tahiti but subsequently has faced competition from *P. otaheitana*, and has now almost completely lost its identity. The *affinis* clade also contains *P. compressa* and *P. laevigata*, which appear to have originated south of the main *P. affinis* range as geograpical and high-altitude isolates.The two *P. affinis* derivatives are allopatric.

The level of mixing between the *P. affinis* and *P. otaheitana* clades is high, with 13% (n=101) of morphologically indentified individuals being in the 'wrong' mitochondrial clade (Fig. 11.9). The *P. otaheitana* clade contains few *P. affinis* morphotypes (3%, n=71) whilst the *P. affinis* clade contains a high proportion of *P. otaheitana* individuals (37%, n=30). This strongly suggests that gene flow between the two clades is not equal, with more introgression of *P. affinis* mitochondria into the *P. otaheitana* clade. Although *P. affinis* has a higher reproductive capacity (20% more young per individual) the difference is most likely to arise from the geographical variation in anatomy. The hybrids are found in the south and west of Tahiti; hybridisation in the north-east is inhibited by pronounced anatomical differences and chiral separation (sinistral *P. otaheitana* and dextral *P. affinis*). Elsewhere anatomical differences are less pronounced but the relatively long *P. otaheitana* penis makes fertilisation likely to be unidirectional, leading to *P. affinis* mitochondrial introgression into *P. otaheitana* morphotypes.

Sinistral *P. otaheitana* appears to have colonised the north and east of Tahiti originally, spreading south and west. It may have originated as a sinistral species on the island, possibly as a result of chiral separation with another species on the ancestral island. In areas not occupied by dextral relatives (from the *P. affinis* clade) chirality has reversed, giving *P. nodosa, P. diminuta, P. producta* and *P. cytherea*. Speciation in this group would appear to be initiated allopatrically with subsequent prezygotic isolation created by anatomical divergence; in each case sympatric *P. otaheitana* and its related species are anatomically distinct. This isolation is not complete as the molecular phylogeny indicates some gene flow in most areas.

The original Moorean *P. mirabilis* clade is ancestral to the Tahitian *P. affinis* which also appears to have colonised Tahiti in the north and/or east of the island. The invasion of Tahiti by this clade probably constrained the main *P. otaheitana* morphotype to a sinistral form. This clade may include *P. compressa* as an ecological specialist.

All Moorea *Partula* share a last common ancestor around 2.8 million years ago (Lee *et al.* 2014), whereas the island dates to only 1.6 million years ago (Uto *et al.* 2007). This indicates that there have been two colonisations of Moorea, one by the *P. mirabilis* clade and one by the lineage leading to all the other Moorean species, both date to 1.4-1.6 million years ago. The former clade may originate on Bora Bora, the home of its basal species *P. lutea*. The main Moorean radiation diverged from other Society island species around 2.9 million years ago, presumably on one of the older Leeward islands. The evolution of dextral *P. mirabilis* on Mt. Rotui and the centre of the island was followed by more widespread forms, splitting into different ecotypes; the large *P. suturalis* group and the smaller *P. taeniata-incrassa.* As described above, where the *P. suturalis* overlapped with *P. mirabilis* its chirality was shifted to sinistrality (*P. suturalis* in the north and west, *P. tohiveana-mooreana* in the south). Where sinistral *P. suturalis* spread into the range of *P. tohiveana-mooreana* its chirality reversed again. In the east chirality in *P. suturalis* is confused by the partial isolation of its dextral daughter species *P. aurantia.* Sperm competition has been intense in these large-shelled lineages, with the *P. suturalis-aurantia* lineage developing notably large caecae and the *P. tohiveana-mooreana* lineage moving towards the extreme vermiform anatomical form.

Fig. 11.8 The Moorean ring species complex, modified figure of Murray *et al.* (1980) with detail of the *mooreana-olympia-tohiveana* ring around Mt. Tohiea. A – *aurantia*; M – *mooreana*; O – *olympia*; S – *suturalis*; T – *tohiveana*. Red – sinistral; purple – mixed; blue – dextral. Gene flow shown as green lines, barriers to hybridisation as black lines.

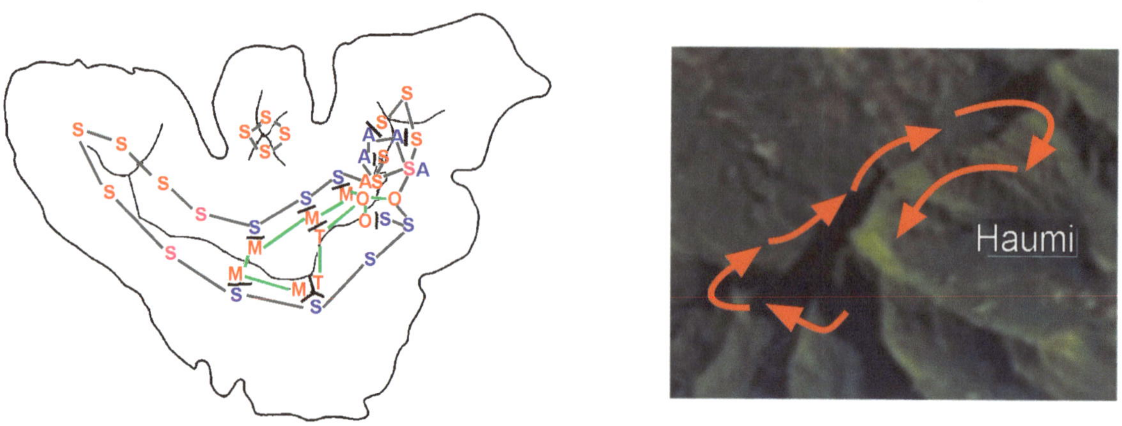

Fig. 11.9 Evidence of hybridisation in Moorean (left) and Tahitian (centre and right) *Partula*. Phlyogenies from (Lee *et al.* 2014), using the original identifications and colour coded with corrected identifications based on dissections and examination of specimens. Unverified identifications in grey, lineages on other islands unlabelled

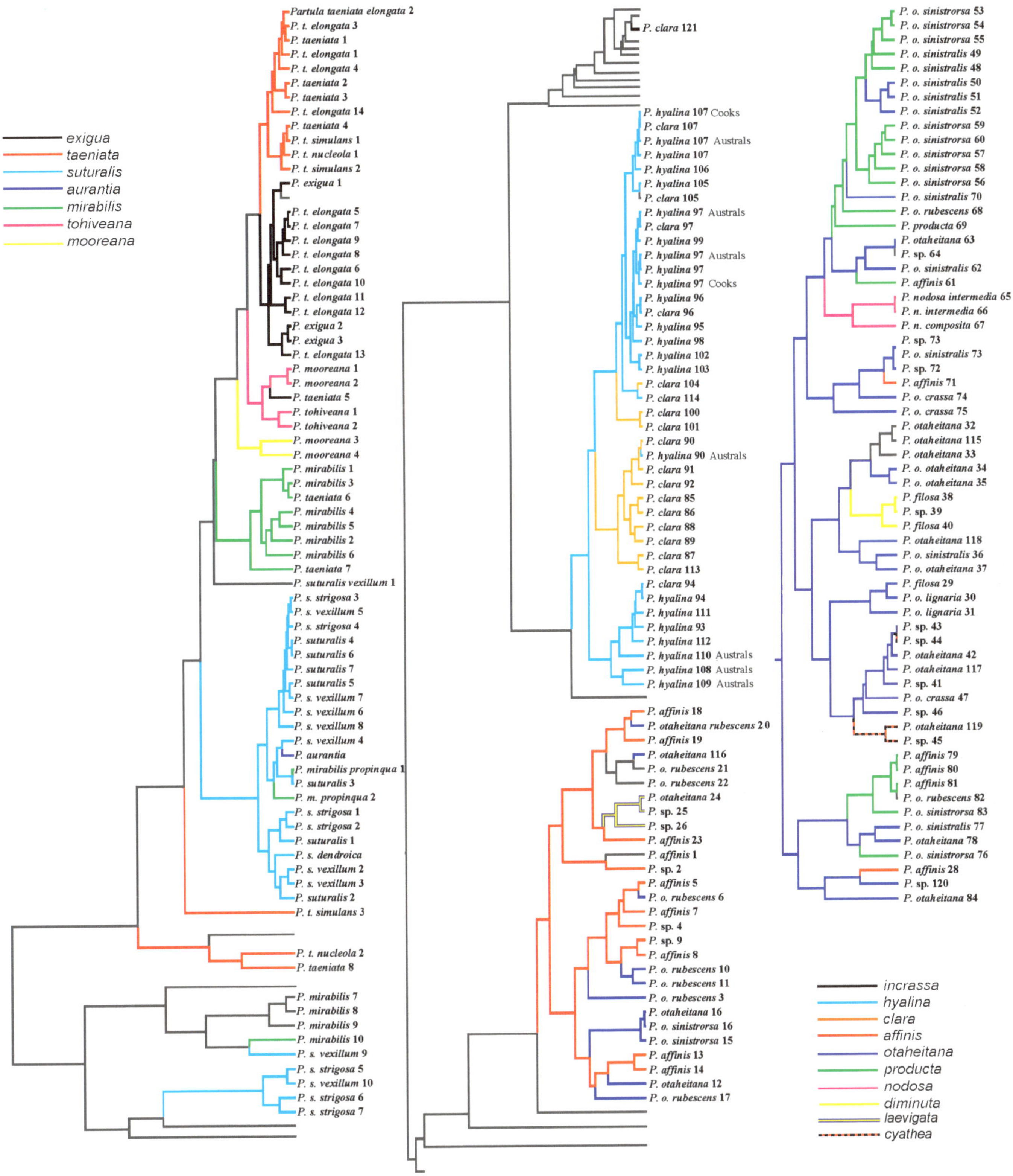

P. tohiveana-mooreana is essentially an ecological specialist lineage, being *Freycinetia* specialists. Speciation within the complex has been parapatric, forming a ring species where populations diverge parapatrically with gene flow between adjacent populations except where the extremes of the lineage overlap, in this case in Haumi valley where *P. tohiveana* and *P. mooreana* are sympatric and do not interbreed (Fig. 11.8). This was first referred to in 1969: "the *mooreana-olympia-tohiveana*

Fig. 11.10 Moorean (top left), Tahitian (lower left), Raiatean (right) *Partula* evolution: drawings – anatomy, arrows – inter-island dispersal, black circles – colonisation, red circles – anatomy change, green circles – ecological specialisation

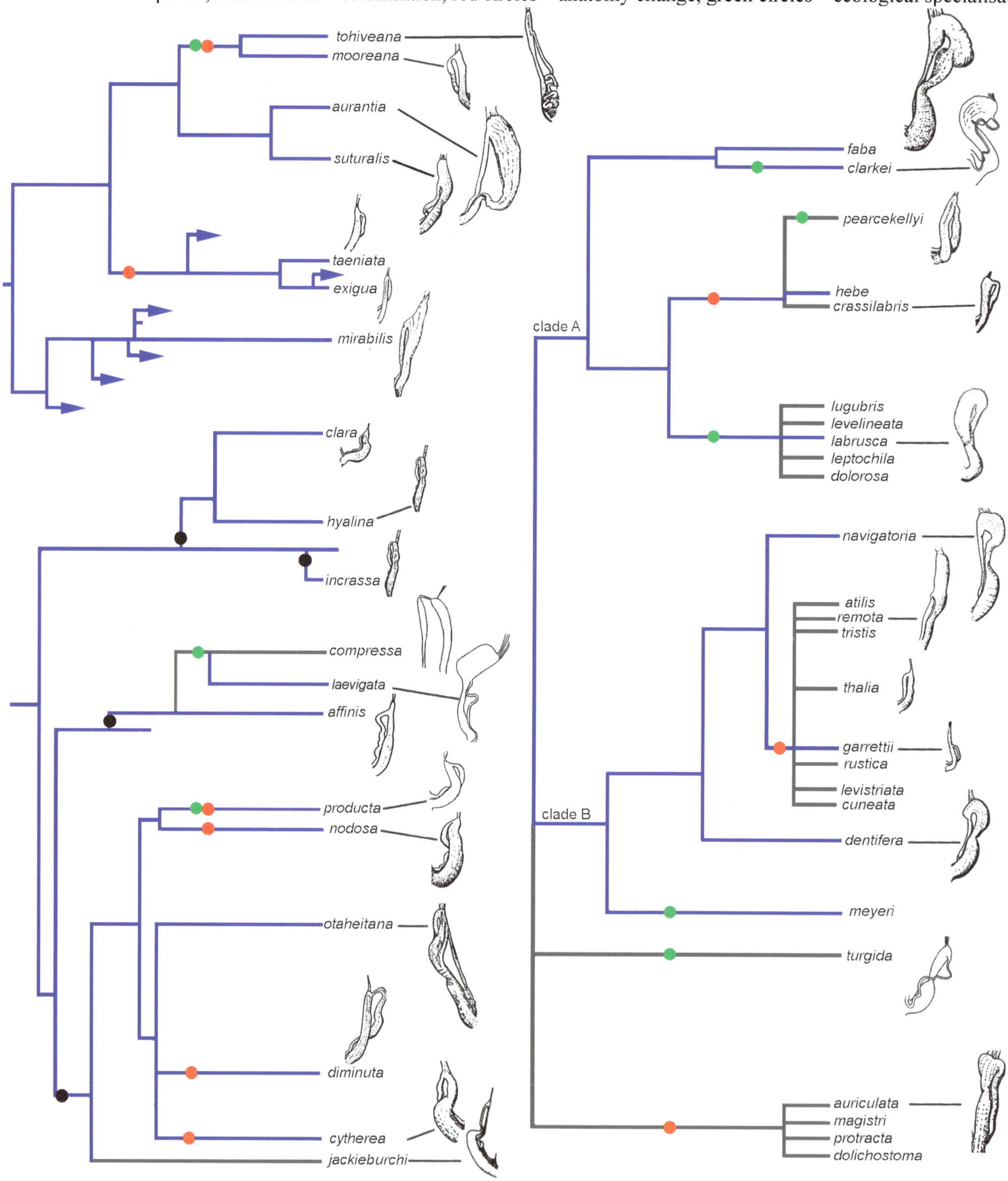

series may represent a ring-species around the vertical (or overhanging) cliffs that join Tohivea and Mouaputa. Since these cliffs are devoid of vegetation they probably constitute a barrier to the movement of snails." (Clarke & Murray 1969). The later extension of this system to include *P. suturalis* (Murray & Clarke 1980; Clarke *et al.* 1984; Murray *et al.* 1991) may be questionable, for whereas *mooreana-olympia-tohiveana* do fit the classic definition of a ring species, *P. suturalis* appears to interact with the ring, rather than being an integral part of it. No other partulid ring species have been identified, possibly due to the scarcity of barriers to movement within the islands.

These patterns were subsequently complicated by hybridisation (Johnson *et al.* 1988) carrying *P. tohiveana* mitochondria into *P. taeniata* populations and *P. mirabilis* and *P. tohiveana* into *P. suturalis-aurantia*, with extensive hybridisation between *P. taeniata* and *P. incrassa*, between *P. taeniata* and *P. mirabilis*, and between *P. suturalis* and *P. aurantia*. Some hybridisation is apparent in recent mitochondrial phylogenies (Lee *et al.* 2014), however this is mostly based on specimens identified from Crampton's (1932) descriptions, without the corrections noted by Kondo & Burch (1983). Re-examining the available samples demonstrates that there are only moderate levels of hybridisation (Fig. 11.9).

P. taeniata-incrassa are also ancestral to Tahiti's *P. hyalina-clara* (as well as that island's *P. incrassa* population). The relationships between *P. hyalina* and *P. clara* are complicated by discrepancies between conchology, anatomy and genetics arising from the very close resemblance between *P. hyalina 'marmorata'* and *P. clara* and some gene flow between them. The molecular phylogeny (with identifications corrected in Fig. 11.9) suggests a split between the white *P. hyalina* and the brown-shelled forms, with subsequent divergence between eastern *marmorata* and western *clara*. However, the Bayesian phylogeny combines the two forms of *P. hyalina* (Lee *et al.* 2014). 6% of the *P. clara* clade samples are phenotypically *P. hyalina* (n=18) and 6% of the *P. hyalina* clades are *P. clara* phenotypes and 13% '*marmorata*' (n=93). These data suggest a significant amount of bidirectional gene flow between the two nominal species, however if the comparison is restricted to native Tahiti samples the only mixed ones are in the *P. hyalina* clades, where 8% resemble *P. clara* (n=71). This may indicate predominant *P. hyalina* mitochondrial introgression into the *P. clara* population, as would be expected from the greater abundance of *P. hyalina*. Despite this gene flow the unstudied genetics of colour maintain two distinct conchological forms.

Phylogenetic studies of Raiatean *Partula* are very few in number (Lee *et al.* 2008, 2014). These group highly divergent species such as *P. faba* with *P. clarkei*, and *P. hebe* with *P. labrusca*. Five species cannot be placed in the phylogeny reliably as their relationships to the genotyped species are not known (*P. turgida* and the *robusta*-group). The other species can be placed in two major clades (Fig. 11.10). These separated around 2.5 million years ago with divergence within the clades starting shortly after. Within clade A there is a split that can be interpreted as an arboreal generalist (*P. faba* lineage) separating from a lower foliage form (*hebe*- and *lugubris*-groups). There is then ecological specialisation within the *P. faba* lineage, giving rise to *P. clarkei*, and high altitude specialisation producing the *lugubris* group as separate from the *hebe*-group. It is notable that in clade B *P. dentifera* is thought to cause a behavioural shift in *P. garrettii* (see p. 228). In the second clade *P. meyeri* is a very early offshoot, presumably as a habitat and geographical isolate. *P. dentifera* splits from the rest of the clade as a widespread arboreal form, and *P. navigatoria* and *P. garrettii* separate recently (within the past 500,000 years) allopatrically. Other speciation events are all allopatric and appear to have occurred within the past 400,000 years.

Tahaa species all appear to be related to those of Raiatea and would have been able to move between the islands at periods of low sea level. Therefore the isolation of these species must have occurred since the last glacial maximum, or within the past 25,000 years. Four of the Tahaa species are clearly close relatives of Raiatean species with one, *P. faba subangulata*, being distinct only at the subspecies level. Two species are more distinct and their exact affinities are unknown: *P. sagitta* and *P. eremita*.

From this evaluation of the evolution of the partulids the relative significance of allopatric drift and adaptive radiation can be deduced. 81% of species appear to have arisen largely allopatrically or parapatrically and only 19% through ecological speciation. Of the species arising allopatrically, speciation of 15% was reinforced by some form of character displacement, effectively an adaptation to the presence of a potentially interfertile sympatric congener. Thus the main mode of speciation in Partulidae is allopatric drift, albeit with local exceptions. This is to be expected in island species, where the primary barrier to gene flow will always be physical isolation on different islands. Ecological speciation was most apparent in the Society islands.

Speciation

The idea of a 'taxon cycle' was proposed by Wilson (1959), suggesting that colonising species adapt and expand their range, but the larger ranges become unstable and fragment into more localised species. These adapt away from the phenotypes that promoted expansion. This implies a regular cycling process underlying biogeography. This idea is not easily quantified and was largely replaced by the equilibrium theory of island biogeography (MacArthur & Wilson 1963). In the case of ants it has been proposed that the initial stage in a taxon cycle is a shift from interior primary forest habitats to more disturbed, marginal ones (e.g. coastal areas) possibly as a result of competition (Wilson 1959; Economo & Sarnat 2012). The coastal taxa are most likely to colonise new islands by rafting, due to their location and physiological adaptations. These also make them relatively successful invaders and they may outcompete resident species in the newly occupied areas (Wilson 1959). Subsequent adaptation to the local conditions leads to fragmentation and decline, ending with the endemic characteristics of low densities and ecological specialisation. This may be due to repeated colonisation forcing older colonists into the interior habitats through competition or pathogen transfer (Wilson 1959; Ricklefs & Cox 1972). As they move away from the margins, populations are more isolated and gene flow limited. A recent analysis of Fijian ants (Economo & Sarnat 2012) found support for the taxon cycle in that invasives were found in the most disturbed habitats, widespread natives in lowland habitats, endemic allospecies in upland forest and deep endemics in the highest altitude sites and at lowest densities: "increasing endemism associated with increasing confinement inland and upland. This is strong support for at least one component of the taxon cycle; that lowland marginal habitats facilitate dispersal, colonization, and geographic connectivity."

Evidence of similar patterns in partulids is limited; in *Partula* the only species found on more than two islands (*P. hyalina, P. grisea, P. similaris, P. micans, P. flexuosa*) are lowland species, often associated with anthropogenic habitats. There is no clear association between lineage age ('deep endemics') and distribution. Colonisations by partulids may be too infrequent for this pattern to be detectable on a broad scale. This may also account for the lack of clear species number – island area associations; the classic island biogeography species-area and height plots are shown in Fig. 11.11. In both cases the Society islands have more species than comparable sized islands in other groups.

Within the Society islands there is a relationship between island age and the number of partulid species (Fig. 11.12a), excluding Raiatea which has almost four times as many species as predicted. As erosion has led to a reduction in land area, the older isalnds would be expected to retian few species, whilst Tahiti and Moorea may be too young for much speciation. Raiatea could therefore be expected to have more species than the other islands. To some extent this is borne out by the origins of the islands' species: 70% of the Society island species have speciated in situ, this proportion reaches a maximum of 92% on Raiatea. However, the total number of species on these islands is largely irrelevant as the species are not distributed evenly across the island, so species number is influenced by topography as well as area and age. The best predictor of species number is the relationship between island age and the maximum number of species per valley (Fig. 11.12b). Raiatea continues to be a discrepancy, although by only by one species.

Fig. 11.11 Biogeographical patterns: a – species-area; b – species-height. Society islands are marked in red

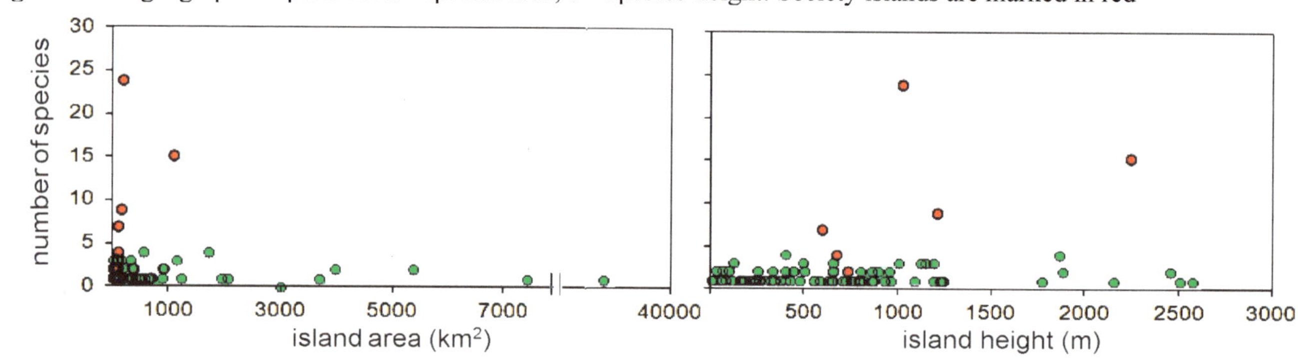

Fig. 11.12 Predictors of species number in the Society islands. B – Bora Bora; H – Huahine; M – Moorea; Mp – Maupiti; R – Raiatea; Ta – Tahaa; Th – Tahiti

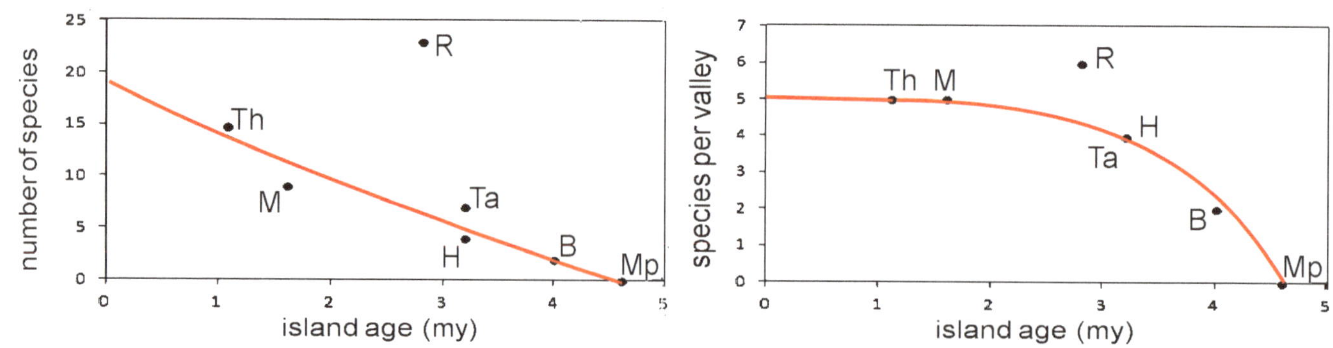

Fig. 11.13 Topography of Raiatea as illustrated by the scale models that Crampton had made in 1946 (Crampton archives, ANSP). Note the deeply dissected mountains of this eroding island. Lower right figures show proximity of Tahaa

Following initial colonisation speciation occurs allopatrically as the islands becomes fully occupied. Within each valley the species diverge progressively in an adaptive radiation, reaching the maximum 5-6 species per valley. These are the ecological niches discussed on p. 224. Raiatea is near the end point of this process, with a small number of widespread generalists (*P. faba*, and fragmenting species such as *P. dentifera* forming subspecies) and a large number of largely allopatric specialists (81%). The latter have restricted ranges (all <60 km^2, and 59% <5 km^2), resulting in an elevated risk of extinction. As the island erodes (Fig. 11.13) and ranges continue to contract the extinction rate rises, ultimately leaving only a small number of widespread forms, as are seen on Huahine and Bora Bora. Ultimately the island become too small to support a viable partulid population; Maupiti marks this point (no species on its 5 km^2 core, whereas Bora Bora at 25 km^2 is still able to support two species).

Adaptive radiations require the existence of niches (physical opportunity) that are not constrained by competitors (ecological opportunity) and that can be filled by phenotypic change (evolutionary opportunity) (Simpson 1953). In partulids these opportunities are limited in the western Pacific where competitors exist in some areas and in *Samoana* which is constrained by the presence of *Partula*. In the east of the range the conditions for radiation are met for *Partula* on the Society islands and *Samoana* on the Marquesas. Only a relatively small part of the partulid radiation can be considered adaptive; the allopatric and parapatric speciations are non-adaptive radiations. Such replacement 'allospecies' are present in a wide variety of taxa (Rundell & Price 2009), including island radiations of snails (Cameron *et al.* 1996; Cook 2008).

A pattern of specialisation, extinction and replacement by generalists has been considered on a theoretical basis and detected in butterflies (Dennis *et al.* 2011). In organisms with a very high potential for specialisation (such as in phytophagous butterflies) the conditions that create a particular specialisation may be unlikely to recur precisely, thus there would be a pattern of recurrent specialisation only in general terms. In the case of partulids, the specialisations are much broader, meaning that the underlying selection pressures change little. Consequently, the evolutionary pattern is closer to what could be considered a 'speciation cycle': generalists specialising, declining, being replaced by generalists, leading to specialisation, and so on. A point of caution that has been raised in discussion of ecological specialism/generalism is the presence of 'faux specialists' (Brooks & McLennan 2002). These are not true specialists but are constrained to a particular niche by competition. This does not affect the speciation cycle, as they function as specialists during part of the cycle, but may escape extinction by reverting to generalists when the opportunity arises. These species are only of significance to the system if there are no genuine specialists; in that case no cycle could develop and no pattern would be detectable. In the Partulidae this does not appear to be the case. It can be argued that some, such as *Samoana attenutata*, are faux specialists as anecdotal evidence suggests that this species may have become more abundant and more generalist as sympatric *Partula* have declined. The significance of this is impossible to evaluate as the cycle has now collapsed; what we observe in the Society islands today is analogous to expansion of early colonists.

In the Society island partulid speciation cycle there is an oddity in that the terrestrial niche appears to be particularly vulnerable, and restricted to only two islands: Tahiti and Raiatea. Of the species on those islands only three are exclusively ground living (*P. atilis, P. compressa* and *P. crassilabris*), all of which are geographically highly restricted. As described above,

Fig. 11.14 Crampton's scale model of Tahiti (published in Crampton 1916), note the more extensive area of high land than on the older island of Raiatea (Fig. 11.13)

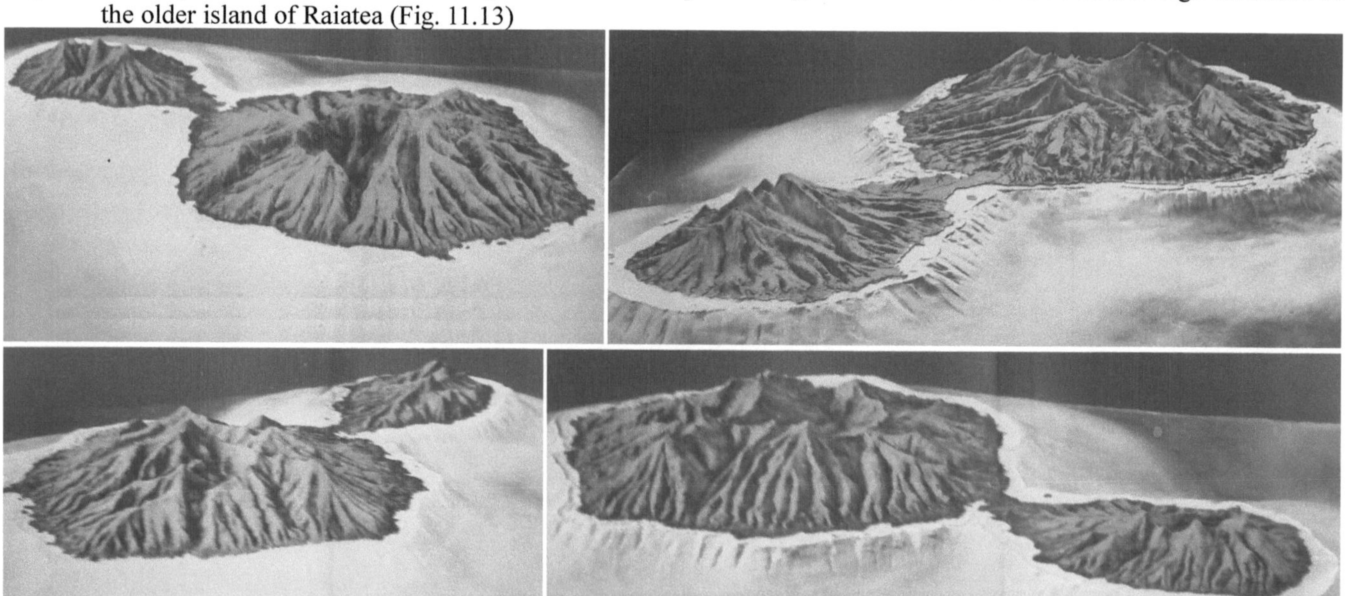

the other species are constrained to terrestriality by competition with other species and tend to become more arboreal when competitive pressures are relaxed. It appears that the terrestrial niche is not one that is favoured by partulids. Even if terrestriality is fully available as a specialism it will be a particularly vulnerable part of the speciation cycle in the Society islands. Terrestrial species will be limited to forest areas with suitable ground cover, this excludes them from most mountain ridges and so limits gene flow to the mouths of valleys. This means that terrestrial species are likely to have elevated extinction rates. Elsewhere *P. guamensis, P. rufa, Eua globosa* and *Sphendone insolita* are fully terrestrial. The first three may now all be extinct.

Plots of the number of lineages through time shows a continued but slowing rate of diversification. This is most apparent in the low diversity genera (*Eua* and *Palaopartula*); the slight slowing of rates (Fig. 11.15) indicates only the very early stages of what has been termed "diversity dependence" (Rabosky & Lovett 2008), in contrast to classic adaptive radiations such as Madagascan vangas (Jønsson *et al.* 2012). Speciation rates for the family seem to be erratic, but tend to increase as the family diversifies. The lack of any significant diversity dependence probably arises from ongoing colonisations. The Raiatean and Tahaa system is one exception; allowing for the uncertainty arising from the large number of undated speciation events, diversity dependence is apparent. There are two notable bursts of speciation: 2.75-3 million years ago shortly after Raiatea was colonised, and 250,000-500,000 years ago when the two islands were last separated by sea-level change.

Comparison with other taxa

Although superficially similar and equally polytypic, Hawaiian achatinelline tree snails appear to have a simpler evolutionary history. Speciation largely follows an island age progression rule, with sequential colonisation from oldest to youngest islands (Holland & Hadfield 2004) and the molecular phylogeny shows a good correspondence with traditional morphologically defined species. This radiation is even more recent than the partulid one, speciating within the past million years (Holland & Hadfield 2004).

Within achatinelline species there is a correlation between population genetic structure and mountain topography

Fig. 11.15 Partulid lineage diversification and speciation rates: lines – lineage diversification (lineage per 250,000 years); bars – speciation events (log scale). Raiatea & Tahaa figure include probable diversification (coloured bars and solid line) and possible ranges (dashed lines and grey areas)

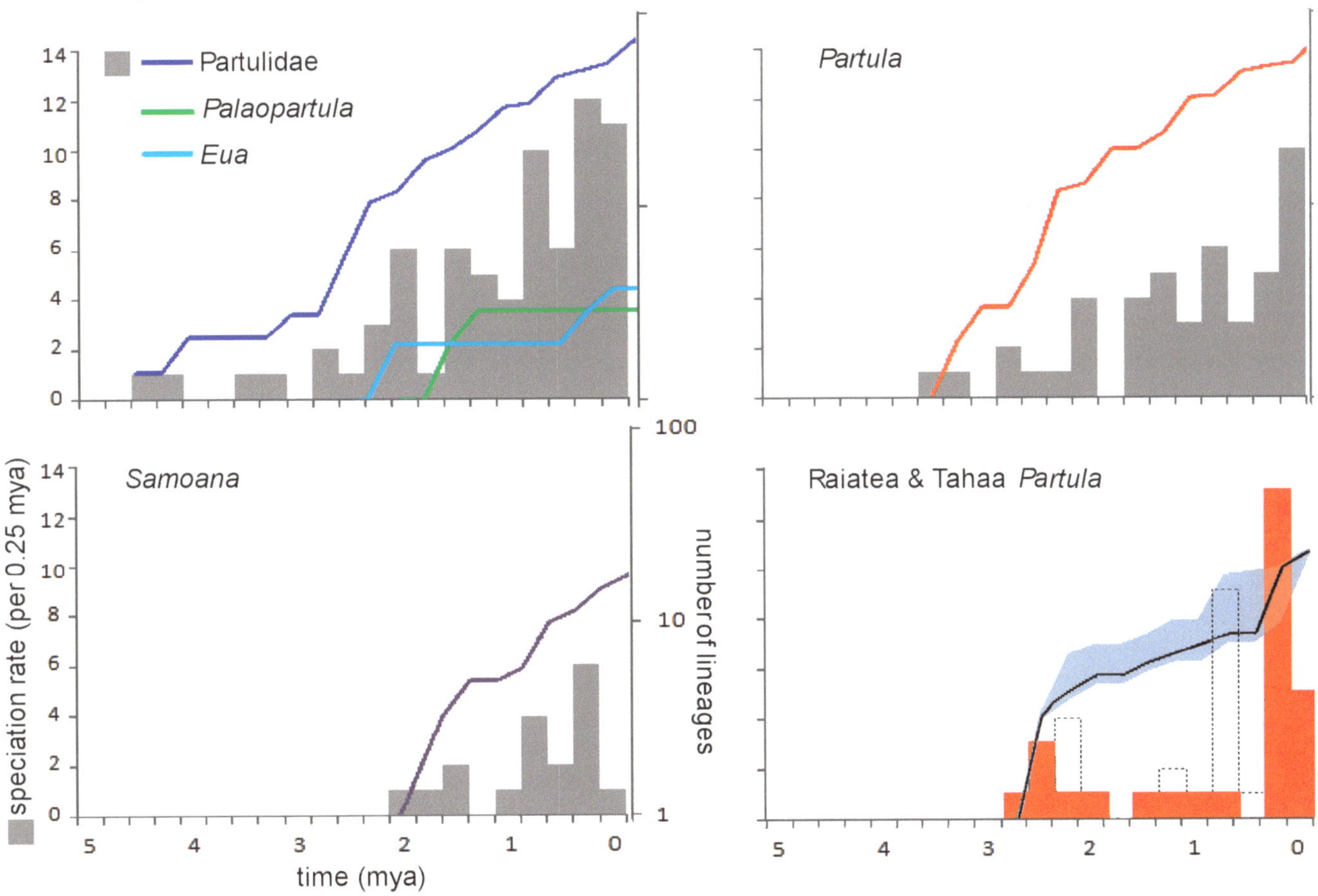

(Holland & Hadfield 2002). Their sensitivity to habitat isolation arises from low dispersal abilities with mean movement distances of 4.95 m in 7 months (Hall & Hadfield 2009). Partulid dispersal is similarly limited (4 m per generation – Murray & Clarke 1984). In the case of the achatinelline *Partulina proxima,* tiny populations (2-4 adults) exist on isolated trees (Hadfield & Miller 1989). The reproductive potential of these populations is extremely low, with an annual fecundity of 6.2 per adult and 98% juvenile mortality. These populations can only be stable if reproductive life is at least 12 years (estimated lifespan is 18-19 years) (Hadfield & Miller 1989), equating to an annual adult morality of approximately 6%. Thus, ecologically achatinellines and partulids are similar, and the differences in their evolutionary patterns are likely to arise from differing ages of radiations. Other superficially similar tree snail genera may have very different ecologies and adaptive strategies (e.g. the oviparous *Liguus* [Tuskes 1981] and *Pachnodus* [Gerlach 1999]).

The low fecundity of partulids and achatinellines is likely to have evolved in conjunction with a low adult mortality (and hence long adult life). Data for wild partulids are very limited; adult mortality is probably no more than 16% (see p. 232). Not all achatinellines have low mortality rates: relatively high rates have been found in *Achatinella sowerbyana* and *A. mustelina* (56-75% per year in juveniles and 20-52% in adults) (Hall *et al.* 2010), suggestive of a recent change in adult mortality factors as other aspects are comparable to *Partulina proxima* (Kobayashi & Hadfield 1996).

Inheritance and selection of polymorphism in other snails is similar. Chirality reversals are frequent in *Albinaria* populations, arising from population fragmentation leading to genetic drift (Kornilios *et al.* 2015). Associated with this drift, speciation in the genus is largely allopatric (Douris *et al.* 2007). In the case of colour patterns in *Liguus fasciatus* the degree of polymorphism is positively correlated to habitat area (Rex 1972), meaning that larger and more diverse areas support more variants. These comparisons all suggest broadly similar evolutionary processes in the different polymorphic tree snails, with slow rates of population growth and dispersal contributing to the creation and maintenance of local variation.

The have been many other studies of snail polymorphism, realting this to phylogeography and attempting to elucidate the underlying evolutionary processes (Berovides Alvarez & Valdes Zamora 1986; Chiba 1999; Gould & Woodruff 1990; Hayashi & Chiba 2000; Holland & Hadfield 2002; Hugall *et al.* 2002; Pfenninger & Magnin 2001; Schilthuizen 2012; Schilthuizen *et al.* 1999; Shimizu & Ueshima 2000; Thacker & Hadfield 2000; Watanabe & Chiba 2001; Woodruff 1989).

Various selective pressures leading to snail polymorphism have been proposed, including avian predators, habitat variation and climate factors (e.g. in Cuban *Polymita* – Berovides Alvarez & Valdes Zamora 1986). In some species shell colours and patterns are strongly correlated with habitat; in *Theba pisana* darker, more banded forms are associated with closed habitats (Johnson 2011, 2012) which may be due to a combination of crypsis and thermal properties (Ożgo 2011; Silvertown *et al.* 2011).

Most of the studies have been on *Cepaea*, with evolutionary research dating back to the largely unpublished studies of Diver (1939). Diver considered variation in *C. nemoralis* to be non-adaptive but this was countered by Cain & Sheppard (1950) proposing selective advantages such as avoiding predators or, more recently, thermal adaptations (Jones 1973; Heath 1975). Visual conspicuousness is an obvious explanation for many species where dark or transparent morphs are usually found in darker sites and white shells in more exposed areas, and in the latter banding is restricted to the lower parts of the shell where it is most inconspicuous (Cain 1977). Divergent selection pressures then lead to the intermediate, visually heterogeneous habitats supporting the most conspicuously polymorphic populations.

The theoretical role of predators in influencing polymorphism was expanded by the proposal of 'apostatic' selection (Clarke 1962), whereby visual predators develop a search-image for common forms of prey, leading to disproportionately high predation levels on those forms, and a selective advantage for the rarer forms (the 'apostates'). This negative frequency dependent selection should lead to cycles in morph frequency as the predators switch their search images, altering the balance of selective advantages. Although apostatic selection and other fluctuations in selection appears to explain the changes in some *Cepaea* populations (Clarke & Murray 1962; Wall *et al.* 1980), others colonies show very high morph ratio stability, with little change in morph frequencies over decades, despite variation in weather and predation levels and significant habitat change (Goodhart 1987; Cameron *et al.* 2013; Cameron & Cook 2013). These patterns of local forms have been termed 'Area Effects' (Cain & Currey 1963) where groups of similar populations are separated by sharp transitions unrelated to habitat. It should be noted that these area effects may not affect conspicuous polymorphisms alone, and that 'molecular area effects' may exist (Ochman *et al.* 1983). Cain & Currey (1963) suggested that environmental factors were unrecognised and that transitions were selectively maintained clines. In at least some cases historical landscape changes can explain area effects (Cameron *et al.* 1980). Goodhart (1963) and Wright (1965) argued that the steep clines at the borders of area effects resulted from hybrid disadvantage between different co-adapted genomes, in addition to any environmental selection pressures. This is supported by close examination of some localities indicating that selection acts at the boundaries of the areas through disruptive selection acting against intermediates, rather than stabilising selection within the populations (Cook & Ożgo 2015). Conversely, hybridisation between genotypes may create new morphs, resulting in higher levels of polymorphism in hybrid zones (Chiba 1997, 1998). Even if the underlying selection pressures may be the same, the area effects may create different geographical patterns in closely related species (e.g. *Cepaea hortensis* and *C. nemoralis* – Cameron 2013). In other taxa hybridisation has been suggested to generate novel adaptive elements (Palmer & Kronforst 2015).

Partulid characteristics

There have been several studies of phenotypic evolution and speciation of Partulidae (much of it summarised by Johnson *et al.* [1993b]). The recent molecular phylogeny (Lee *et al.* 2014) provides a good basis for evaluating the phylogenetic aspect of phenotypic evolution, specifically for reproductive anatomy, radulae, colour patterns and apertural dentition.

In reproductive anatomy there is an apparent progression towards simplification, from the complex internal pilaster structures of *Eua* to the simpler arrangement of *Samoana*, and their reduction into low ridges in *Partula*. Although the pilasters have been described (Pilsbry & Cooke 1934; Kondo 1955), their functions have not been considered. Following penial protrusion the pilasters would be presented on the outside of the penis and stretched into broad external ridges through the mechanics of the inversion process (Fig. 11.1). In this form their most likely function would be stiffening of the penis. The development of two unequal pilasters (in *Samoana*) would result in curvature of the everted penis. Their reduction into ridges in *Partula* and *Palaopartula* results in a more uniform mechanical support, allowing for more flexibility. This may facilitate mating where the oviduct is longer or more convoluted. In species where the ridges are further reduced to mere papillae mechanical support is presumably lost.

The development of the distinct penial caecum may involve a modification of the retractor musculature. Expansion of the distal end of the penis into a caecum would cause a rearrangement of the muscle, leading to tendency for a split into two portions, the main retractor attaching onto the caecum and what Kondo (1955) called a 'vestigial' retractor on the penis proper. This is normally seen in western *Samoana* species (except *S. stevensoniana*) and the Society islands species (except *S. diaphana*) where the caecum is very elongate. In *Partula* this 'vestigial' retractor has been lost, except in *P. jackieburchi* and some *P. othaeitana.* Evertion of the penis leads to protrusion of the caecum within the oviduct where it would act as an anchor, and this is likely to increase the potential of successful sperm transfer. Mating between different anatomical forms may be problematic; a species with a large caecum may be mechanically unable to mate with a species with a short oviduct. Conversely, a species with a short penis and reduced caecum may have very low mating success with a species with a very long oviduct. Penial anatomy may provide another alternative means of maximising the chances of fertilisation through alterations to penis length; in some species the caecum is well developed but not clearly distinguished from the penis. On eversion this does not form an anchoring 'head' but remains a simple tube (Fig. 11.16). Where the entire penial apparatus is greatly elongated sperm transfer can be achieved during a mating of short duration. It may be expected that the degree of caecum development and penis length may correlate with the level of potential sperm competition: large caecae and elongate penises may be found where numerous closely related species coexist, providing some degree of mechanical isolation where genetic isolation is limited.

Palaopartula also shows the two competitive penis anatomies: a well developed caecum in *P. calypso* and a reduced caecum but elongate penis in *P. thetis* and *P. leucothoe*. In *Samoana* there is a phylogenetic split between the basal species (unmodified *S. thurstoni* and *S. fagilis* with its reduced caecum) and the main clade. The latter typically have an elongate,

Fig. 11.16 Penis shape change following eversion – internal position and everted: a – *P. navigatoria* showing the expanded caecum during eversion; b – *P. hyalina*; c – *P. affinis*; d – *P. rosea* × *varia*. b & c full figures after Kondo (1955), *P. affinis* everted after Kondo & Burch (1980)

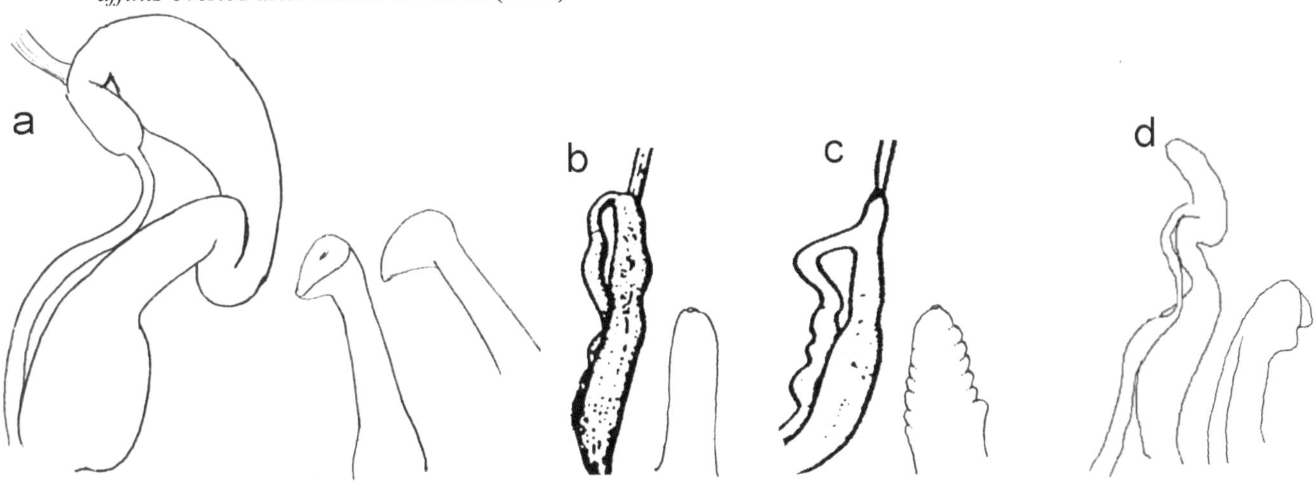

thin-walled caecum (reduced in some *S. medana*). There is very little variation within these clades, but also relatively little sympatry (two Samoan islands with two sympatric species, three Marquesas islands with three species and one Societies island with three species, and only two potentially hybridising species on each island if the anatomically divergent species *S, medana* and *S. diaphana* are excluded). Microsympatry appears to be very limited on each island (data from Kondo MS, T. Coote pers. comm.).

Sympatric *Partula* tend to have different anatomical types, probably as a result of selection for prezygotic isolation whereby the form of the caecum inhibits cross-fertilisation. This may apply to the vermiform penis of *P. tohiveana* which is unique in Moorean species and, less obviously, to the less distinct forms of *P. suturalis, P. aurantia* and *P. mirabilis*. On Tahiti *P. otaheitana* forms differ from all congeneric species, the presence of different *otaheitana* anatomical geographical forms indicates the convergent evolution of anatomical distinctiveness. The convergence of *P. jackieburchi* with *Samoana* is noteworthy, but is only superficial; the penis remains typical of the genus, with no *Samoana*-like pilasters, and the similarity of the caecum arises only from its reduction. This implies an associated reduction in sperm competition for *P. jackieburchi*. However, this species was found in the same valleys as *P. otaheitana*, with which it was able to hybridise (Johnson *et al.* 1993a). *P. jackieburchi* can perhaps be regarded as a relict species, largely displaced by *P. otaheitana* and only surviving in rare, scattered populations where *P. otahetiana* was very scarce, or absent. In allopatric island radiations anatomical divergence is less apparent (e.g. species of Melanesia).

Based on the phylogenetic pattern in the family as a whole and the relatively large penis of the basal *Samoana, Sphendone* and *Palaopartula*, it appears that the partulid penis starts as a small, pilaster-supported structure as seen in *Eua*, and subsequently develops a distinct caecum (Fig. 11.17). Where sperm competition is low (single species in isolation) the caecum tends to become reduced, either passing through a thin-walled, non-functional phase (*Samoana*-like) or simply reducing in length (many *Partula* species). Where multiple potentially interbreeding species coexist the caecum tends to diverge into an enlarged form (the '*faba*-type' penis referred to by Kondo [1955]) or merging into a vermiform penial structure ('*tohiveana*-type').

The very small caecum found in the rarely sympatric *Eua* species is probably not plesiomorphic; it is a ridged cavity with a minute aperture, incapable of inversion and has a completely different function from the caecum of all other partulids. It cannot be inverted to function in sperm competition, nor can it be used to form the spermatophore as in a true epiphallus as the aperture is too small to allow passage of a fully formed spermatophore. With these constraints its only function must be as a sperm reservoir prior to mating, the complex ridge arrangement facilitating release of sperm when contracted.

Radula variation shows no clear pattern, other than the distinctiveness of the *Palaopartula* form and the reduced number of laterals in *Samoana*. Variation in tooth shape is minimal but numbers of teeth vary considerably. This variation is not completely random as different species do have different, but overlapping, ranges and there is a genetic component to this variation (see p. 26). It does not follow any phylogenetic pattern, nor is there a significant association with ecological grouping; there is no pattern to lateral teeth, generalists have the lowest marginal (and hence total) counts, followed by plant feeders (generalists and specialists), detritivores and fungus feeders. Plant generalists have significantly fewer teeth than fungal generalist (exact $t=2.423$, $P_{2,108}=0.017$) or detritus generalists ($t=2.920$, $P_{2,208}=0.04$). Fungal and detritus generalists are

Fig. 11.17 Evolutionary steps in the transition from the *Eua*-type penis to *Palaoartula*, *Samoana* and *Partula*

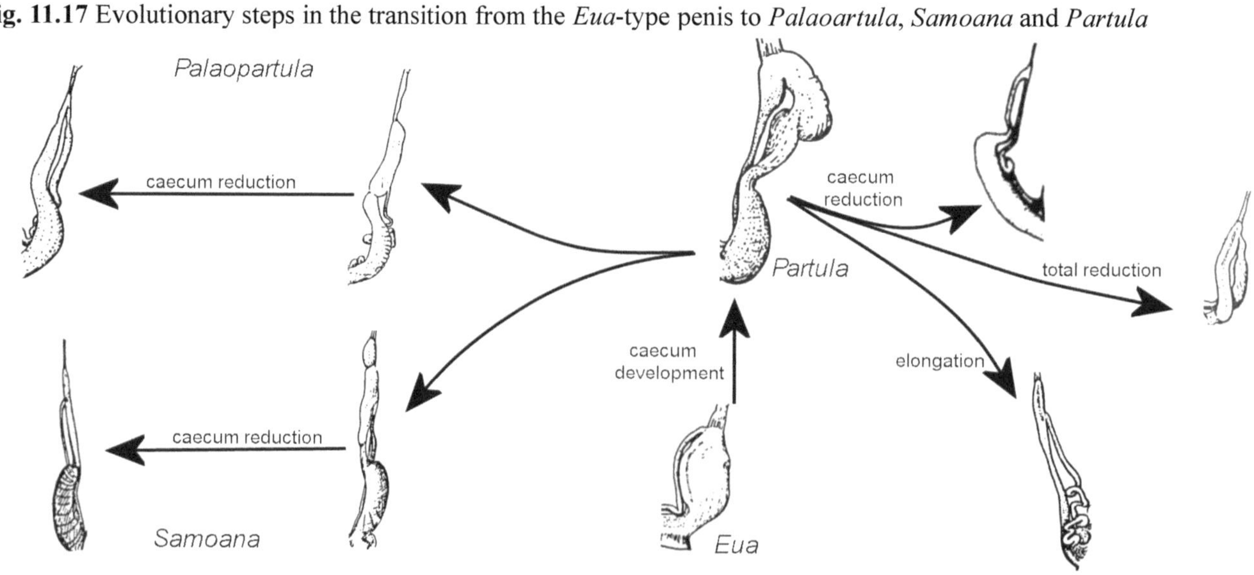

not significantly different, nor are generalists and plant feeders. This suggest that there is a limited functional aspect to radula tooth number, which leads to convergence and divergence in numbers, but with considerable overlap.

Apertural dentition is well developed in a small number of species, and only in the Society islands (Raiatea and, to a lesser extent, Tahiti and Moorea). The reason for the development of apertural dentition is not known; complex barriers are generally thought to be anti-predator defences in small species (Boettger 1921, 1935; Solem 1972; Gittenberger 1996; Wada & Chiba 2013) or to minimise water loss (Boettger 1921, 1935), but the simple barriers in the large apertures of *Partula* would not be effective defences and their distribution does not correspond to areas with higher potential water loss than in denticle-less species. The apparent ineffectiveness of denticles as anti-predator defences in *Partula* was noted by Goodfriend (1986) who concluded that their function remained a mystery. A non-adaptationist explanation can be proposed, suggesting that they are not directly selected for but result from a cessation of growth before the cessation of calcium deposition. This could explain the development of thickened keeled lips, but would not produce distinct denticles. The denticles take two forms; the simplest is a partietal ridge that may develop into a denticle in the aperture, the more obvious form is a discrete denticle, not associated with any ridge. The ridge may be structural, thickening the parietal region, but this is not the case for the denticle. The thickening of the lip could be structural or a result of continued calcium deposition. The nodular denticle appears to be too discrete a structure for a mere calcification.

Denticles are normally present or absent in a species, although there are few species where they are variable. This variation may be the result of introgression, and variability is particularly apparent in the Moorean species *Partula suturalis* and the Tahitian *P. otaheitana*, both species that form parts of hybrid complexes, suggesting that the denticle may have appeared in the species as a result of introgression from a normally dentate species (*P. tohiveana* and *P. mooreana* on Moorea and *P. affinis, P. cytherea, P. diminuta, P. nodosa* and *P. producta* on Tahiti).

Denticulate species are almost always terrestrial or inhabitants of low vegetation whereas canopy species are always edentulous (Table 11.1). Excluding species known from a single specimen, those of unknown ecology and the cloud forest and burrowing specialists, denticulate or auriculate species are only found on, or near the ground (18 species), species with simple apertures are largely found higher above ground (18 species). Variable species are more problematic: *P. suturalis, P. mirabilis* and *P. otaheitana*. All three species form hybrid complexes with other species and the genes for denticle formation may have originated in more terrestrial taxa. In addition, some populations of *P. otaheitana* are at least partially terrestrial, adding further complexity. The exceptions to the pattern are the simple terrestrial and low vegetation *P. lugubris, P. leptochila* and *P. meyeri*, all of which are at least largely high-altitude species. This indicates that there is a function for the denticle that is associated with the environment. In conjunction with the expanded lip, and especially when associated with auriculation of the lip, it would act to increase the surface area in contact with the substrate. This would facilitate the formation of a complete mucus seal around the aperture during aestivation. As the denticle is slightly (occasionally deeply) recessed this explanation would only apply at the subadult stages where the lip is forming at the level of the denticle (Fig. 11.18). It can be hypothesised that the formation of an effective seal would be of limited benefit to juveniles that are unlikely to survive prolonged aestivation and should concentrate their resources on rapid growth. Where maturity is approached a mucus seal may minimise water loss and prevent the entrance of potential predators. Water loss would be expected to be more of an issue for lowland species than

Fig. 11.18 Development of the denticle in *Partula hebe*, showing the development of the denticle just behind the line of the early stage lip in subadults and the completion of the keeled lip serving to seal the shell against the substrate

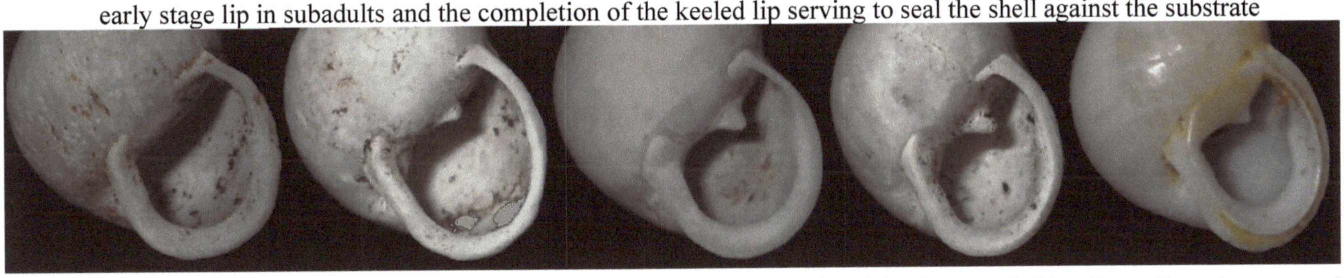

Table 11.1 Relationship between ecology and aperture characteristics in Society islands *Partula*. Numbers of species grouped into micro-habitat preferences. Labiate forms are only found in denticulate species.

	terrestrial	low	mid-high	specialised	canopy	?	notes
denticulate	5	8	1[1]	2[2]			1 – *sagitta*; 2 – *tohiveana, mooreana, Freycinetia* specialists
auriculate						1[3]	3 – *P. magistri*
denticulate & auriculate	1	4					
simple	1	3[4]	17		2	2[5]	4 – *lugubris, leptochila, meyeri*; 5 – *levistriata, eremita*
variable		1[6]	2[7]				6 – *otaheitana*; 7 – *suturalis, mirabilis*

those at higher altitudes, and more for arboreal species than for terrestrial ones, whereas predators such as ants may be more problematic in or near the leaf-litter. In this context it is notable that *P. tohiveana* and *P. mooreana* are some of the very few high denticulate species and both are associated with *Freycinetia*. This plant collects litter in its axils, which is frequently inhabited by colonies of ants (pers. obs.). The only other high denticulate species is *P. sagitta*, which is only known from a single collection with limited data.

The scarcity of denticles in species from outside the Society islands (Fijian *Partula* sometimes have a denticle) may be a result of the almost complete absence of terrestrial species. The only exceptions are the very distinct *P. guamensis* and *P. rufa*, the partially terrestrial *Eua zebrina* and *Sphendone insolita*. The simplicity of apertures outside of the Society islands may reflect this paucity of terrestrial partulids, arising from the restriction of species to the generalist niche due to the existence of terrestrial competitors or other conflicts, or the presence of different terrestrial predators.

Inheritance of characters

Crampton's study, and Meyer's that preceded it, aimed to determine the Mendelian basis of inheritance of chirality and colour pattern in partulids. At the same time chirality was being investigated by Boycott & Diver (1923) working on *Lymnaea*. They determined that inheritance was Mendelian, with a dominant dextral allele and a recessive sinistral one, but delayed by a generation through a maternal effect. Crampton (1916) concluded that the same pattern held true for *Partula* based on his embryonic data and this was later confirmed by laboratory crosses (Clarke & Murray 1969; Johnson 1982; Johnson *et al.* 1987, 1990; Davidson *et al.* 2009). *Partula* data have been interpreted to mean that switching between chiral forms in partulids represents character displacement for reproductive isolation (Johnson 1982). The genetic basis of chiral variation in *Lymnaea stagnalis* has now been determined to be related to a variant of the Ldia2 formin developmental gene. Although the pattern of inheritance is very similar in *Partula*, and chiral variation probably arises from mutations affecting development involving formin genes, the precise mechanism is different (Davison *et al.* 2016).

In their investigation of the inheritance of coiling direction Murray & Clarke (1966, 1976b) found that chirality in *P. suturalis* (Fig. 11.19) is inherited independently of the colour genes. It also affects shell shape (Johnson 1987; Goodin & Johnson 1992). The sinistral allele was dominant to dextral in contrast to other Moorean partulids: crossing dextral *P. mirabilis* with sinistral *P. tohiveana* indicates that dextrality is dominant in these (Johnson *et al.* 1993b). Given that most partulids are dextral it is probable that the *P. suturalis* condition arose as a reversal of the normal dominance pattern.

Under this hypothesis the original colonist of Moorea would have been dextral (such as *P. mirabilis*), diverging into different allopatric species. Where the ranges of these overlapped in the south the chirality of one group shifted, giving rise to the sinistral *P. mooreana-tohiveana* and *P. suturalis*. Where the ranges of these two groups overlapped *P. suturalis* reverted to

Fig. 11.19 Variation in coil in *Partula suturalis* within one population (photo: D. Clarke)

dextrality, giving a dextral population with sinistrality being dominant. Some populations became isolated in the north-east of the island, isolating *P. aurantia* from the main *P. suturalis* group.

Mixed chiral broods have been recorded on Moorea (Crampton 1925b, 1932). Crampton (1932) considered these mutations but as Johnson *et al.* (1993b) pointed out, this is theoretically impossible as the maternal effect should mask any mutation for the first generation, and it is more probable that these are developmental errors. Such occasional mutations or developmental errors do occur but are extremely rare. This is indicated by the extremely rare reversed chirality shells in normally invariant species – 0.004% of Crampton's *P. taeniata* shells (n=28,530), 0.167% of *P. mirabilis* (n=4,196). No such reversals have been recorded in any captive bred snails (n=98,912, ZSL data). Crampton recorded mixed chirality in 0.1% of twin broods in *P. suturalis* (n=4,319) but only in Faamaariri valley. This population hybridised with *P. aurantia* (Murray & Clarke 1968) and this may represent hybridisation induced developmental errors. Similarly, single broods of *P. mooreana* and *P. mirabilis* contained mixed chirality offspring, both in Faataofe valley. This valley was the only area where reversed chirality was recorded in these species and Johnson *et a*l. (1993b) suggested that this may also be a case of hybridisation. No mixed broods were found in the large samples from Tahiti, even though hybridisation is also widespread on that island. This may indicate that hybridisation on Tahiti was between more closely related species than on Moorea, resulting in fewer developmental problems. This would be in accord with the more recent colonisation of Tahiti by *Partula*.

Matings between mixed chiral pairs have been found to have reduced success, acting as a partial prezygotic barrier to gene flow (Johnson 1982). Davison *et al.* (2009) found no such effect and concluded that Johnson's data reflected long-term sperm storage, whilst their data came from recent matings. This suggests that the positive frequency dependence effect of chirality mainly arises in low density populations where mating may be infrequent.

On Tahiti, it would appear that the island was first occupied by the largely dextral *P. affinis* in the east of the island. This species has a low frequency of sinistrals (<5%) in all populations except in the south-east Tahiti-Iti peninsula where Crampton recorded 5% sinistrals in Haavini and 25% in Vaiau. Surviving populations in the region are now fully sinistral. This is the only area where *P. affinis* and its dextral relative *P. laevigata* were sympatric and hybridised. The chirality shift in *P. affinis* probably represents character displacement, giving some degree of reproductive isolation. The west of the island was occupied by the dextral *P. nodosa-diminuta-producta* clade, of which *P. diminuta* is fully dextral and *P. nodosa* dextral except in Maruapoo (93% sinistral) and in Atehi/Hapaa (fully sinistral). *P. producta* was originally dextral but has shifted in two valleys in the south-east where it is sympatric with *P. affinis* (being 90% and 32% sinistral). Subsequent *P. otaheitana* colonisations would have been forced to become sinistral and this is the case in almost all valleys (>99% sinistral). Where sinistral *P. otaheitana* colonised valleys not occupied by close relatives the sinistral form lost its advantage and started reverting to the dextral form. These partial reversals were found in the north (Fautau 55% sinistral, Hamuta 64%, Mt. Atara 75%). In the south a complex mosaic of habitats led to isolation of *P. otaheitana* and *P. producta* in some valleys and sympatry in others. This resulted in mixed chirality in *P. producta* in Papehue, Atehi and Vaihiria (53-54%). *P. producta* are also sinistral even in isolation: Vairaharaha, Moaroa, Taharuu, Teohu. Sinistral *P. producta* and *P. otaheitana* are found in the same valleys but not in microsympatry in Papeiti and Temarua.

Crampton's attempts to identify the Mendelian basis of colour patterns were less effective. For the Tahitian species he gave a long and complicated analysis of inheritance of the relatively simple red and yellow colours in *P. otaheitana* (Crampton 1916). This indicated that red was dominant in most valleys, but not all. Even in the valleys where red dominance is most plausible there was an excess of yellow embryos in red adults, suggesting that colour may change during development. By the time he published the Marianas volume (1925) he seems to have abandoned this approach; it is not repeated in the Moorean volume (1932) and there are no notes on inheritance in his archives. Crampton had intended to undertake experimental crosses

Table 11.2 Pattern alleles (from Clarke & Muray 1976a, b), equating *P. taeniata* and *P. suturalis* and other captive taxa

taeniata genotype	pattern	***suturalis*** genotype	pattern	***mirabilis***	***mirabilis x taeniata***	***tohiveana***	***clara***	***affinis***	***varia***	***rosea***	***faba***
$B1^{-}B2^{-}$, $B1^{-}B2^{-}$	plain	M^{A}, M^{A}	atra	plain		plain	plain	plain	plain	plain	plain
$B1^{f}B2^{-}$, $B1^{-}B2^{-}$	frenata	M^{F1}, M^{A}	frenata	frenata**		frenata		frenata	bisecta	bisecta	
$B1^{f}B2^{-}$, $B1^{f}B2^{-}$	frenata	M^{F1}, M^{F1}	frenata	frenata**		frenata		frenata	bisecta	bisecta	
$B1^{f}B2^{-}$, $B1^{-}B2^{z}$	lyra	M^{F1}, M^{C}	lyra**	lyra**							
$B1^{f}B2^{z}$, $B1^{-}B2^{-}$	lyra	M^{F2}, M^{A}	bisecta	bisecta			bisecta	bisecta*	bisecta	bisecta	bisecta
$B1^{f}B2^{z}$, $B1^{f}B2^{-}$	lyra	M^{F2}, M^{F1}	frenata	lyra**							
$B1^{f}B2^{z}$, $B1^{f}B2^{z}$	lyra	M^{F2}, M^{F2}	frenata	lyra**			bisecta	bisecta*	bisecta	bisecta	bisecta
$B1^{f}B2^{z}$, $B1^{-}B2^{z}$	lyra	M^{F2}, M^{C}	bisecta	lyra**							
$B1^{-}B2^{z}$, $B1^{-}B2^{-}$	zonata	M^{C}, M^{A}	atra	zonata**	zonata			cestata**	zonata	zonata*	
$B1^{-}B2^{z}$, $B1^{-}B2^{z}$	zonata	M^{C}, M^{C}	cestata	zonata**		zonata**	cestata**	cestata**	zonata	zonata*	
$C^{N/Y/W}$	colour	$M^{A/S}$	colour*	$C^{N/Y/W}$	C^{N}	$C^{Y/W}$	$C^{N/Y/W}$	$C^{N/Y/W}$	$C^{N/Y/W}$	$C^{N/W}$	$C^{N/Y}$
$L^{P/W}$	lip	L^{W}	lip								
$S^{N4/-}$	spire	M^{X}	apex	mitella	-	-	-	-	bicolor		-

**P. suturalis 'dendroica'* retains the colour forms controlled by $C^{N/Y/W}$ alleles; ** wild patterns not produced in captive populations of colour forms (Crampton 1929) and kept some species alive for several months (Crampton 1916) but Mendelian crosses were not undertaken until the work of Murray & Clarke (1966, 1976a, b) on *P. suturalis* and *P. taeniata*. They identified several different loci controlling different aspects of colour, each with several alleles and combining to act as supergenes.

Crampton defined a wide range of forms, but was not consistent when considering the different species. Murray & Clarke (1976a, b) used a simpler categorisation of four banding types, defining frenata as having two dark lines, zonata with a broad central dark band flanked by lighter areas, lyra with three bands, bisecta being dark with a broad central light band and atra as being the unbanded dark form. There is also a cestata form with two pale bands which form a continuum with zonata. They investigated these patterns in *P. taeniata* and *P. suturalis*, finding different systems of inheritance in the two species (Table 11.2). The differences in banding pattern and inheritance suggest that *P. taeniata* and *P. suturalis* evolved their patterns convergently, from unbanded ancestors (Murray 1972). This resulted in different genetic controls and different patterns. Where patterns are shared frenata appears to be very similar in both species whereas zonata and bisecta differ in detail.

The hypothesis of different genetic systems for banding in *P. taeniata* and *P. suturalis* is made problematic by the recognition of signficant hybridisation between Moorean species. The presence of the typical banding forms in the hybrids suggests that there may be an underlying similarity. The *P. suturalis* supergenes can be related to those of *P. taeniata* if it is hypothesised that the two linked banding loci in the latter (B1 and B2) have become more tightly linked, effectively forming a single locus (M) in *P. suturalis* (Table 11.2). The *P. taeniata* frenata allele ($B1^f$) in isolation is equivalent to the *P. suturalis* M^{F1}, zonata forms ($B2^z$ and M^C) are equivalent (*P. suturalis* cestata intergrading into zonata). The paired alleles ($B1^f\ B2^z$) are equivalent to M^{F2} and the paired recessive $B1^-\ B2^-$ is equivalent to atra M^A. The banding alleles lighten the base colour; in the absence of these the base is dark in *P. suturalis* (M^A) but varied in *P. taeniata.*

Wild population data also provide support for a common pattern. The presence of strigosa in *P. taeniata/mirabilis 'propinqua'* is suggestive of introgression of the *P. suturalis* M^S allele. Crampton's *P. taeniata propinqua* was described as having lyra and zonata forms, but no frenata. The absence of frenata suggests that the allele needed for lyra is maintained by frequent crossing with *P. taeniata* frenata individuals. Unless *P. mirabilis* and *P. taeniata* were interbreeding freely in this area

Fig. 11.20 Phylogeny of Society island *Partula* with inferred colour and banding alleles. Allelic dominanace marked from black (dominant) to yellow (recessive). Modified alleles producing variant patterns are shown as red points and codominant alleles as merged green points. The phylogeny is coloured with puative evolutionary transitions for banding genes: green - development of the *P. taeniata* type supergene; red - modification of an allele; dark blue - loss of alleles; light blue complete loss of banding. Shells show *P. suturalis* patterns, including the transition between zonata and cestata

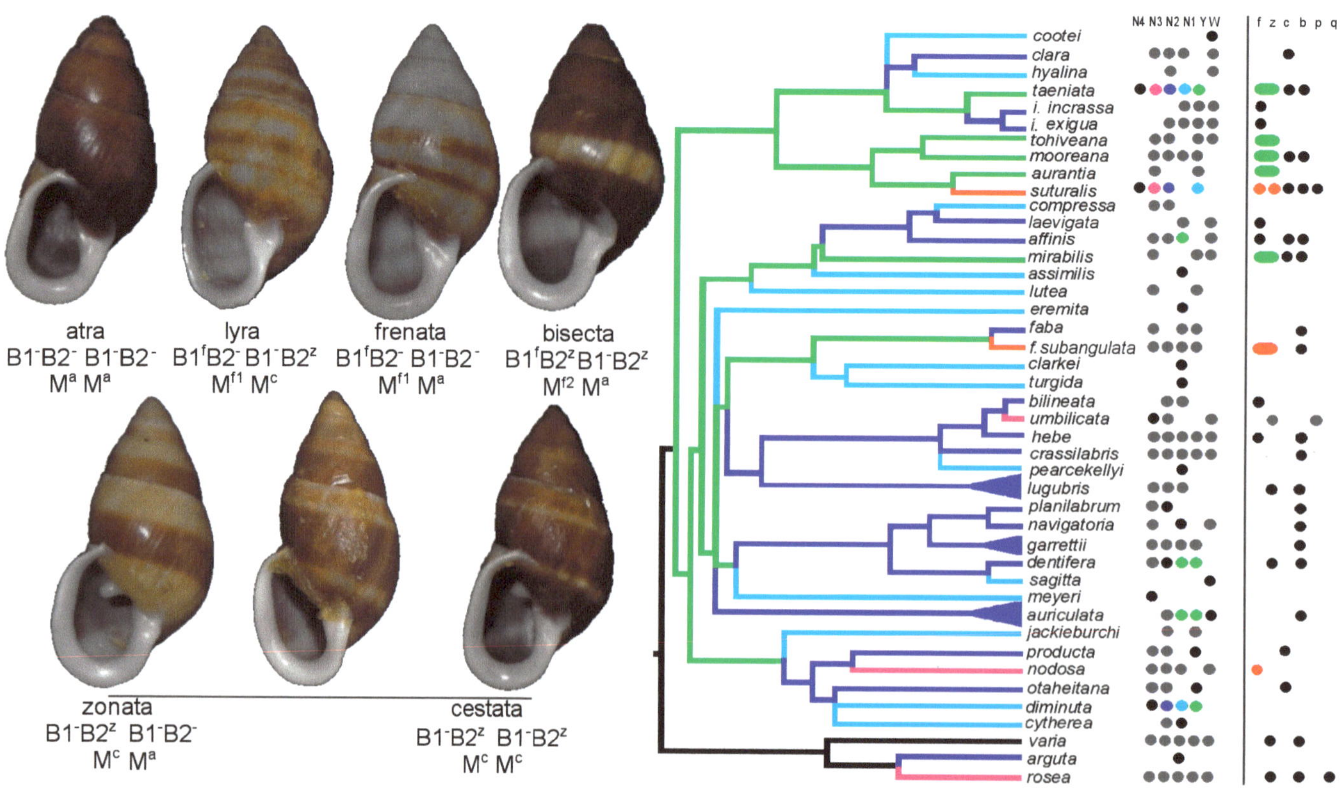

lyra would be expected to have been rare.

No experimental crosses have been carried out on other species, but the conservation breeding populations do provide some information which can be compared to the known inheritance patterns to infer genotypes (Table 11.2), as do Crampton's (1916, 1932, MS) embryo data. Using a null hypothesis that the alleles on the two *P. taeniata* banding loci and the separate colour loci produce the observed patterns allows embryonic colour frequencies to be predicted and compared to Crampton's dissections of embryonic material from 17 Society island taxa: *P. affinis, P. aurantia, P. auriculata* (and '*P. robusta*'), *P. faba subangulata, P. dentifera imperforata, P. hebe, P. leptochila, P. mirabilis, P. navigatoria, P. otaheitana, P. planilabrum, P. producta, P. robusta, P. garrettii thalia, P. t. tohiveana, P. t. olympia* and *P. umbilicata.* Most data are consistent with the *P. taeniata* system but all null hypotheses were rejected in *P. dentifera, P. garrettii* and *P. leptochila*, in addition all *P. hebe* embryos were white (Appendix III). A full frenata-zonata-lyra system was supported in *P. faba subangulata* (with modification of the lyra pattern), some *P. umbilicata* (4/12 populations) and in the only *P. affinis* populations with the full range of banding patterns. Simplified banding patterns consistent with the loss of one or more banding alleles appear to be dominant to plain shells in *P. navigatoria, P. auriculata, P. tohiveana*, the remaining *P. affinis* and most *P. otaheitana* (5/9).

Colours generally supported dark shades being dominant to light (as in *P. taeniata*): *P. diminuta,* some *P. faba subangulata* (2/5) and *P. umblicata* (1/9), but the reverse was the case in *P. navigatoria, P. auriculata, P. planilabrum* and one *P. faba subangulata* population. Colours in *P. otahetana* are different from most other species and their inheritance seems to follow two different patterns. For the majority of populations (5/9) yellow is dominant, two (*P. o. rubescens*) have red dominance and two (*P. o. otaheitana*) have light and dark shades being codominant. These may reflect local divergence.

These limited data do provide some support for all Society islands *Partula* patterns being controlled by a common genetic system with a modification in *P. suturalis* and the loss of alleles in several taxa (Fig. 10.20), probably as a result of drift caused by repeated founder effects. It has previously been noted (Cowie 1992) that the apparent different genetic bases for polymorphism in *Partula* species contrasts to the situation in *Cepaea* where different species share the same colour and pattern genes, despite a much earlier divergence of *Cepaea* species (Johnson *et al.* 1993b), estimated at 18 million years ago (Neiber & Hausdorf 2015) whilst Moorea is only 1.6 million years old (Uto *et al.* 2007). The common pattern of inheritance outlined here does not require rapid convergent evolution in the *P. suturalis* and *P. taeniata* lineages.

The failure to support any null hyptohesis in *P. dentifera, P. garrettii* and *P. leptochila,* and some populations of *P. affinis* (Papeiha, Utuufai, Faone), *P. auriculata* ('*robusta*'), *P. otaheitana* (Papenoo, Tenaire, Vaipuarii, Tiamao), *P. tohiveana* (*P. t. tohiveana*) and *P. umbilicata* (Hamene, Hurepiti, Patio, Tapuamu) indicates that these populations may not be at equilibrium, with allele frequencies affected by selection or drift, or that they may be produced by other inheritance systems. There may also be errors in the scoring of embryonic colours; ontogenetic colour change may be indicated in *P. leptochila, P. auriculata 'robusta', P. garrettii* and *P. tohiveana* where dark coloured and banded embryos are under-represented (Table 11.3), suggesting darkening and banding developing over time. An alternative explanation would be that the light snails are selected against in these species, but this is unlikely in *P. leptochila* at least as adults are mostly light (68%). Some ontogenetic colour change has been recorded, where colour is slightly dilated by faster growth in *P. taeniata* (Murray & Clarke 1976a). More complex ontogenetic change may occur in *P. umbilicata* which has more unbanded medium shaded embryos, suggesting both ontogenetic lightening and darkening as well as development of banding patterns. Alternatively there may be selection against unbanded and medium individuals. A similar situation occurs with *P. dentifera* which has more medium shaded adults and fewer banded embryos, these differences are insignificant if the median and banded categories are combined, suggesting that median embryos become banded with time. Alternatively there may be selection against unbanded and extreme shades.

In the case of the *P. otaheitana* and *P. affinis* populations there is a geographical component to the populations that do not fit null model expections (Fig. 11.21). The excess of yellow embryos compared to red suggests that the red colour may develop with age (as suggested by Crampton 1916), or that there is selection against yellow individuals in some valleys. The under-representation of banding in southern embryos indicates a progressive development of banding as well as colour. Chirality in Fautaua showed an excess of dextral embryos, probably indicating that the population was not at equilibrium. This was an area with a very mixed population, apparently the result of the overlap of several different genotypes (including input from *P. cytherea*). In this population 71% of sinistral adults bore young of the same coil, as did 78% of dextrals. These discrepancies may arise from hybridisation, as their distribution matches the areas where introgression has been detected

Table 11.3 Significant differences between adults and their embryos (data from Crampton 1916, 1925, 1932, MS)

Taxon		χ^2	df	P	Taxon		χ^2	df	P	Taxon		χ^2	df	P
P. affinis	Papehia	491.89	1	<0.01	*P. leptochila*		4.48	1	<0.05	*P. tohiveana*		11.78	1	<0.01
	Utuufai	35.22	1	<0.01	*P. otaheitana*					*P. umbilicata*				
	Faone	548.45	1	<0.01		Papenoo	44.93	1	<0.01		Hamene	112.50	3	<0.01
P. auriculata 'robusta'		138.67	2	<0.01		Tenaire	6.40	1	<0.05		Hurepiti	14.10	2	<0.01
P. dentifera imperforata		19.97	3	<0.01		Tiamo	783.61	3	<0.01		Patio	1497.99	2	<0.01
P. garrettii thalia		16.13	1	<0.01		Vaipuarii	410.31	1	<0.01		Tapuamu	751.13	2	<0.01

(Lee *et al.* 2007b), with *P. producta* and other species. Data for *P. affinis* are limited to six valleys with plain embryos under represented in three. These also had a lower frequency of banding in adults than elsewhere (14% [n=1,097], compared to 24% [n=993]), supporting the idea banding was selected against in this area.

The selection pressures that led to the evolution of the banding polymorphism have been speculated to include predation (Johnson *et al.* 1993b) although shell banding and colour have also been noted to be correlated with other factors in other snails; shell strength (Rosin *et al.* 2013), thermal properties (Heath 1975), neural activity (Ermentrout & Campbell 1986) and resistance to parasites (Scheil *et al.* 2013). For predation to be the driver behind the evolution of colour polymorphism in *Partula*, the predator must be visual. No significant visual predators have been identified although extinct *Aplonis* starlings and native kingfishers have been hypothesised as predators (Gerlach 2014) as well as warblers and introduced chickens and rats (Johnson *et al.* 1993b). If any of these were the natural predators then a correlation would be expected between predator distribution and banding. However, this is not apparent; although the colour patterns are found in islands which would have been occupied by such predators, most partulid populations are unbanded and not polymorphic, despite the presence of the similar predators. Similarly, there is no apparent reason why thermal selection pressures should be much greater in the Society and Marianas islands than elsewhere. Whether parasitism has a role is completely unknown, and seems unlikely to have such a strong geographical component. Thermal or parasitic selection pressures may lead to the evolution of banding patterns or specific colours, but would not be expected to lead to the evolution of conspicuous polymorphism.

Even within the Society islands there is variation in polymorphism. There is a clear distinction between genera and ecotypes. Although Society islands *Partula* are highly polymorphic, *Samoana* are not. This is typical of the genus, with the exception of the Marquesas where *S. ganymedes* is polymorphic. In Society island *Partula* arboreal species are more polymorphic than terrestrial ones, although this is not the case for the canopy-dwelling species which resemble *Samoana* in shell thickness and colour. This supports the predation hypothesis, where arboreal species would be more detectable to predators than those living in the shaded litter, where a dark cryptic patterning might be most adaptive. Although many of the coloured morphs are conspicuous they may survive if polymorphism confers an apostatic selection advantage, with a visual predator feeding preferentially on the commonest morphs. This would also suggest that where different species' ranges overlap (resulting in a higher level of polymorphism) only polymorphic species would be able to be abundant, with monomorphic species surviving only when rare, as they have no apostatic advantage. Monomorphic species would thus only be common in the absence of the predatory pressure. Conversely, the more polymorphic species that coexist, the stronger the apostatic advantage. So selection will push towards sites being occupied by numerous polymorphic or few monomorphic species. However, many monomorphic species may coexist but with fluctuating populations. This would favour the evolution of polymorphism in those species.

The positive correlation of phenotypic and species diversity in partulids is not a general pattern in molluscs. In Galapagos *Bulimulus* tree snails phenotypic variation is negatively correlated with the number of sympatric congeners, suggesting restriction due to competition, whereas environmental variation is associated with greater phenotypic variability (Parent & Crespi 2006). The greatest species diversity is found on the older islands with more plant diversity (Parent & Crespi 2006, 2009). Similarly, there is no clear pattern in *Achatinella* species (M. Hadfield, pers. comm.).

Colour variation is not evenly distributed across the Society islands (Fig. 11.21); Moorea and Tahaa have exceptionally high variability, as do parts of Huahine. On the other islands moderate levels of variability are found on Bora Bora but only on parts of Raiatea and Tahiti. There are no obvious explanations for these observations and the diversity of

Fig. 11.21 Distribution of colour variation in Society island *Partula* shown as number of colour forms per species. Points mark populations with signficant differences between adult and embryonic colour frequencies

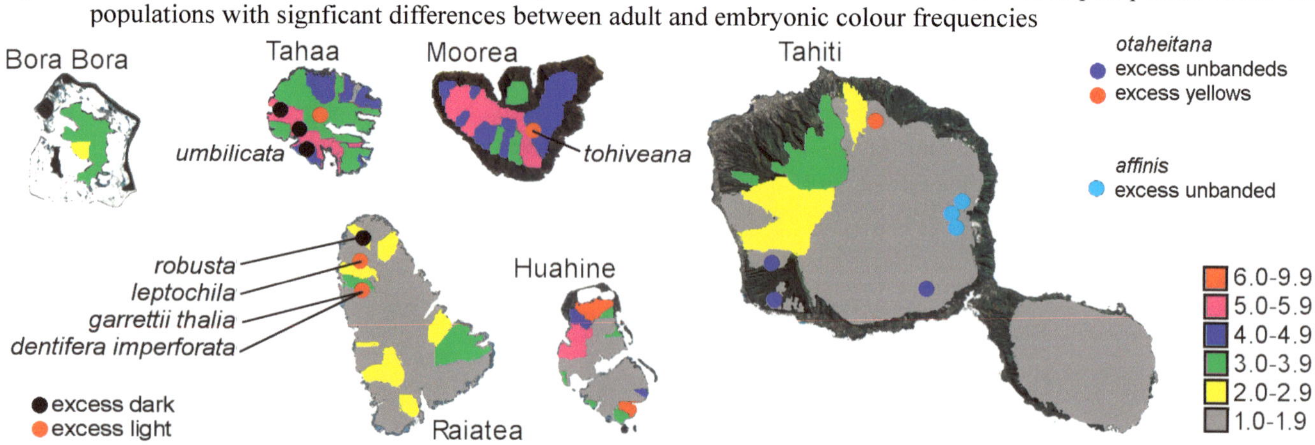

colour forms on Huahine is of particular interest as there the two most variable of all partulids were found in sympatry. This would suggest that selection pressures for polymorphism were most intense on Huahine, Moorea and Tahaa. The highest levels of polymorphism show some association with relative rainfall levels: the valleys with the highest levels of polymorphism on Tahiti are the driest, the west of Moorea is drier than the east, the north-west of Huahine has large areas of dry scrub and Tahaa, as a low island, is relatively dry. The reasons for this association are obscure; possibly the natural habitats of the drier valleys are more open, facilitating foraging by visual predators. This cannot now be confirmed due to the invaded and degraded nature of the present-day forests.

Selection favouring expanding polymorphism may be countered by certain costs, such as local adaptation of specific forms (and consequent selection against others), the polymorphism thus contributing to genetic load. A cost to polymorphism would be expected to lead to its reduction when its advantages are reduced. For example, if *Aplonis* starlings were the native predator, then their extinction would have relaxed the pressure on the snails, leading to loss of polymorphism. The colour forms might be maintained for a considerable period due to their inheritance, although drift could lead to fixation of one of the alleles. This might be seen as a mosaic of monomorphic (homozygotic) populations with polymorphic zones between them. Alternatively, polymorphic populations may be at a high risk of extinction due to correlated sexual, behavioural, life-history and ecological traits, with change leading to conflicts at genetic and population levels (Bolton *et al.* 2015). Loss of polymorphism in most remaining wild and captive populations may be attribute to genetic drift in these relict or inbred populations.

The almost complete lack of polymorphism in *Samoana* may be explicable if all its radiations are either relatively old (the western islands) or canopy dwellers (Societies). The only polymorphic species is *S. ganymedes* (and to a lesser extent its sibling species *S. inflata*). In this species the polymorphism is limited but conspicuous, with three distinct morphs: unbanded, cestata and the fasciata form which has the sutural third of each whorl paler than the base. There are no data on the distribution or habitat preferences of these forms. In *Eua* the radiation of Samoan species is recent, but these are almost entirely allopatric single island taxa and a lower rate of radiation is generally apparent. Polymorphism is only notable in *E. zebrina.*

Crampton (1908) followed Gulick in thinking that random changes explained pattern, what Wright (1955) later called drift. As summarised by Johnson *et al.* (1993b) the subsequent work on Moorea indicated otherwise, finding marked local differences in frequencies. They discussed this in the context of 'area effects' as in *Cepaea nemoralis* which may be due to past vegetation history creating isolated populations in refuges. Local divergence could be due to predation, microclimates creating different selection pressures, or drift. They noted the presence of steep clines which may be due to recent contact between isolated populations, Clarke (1966) thought this could be due the coadaptive processes of local adaptation, without the need for isolation, giving rise to parapatric speciation. Strong clines were recorded over a 200 m transect in relatively homogenous habitat in Faatoai where frequencies of dark shells varied from 0-55% and bandeds from 50-0%. This area had the highest frequency of banding of anywhere in the northwest of Moorea. They proposed that "the increase in banding is due to the special conditions of the 'genetic environment' produced by hybridization between two coadapted gene complexes." They also noted that the banding was unusual, with intermediates between unbanded and banded, "suggesting some breakdown in the control of gene expression." There was also an abrupt increase in length and decrease in width at the point where the frequency of dark shells dropped. There was a change from 1.1 eggs and young per snail to 3.0 "These observations support the view that the different proportions of shell colors are associated with different coadapted gene complexes." In another transect in the same area there was no such sharp transition. In Urufara valley the clinal colour change was the opposite of the Faatoai pattern, but with no clear pattern to banding or egg numbers. These differences indicate that there is no general patternvof adaptation, just local ones.

Clarke (1968) and Murray (1972) argued that as all Moorean *Partula* show area effects and the locations of these do not coincide allopatric divergence would require hundreds or thousands of refuges, so it is more likely that the effects arose parapatrically. Johnson *et al.* (1993b) note that lack of information on ancient human impacts means that neither allopatric nor parapatric mechanisms can be ruled out. "In the shell color and banding polymorphisms of *Partula*, there are regional variations in the frequencies of morphs that may be due to selection by external physical and biotic factors, and influenced by the internal genetic environment, but the strongest determinant of morph frequencies in a species seems to be the array of morphs in its sympatric congeners." This latter point is particularly apparent in the frequency of the frenata morph in *P. taeniata* which is positively correlated to the number of sympatric *Partula* species (Johnson *et al.* 1993b) (Fig. 11.22). A possible cause may be found in other correlations; *P. taeniata* become more slender in sympatry (Johnson *et al.* 1993b) which has been suggested to arise from *P. taeniata* exclusion from *Freycinetia* causing wider phenotypic changes. Exclusion from the enclosed axils of *Freycinetia* may expose *P. taeniata* to predation risks, and if banding offers some form of protection from predation, through crypsis or apostatic selection, then selection would be expected to lead to an increase in banding frequency and diversity.

Where banding is common in *P. taeniata*, the frequencies of the different morphs are rarely as expected simply from their inheritance patterns; frenata is generally more common than zonata but this is reversed in the inner valleys, which also lack lyra (as do the south and outer west valleys), whereas lyra is the only banding form in the north-east. This is particularly interesting as lyra is the heterozygote of the frenata-zonata system and so all three should be present. For the different subspecies Crampton (1932) gave the ratio of frenata:zonata:lyra as: *P. t. elongata*: 1:1.4:0.1; *P. t. simulans* 1:0.3:0.1; *P. t. striolata* 1:3:102.2; *P. t. nucleola* 1:1.5:2.6; *propinqua* 0:1.0:3.7; *spadicea* 1:10.2:13.0. This supports the idea that the banding patterns are not merely maintained by drift, but are selected differently in each area.

Crampton (1932) combined the *P. taeniata* embro data into regions and subspecies, rather than analysing them by individual valleys. In these banding patterns (all patterns combined) are under-represented in all populations. Crampton (1932) noted the difficulty in determining the colour pattern in embryos of *P. taeniata*, observing that many pale or striped adult shells had darker apices, and so would have been placed into one of the darker categories as embryos. This may account for the under-representation of pale and striped embryos.

Carnivory

Four species have been recorded eating other snails: *Eua zebrina, Partula clara, P. incrassa* and *P. hyalina*. It is notable that the three *Partula* species are all sibling species, thus partial carnivory has evolved only twice in the Partulidae. Snail eating may evolve to provide a high-protein diet or increase access to calcium through digestion of their prey's shells. Although the three *Partula* species have thin shells for their genus, there are thinner shelled non-carnivorous partulids.

Egg-shells

Egg-shells are uncalcified in *Eua globosa, Partula clara, P. hyalina* and some *P. taeniata* populations. Murray & Clarke (1988) noted that *P. taeniata* eggs were uncalcified in populations sympatric with *P. incrassa* and considered this a case of character displacement. Hypothetically the case the underlying selection pressures could arise from *P. taeniata* being excluded from some unidentified resource. The similarity between the list of semi-carnivorous partulids and those with uncalcified eggs may suggest a calcium limitation factor underlies both, although the correspondence is not exact. If a mutation resulted in a reduction in calcium metabolism in the eastern Moorean populations, this could have driven a selection for shell consumption (*P. incrassa*) or a reduction in calcium deposition around the egg (*P. taeniata*). *P. clara* and *P. hyalina* being descendants of the *P. taeniata* lineage may simply have inherited the reduced egg calcification if derived from this population. However, this does not explain why *E. globosa* is the only other taxon to have lost calcification, this species having a robust shell, unlike the other three species.

Changes in distribution

Crampton (1916, 1932) considered changes in the distribution between Garrett's collections and his own to be be particularly notable. For example, Garrett had found *P. clara* only in the upper parts of south-west Tahiti whereas Crampton collected it everywhere except in the north and east. More notably *P. nodosa* had been found in only one valley by Garrett whereas Crampton recorded it in seven and thought this difference to be due to a mixture of limited exploration in the west and to range expansion (Crampton 1916). Garrett collected it in Orofere valley in the 1860s but Crampton did not find it there in 1907. In 1908 he "found a few individual *in exactly the same locality which was visited earlier* [his italics]." He thought its rarity there was due to recent colonisation, but conversely attributed its rarity in Atehi to decline. In Punaruu it seemed to

Fig. 11.22 Selection pressures on *P. taeniata*. Left – degree of *Freycinetia* specialisation and banding patterns for Moorea (plant data from Murray *et al.* 1993, banding data from Crampton 1932). Points coded by the number of main banding patterns in that species: red – 4, yellow – 3, green – 1. Right – effect of sympatric species diversity on banding patterns and shape (data from Johnson *et al.* 1993b)

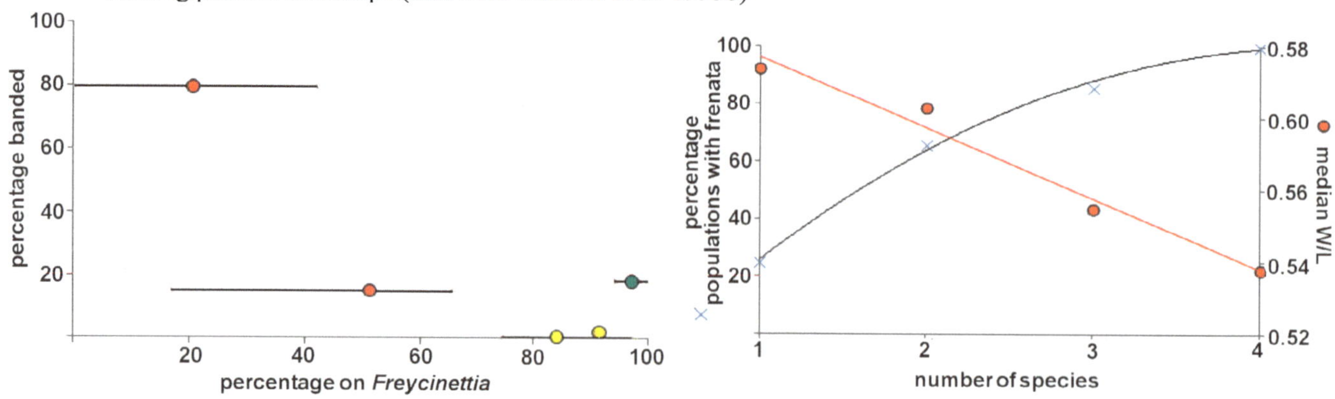

have spread across the river since Garrett's time. Crampton's observations do suggest some change but the significance of his observations are impossible to evaluate.

Of the widespread and abundant species found by Crampton, only *P. mirabilis* had not been collected by Garrett earlier (Fig. 11.23). "From the observations of the present writer during the years from 1907 to 1923, it is certain that the species has spread into this wider territory from a central region that must have been so small in Garrett's time as to escape the scrutiny of that careful explorer and observer…" (Crampton 1924). "Certain obvious features of the general distribution of *mirabilis* make it plain that the species must have spread recently from a smaller area of occupancy. Whereas 5 per cent of the Vaianai series of 1923… were *mirabilis*, none were found in that valley in prior years; it is true that the last year's collections were taken mainly in the higher part of Vaianai, but *mirabilis* was found in the middle level where it had not been seen when the substantial earlier lots were collected. However, the fact that it was more abundant in the higher valley near the pass, beyond which the species exists in abundance, is itself proof that *mirabilis* had only recently migrated across the ridge. The single specimen from Oio, taken in 1919, proves that a tongue of penetration had diverged from the main stream into Vaianai almost at the pass itself, to push farther downward to the west of the ridge which separates Vaianai and Oio.

"The next and most prominent feature of distribution is the abundance of *mirabilis* in Matapoopoo Valley, in the sector dominated by Mount Rotui. Here there is no question that its numbers have rapidly increased recently, for even when all allowances are made for local variations in numerical abundance, the high percentages of 83 and 97 for the eastern and western series of 1923 are in marked contrast with 33 per cent and 24 per cent for 1909 and 1919, respectively. The associates in the way of *P. taeniata* and *P. dendroica* have not markedly diminished in actual numbers in the interval; the differences indicate a prodigious increase of *mirabilis*. Again, the fact that *mirabilis* occurs in the Rotui sector in Matapoopoo Valley proves its late incursion into this territory. The ecological conditions at the contacts of Matapoopoo with Vaihiaiia on the east and Tepatu on the west are such as to check the tides of migration to some extent, it is true, but they can not be regarded as insuperable obstacles when *taeniata* and *dendroica* have overpassed them." (Crampton 1932). Murray & Clarke (1980) disagreed with these interpretations, attributing all apparent changes in distribution to differences in sampling areas, with Crampton collecting at lower altitudes than they did, and often in areas that were not visited by Garrett.

From the available data it appears that partulid evolution is largely a consequence of occasional dispersal. This creates broad distribution patterns of the taxa and the starting point for allopatric speciation by drift. The phenotypic and taxonomic diversity patterns studied by Crampton and later by Clarke, Murray and Johnson arise from the interactions between these lineages and with the environment. Much of the difficulty in interpreting the recent diversity has resulted from the prehistoric modification of the environment, including the probable extinction of native predators and changes in vegetation.

Fig. 11.23 Distribution maps of *Partula mirabilis*: left Crampton's (1924) conjecture of dispersal from point Y, right Murray & Clarke's (1980) distribution. Matapoopoo, Oio, Tepatu, Vaiani and Vaihiaiia marked as abbreviations

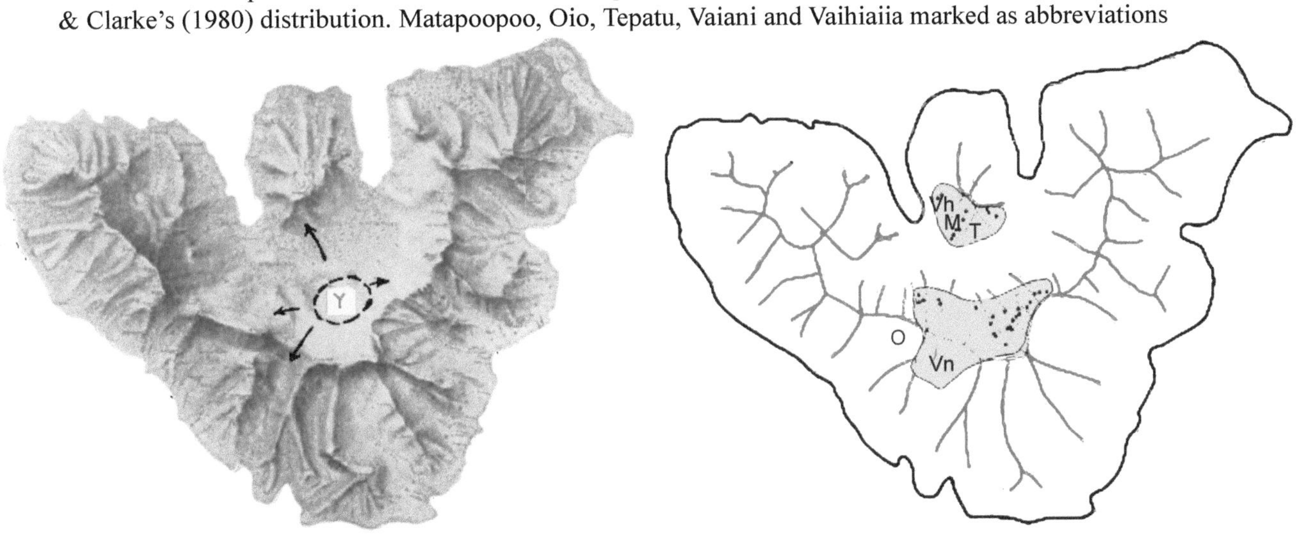

12. Status and conservation of Partulidae

The Partulidae are one of the most threatened families of animal (32-39% of species extinct, 9% extinct in the wild, 50% threatened, n=104) (Table 12.1), compare to an overall extinction rate of 7% for land snails (Régnier *et al.* 2015a). The only family thought to have a higher extinction rate is the Amastridae (at least 31%, possibly 95%, n=282: Régnier *et al.* 2015b). The high levels of partulid extinction result from habitat loss and modification through clearance and plant invasions, and the impacts of introduced predators (principally the carnivorous snail *Euglandina rosea* and, most recently, the predatory flatworm *Platydemus manokwari*). Here the status of the wild populations is reviewed through evaluating the threats present on each of the islands or island groups where partulids have been recorded. This is followed by a synthesis of the threats facing the family as a whole.

Melanesia

New Guinea – It has been reported that around 80% of Manus island in the Admiralties remains forested (although this includes secondary forest) (Rannells 1995). Many other islands were cleared for coconut plantations in the past. Currently deforestation is occurring due to timber extraction and shifting agriculture (Stattersfield *et a*l. 1998). The partulids known from these islands are adaptable species found in secondary habitats and appear not to be threatened.

Solomons – There has been extensive deforestation in past 30 years through logging and copra and oil plantations, and subsistence agriculture. The area under cultivation doubled between 1972 and 1992 (Thistlethwait & Votaw 1992). As with the New Guinea islands, many of the Solomons partulids are adapted to secondary habitats and appear not to be threatened, although little is known of the larger species *P. cramptoni.* Rennel island retains 90% of its primary forest but logging operations started in 2008. Clearance of forest and opening up of roads has led to an increase in invasive plants (Dingwall 2013).

Platydemus was first recorded from Guadalcanal in 2014, although some genetic diversity in the populations suggests it may be an ancient population (Justine *et al.* 2015).

Vanuatu – Around 70% of Vanuatu remains vegetated and half of this is forest cover. Significant native forest remains on Espiritu Santo, Malekula, Erromango and Efate (Bakeo & Qarani 2005). Much of this is heavily invaded by introduced plant species. *Euglandina* was deliberately introduced in 1982 or 1983 (Gargominy *et al.* 2011), it is known to be present on Espiritu Santo, Aorea and Ambrym (Bowden 2000). *Platydemus* was recorded in Vanuatu in 2002 (FAO-SAPA 2002) but at present there is no information on its distribution in the islands. Partulids appear to be very rare in Vanuatu but there are few historical records for comparison. In the main islands, *P. auraniana* was present on Aore island in the past but could not be found in 2006. The only partulid to be found in 2006 was a single *P. pyramis* shell on Espiritu Santo (Fontaine *et al.* 2011). *P. auraniana* was still present in the Torres islands of Hiw, Metoma, Tegua, Loh and Toga in 2007 (Ó Foighil 2012c).

Fiji – There is evidence of habitat modification going back 2,500 years in Fiji (Gibbons 1985). Most forest has been converted to plantations (trees and sugar cane) and grazing land, although 40% of Fiji's land retains tropical moist forest cover (Kirkpatrick & Hassall 1985). Most systems are heavily modified and primary forest is restricted to small patches in the more inaccessible areas (Ash 1992). Partulids have only been recorded from the smaller islands which retain a relatively large proportion of natural forest (Brodie *et al.* 2016). There is no evidence that partulids ever occurred on the larger islands.

On Rotuma island deforestation has been extensive, but a large proportion of the island retains secondary forest cover (Stattersfield *et al.* 1998). When *P. leefei* was first collected in 1897 J.S. Gardiner (in Smith 1897) stated "There is no indigenous forest left anywhere on the island". *P. leefei* was tolerant of the degraded habitat, being collected in large numbers in 1938 (525 specimens in BPBM – Brodie *et al.* 2014). The species is currently scarce, with just 5 shells located in 2012 (Brodie *et al.* 2014, 2016) and a live animal found in 2016 (G. Brodie pers. comm.). *Euglandina* has not been introduced to Fiji but *Platydemus* was found at three sites in the west of Rotuma in 2012 (Brodie *et al.* 2014, 2016). It is probably too localised to have caused a decline in *P. leefei.* It is also present on the main Fijian island of Viti Levu (Brodie *et al.* 2014).

Micronesia

Palau – Around 25% of the original forest cover on Palau has been converted to grassland or secondary vegetation (Cole *et al.* 1987). However, the pressures on the remaining forest are currently not severe outside of Koror and Babeldaob islands due to a relatively low human population density. *Euglandina* has does not appear to have become established in Palau but *Platydemus* has been present on Koror since 1991 and Ulong since 1992. Of the *Palaopartula* species, *P. leucothoe* has always been considered rare (Kondo 1955) and was not located by in the known localities on Peleliu and Angaur, in 2003 (Rundell

Table 12.1 Conservation status of Partulidae. EX – extinct, EW – extinct in the wild, CR – Critically Endangered, EN – Endangered, VU – Vulnerable, LC – Least Concern. *Er – Euglandina; Pm – Platydemus manokwari*

	status	threats
Eua expansa	EN	*Pm*, habitat
Eua globosa	CR (EX?)	habitat
Eua montana	EN	*Pm*, habitat
Eua zebrina	EN	*Er*, *Pm*?, habitat
Eua sp.	EN	*Pm*
Sphendone insolita	VU	habitat
Palaopartula calypso	CR	*Pm*, habitat
Palaopartula leucothoe	CR (EX?)	habitat
Palaopartula thetis	EN	*Pm*, habitat
Partula affinis	CR	*Er*
Partula arguta	EX	*Er*
Partula assimilis	VU	habitat
Partula atilis	EX	*Er*
Partula auraniana	EN	*Er*
Partula aurantia	EX	*Er*
Partula auriculata	EX	*Er*
Partula bilineata	EX	*Er*
Partula clara	CR	*Er*
Partula clarkei	EX	*Er*
Partula compressa	CR	*Er*
Partula cootei	EX	habitat
Partula cramptoni	VU	habitat
Partula crassilabris	EX	*Er*
Partula cuneata	EX	*Er*
Partula cytherea	CR (EX?)	*Er*
Partula dentifera	EX	*Er*
Partula desolata	prehistoric EX	-
Partula diminuta	EX	*Er*
Partula dolichostoma	EX	*Er*
Partula dolorosa	EX	*Er*
Partula emersoni	CR	*Pm*
Partula eremita	EX	*Er*
Partula faba	EX	*Er*
Partula flexuosa	LC	habitat
Partula garrettii	EW	*Er*
Partula gibba	CR	*Pm*
Partula grisea	LC	habitat
Partula guamensis	EX	*Pm*
Partula hebe	EW	*Er*
Partula hyalina	VU	*Er*, *Pm*
Partula incrassa	E	*Er*, *Pm*, habitat
Partula jackieburchi	EX	*Er*
Partula labrusca	EX	*Er*
Partula laevigata	CR	*Er*
Partula lanceolata	VU	habitat
Partula langfordi	EX	*Pm*, habitat
Partula leefei	CR	*Pm*, habitat
Partula leptochila	EX	*Er*
Partula levistriata	EX	*Er*
Partula lirata	VU	habitat
Partula lugubris	EX	*Er*
Partula lutea	EX	*Er*
Partula sp. (Rota)	CR	*Pm*, habitat
Partula magistri	EX	*Er*
Partula makatea	prehistoric EX	-
Partula meyeri	CR	*Er*, climate
Partula micans	LC	-
Partula mirabilis	EW	*Er*
Partula mooreana	EW	*Er*
Partula navigatoria	EW	*Er*
Partula nodosa	CR	*Er*, *Pm*
Partula obesa	CR	*Er*, *Pm*, habitat
Partula otaheitana	CR	*Er*, *Pm*
Partula pacifica	CR	*Er*
Partula pearcekellyi	EX	*Er*
Partula planilabrum	EX	*Er*
Partula producta	EX	*Er*
Partula protracta	EX	*Er*
Partula pyramis	CR	*Er*
Partula radiolata	CR	*Pm*, habitat
Partula radiosa	VU	*Er*
Partula remota	EX	*Er*
Partula rosea	EW	*Er*
Partula rufa	CR (EX?)	habitat
Partula sagitta	EX	*Er*
Partula salifana	EX	*Er*
Partula similaris	LC	habitat
Partula suturalis	EW	*Er*
Partula taeniata	CR	*Er*, *Pm*
Partula tohiveana	EW	*Er*
Partula tristis	EX	*Er*
Partula turgida	EX	*Er*
Partula umbilicata	EX	*Er*
Partula vanikorensis	VU	habitat?
Partula varia	EW	*Er*
Samoana abbreviata	CR	*Er*, *Pm*?, habitat
Samoana alabastrina	VU	habitat
Samoana annectens	CR	*Er*
Samoana attenuata	CR	*Er*, *Pm*
Samoana bellula	CR	climate
Samoana burchi	CR	*Er*, *Pm*
Samoana conica	EN	*Er*, *Pm*, habitat
Samoana cramptoni	CR(EX?)	habitat
Samoana decussatula	CR	*Er*, habitat
Samoana diaphana	EN	*Er*, *Pm*
Samoana dryas	EN	habitat
Samoana fragilis	CR	*Pm*
Samoana ganymedes	CR	*Er*, habitat
Samoana gonochila	CR	*Er*, habitat, climate
Samoana inflata	CR	habitat
Samoana margaritae	VU	habitat
Samoana medana	CR	*Er*, habitat
Samoana minuta	CR (EX?)	*Er*
Samoana pilsbryi	CR (EX?)	*Er*
Samoana stevensoniana	CR	*Pm*, habitat
Samoana thurstoni	EN	habitat

2005, 2010). One dead shell was found on Klaltebechel in 2007 (Ó Foighil & Rundell 2012c). On Babeldaob and Ngeruktabel Rundell (2005) found a *Palaopartula* (suggested to be *calypso*) to be 'thriving' but highly localised. The Palau species have so far not suffered significant declines, but may be at risk from future habitat loss or the spread of *Platydemus*.

Palaopartula calypso and *P. leucothoe* have always been rare species but *P. thetis* has declined since the 1930s and now occurs only as isolated individuals in small forest patches on six islets. Habitat deterioration and rat predation are suspected to be the main causes of decline. *Sphendone insolita* is currently relatively abundant (Slapcinsky & Krauss 2016).

Marianas – The Marianas islands were settled by humans around 6,000 years ago. Partulids have been heavily exploited in the islands for a long time, being used for food and for their shells. Despite exploitation they remained extremely abundant until recently. Habitats on the Marianas have been extensively modified since the first human settlement but forest cover remained extensive on most islands until the 19th and 20th centuries. Crampton (1925) recorded partulids in almost all forested areas and *Partula gibba* and *P. radiolata* had been able to adapt to use invasive plants. *P. salifana* and *Samoana fragilis* were more restricted and may have been less adaptable. Declines in partulid abundance were first observed in 1989 (Hopper & Smith 1992).

Macrogonaxis kibwezensis and *M. quadrilateralis* were introduced to Guam in 1954 and 1967 respectively, and the former to Rota. *Euglandina* was introduced to Guam in 1957. It is not present on Rota. *Platydemus* appeared on Guam in 1978 (Hopper & Smith 1992) and was deliberately introduced to Saipan in 1981. It arrived on Tinian in 1984, Rota in 1988 and Aguiguan in 1992 (Bauman 1996).

On Guam *Partula gibba* was restricted to few small isolated populations, although these seemed healthy in 2005 (57-60% immature) with high population densities (4.7-17.2m^{-2} in 1993-5) (Smith *et al.* undated). Population densities in one of these sites declined to 2.2 m^{-2} in 2007, which may be a result of changes in forest structure (Smith *et al.* undated). Currently only one surviving population is known. *P. radiolata* also has some viable populations (70% immature) (Smith *et al.* undated) and is now the most abundant partulid on Guam, present in all sites with surviving partulids (Hopper & Smith 1992; Smith 1995). There was some indication of a decline in abundance in 2008 and the largest site was lost to development and habitat damage caused by typhoons in 1990-2, and one was destroyed by fire in 1991-2 (Smith & Hopper 1994; Centre for Biological Diversity, undated). *S. fragilis* was present in six sites in 1989 (Hopper & Smith 1992), all of which had disappeared by 1996. A single small colony was found in 2007 (Smith *et al.* undated) and the population estimated at 300 individuals (Centre for Biological Diversity, undated). *P. salifana* was last recorded in 1946 (USNM 606976). Dead shells were found in the historic range of the species in 1989 (Hopper & Smith 1992). Kerr (2013) thought that *Platydemus* had eliminated *Macrogonaxis* and *Euglandina*, as none were found in 3 years of surveys, and none were found in 1992 (Hopper & Smith 1992).

On Aguiguan *Partula gibba* is extinct probably following extensive deforestation from the 1940s onwards (Hadfield 2015). *P. langfordi* was found in five sites in 1985, only one in 1992 and in 1995 only shells could be found (some fresh). Since then only old shells have been found (Centre for Biological Diversity, undated; Kerr 2013). On Tinian *P. gibba* is still present in a largely inaccessible areas where small patches of good habitat remains (B. Holland pers. comm.). On Rota *Partula* were abundant and widespread in the past (as indicated by museum specimens) up until sometime after 1969 (Bauman 1996). Extensive habitat remains but recently *Partula* has only been recorded in five sites, suggesting that it has declined, although it was still present in 2010 (Hadfield 2010). *S. fragilis* was present at Sabana in 1959 but the area was lost to agriculture in 1995 and the snails could not be found (Bauman 1996); one population was found in 1996 but again lost, and a further rediscovery was made in 2015 (A. Kerr pers. omm.). On Saipan in 2010 *P. gibba* was locally abundant in a lowland site (Hadfield 2015). On Sarigan feral ungulates were removed by 2000, recovery of vegetation also saw a recovery of the *P. gibba* population (Smith 2008). The species was present in great abundance in 2010 (Hadfield 2015). On Pagan large numbers of feral cattle degraded the woodland in the north of the island by eating much of the undergrowth. In areas where cattle had increased in abundance since 1949 *P. gibba* had disappeared by 2010 (Hadfield 2015). In the south of the island *P. gibba* could still be located and in this area cattle were absent, possibly excluded by the terrain and lack of fresh water. Goats and pigs were present in all areas but did not affect the understory to the same extent as the cattle (Hadfield 2015).

These different islands suggest that relict populations of *S. fragilis, P. gibba* and *P. radiolata* may survive in the presence of introduced predators, although the situation for *P. gibba* on Guam is critical. The data also suggest that *Platydemus* may eliminate *Euglandina*. Although at least two Marianas partulid species are able to adapt to invaded habitats, they are not able to tolerate extensive removal of undergrowth.

Carolines – The Caroline islands have been modified by human occupation for at least 2,500 years. Most lowland forest has been converted to grassland or degraded. Remaining forests are heavily invaded by introduced plants. On Kosrae and Pohnpei montane and cloud forest areas remain (Stattersfield *et al.* 1998). Upland forest is being converted to agriculture and invaded by a variety of introduced plants. *Euglandina* was introduced to Pohnpei in 1960 but appears not to have become established

(Schreiner 1989). *Platydemus* has been present on Pohnpei since 1993.

On Pohnpei *P. guamensis* and *P. emersoni* were relatively abundant in 1936 when Kondo collected 1400 and 500 respectively. Surveys in 2005-6 and 2008 failed to find any *P. emersoni* and only one shell of *P. guamensis* was recorded on the main island and three shells on Temwen islet. No *Euglandina* were found but *Platydemus* was abundant in many sites (Pelep & Hadfield 2011). Since then one *P. emersoni* has been found, in 2011 (B. Holland pers. comm.). The lack of empty shells suggests that the loss of the main populations must have occurred considerably earlier. *P. guamensis* was a large, terrestrial species and may have been particularly vulnerable to *Platydemus* and the abundant Polynesian and black rats on the island.

On Kosrae *P. rufa* was localised in 1936 when Kondo collected several specimens. The next survey, in 2005, failed to locate any evidence of partulids (Rundell & Czekanski-Moir 2015). *Euglandina* and *Platydemus* remain absent from the island and the apparent extinction of *P. rufa* can probably be attributed to habitat loss, almost all lowland forest being cleared or modified. Extensive areas of upland forest remain but all partulid records were from lowlands (Kondo field notes, BPBM).

Fig. 12.1 Changes in relative abundance of native partulids in the past 100 years. Arrows mark predator introduction: yellow – *Macrogonaxis*, red – *Euglandina*, black – *Platydemus*. X – extinction

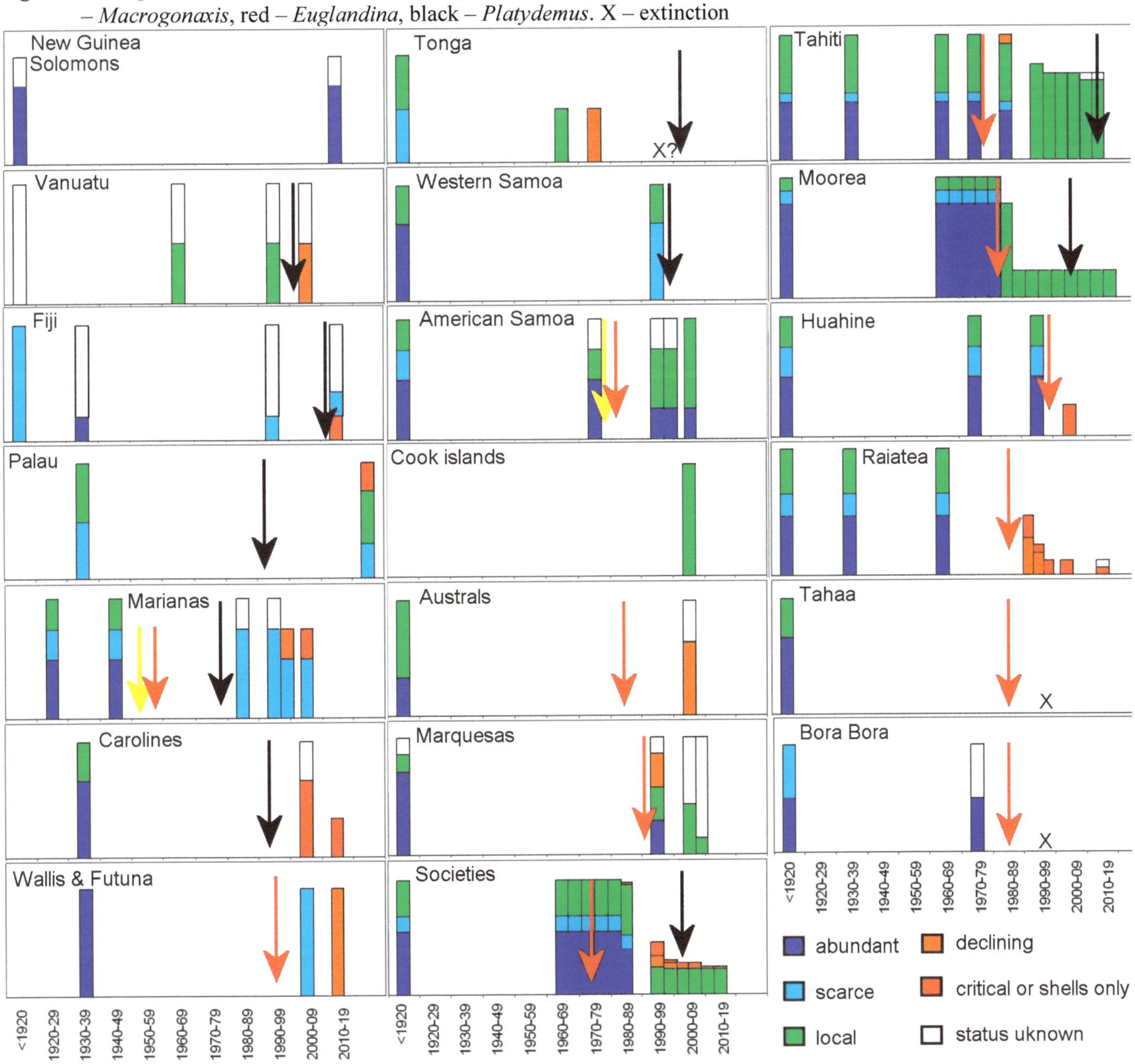

Polynesia

Tonga – Human population growth on Tonga is rapid, resulting in increasing forest loss (Rinke 1986b). Partulids are only known from 'Eua island, much of which has been deforested (Drake *et al.* 1996) although some areas of good habitat remain.

Eua globosa was collected in 1932 in one site. In 1967 it was found in a second site but in 1970 that area was cleared by bulldozer. At that time 60 snails were found. In 2000 only empty shells could be found, although not all areas were surveyed (Ó Foighil 2012a). *Samoana cramptoni* is known from a single locality in 1933. Searches for it in 1967 and 1970 failed to located any snails. In 1967 Kondo suspected that it was extinct: "No *Samoana cramptoni* despite much effort… [it] may have disappeared altogether" (Kondo field notes, BPBM). These species declined long before *Platydemus* appeared in Tonga (in 2002) and their decline can be attributed to habitat loss.

Western Samoa – The Samoan islands have been occupied by humans for around 3,000 years and only 20% of the original lowland rain forest remains as a result of clearance for timber and agriculture. Most people live on 'Upolu and all accessible forest has been lost from that island. Savai'i retains areas of high forest as it is one of the highest islands in the Pacific, but this is at risk from logging and clearance for plantations (Merlin & Juvik 1983). *Lissachatina* was introduced to 'Upolu in about 1990 and to Savai'i in 1996 raising the possibility that predators may be introduced as happened when *Platydemus* was introduced to Savai'i by the Samoan Ministry of Agriculture and fisheries (1996-8, and again in 2001 – Brook 2012). The flatworm was first recorded on 'Upolu in 1998 (Purea *et al.* 1998).

On Savi'i *Eua expansa* and *Samoana conica* were widespread but uncommon in early records, they remain widespread but *S. conica* appears to have declined by 1994 (Cowie & Robinson 2003). In 2012 *E. expansa* and *S. stevensoniana* were found in one site, at 880 m above Asau (Brook 2012), *S. conica* had a wider but patchy distribution. *Eua* sp. is known only from one locality.

On 'Upolu *E. expansa* and *S. conica* were widespread and uncommon in the past but are both now very scarce (Cowie 1993). *E. montana* was always highly localised. The partulids declined before the arrival of *Platydemus* on 'Upolu, probably due to habitat degradation.

American Samoa – As with Western Samoa, these islands have a long settlement history and have lost almost all the original forest cover. *Euglandina* was introduced to Tutuila in 1980, three years after the arrival of *Lissachatina fulica* (Cowie 19988b), and arrived on Ta'ū by 1992 (Cowie *et al.* 2002). It was well established on Tutuila by 1984. *Platydemus* may also be present on Tutuila. *Macrogonaxis kibwezensis* had been introduced earlier, as early as 1977 on Tutuila.

In 1975 *Eua zebrina* and *Samoana conica* were widespread and abundant on Tutuila, but *S. abbreviata* was not located and was suggested to be extinct (Solem 1975, unpublished report quoted in Cowie & Cook 2001). In 1992 (12 years after the introduction of *Euglandina*) the only partulids to be located were on Tutuila were *S. conica* in one site, *S. abbreviata* in one site and *E. zebrina* in two sites (Cowie & Cook 1999). All three species were scarce, with only a single *S. abbreviata* found. In 1998 more extensive surveys were carried out (Cowie & Cook 2001). On Tutuila 15 live *S. abbreviata* were found in a single site, 288 *S. conica* and 1,102 *E. zebrina*, indicating that the latter does retain some significant populations despite the general decline. For all three there were several sites where only empty shells could be found (Cowie & Cook 2001). *S. abbreviata* was still present in 2001. In 1992 Smith (1993) recorded a population of *E. zebrina* on Nu'usetoga island, off Tutuilia. Many rat-eaten *E. zebrina* shells were noted, but no *Euglandina* were present.

S. thurstoni appears always to have been a scarce species (Cooke & Crampton 1930) despite Solem's 1975 description of it being present 'in fair numbers' on Ofu in 1975 (Solem 1975, unpublished report quoted in Cowie & Cook 2001). On that island four *S. thurstoni* were found in undisturbed upland habitat, and 88 *E. zebrina* in 1988 (Cowie & Robinson 2003). *S. thurstoni* was still present on Olosega in 2001. The decline in the partulids on Tutuila occurred after the introduction of *Euglandina*, but the predator has not caused complete extinction of any of the species so far. Although partulids are rare on Ofu there is no clear evidence of a decline.

Wallis & Futuna – Wallis and Futuna have been occupied by humans for at least 3,000 years. Over this period most forests have been cleared, with 30% remaining on Futuna, mostly above 400 m. In contrast, Alofi is one of the least degraded high islands in the Pacific (Mueller-Dombois & Fosberg 1998), and remains uninhabited, although some land is cultivated by residents of Futuna. Black rats have only recently reached Futuna and are restricted to human habitations, they are still absent from Alofi. *Euglandina* was introduced to Futuna in 1993 and is also present on Alofi (Meyer *et al* 2008; Gargominy *et al.* 2011). Small populations of *Partula obesa* remained on Futuna in 2008, and more extensive ones on Alofi, with abundant empty shells in some areas indicating significant population declines. In 2011 small, scattered populations were found in several areas of Alofi (Ó Foighil 2012b). It is probable that partulids in Wallis and Futuna are declining as a result of predation by *Euglandina*.

Platydemus was first recorded from Futuna in 2008 (Justine *et al.* 2015).

Cook islands – Human occupation dates back 700-900 years, although palaeontological data may suggest human arrival 2,500 years ago (reviewed in Brook 2010). Development and clearance for agriculture were extensive in the 19th century, although upland forest areas have survived. Lowland forests have largely been turned into coconut plantations. Native forest remains on the upper parts of Rarotonga, including some of the best remaining primary montane rain and cloud forests in the eastern Pacific. *Partula assimilis* apparently declined as a result of habitat loss in the 1800s; Garrett (1881) noted that the presence of abundant dead shells in some valleys indicated that these populations had died out. At that time it was still locally abundant inland. This remained the case in 2005-7 (Brook 2010).

Australs – The Austral islands are extensively degraded with lowland habitats mostly having been cleared for agriculture or lost to fire; Rapa is mostly introduced vegetation (64% of the area) with *S. margaritae* surviving in the isolated native and mixed vegetation areas. Remaining forest areas are heavily degraded by invasive plants and the effects of goats and pigs. Forest fragments remains at high altitude on Rapa and Rurutu, and on limestone plateau areas on Rimatara. *Euglandina* was introduced to Tubai in 1985 and is now widespread (Gargominy & Fontaine 2003, 2014).

Partula cootei was spread across the north of Rurutu in 1934 but was recorded from only a single locality in 2005 (as old shells only). It is only know from subfossils on Rimatara. It is probably extinct as a result of habitat degradation and a combination of competition and hybridisation with *P. hyalina*. There is some possible evidence of competition as dietary data (p. 226) suggest that *P. cootei* may have been forced away from a generalist niche to be constrained to detritus feeding.

P. hyalina was found throughout Tubai, Rurutu and Raivavae but in 2003 live snails on Tubai were found only on the littoral plain in the west. On Rurutu and Raivavae they were highly abundant in some areas (Fontiane & Gargominy 2002; Gargominy & Fontiane 2003). It is the most adaptable partulid and the only cause of its decline on Tubai was predation by *Euglandina*.

Since its discovery on Raivavae in 1934 *Samoana dryas* has been restricted to the fragments of high altitude forests. This species appears to be continuing to deline as the 2003 MNHN expedition found live populations in the *dryas* in the type locality, but only old shells in the *oreas* and *hamadryas* localities.

Marquesas – The Marquesas have been settled for 1-2,000 years. Lowland forests have been completely lost to invasive species (Florence & Lorence 1997) or converted to grass in drier areas as a result of frequent fires (Rolett 1993). High-altitude forest remains on Ua Pou, Nuku Hiva, Fatu Hika, Ua Huka, and Fatu Hiva. *Euglandina* was introduced to Hiva Oa and Nuku Hiva in 1992, and Fatu Hiva after 1995, but has not yet been recorded from the other islands.

S. gonochila was always a relatively scarce species on Nuku Hiva. Kondo (MS) commented on the Nuku Hiva population: "Lebronnec's tiny sample of five adults is explainable in this way: The coffee plants intended from commercial usage had escaped and spread like wildfire, killing off the native plants in Tahauku, Atuona, and other valleys. At the height of the coffee explosion, decussatula's [i.e. *S. gonochila*] population dropped to near zero, therefore Lebronnec's sampling was impoverished.

"In 36 years, decussatula adapted to coffee plants and recovered to a small degree. In 1969, when Kondo collected in Tahauku Valley with Lebronnec's grandson Garcienne Frebault, the Tahauku population was able to furnish 33 adults in three trips." (Kondo MS).

By 1995 *Euglandina* had eliminated all but one colony, but *S. gonochila* still survived on the Toovii plateau in 1997 and on Mt. Tekoa in 2004 (Lee *et al.* 2014). The species was better established on Ua Huka where it remained abundant in 1995 but has since become restricted to the highest peaks; it was still present in the Vaikivi reserve in 2006 (Meyer 2006).

On Ua Pou *Samoana bellula* was common in 1995 but appears to have been restricted to Mt. Poumaka and Teavakaakiti by 2003. It ws still present on Mt. Pouteainui in 2005 (Lee *et al.* 2005) although only empty shells were found in 2003. The surviving animals were found on native plant species, but also on *Coffea arabica* (Meyer 2003). No *Euglandina* have been reported. This species has a very erratic collection history; Garrett only collected two in the 1880s but subsequently it was found plentifully.

On Hiva Oa *S. decussatula* has declined in most areas, but in 2007 it still survived in one site. *S. ganymedes* was restricted to the Puamau plateau in the east of the island in 1995. This area appeared to be separated from the range of *Euglandina* by a large area of grazing land. Small numbers were still present on Mt. Temetiu in 2007, and a *Samoana* was present in 2011 (Coote 2009; Coote *et al.* 2015). *S. decussatula* appears to have been rare on Hiva Oa since the early 20th century. A more recent decline may have taken place in *S. ganymedes* with the declining numbers being taken by Kondo (MS) to be an indicator of the declining forests over this time. *S. medana* collections are highly erratic, and seem to have declined considerably, although some records of *S. decussatula* may be this species.

Tahuata remains free of *Euglandina* but partulids are highly restricted; in 2009 J.-Y. Meyer located *S. decussatula* in one site (12 individuals at 400 m in Vaitahu valley, with an isolated individual at 830 m) and found one *S. inflata* shell on Mt.

Haaoiputeomo. Of the main collections, only LeBronnec collected *S. decussatula* and *S. medana* on Tahuata and only 14 and 41 individuals respectively.

On Fatu Hiva *S. minuta* and *S. pilsbryi* were widespread and abundant in the 20th century, *S. medana* was localised and much rarer. 1995 and 2000 surveys found *S. minuta* to be relatively abundant but *Euglandina* was introduced in around 2001; in 2003 no partulids were found (Taputuara & Meyer 2007).

Societies – Human first occupied the Society islands 1-2,000 years ago. Lowland forests have been largely replaced by coconut plantations. In most areas natural forest is restricted to areas above 500 m. There is also extensive invasion by introduced plants, especially *Miconia calvescens* which has been spreading across the islands since 1937. *Lissachatina* was introduced to Tahiti in 1967 and quickly spread to the other islands. *Euglandina* were released on Tahiti in 1974 and Moorea in 1977, with unofficial introduction to other islands in the 1980s and 1990s. *Platydemus* was first recorded on Tahiti in 2006 (T. Coote pers. comm.) and on Moorea in 2009 (Lovenburg 2009).

Partulid declines were already detectable on Tahiti in 1909; Crampton (1916) found that in that year the range of *Partula diminuta* had contracted compared to the previous year due to agricultural clearances reducing habitat at low levels. Other declines are likely to have occurred, but all species remained abundant in 1970 when J. Burch collected extensively. By 1984 *Euglandina* had spread across much of the south of the island and by 1987 it was present in most areas. In that year partulids were found only in Tiarei and Mahaena (*P. otaheitana, P. affinis, P. hyalina* and *P. clara*) (Murray *et al.* 1988). Subsequent surveys (from 1995) found that *Euglandina* had not invaded these populations in Tahiti-Iti until 2001 (Coote 2007; Coote *et al.* 1999). *P. affinis* and *P. hyalina* still survive there in reduced numbers. Elsewhere *P. otaheitana, P. compressa, P. laevigata* and *Samoana* species survived at high altitudes, above the main range of *Euglandina* (Coote *et al.* 1999). Surprisingly, some *P. hyalina, P. clara* and *P. incrassa* populations survived the invasion of *Euglandina* and scarce populations have reappeared in several valleys once *Euglandina* populations declined to low levels.

Moorean *Partula* were abundant until the early 1980s. Their rapid decline following the introduction of *Euglandina* in 1977 had started by 1982 and has been well documented (Clarke *et al.* 1984; Murray *et al.* 1998). The first losses were in the northeast of the island and by 1982 *P. aurantia* was thought to be extinct in the wild (Clarke *et al.* 1984); the only survivors in this area were some *P. incrassa* (Wells 1995). Some *Partula* were found in Haapiti valley in 1986 but in 1987 no partulids could be found on Moorea and it was concluded that all were extinct in the wild (Murray *et al.* 1988). In 1996 a small

Fig. 12.2 *Samoana* population changes based on collection records. a) decline in Marquesas *Samoana* abundance associated with 20th century habitat degradation; b) equivalent data for Society islands, showing lack of pattern. Data are given as proportional abundance (number of specimens divided by maximum number from that collector, to account for differing collection effort). Collections are from 1880s (Garrett), 1906-9 (Crampton), 1929-34 (LeBronnec and Mangareva expedition), 1968-80 (Kondo *et al.* Marquesas; Clarke & Murray and Burch *et al.* Societies), recent (PILSG 1995 Marquesas, Coote Societies).

a: bel. – *S. bellula;* dec. – *S. decussatula;* gan. – *S. ganymendes;* gon. – *S. gonochila;* inf. – *S. inflata;* med. – *S. medana;* min. – *S. minuta;* pil. – *S. pilsbryi.* FH – Fatu Hiva; HO – Hiva Oa; NH – Nuku Hiva; Ta – Tahatua; UH – Ua Huka; UP – Ua Pou

b: ann. – *S. annectens;* att. – *S. attenuata;* bur. – *S. burchi;* dia. – *S. diaphana*

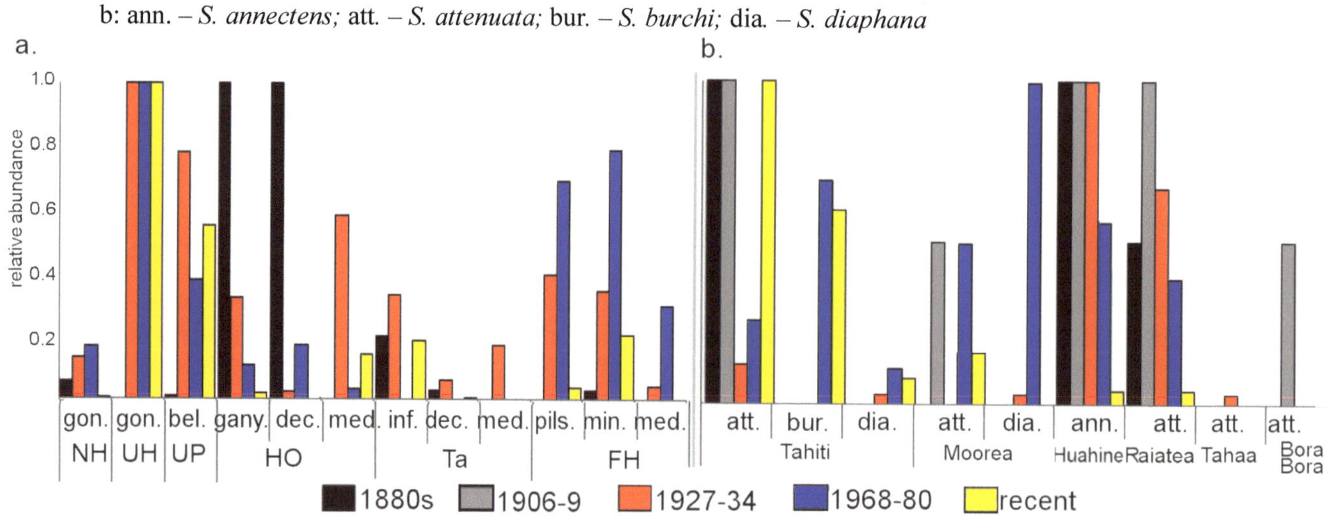

population of *S. attenuata* was found and several of *P. taeniata* in 2000 (Coote 1999; Coote & Loève 2003; Lee *et al.* 2009) and currently four populations are known to survive. No other *Partula* species remains on Moorea.

There may have been 19th century declines in *Partula* on parts of Raiatea. Crampton (MS) observed that for *P. dentifera callifera* 'The wide, open and nearly level floor of the valley must have been well wooded in Garrett's time but extensive areas have since been cleared for agricultural purposes and have become unsuitable for Partulae. Such changes probably account for the fact that so few examples of *P. callifera* were collected by the writer in a region where, as Garrett wrote, this species was "not uncommon"." Despite the possibility of local declines, partulids were still common in most areas when the island was surveyed by in 1970 (Burch) and in 1908 (Murray and Johnson). Unofficial introductions of *Euglandina* were made to Raiatea in around 1986 (Coote & Loève 2003). The pattern of spread and extinction has been documented, leading to the conclusion that all partulids were extinct on the island by 1998 (Gerlach 1994; Coote & Loève 2003). *P. meyeri* was discovered on Mt. Tefatua in 2005 (Burch 2007). No *Euglandina* were found on the summit of the mountain and it was thought to be above the *Euglandina* altitude range (as predicted earlier – Gerlach 1994). However, J.W. Moore collected several species of *Partula* from the summit of Mt. Tefatua in 1927 and their absence in 2005 indicates that *Euglandina* had already eliminated these populations. It is probable that a relict *P. meyeri* population survived the invasion, although no other *Partula* are thought to survive on Raiatea. Some species are thought to have been rare canopy species (e.g. *P. turgida* and *P. pearcekellyi*) and it is not impossible that these may still survive in overlooked sites.

Tahaa was not surveyed by any partulid collector after Crampton in 1909 until 1994, by which time all partulids were extinct (Coote & Loève 2003). The island has been extensively degraded by agriculture and plant invasion, and is invaded by *Euglandina*. Bora-Bora supported partulids when surveyed by the Mangareva Expedition in 1934, but none were present by 1991 (Coote & Loève 2003).

On Huahine partulids were present in 1991, at which time *Euglandina* was still absent (Coote & Loève 2003). By 1994 *Euglandina* were established and partulids in decline; none could be found in 2000 (and no live *Lissachatina* or *Euglandina*) (Coote & Loève 2003).

The three Tahitian *Samoana* still survive. *Samoana attenuata* is still present at Haapupuni, Hamuta, Apirimaue, Tereia, Vaipoe, Tiapa, Papehue and Ururoa, and on Moorea at Mt. Tohiea Belvedere, in Vaianae, Afareaito, Haumi and Tefeo valleys. On Raiatea it was present on the Temehani plateau in 2005 (Lee *et al.* 2008) and in Averarahi valley in 2011 (F. Jacq pers. comm.); it is probable that relict populations persist. The number of empty shells found on Tahiti support Garrett's (1884) view that it was more abundant than the scarce historical records indicate: "It is more abundant in Toloa and Hapai valleys than elsewhere. Owing to its peculiar habit of living on the foliage near the tops of trees, it easily escapes observation. It occurs more rarely at Tahiti, where it has, also, a wide range, and is confined to the upper portions of the valleys." Recently there have been an increasing number of sightings of this species and of *S. diaphana* on Tahiti, which may suggest the species is expanding, possibly as a result of competitive release following the decline of the *Partula* species, alternatively this may be due to more thorough surveys of higher altitude areas. *S. burchi* is still present on Tahiti. Huahine's *S. annectens* was known from 65 specimens in 1934, some in 1970 and one sighting in 2005.

Threats to partulids

Habitat loss and degradation

Although 46% of partulid species have been affected by habitat deterioration, it does not appear to have had a major impact on most partulid populations, except for the probable extinctions of *E. globosa* and *S. cramptoni* on 'Eua island in Tonga and possibly *P. rufa* on Kosrae. It may be a contributing factor to declines (Pippard 2012), particularly in the Marquesas. An increase in invasion by alien plants may reduce habitat suitability but there are no direct data on this.

Euglandina 'rosea'

Euglandina predation is the single most significant threat to partulids, with 69% of species affected by its presence and 72% of population declines attributed to its presence. Of the 51 probable extinctions (including species extinct in the wild), 92% are primarily attributable to *Euglandina* predation. The impact of *Euglandina* has been disproportionate on terrestrial species; six of the seven largely terrestrial species affected by *Euglandina* are extinct, compared to 58% of the arboreal species (n=62). The surviving species largely exist in higher altitudes lacking established *Euglandina* populations, or are found in areas that have not been invaded to date. There are some species (notably *P. taeniata, P. hyalina* and *P. clara*) which do survive in the presence of *Euglandina*, albeit in very reduced numbers.

An altitude limit to the spread of *Euglandina* was predicted (Gerlach 1994, 2001) based on its temperature requirements. The exact predictions have proved to be incorrect but currently *Euglandina* does not establish permanent populations at higher altitudes, e.g. above 1,200 m on Tahiti.

Data on the establishment of *Euglandina* populations and the early stages of their impacts on partulids are very poor, as no monitoring of releases was carried out. Field surveys shows that introductions could be considered established within three years in some islands (Huahine, Hiva Oa) or four (Tutuila) by which time declines in partulids were apparent. Extinctions have been recorded within nine years of introduction.

Recent research has demonstrated that introduced *Euglandina* comprise at least two species (J. Slapcinsky pers. comm.); whether these have different impacts on partulids is not known at present.

Platydemus manokwari

The New Guinea flatworm appears to be native to New Guinea and Australian populations are thought to be introduced due to their recent identification (Queensland in 1976 and Northern Territory in 2002) (Justine *et al.* 2015). It was first introduced to the Maldives and Philippines in the 1980s and started appearing in islands occupied by partulids in 1978 (Guam on the Marianas: Eldredge 1994).

From the first introduction to Guam, the flatworm spread to the main Marianas islands (Saipan in 1982, Tinian 1984, Rota 1988, Guam and Aguiguan by 1992). New island groups were invaded from 1991: Palau (Koror 1991 and Ulong 1992), the Carolines (Pohnpei 1993), Samoa (Alafua and 'Upolu 1998), Tonga (by 2002), Vanuatu (2002), Societies (Tahiti in 2006 and Moorea in 2009) and Fiji (Rotuma 2012). A new phase in its expansion started with its arrival on continental areas (a greenhouse in France in 2013, and in Florida in 2015) (Justine *et al.* 2015).

The New Guinea flatworm is known to be a major predator of snails in Japan (Sugiura 2009). Currently 23 species of partulid are affected by *Platydemus*, and it is thought to have had significant impacts on four species. In the Caroline islands it appears to have been the only significant factor in the extinction of the terrestrial *Partula guamensis*, but the high-level arboreal *P. emersoni* survives (Pelep & Hadfield 2011; B. Holland pers. comm.). On Guam the proliferation of flatworms is associated with the disappearance of *Euglandina*, and it has replaced this species as the main predator of partulids. *P. gibba* has become extremely rare and survives precariously on the island and is now rarer than *P. radiolata. Samoana fragilis* also survives in one site. All three species are heavily predated by the flatworms (C. Fiedler & A. Kerr pers. comm.).

As *Platydemus* continues to establish its presence in new areas it is likely to replace *Euglandina* as the main predator of partulids. This may change the dynamics of the predator-prey system. In the Japanese Bonin islands the decline of *Euglandina* from a high level of abundance in 1985 to extreme rarity in 1995 was attributed to the introduction of *Platydemus* in the 1990s (Ohbayashi & Takeuchi 2007). *Platydemus* has been found to be a major predator of ground snails (particularly small snails, but being able to kill all sizes). It is known to prey on arboreal snails at least 2 m above ground but the predation rate is reduced on trees, with field experiments recording approximately 40-67% of ground snails consumed in a day, 20-25% 0.5 m above the ground and no more than 20% at 1 m (Sugiura & Yamaura 2008). There are no data for 2 m; extrapolating the limited data suggests that around 40% of trails will be followed 2 m above ground and that arboreal predation will be insignificant above 2.5 m above ground. Largely terrestrial partulids are likely to suffer very high levels of predation whilst largely arboreal species may persist. *Euglandina* has been calculated to have a similar attack rate (34%) (JG data), but may be a more effective arboreal predator. *Platydemus* activity is restricted to damp conditions and the drying effect of air circulation may restrict arboreal foraging. *Euglandina* is more tolerant of dry conditions and can often be found roosting on trees. Thus it

Fig. 12.3 *Euglandina 'rosea'*: left – distribution, right – attacking a snail. Map: green – partulid localities; purple – *Euglandina* populations without partulids; red – partulid population invaded by *Euglandina*

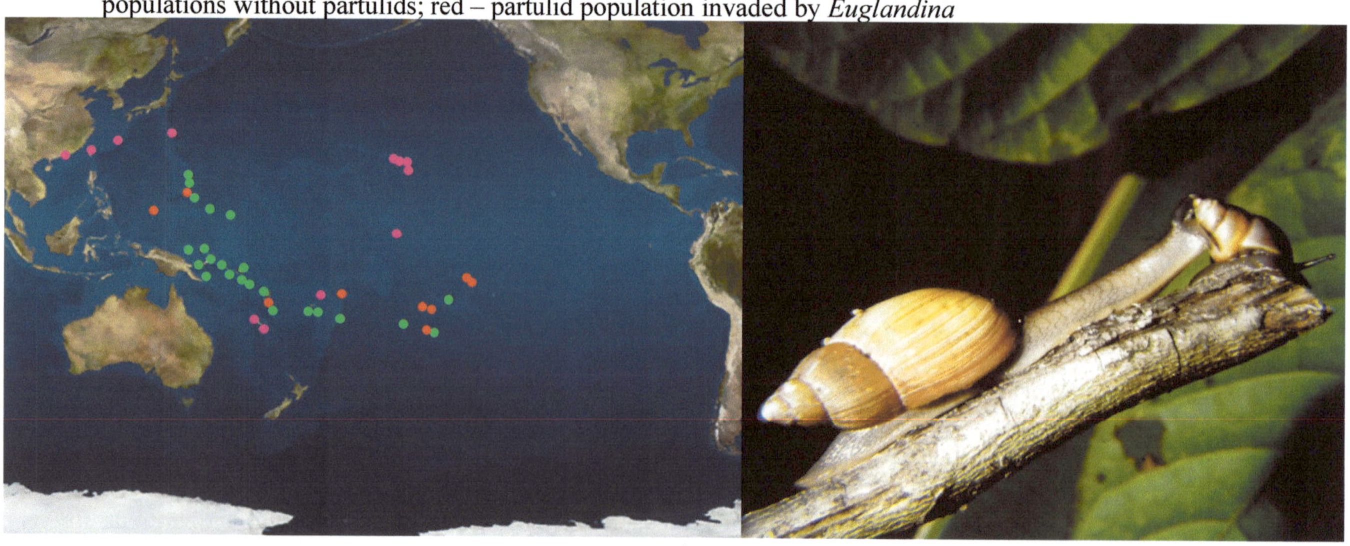

may be the case that *Platydemus* may be less problematic for arboreal partulids than *Euglandina* has been.

It has been estimated that ambient temperatures of 10°C are too low for population establishment (with reproduction occurring above 15-18°C and no juvenile survival at 10°C) (Sugiura 2009). This suggests that all partulid habitats are vulnerable to *Platydemus*, although invasion at higher altitudes may be slow.

As with *Euglandina*, introductions of *Platydemus* have not been monitored. The limited data available suggest that *Platydemus* reduces partulid populations within 12 years of introduction and in Guam it displaced *Euglandina* within 14 years. On most islands it has not been present for long enough for its impacts to become apparent. The colonisation of Tahiti provides some information, with its first record in 2006, isolated records since then and a marked increase in sightings in 2015.

Other invasive animals

Rats are probably the most widespread introduced snail predators. They have been suggested to be predators of partulids (Johnson *et al.* 1993b). Few rat-damaged partulids have been recorded and rat predation does not appear to be significant in most areas. The only notable exception is Fiji where a relatively high proportion of shells show rat damage (G. Brodie pers. comm.).

Little fire ants (*Wasmannia auropunctata*) have been present on Tahiti since 1977 and could be problematic (Coote *et al.* 2015). They have long been established in Wallis and Futuna (Alofi since 1997), Vanuatu (including Gaua, Mota and Vanua Lava from 1998), Marianas (Guam from 2011) and parts of the Solomon islands (including Guadalcanal from 1974) (Wetterer & Porter 2003; Raymundo & Miller 2012). So far no partulid declines have been specifically linked to their presence, and severe impacts may be limited to situations combining habitat disturbance and invasion.

Climate change

The decline of the Marquesas species *Samoana gonochila* on Ua Huka and *S. bellula* on Ua Pou in the absence of *Euglandina*, *Platydemus* or significant habitat loss have been suggested to be due to climate change (Coote *et al.* 2015). This is speculative at present but many partulid populations are restricted to small habitat islands which would be vulnerable to any changes.

In the Society islands it is projected that rising temperatures will facilitate *Euglandina* invasion at higher altitudes (Petit & Prudent 2010; Sauter *et al.* 2013). This is also likely to be true for *Platydemus manokwari.*

Conservation

Conservation management of the surviving partulids has relied on conservation breeding and reintroduction. Habitat management and protection have been relatively minor aspects due to the predominance of the *Euglandina* (and now *Platydemus*) problem. Conservation breeding has been seen as the priority for partulids since 1986 (Clarke & Wells 1992; Wells 1995). Conservation concerns were published in 1980 (Tillier & Clarke 1980), and conservation breeding started in 1981, building on the techniques developed during the genetics research (Bloxham *et al.* 1984; Bloxham & Tonge 1986; Tonge & Bloxham 1991; Pearce-Kelly 1994). Partulid breeding has led to research on invertebrate pathology, diet, population

Fig. 12.4 *Platydemus manokwari*: left – distribution, right – eating *Samoana fragilis* (photo G.C. Fiedler). Map: black – original range; purple – introduced populations; green – partulid populations; red – partulid populations with *Platydemus*

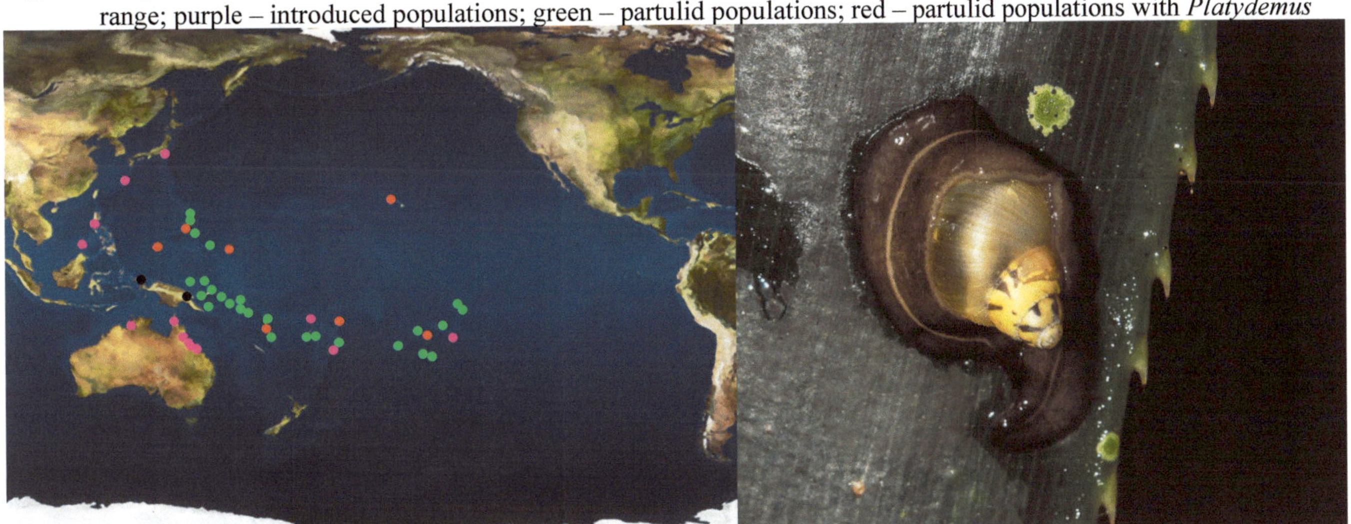

management and colony data management (Burlingham-Johnson *et al.* 1994; Amin 1997, 1998; Clarke *et al.* 1998; Stallard & Gilbert 2003; Gouveia *et al.* 2006, 2007, 2011).

Conservation breeding has been highly successful for some species but not others (Mace *et al.* 1998). Even where populations have been established mortality has been high, possibly attributable to environmental fluctuations (Gouveia *et al.* 2006; Gouviea 2012) and bacterial or microsporidial parasite outbreaks (Cooper & Knowler 1992; Cunningham *et al.* 1996; Cunningham & Daszak 1998; Selman *et al.* 2004; Cwiertna *et al.* 2010). Problems in conservation breeding of *Achatinella* snails have been attributed in part to inbreeding (Price & Hadfield 2014). In *A. lila* a decrease in fecundity in the second filial generation was accompanied by a decrease in juvenile survivorship. Fecundity, survivorship and heterozygosity started to increase in the third generation, suggestive of some degree of allele purging. A similar pattern is apparent in some partulids, with second or third generation declines in fecundity in the species with the most severe bottlenecks (Fig. 12.5-6). The decline of the captive *A. fuscobasis* populaiton may be due to disease susceptibility of the captive stock founded from genetically uniform populaiton animals reflecting a historical bottleneck (Sischo *et al.* 2016).

Beyond the failure of any *Samoana* species to become established in captivity, no generalisations can be made about adaptation to captive conditions. *S. ganymedes* survived 11 years, but only reached the third filial generation (F_3). Of all *Samoana* kept, this was the most *Partula*-like in its robust shell, with the exception of *S. inflata*. Only a small number of founders were collected for this latter species (4 compared to 78 for *S. ganymedes*). All other *Samoana* rapidly declined to extinction, without any clear external cause. *S. attenuata* showed exceptionally high fecundity and high self-fertilisation rates

Table 12.2 History of partulid conservation breeding to 31st December 2015

Species	duration	2015 population	Species	duration	2015 population
Partula affinis	1995-	659	*Partula mooreana*	1985-	1,427
Partula arguta	1991-5	-	*Partula navigatoria*	1991-	506
Partula auraniana	2007	-	*Partula nodosa*	1984-	3,736
Partula aurantia	1982-92	-	*Partula obesa*	2009-11	-
Partula clara	2000-12	-	*Partula otaheitana*	1995-2007	-
Partula clarkei	1991-5	-	*Partula pacifica*	1996-2001	-
Partula compressa	1997-9	-	*Partula producta*	1984-9	-
Partula diminuta	1984-9	-	*Partula radiolata*	1993-2009	-
Partula dolorosa	1992-3	-	*Partula rosea*	1991-	373
Partula faba	1991-2016	1	*Partula suturalis*	1980-	737
Partula garrettii	1991-	179	*Partula taeniata*	1981-	420
Partula gibba	1972-2015	-	*Partula tohiveana*	1982-	1,305
Partula hebe	1991-	271	*Partula varia*	1991-	672
Partula hyalina	1995-	431	*Samoana attenuata*	1995-9	-
Partula incrassa exigua	1982-87	-	*Samoana bellula*	1995-8	-
Partula labrusca	1992-2001	-	*Samoana burchi*	1996	-
Partula laevigata	1997-2001	-	*Samoana ganymedes*	1995-2006	-
Partula leptochila	1991-3	-	*Samoana gonochila*	1995-7	-
Partula lirata	1999-2002	-	*Samoana inflata*	1995-6	-
Partula lugubris	1992-3	-	*Samoana minuta*	1995-6	-
Partula mirabilis	1984-	288			

Fig. 12.5 Fecundity declines in captivity. Left – effect of bottleneck severity on fecundity as a proportion of parental fecundity ($y=0.324x^{0.135}$, $R^2=0.415$; regression excluding frequent self-fertilisers, shown as blue points). Right – generational fecundity as a proportion of parental fecundity in bottlenecked species: *P. garrettii* (n=8, no inbreeding depression); *P. hebe* (n=8, prolonged moderate depression; *P. rosea* (n=7, pronounced depression, recovery after 8 years); *P. tohiveana* (extreme bottleneck in F_1 [n=4] with inbreeding followed by recovery)

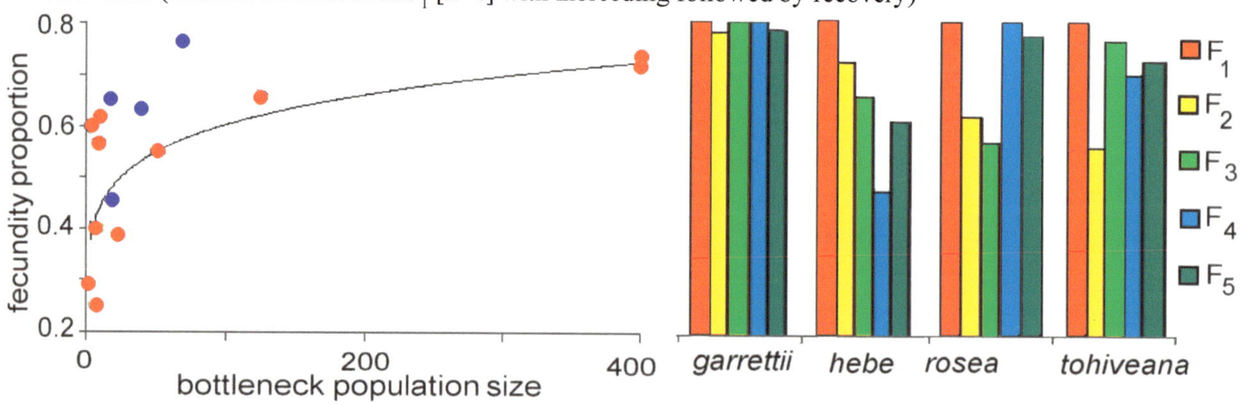

in the parental generation but fecundity dropped by 90% in the F_1 generation, with no recruitment to the F_2 generation. The cause of the sensitivity of *Samoana* species is not known.

Within *Partula* several apparently similar species have adapted very differently (Fig. 12.7); *P. affinis* has proven adaptable, with sustained fecundity and low mortality whereas reproduction in *P. otaheitana* has been extremely poor and populations failed to recover after the loss of the parental generation. Of the two large Raiatean species, *P. navigatoria* has shown sustained population growth with the exception of temporary drops in fecundity in the F_3 generation. In contrast, *P. faba* failed to adapt to captivity; fecundity of the filial generations has never matched that of the wild-collected animals and neonatal mortality was high. Increased mortality of all age stages and a collapse in fecundity in the F_4 generation was only the tail-end of a long process of decline. The population graph shows that the apparent early success of this species was due to rapid reproduction of the parental generation and relatively low adult mortality. The decline from 1995 has been due to declining fecundity, a long-term problem masked by the longevity of this species. The almost complete failure of the F_4 generation may be due to inbreeding or to unidentified issues. The reasons for the failure of these populations are not known; the *P. otaheitana* founders were collected at high altitude and the captive conditions that suit low- to mid-altitude species may not have been suitable for them. Most *P. faba* were collected from the same site as *P. navigatoria* and the species would have been expected

Fig. 12.6 Demography in three *Partula* species with population bottlenecks (a – *P. garrettii* [34 founders] b – *P. hebe* [8], c – *P. rosea* [9]) and one without (d – *P. varia* [190]). Bottlenecks marked with arrows and number of adults. Left – weekly mortality (left axis and lines) and weekly fecundity (right axis and bars) in different captive generations (P – parental generation, F – filial generations, X – mixed filial generations). Right – adult population changes by generation

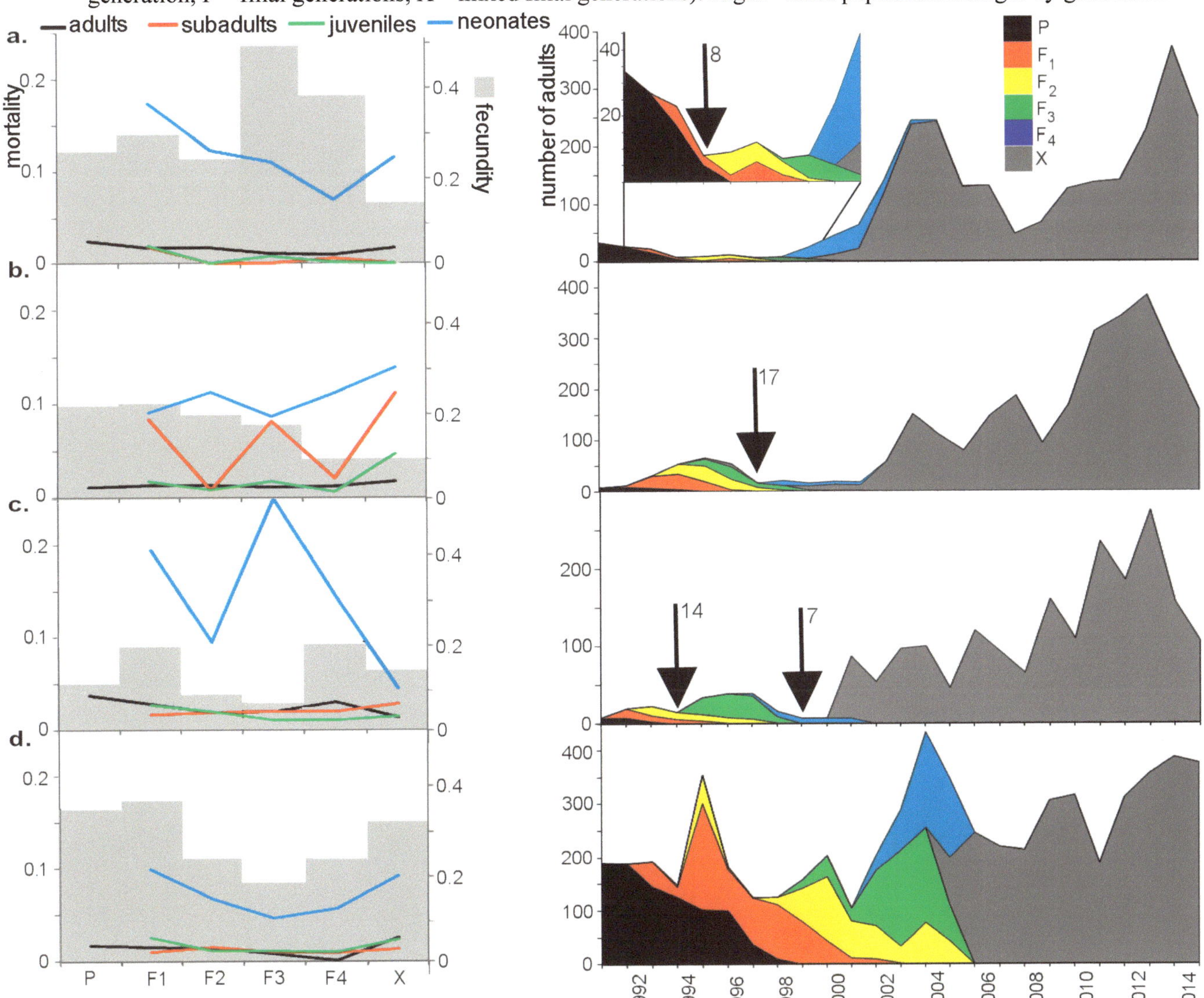

to be adaptable, the high neonatal mortality may indicate a specific ecological requirement of that developmental stage.

The loss of other captive populations may be attributed to disease, as in the case of *P. clarkei* which is suspected to have been eliminated by infection of the parasite *Steinhausia* (Cunningham & Daszak 199). Prior to its sudden loss this species was adapting well to captivity (Fig. 12.8), with improving fecundity (in peak and mean fecundity of each captive generation). Mortality rates were relatively high for subadults and juveniles and increasing, but this was offset by stable or declining neonatal mortality. The result of the interaction of these factors was a rapid population growth for three years. The sudden tripling of adult mortality and collapse in fecundity resulted in a population crash and extinction in just 70 days. A similar pattern was seen *P. labrusca*, although a pathogen was not identified. In this species fecundity was low in all filial generations but survivorship was improving, with the exception of a population crash in 1997 (77% decline in 126 days) and extinction in 2001, 217 days after the population peak. Both species may have been particularly vulnerable due to their narrow genetic base, being founded by 18 and 7 founders respectively.

Reintroduction

Reintroductions were first attempted in the Society islands in 1994 (Coote *et al.* 2004). This followed an experimental

Fig. 12.7 Differing fates of *Partula* species: a – *P. affinis*, b – *P. otaheitana*, c – *P. navigatoria*, d – *P. faba*. Left – mortality and fecundity, right – adult populations divided by generation. Key as in Fig.12.5

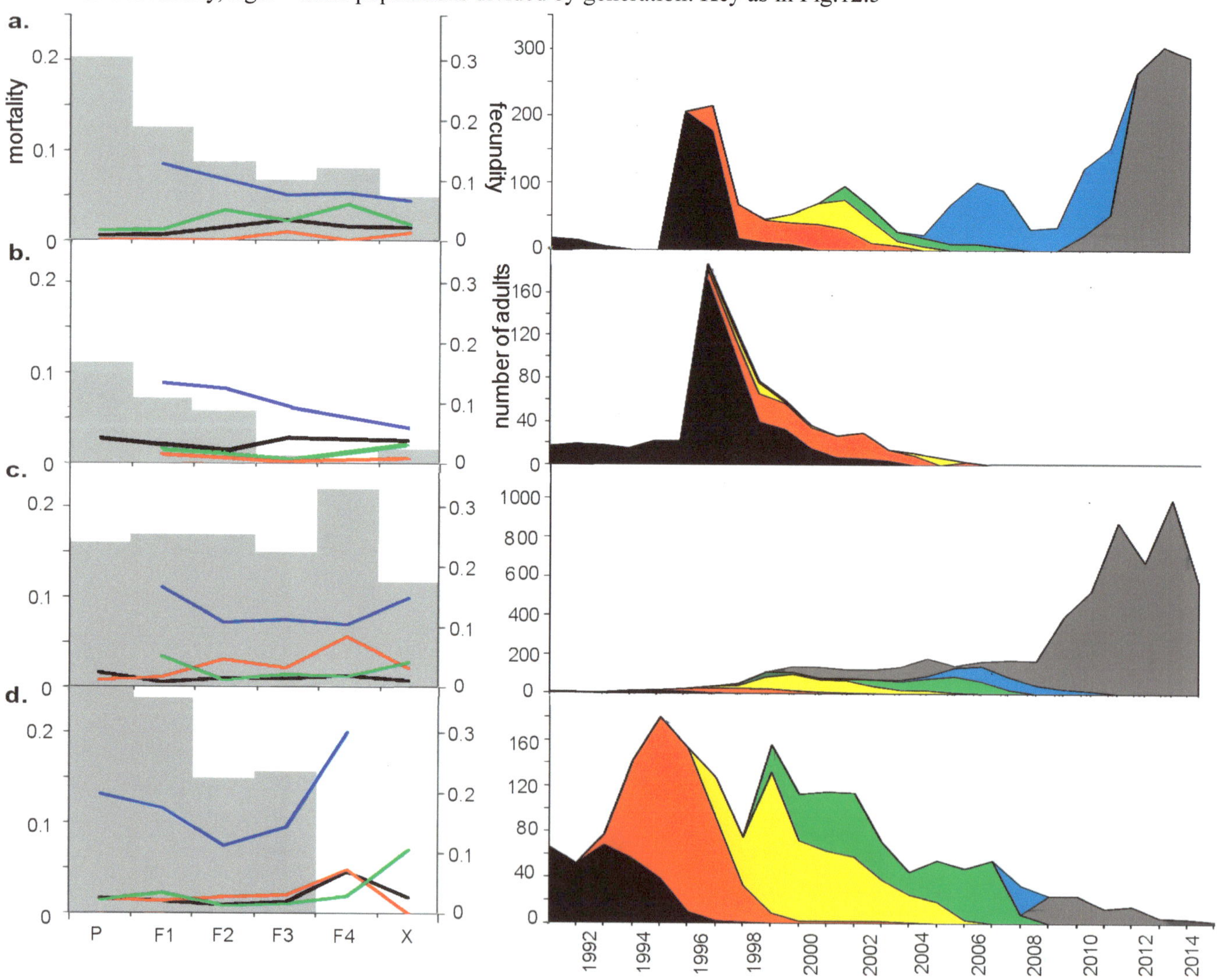

release in the confined greenhouse environment (Pearce-Kelly *et al.* 1995). As *Euglandina* persists on all of the Society islands reintroductions could only be undertaken within predator-proof exclosures (Clarke & Wells 1992). 300 snails (*P. tohiveana, P. suturalis* and *P. taeniata*) were released into an exclosure of 20×20 m on Moorea in 1994 (Coote *et al.* 2004). This reserve was within natural forest cover and excluded predators with a metal fence protected with an electric fence and a salt filled trough. Within a year it was found that maintenance problems had enabled *Euglandina* to invade the exclosure, killing all but three *Partula*. The reserve was restocked with 282 snails in 1996 (Clarke *et al.* 1996; Coote *et al.* 2004). Further maintenance problems occurred and the experiment was halted 4 years after the original release, at which time eight snails remained, all wild-born (Coote *et al.* 2004).

Although the release did not establish a population, it did demonstrate that captive-bred partulids can adapt to natural habitats and breed successfully. Further releases were planned for Tahiti and a reserve constructed on Tahiti-Iti in 2002 (Coote & Loève 2003). However, this was destroyed by a rock fall and the difficulty of accessing the site led to plans for new reserves in the more accessible Te Faaiti valley in 2008 (Coote *et al.* 2015). The reserve was constructed in 2012 within a forest restoration area. The felling of invasive trees meant that most large plants were removed from the area, minimising the risk of damage to the enclosure, but increasing its exposure. When the snails were released in 2015 the reserve contained a good growth of shrubs and small trees. 119 snails (*P. hyalina, P. nodosa* and *P. affinis*) were released but exceptionally hot weather, combined with the lack of tree cover, resulted in very high mortality.

Fig. 12.8 Population decline in *Partula clarkei* (a-c, g) and *P. labrusca* (d-f, h). a & d – weekly fecundity in different captive generations; b & e – weekly mortality and fecundity; c & f – annual adult population changes by generation; g – population changes by age group in *P. clarkei*; h – population decline by age group in *P. labrusca*. Key as Fig. 12.6

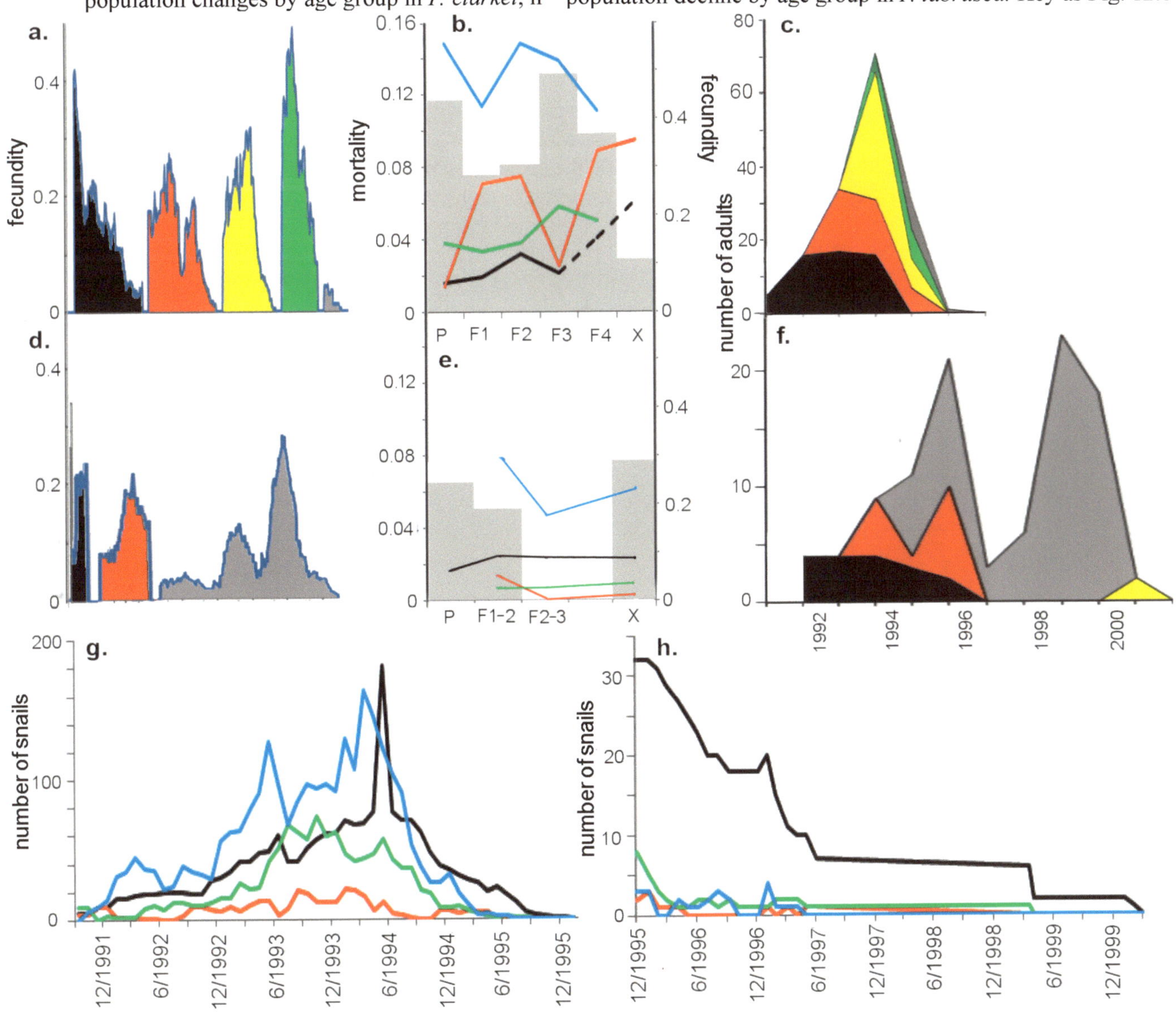

At the same time as the reserve was stocked an experimental free release of *P. nodosa* was undertaken in Papehue valley. 150 snails were placed on *Inocarpus fagifer* trees. In some valleys these trees have retained *Partula* (*P. clara* and *P. hyalina*) populations despite the presence of *Euglandina* and it has been hypothesised that the predators may not forage efficiently on the plants. The released snails adapted well with many births starting three weeks after release. Releases of other species are scheduled to take place in 2016 and 2017.

Potential for future conservation and research

The releases on Tahiti are being used to evaluate the practicality of further releases of these and other species. Releases onto trees, rather than into reserves, raises the possibility of reintroduction to islands where long-term maintenance of reserves may be impractical.

Although the mass extinction of partulids caused the cessation of the genetic research on the family, new methods and events create new opportunities. Next-generation DNA sequencing should enable expansion of the molecular phylogeny to include at least some of the extinct species. These methods could ultimately allow identification of genes for phenotypic polymorphisms in different species but this will be of limited value beyond phylogenetics as genetic interactions could not be studied in extinct lineages. Population genetics in some of the extinct species could be studied as significant numbers of samples exist for different populations in Crampton's collections of the Society islands. Samples from other extinct populations are more restricted, but do exist.

Captive populations are almost all inbred, monomorphic lines and although polymorphism is retained in captive *Partula suturalis*, the inheritance of polymorphism has already been elucidated in that species. This could still be combined with gene sequencing to provide more detailed information on the evolution of the inheritance pattern. The relict populations in the Society islands offer limited potential for evolutionary research; almost all populations are monomorphic (the exceptions being *P. taeniata/mirabiis propinqua* of Opohunu bay and, to a lesser extent, *P. otaheitana*). They are also isolated from one another and there is little scope to study gene flow.

Despite these limitations, there is a new opportunity to take the early genetic studies in a new direction. Any future re-established wild populations will effectively be new colonisations, offering the opportunity to study range expansion and species adaptation. In the case of *P. suturalis* this will include the possibility of monitoring changes in polymorphism where the environmental variables and polymorphic inheritance are known.

Fig. 12.9 An individual from the introduced *Partula hyalina* population on Rurutu (photo: T. Coote)

13. In memoriam: *Partula faba*

On 21st February 2016 the last surviving *Partula faba* died in Edinburgh Zoo. This was a significant event in *Partula* history: the 34th partulid extintion and the loss of the first species to be described, 232 years ago.

Partula faba was the first live partulid that I encountered, in Vaiapu valley on Raiatea in Setember 1992. *Euglandina* had passed through that valley just days before, leaving a carpet of empty shells of *P. faba* and *P. garrettii.* Just a single live *P. faba* remained. Over the next week I did find more in valleys that had not been devastated by *Euglandina* at that time, but those I collected on the Temehani plateau on 8th September 1992 were the last sighting of the species in the wild.

The single Vaiapu snail lived on in London Zoo for six months, during which time it produced several offspring. Initially it seemed as though *P. faba* had been saved, as the 65 that I had collected and the 89 brought back by the Operation Partula Expedition the year before survived and bred. However, although the wild collected snails did well, subsequent generations never prospered and gradually the population dwindled. As different zoo populations declined the different colonies were brought together at London Zoo, but the declines continued. They were then moved to Bristol Zoo which had fewer *Partula* and was better able to concentrate on this critical species. This also did nothing to slow the decline and in 2015 the last two were moved to Edinburgh Zoo which had experience of pulling *Partula* back from the brink, having rescued *P. tohiveana* from a bottlenck of four. By the end of the year just one remained. Self-fertilisation had occurred once when the species was more abundant, so even then there remained a faint hope of recovery. Sadly the last one died before that could occur.

The loss of *P. faba* epitomises the problems facing partulids: their vulnerability to extinction in the wild and the difficulty of maintaining them in captivity. They coexisted with humans for about 1,000 years, and most have been capable of withstanding habitat changes and intensive collection, but have been unable to survive *Euglandina* introduction. Being small and comparatively simple to keep, it is relatively easy to keep large populations in captivity. However, identifying exactly what they need for sustained breeding, or the causes of decline have proved extremely difficult. For *P. faba* my research into the diet of the specimens that Crampton preserved failed to identify any key component that could have saved them. Similarly, pathological studies have failed to identify any specific cause of death.

For *P. faba* the story is probably over. As a large and highly conspicous tree-snail there is no realistic possibility of its continued, undetected survival in the wild. In recent years there has been much talk of resurrecting species through techniques such as cloning. This is well advanced for mammals, but we are a long way from developing the necessary techniques for an ovoviviparous snail. However, *P. faba* has been included in the 'Frozen Ark', the repository of material that may one day allow this to move beyond science-fiction. It is not a coincidence that *Partula* are well represented in the Frozen Ark, set up by Bryan and Ann Clarke, and Ann McLaren. Initially based on preservation of the DNA of the Moorean *Partula* species, the Frozen Ark is yet another example of how *Partula* have inspired developments in science and conservation.

Fig. 13.1 232 years of *Partula faba:* 1784 – Thomas Martyn's *Helix faba*; 1992 – the last wild *P. faba* on Temehani, February 2016 – the world's last *P. faba* (photo: G. Bennett)

Acknowledgements

I am grateful to many people for sharing their information, especially to Trevor Coote, Jean-Yves Meyer, Paul Pearce-Kelly, Dave Clarke, Bryan Clarke, Jim Murray and Michael Johnson, and to Jack Burch, Darmaid Ó Foighil, Cindy Bick, Tahewan Lee, Mike Hadfield, David Sischo, Curt Fiedler, Brenden Holland, Alex Kerr, John Slapcinsky, Gilianne Brodie and Fred Brook. Useful information was also provided by Anita Eschner, Angus Davidson, Fred Jacq, Jean-François Butaud, Bill Sutherland, Nate Shoobs and André Delsaerdt.

Access to collections was greatly facilitated by Gary Rosenberg and Paul Callomon (ANSP), Norie Yeung and Regie Kawamoto (BPBM), Jon Ablett and Fred Naggs (NHMUK), Richard Preece (UMZC), Elaine Brown (ZSL). I am grateful to the following for the loan of specimens: Mark Siddall (AMNH), Regie Kawamoto (BPBM), John Slapcinsky (FLMNH), Jochen Gerber (FMNH), Adam Baldinger (MCZ), Olivier Gargominy, Virginie Héros, Philippe Maestrati and Benoît Fontaine (MNHN), Jon Ablett (NHMUK), Bernhard Hausdorf (ZMH) and Tom Schiøtte (ZMUC).

Photographs of museum specimens were provided by Francisco Borrero and Paul Callomon (ANSP), Tim Pearce and Charlie Sturm (CMNH), Jochen Gerber (FMNH), Manuel Caballer and Philippe Maestrati (MNHN), Harriet Wood and Ben Rowson (NMW), Darryl Potter and John Healy (QM), Brian P. Murphy (USNM), Barbara Oberholzer (UZH), Christine Zorn (ZMB), Harry G. Lee and Bill Frank. Photographs of live snails were provided by Simon Aiken (Simon's Specimen Shells Ltd.), Gareth Bennett, Ric Brewer, Dave Clarke, Robert Cowie, Trevor Coote, G. Curt Fiedler, Olivier Gargominy, Fred Jacq, Jean-Yves Meyer, Brenden Holland and David Sischo.

The project was made possible through the financial support of Peter Kistler, ZSL, ANSP and the Philadelphia Shell Club, UMMZ and BPBM. For arranging this support I am grateful to Paul Pearce-Kelly, Gary Rosenberg, Darmaid Ó Foighil, Ken Hayes and Nori Yeung. Publication is supported by Ada, D. Baker, R. Brewer, S. Butcher, A. Collette, Craig, J. Dimatos, J. Gamble, C. Howe, Jarrah, Julie K., J.A. Lara, J. McCormick, A. Minelli, E. Lamont, H.G. Lee, G. Macaulay, B. Oppenheim, K. Pan, A. de Rienzo, B.W. Skakle, R. van Steenselen, B. Sumner, R. Thoma, B. Tokic, M.S. Uhle, O. Vigneresse, J. Walewski and A.L. Wood.

Finally I am grateful to Sam Aberdeen for asking a question that caused me to re-examine my old notes and revive a 20 year old incomplete project.

Partula tohiveana (photo: D. MacCaffrey)

Appendix I – Measurement data

Sources: 1 – Crampton 1916; 2 – Crampton 1925; 3 – Cook & Crampton 1930; 4 – Crampton 1932; 5 – Crampton 1956; 6 – Crampton & Cook 1956; 7 – Crampton & Cooke 1953; 8 – Kondo 1970; 9 – Solem 1972; 10 – Johnson *et al.* 1986; 11 – Bauman & Kerr 2013; 12 – Kerr & Fiedler 2015; 13 – Slapcinsky & Kraus 2016; 14 – Crampton MS; 15 – museum specimens; 16 – JG specimens

Measurements and percentages given as mean values with standard deviations. Valleys clockwise from north

			N	Length	Width	Mouth H	W/L	Mouth H/L	Source
Eua	*globosa*		25	21.22±0.558	18.98±0.962	13.31±0.346	89.57±2.637	62.74±1.677	14, 15
	expansa	Savai'i	20	17.45±0.621	13.28±0.536	11.30±0.280	76.05±2.550	64.55±8.550	14
		'Upolu	7	17.15±0.831	19.10±0.777	11.90±0.784	73.50±2.611	68.99±8.231	14
	Eua sp.	Savai'i	2	19.70	12.15	12.15	61.60	59.60	15
	zebrina	(total)	7077	18.90±1.171	12.10±0.698	11.78±0.856	63.98±2.728	62.27±2.388	14
	montana		17	18.06±0.975	12.30±0.559	9.43±1.572	69.43±4.276	51.09±5.096	15
Samoana	*fragilis*	Guam D	36	14.4±0.753	9.77±0.525	8.360±0.441	67.78±2.388	57.97±1.462	2
		Guam S	1	12.00	9.80	7.90	82.00	66.00	15
		Rota	12	17.35±0.783	11.78±0.850	9.88±0.697	67.87±2.423	56.88±2.074	15
	alabastrina	Matuku	10	22.55±0.555	13.21±0.675	12.75±0.315	58.64±2.416	56.79±1.030	14
		Moala	10	21.07±0.946	12.53±0.498	12.04±0.699	59.37±1.863	57.54±1.392	14
	cramptoni		9	18.68±0.262	11.28±0.371	9.080±0.113	60.17±2.108	48.89±0.737	14
	conica	Savaii	21	24.25±0.799	14.70±0.435	14.10±0.665	60.50±1.498	57.50±1.591	14
		'Upolu	69	24.65±1.343	14.30±0.678	13.70±0.842	57.70±1.877	55.50±1.354	14
		Tutuila	801	22.90±1.189	14.61±0.701	13.86±0.678	63.82±2.581	60.39±1.824	14
	abbreviata		350	19.63±0.967	13.68±0.927	12.53±0.796	69.64±2.375	63.73±1.706	14
	stevensoniana		29	22.94±0.646	13.54±0.614	11.93±1.004	59.27±2.053	52.20±4.200	15
	thurstoni		4	19.00±1.824	12.95±1.066	10.30±1.453	68.50±7.454	54.12±9.451	15
	dryas	Total	83	14.95±0.258	9.61±0.201	9.11±0.193	65.86±0.455	60.68±0.370	6, 15
	margaritae	Rapa	245	22.32±0.276	12.61±0.210	8.22±0.266	56.28±0.266	52.71±0.315	6
		'Rurutu'	1	21.30	12.50	10.90	58.69	51.17	8
	gonochila	Nuku Hiva	48	15.44±1.129	9.71±0.573	9.25±0.716	63.02±2.402	59.93±2.060	15
		Ua Huka	242	18.74±1.152	11.57±0.805	10.68±0.835	56.97±2.271	57.03±1.995	15
	bellula	total	225	15.34±1.470	9.24±0.818	8.68±0.810	60.28±2.222	56.65±2.735	15
	decussatula	Tahuata	14	15.36±1.165	10.76±0.896	10.02±0.696	70.07±3.466	65.35±3.516	15
		Hiva Oa	24	15.85±0.638	11.06±0.447	10.44±0.478	69.81±3.091	65.90±2.308	15
	pilsbryi		93	18.79±0.617	11.47±0.435	11.21±0.451	60.70±2.182	59.39±1.797	15
	medana	Tahuata	30	12.07±1.362	8.39±0.863	7.82±1.121	64.58±2.916	64.90±2.995	15
		Fatu Hiva	6	11.98±0.436	8.45±0.414	7.75±0.501	70.56±3.515	64.67±3.365	15
		Hiva Oa	59	11.75±0.665	7.46±0.381	7.76±0.456	68.61±2.197	63.74±2.212	15
	inflata		117	21.05±0.762	15.15±1.001	12.91±0.659	71.96±2.893	61.37±2.305	15
	ganymedes		281	21.38±0.978	13.38±0.947	11.91±0.782	62.74±3.065	55.56±2.396	15
	minuta	Fatu Hiva	70	13.43±0.672	10.63±0.592	9.52±0.491	79.24±1.960	70.60±1.610	14, 15
	attenuata	Bora Bora	2	15.35	7.80	8.50	50.60	55.85	14, 15
		Moorea	9	14.37±0.381	6.47±0.122	8.08±0.432	45.04±1.469	56.22±1.894	10
		Raiatea	240	15.67±0.544	8.27±0.305	8.77±0.298	52.57±1.679	55.78±1.270	14
		Tahaa	25	15.01±0.727	7.99±0.474	8.25±0.561	53.14±1.622	54.70±1.770	14
		Tahiti	40	15.73±0.471	8.25±0.362	8.78±0.413	55.79±1.885	52.43±1.610	15
	annectens		58	13.92±0.390	8.08±0.288	8.25±0.245	57.91±1.300	59.20±0.989	14
	burchi		19	17.59±0.756	9.05±0.734	9.52±0.384	51.50±4.104	53.98±1.365	15
	diaphana	Moorea	24	13.60±0.813	7.82±0.622	7.29±0.731	57.59±4.278	55.65±3.580	15
		Tahiti	5	15.45±0.752	9.35±1.207	9.45±0.052	60.35±5.220	58.55±0.045	15
Palaopartula	*calypso*		38	27.65±1.227	13.59±0.886	11.79±0.914	43.0±2.673	54.85±1.442	15
	thetis	Angaur	12	26.90±0.870	13.18±0.539	15.02±0.500	49.00±1.893	55.16±1.434	14
		Babelthaup	51	27.74±0.817	13.58±0.434	15.30±0.444	48.89±1.300	55.19±1.421	14
		Koror	38	26.05±1.175	13.39±0.507	14.11±0.724	51.47±2.136	54.03±1.428	14
		Peleliu	5	27.59±0.814	13.58±0.392	15.26±0.774	48.90±1.625	55.50±2.191	14
	leucothoe	Angaur	38	18.84±0.513	10.13±0.377	10.63±0.371	53.76±1.370	56.37±0.978	14
		Koror	1	19.55	10.70	10.50	54.50	53.50	14

			N	Length	Width	Mouth H	W/L	Mouth H/L	
Palaopartula	*leucothoe*	Peleliu	62	20.75±0.605	11.35±0.342	11.76±0.304	54.68±1.700	56.16±1.114	14
Sphendone	*insolita*		56	21.1±0.8	14.0±0.4	11.3±0.4	63.6	54.5	13
Partula	*grisea*	Lihir	2	19.65	11.05	9.25	56.25	49.65	15
	similaris	Woodlark	110	19.27±0.548	11.05±0.384	10.44±0.353	57.38±2.585	54.20±1.920	15
	flexuosa	Guadalcanal	9	15.53±0.548	8.71±0.293	8.03±0.255	56.11±1.694	51.75±1.925	15
	micans	total	23	15.34±0.813	8.47±0.430	7.30±0.341	55.11±2.081	45.63±2.382	15
	cramptoni	Bellona	22	23.30±0.692	13.68±0.286	12.98±0.384	58.77±1.922	55.70±1.331	15
		Rennel	1	28.80	15.00	12.30	52.00	43.00	15
	vanikorensis		25	21.84±1.208	13.06±1.187	10.20±0.553	59.763±.382	46.74±1.858	15
	radiosa		2	21.85	12.3	10.75	56.50	49.2	15
	pacifica	Tanna	10	23.45±0.450	13.95±0.551	12.74±0.754	57.20±1.871	52.80±1.931	15, 9
		Efate	2	21.90	11.63	10.65	53.09	48.61	15
		Anetiyum	3	22.66±0.802	13.40±0.693	12.20±0.755	59.10±1.469	53.82±2.541	9
	pyramis	Malekula	3	24.03±1.124	12.63±0.379	12.23±0.551	52.65±3.103	50.91±0.926	15
	auraniana		49	18.56±0.594	11.54±0.602	8.68±0.707	61.78±2.599	46.38±3.530	15
	gibba	Aguiguan	48	16.85±0.095	12.30±0.055	-	-	-	8
	gibba	Guam D	2365	16.99±0.736	11.79±0.565	9.69±0.415	69.38±2.336	56.92±1.550	2
		Guam S	7	16.87±1.05	12.60±0.85	9.89±0.63	-	-	12
		Rota	21	16.60±0.792	11.20±0.681	9.40±0.590	67.47	56.63	11
		Saipan D	2586	17.08±0.770	12.03±0.573	9.70±0.434	70.37±2.386	57.19±1.692	2
		Saipan S	117	16.50±0.699	13.70±0.821	8.90±0.552	83.00±0.992	54.00±1.232	15
	radiolata	Guam	1416	16.62±1.051	10.19±0.626	9.30±0.605	61.22±2.215	55.80±1.962	2
	salifana	Guam	105	18.05±0.920	11.01±0.344	10.20±0.337	60.85±1.908	56.62±1.409	2
	langfordi	Aguiguan	73	13.60±0.240	10.00±0.250	8.00	72.06±1.011	57.14	8
	sp.	Rota	6	16.77±1.078	11.11±0.800	9.53±0.725	66.27±2.812	56.83±2.208	15
	desolata	Rota	4	18.70±1.300	14.40±0.513	11.30±1.170	80.50	60.75	11
	sarissa		18	20.07±0.754	10.49±0.326	10.51±0.330	52.11±1.380	52.33±1.343	14
	lirata	Cicia	1	20.90	11.00	11.00	52.63	52.63	15
		Vanua Malavu	323	19.22±1.224	10.21±0.595	9.80±0.575	50.93±2.341	53.22±2.273	14
	lanceolata	Mago	55	19.59±0.869	11.01±0.432	10.28±0.376	56.26±1.964	52.44±1.391	3
	leefei		209	13.40±0.685	8.45±0.342	7.51±0.318	62.83±1.698	55.82±1.386	14, 15
	rufa		82	22.44±0.283	14.98±0.917	12.24±0.802	66.67±2.867	54.50±2.048	14
	guamensis		111	26.09±1.106	17.02±0.694	12.61±0.707	65.31±2.567	48.32±2.216	15
	emersoni	D	195	20.15±0.826	10.10±0.995	10.10±0.338	49.50±1.618	49.50±1.294	14
		S	4	19.67±0.419	10.80±0.245	9.45±0.252	54.90±1.283	48.05±1.808	15
	obesa	Alofi	46	15.79±0.776	9.68±0.046	8.51±0.335	61.11±1.696	54.02±1.732	14, 15
		Futuna	118	15.34±0.735	9.54±0.228	8.26±0.394	62.32±2.506	53.93±2.923	15
		Futuna S	5	15.41±0.720	10.06±0.384	8.26±0.382	59.97±2.141	52.49±1.623	14, 15
	assimilis	Rarotonga	25	16.70±0.943	10.07±0.425	8.67±0.358	50.00±1.805	51.92±1.888	14, 15
	o. otaheitana	Mt Pitamateta D	1	18.60	9.30	6.70	61.18	51.68	15
		Mt Pitamateta S	8	20.03±0.865	12.26±0.821	10.34±0.709	60.16±1.209	51.65±1.775	15
		Mt Marau D	3	20.23±0.757	12.16±0.208	10.63±0.543	62.75±2.135	52.54±1.776	15
		Mt Marau S	2	20.15	12.65	11.25	58.20	52.11	15
		Mt. Aorai S	2	19.25	11.10	8.90	54.62	49.66	15
		Fautaua D	421	19.32±0.926	10.97±0.522	9.89±0.497	56.74±2.266	51.12±1.960	1
		Fautaua S	506	19.10±0.894	10.99±0.466	9.83±0.477	57.43±2.649	51.41±1.977	1
		Hamuta D	166	18.50±0.907	10.78±0.500	9.65±0.483	58.41±2.341	52.05±1.979	1
		Hamuta S	289	18.33±0.836	10.71±0.440	9.56±0.532	58.30±2.398	51.69±1.839	1
		Pirae S	939	18.28±1.010	10.67±0.505	9.46±0.416	59.62±1.972	52.59±1.844	1
		Pohaitara S	379	17.56±0.790	10.48±0.451	9.26±0.477	61.50±2.342	55.50±1.885	1
		Ururoa D	1	17.45	10.90	9.90	59.34	52.66	1
		Ururoa S	340	17.45±0.816	10.36±0.458	9.21±0.543	58.75±2.410	51.82±1.776	1
		Tipaerui D	898	18.00±0.846	10.46±0.422	9.47±0.389	58.46±1.541	52.95±1.290	1
		Tipaerui S	24	18.26±0.684	10.68±0.387	9.71±0.497	56.74±2.266	51.11±1.960	1
		Vaitupa S	317	17.17±0.724	9.68±0.358	9.07±0.400	56.31±2.112	52.74±1.757	1
		Atitara D	1	16.25	9.59	9.30	58.02	53.85	1

		N	Length	Width	Mouth H	W/L	Mouth H/L	
o. crassa	Atitara S	154	16.72±0.781	9.71±0.392	9.02±0.395	56.02±3.769	47.95±1.837	1
	Orofero S	196	17.54±1.018	10.47±0.629	9.47±0.463	57.88±2.363	52.11±1.940	1
	Aoua S	635	17.16±0.859	9.95±0.433	9.13±0.535	56.45±2.495	51.90±1.976	1
	Papehue S	137	17.84±0.873	10.11±0.455	9.27±0.446	57.95±2.263	51.63±1.777	1
	Atehi S	408	17.23±0.872	9.99±0.440	8.90±0.395	58.42±2.349	51.48±1.821	1
	Maruapo S	133	17.00±0.667	9.95±0.368	8.77±0.488	58.11±2.611	52.74±1.856	1
	Punaruu S	522	17.19±0.893	9.99±0.516	9.08±0.455	57.91±2.184	53.41±1.812	1
	Taapuna S	219	18.06±0.912	10.47±0.437	9.66±0.395	56.51±3.769	49.66±1.837	1
o. rubescens	Tuauru S	8	19.50±0.559	10.55±0.321	9.70±0.451	55.91±2.016	48.99±1.533	1
	Ahonu	145	20.66±0.952	11.32±0.529	10.28±0.531	55.37±2.350	50.66±1.940	1
	Papenoo	336	20.38±0.925	11.30±0.563	10.33±0.445	57.66±3.376	52.24±2.022	1
	Faarumai	19	20.54±0.893	11.86±0.586	10.76±0.309	58.30±2.948	51.30±1.327	1
	Tiarei	15	19.98±0.727	11.65±0.429	10.27±0.243	60.00±1.979	53.00±0.500	1
	Mahaena	6	18.58±0.471	11.17±0.275	9.87±0.536	56.68±2.281	51.06±1.854	1
	Paraura	89	19.15±1.027	10.84±0.516	9.79±0.508	56.18±2.344	50.59±1.674	1
	Papeiha	111	20.13±0.989	11.29±0.511	10.19±0.550	56.51±2.552	50.50±1.787	1
	Oopu	272	22.83±1.130	12.89±0.542	11.53±0.316	60.50±1.782	51.50±1.692	1
	Apirimaue	1	16.75	10.30	8.70	54.00	49.38	1
	Mt Atara S	3	21.27±0.289	13.03±0.874	10.03±0.498	66.00±2.576	50.00±1.872	15
	Taravao D	1	20.00	13.20	10.00	55.64	50.77	15
	Ahaavini	214	19.68±0.925	10.94±0.494	9.98±0.417	58.76±2.118	51.88±2.255	1
	Tautira	82	20.26±0.711	11.89±0.480	10.51±0.449	57.70±2.536	50.71±1.772	1
	Aiurua	119	20.31±0.997	11.70±0.474	10.29±0.449	57.70±2.536	50.71±1.772	1
o. sinistrorsa	Faarahi S	461	20.10±1.059	10.00±0.576	10.09±0.586	54.70±2.240	50.07±1.913	1
	Tenaire S	364	19.10±0.956	10.68±0.458	9.82±0.508	55.84±2.474	51.36±1.947	1
	Tenaire D	523	17.95±0.841	10.10±0.409	9.24±0.452	56.19±2.294	51.41±1.880	1
	Maaru S	132	18.00±0.805	10.31±0.450	9.13±0.483	57.12±2.353	50.71±1.915	1
	Maaru D	69	18.08±0.829	9.95±0.435	9.11±0.444	54.92±2.446	50.51±2.096	1
	Faurahi S	1	18.10	10.50	7.60	54.83	50.50	15
	Taharuu D	3	18.05±0.845	9.83±0.499	9.17±0.574	48.50±1.700	46.10±1.633	1
	Tearatapu S	30	16.70±0.891	9.71±0.447	8.88±0.802	58.83±2.011	51.33±1.041	1
	Temarua D	2	17.83	10.43	9.13	56.83	52.02	1, 15
	Vaipoo S	61	17.01±0.742	9.67±0.395	8.86±0.425	54.58±2.085	50.63±1.542	1
	Opiriroa S	162	18.21±0.762	9.96±0.380	9.24±0.378	57.50±2.059	52.14±1.764	1
	Maruia D	25	18.24±0.704	10.51±0.360	9.53±0.437	57.02±2.158	51.72±1.895	1
	Maruia S	103	18.02±0.727	10.28±0.352	9.34±0.445	55.40±2.657	51.38±1.830	1
	Otuna S	346	18.55±0.812	10.29±0.439	9.56±0.391	55.89±2.184	50.85±1.887	1
	Tereehia S	69	17.61±0.848	9.86±0.402	8.97±0.434	56.97±2.084	51.45±1.845	1
	Tiamao D	148	17.84±0.811	10.18±0.400	9.21±0.428	56.90±2.159	51.64±1.857	1
	Tiamao S	176	17.81±0.765	10.15±0.351	9.22±0.153	55.75±4.250	46.33±1.529	1
	Vaihonu S	3	18.07±0.321	10.00±0.566	8.37±0.460	56.71±2.152	51.60±1.726	15
	Vaipuarii S	176	18.62±0.825	10.58±0.431	9.63±0.430	57.07±2.080	51.55±1.824	1
	Faaroa S	1	20.40	9.90	9.40	55.67	51.82	15
	Vaiaaia S	137	20.12±0.746	11.22±0.452	10.43±0.501	55.06±2.095	50.30±1.709	1
	Aiavaro S	168	20.13±0.928	11.08±0.432	10.12±0.502	53.43±2.259	49.75±1.673	1
	Vavii S	219	21.4±0.939	11.44±0.518	10.65±0.564	52.41±2.188	49.40±1.773	1
	Vaipoe S	190	21.81±0.987	11.46±0.495	10.79±0.452	56.03±2.309	51.30±1.904	1
laevigata	Mt Atara	2	20.85	12.80	10.90	61.32	47.20	15
	Teihomono	2	20.00	12.00	11.05	59.99	55.76	15
jackieburchi	Aivaro	77	18.55±0.838	11.47±0.596	9.42±0.478	63.76±3.660	51.80±2.473	15
	Aorai	15	19.36±0.976	12.34±0.889	10.03±0.498	58.33±1.534	47.55±1.439	15
	Faone	7	18.76±1.112	10.94±0.730	8.91±0.432	54.63±3.251	47.32±2.073	15
	Faatautia	1	20.50	11.20	9.70	62.08	50.05	15
	Onoheha	19	17.17±0.651	10.65±0.477	8.59±0.494	61.14±2.517	50.90±1.324	15
	Tautira	6	18.47±0.826	11.3±0.851	9.40±0.648	64.04±3.298	50.00±2.309	15
	Tiarei	1	17.80	11.40	8.90	61.86	50.54	15
producta	Orofero D	4	19.45±0.451	10.90±0.852	9.32±0.562	59.63±2.705	53.92±2.002	15

			N	Length	Width	Mouth H	W/L	Mouth H/L	
Partula	*producta*	Faarahi D	2	20.90	11.70	10.70	55.98	51.20	1
		Apirimaue S	284	19.39±0.927	10.7±0.501	9.60±0.495	55.10±2.381	59.40±1.840	1
		Titaviri S	165	19.75±0.919	10.91±0.503	10.13±0.475	55.17±2.716	51.26±1.887	1
		Titaviri D	77	18.12±0.804	10.4±0.364	9.59±0.469	57.32±1.736	52.80±2.000	1
		Vaihiria S	303	18.27±0.775	10.23±0.397	9.24±0.404	55.95±2.098	50.49±1.726	1
		Vairaharaha S	247	18.09±0.759	10.12±0.407	9.25±0.445	55.89±2.098	51.08±1.726	1
		Faurahi D	45	19.02±0.917	10.19±0.354	9.59±0.426	55.38±2.222	50.70±1.785	1, 15
		Moaroa S	432	18.64±0.806	10.34±0.391	9.46±0.477	55.75±1.995	51.03±1.750	1
		Taharuu S	320	18.57±0.763	10.36±0.406	9.49±0.439	57.37±2.158	51.34±1.792	1
		Teohu S	457	17.26±0.772	9.92±0.410	8.88±0.416	55.93±2.354	51.20±1.761	1
		Papeitii S	474	16.67±0.762	9.34±0.419	8.55±0.401	56.70±2.205	51.81±1.651	1
		Temarua S	539	16.85±0.726	9.56±0.394	8.75±0.404	55.53±1.804	50.33±1.583	1
		Tereehia D	60	18.27±0.745	10.14±0.385	9.18±0.456	55.00±2.444	50.83±1.770	1
	cytherea	Mt. Marau D	8	20.39±1.760	11.61±0.810	9.94±0.618	63.33±4.576	52.28±2.622	15
		Mt. Marau S	26	20.06±1.284	12.68±0.968	10.47±0.643	64.05±3.482	54.23±2.799	15
		Orohena S	33	18.33±0.839	11.77±0.608	9.67±0.883	60.04±2.640	51.45±4.429	3
	affinis	Ahonu D	143	17.01±0.832	10.06±0.492	9.02±0.432	60.50±1.414	51.83±1.247	1
		Ahonu S	3	17.45±0.424	10.57±0.410	9.10±0.220	58.28±2.901	51.20±1.136	1
		Orofara	11	16.10±0.323	9.12±0.454	8.33±0.436	59.06±1.987	53.51±1.846	15
		Tuauru D	366	17.32±0.786	10.32±0.428	9.32±0.189	60.83±0.471	55.17±1.247	1
		Tuauru S	3	17.35±0.707	10.57±0.377	9.63±0.444	58.05±1.777	51.41±1.375	1
		Faaripoo	308	16.73±0.719	9.90±0.417	8.97±0.436	58.21±2.276	53.49±1.859	1
		Papenoo D	147	16.63±0.830	9.70±0.434	8.90±0.446	59.83±1.333	52.93±1.050	1
		Papenoo S	7	17.19±1.057	10.16±0.563	9.13±0.477	57.74±2.261	52.73±1.734	1
		Farapa	582	17.83±0.949	10.29±0.435	9.42±0.372	59.58±1.963	52.52±1.871	1
		Faaromai	313	15.54±0.643	9.27±0.347	8.17±0.384	57.14±2.012	51.44±1.645	1
		Tiarei	151	16.22±0.687	9.30±0.365	8.37±0.392	58.06±2.168	52.35±1.831	1
		Mahaena D	137	16.70±0.719	9.70±0.355	8.76±0.555	62.25±1.785	54.75±2.278	1
		Mahaena S	4	17.15±0.512	10.70±0.616	9.45±0.499	63.173.399	55.17±2.749	1
		Paraura	579	16.31±0.730	9.66±0.376	8.66±0.261	57.92±2.060	50.75±0.924	1
		Faatautia	12	16.18±0.605	9.37±0.170	8.23±0.419	58.02±2.103	53.35±1.818	1
		Papeiha	163	16.85±0.672	9.79±0.368	9.01±0.412	58.69±2.122	52.61±1.624	1
		Utuofai	338	16.66±0.758	9.80±0.384	8.79±0.407	58.73±2.090	52.53±1.720	1
		Faaone	587	16.49±0.743	9.70±0.394	8.68±0.453	58.89±1.345	52.86±1.228	1
		Tehoro	303	16.04±0.668	9.42±0.329	8.48±0.502	60.09±2.125	52.53±1.877	1
		Aaavini S	3	15.05±0.424	9.57±0.618	8.17±0.438	59.36±2.183	53.53±1.955	1
		Ahaavini D	56	15.68±0.736	9.32±0.376	8.43±0.419	59.12±2.196	52.95±1.791	1
		Aionifaa	14	16.01±0.546	9.21±0.319	8.33±0.337	58.35±2.183	52.36±1.673	1
		Aiurua	100	15.58±0.646	9.11±0.322	8.18±0.823	59.02±2.011	49.95±2.617	1
		Vaiau D	11	15.54±0.680	9.03±0.410	7.99±0.303	62.30±1.939	53.70±2.315	1
		Vaiau S	5	15.29±0.294	9.5±0.219	8.22±0.443	58.57±1.971	52.21±1.682	1
		Hotutunu	45	16.45±0.740	9.64±0.432	8.61±0.551	58.90±2.154	52.90±0.490	1
		Vaiaaia D	5	16.37±0.960	9.70±0.593	8.70	61.50±1.972	53.50±1.703	1
		Vaiaaia S	1	16.80	10.30	8.90±0.419	59.48	53.95	1
		Vavii	415	16.65±0.732	9.93±0.382	9.00±0.361	58.00±1.954	51.50±1.291	1
		Haoma	44	16.84±0.777	10.13±0.406	8.84±0.413	57.29±1.145	51.78±1.623	1
		Oopu	36	16.76±0.685	9.83±0.422	8.89±0.292	57.17±2.134	51.00±1.348	1
		Apirimaue	6	17.75±0.458	10.13±0.423	9.07±0.400	58.30±2.310	52.20±1.735	1
		Tautira	205	15.99±0.793	9.31±0.404	8.36±0.438	57.60±2.253	52.93±1.974	1
		Vaipoe D	69	16.55±0.792	9.56±0.432	8.78±0.173	57.85±2.554	51.06±1.070	1
		Vaipoe S	4	16.95±0.640	9.80±0.163	8.65±0.471	59.32±2.118	53.49±1.873	15
		Tipaerui D	3	18.27±1.007	10.77±0.493	9.63±0.551	58.97±1.544	52.75±1.439	15
		Tipaerui S	1	16.80	10.80	9.30	64.29	55.36	15
		Fautaua D	5	18.18±0.542	10.72±0.462	9.58±0.387	58.93±1.560	52.69±0.867	15
		Fautaua S	3	17.13±1.007	10.70±0.781	8.83±0.513	62.44±2.114	51.59±2.147	15
	affinis ?	Punaruu S	5	17.76±1.016	10.48±0.665	8.88±0.422	58.73±2.114	52.89±1.767	15
	compressa	Mt Aorai	10	19.53±1.007	10.86±0.601	10.67±0.542	56.20±2.001	52.50±2.566	15
		Orofere	1	18.30	10.30	9.80	56.30	53.60	15

		N	Length	Width	Mouth H	W/L	Mouth H/L	
compressa	Tefaaiti	7	18.34±1.028	9.91±0.540	9.33±0.562	56.90±3.074	54.28±3.157	15
	Viriviriterai	12	18.77±1.000	11.35±0.756	10.45±0.891	60.500±3.071	55.65±3.261	15
	Mt Orofena	2	18.95	10.50	9.70	59.35	57.61	15
	Pic Rouge	1	21.30	12.50	12.00	58.69	56.34	15
	Mt. Marau	1	16.60	10.70	8.60	64.46	51.81	15
nodosa 'exigua'	Atehi D	4	14.08±0.492	8.85±0.166	7.75±0.621	60.90±1.760	53.78±1.592	1
n. 'intermedia'	Orofero D	18	15.37±1.195	9.29±0.638	8.28±0.380	61.94±2.170	54.65±1.577	1
	Aoua D	382	14.42±0.671	8.95±0.408	7.91±0.369	62.33±1.934	54.52±1.500	1
	Papehue D	205	14.76±0.659	9.20±0.398	8.07±0.470	58.93±2.259	52.43±1.624	1
n. 'laeva'	Maruapoo D	28	15.61±0.874	9.22±0.458	8.20±0.403	60.04±2.424	52.24±1.825	1
	Maruapoo S	649	15.55±0.741	9.34±0.406	8.15±0.506	61.39±2.220	54.22±1.715	1
n. 'composita'	Punaruu D	160	15.58±0.878	9.56±0.460	8.42±0.458	60.30±2.102	53.80±1.806	1
	Taapuna D	298	15.99±0.821	9.66±0.435	8.62±0.340	62.30±1.950	54.33±2.794	1
	Taapuna S	6	15.90±0.971	9.90±0.490	8.63±0.260	62.70±1.785	54.50±1.732	1
diminuta		209	14.16±0.634	9.10±0.345	7.63±0.390	64.28±2.207	53.75±1.798	1
clara 'prima?'	Papenoo	5	15.76±0.502	9.14±0.265	8.90±0.417	57.79±2.157	56.14±1.761	1
clara 'prima'	Otuna	9	15.48	9.06	8.63	58.28	55.94	1
	Opiriroa	6	15.35	8.73	8.33	57.00	54.67	1
	Temarua	5	14.81	8.58	8.46	57.50	57.46	1
	Papeiti	9	14.62	8.50	8.30	58.17	56.39	1
	Teohu	31	15.26	9.02	8.75	58.92	57.37	1
	Moaroai	25	15.65	9.14	8.98	58.26	57.10	1
	Faarahi	9	15.68	8.92	8.72	56.83	55.61	1
	Vairaharaha	94	14.72	8.84	8.29	60.01	56.14	1
	Vaihiria	75	15.22	8.97	8.70	58.72	56.98	1
clara 'angusta'	Tereehia	35	15.9±0.783	8.79±0.406	8.76±0.523	55.79±2.250	54.20±2.237	1, 15
clara 'minor'	Orofere	27	14.96±0.446	8.48±0.285	8.32±0.305	56.53±1.895	55.51±1.655	1
	Atitara	32	14.68±0.485	8.27±0.292	8.07±0.368	56.41±1.738	55.85±1.512	1
	Vaitupa	38	15.16±0.807	8.73±0.409	8.29±0.391	57.76±2.520	55.55±1.917	1
	Vaipuarii	20	14.57±0.580	8.67±0.336	8.06±0.320	59.40±1.546	55.05±1.658	1
	Tiamao	21	14.81±0.440	8.58±0.243	8.29±0.352	58.02±1.789	55.64±1.959	1
clara 'parva'	Punaruu	6	14.75±0.693	8.50±0.416	8.13±0.468	57.83±2.055	54.83±1.106	1
	Maruapoo	3	13.25±0.245	7.50±0.163	7.37±0.094	56.17±1.247	55.17±0.471	1
	Atehi	20	15.04±0.429	8.25±0.275	8.23±0.270	54.65±1.459	54.55±1.936	1
	Papehue	30	14.93±0.649	8.41±0.375	8.41±0.461	56.10±1.200	56.00±1.384	1
hyalina	Mangaia (Cooks)	46	15.92±0.819	8.96±0.298	9.30±0.407	55.46±1.737	57.14±1.429	1
	Moki (Cooks)	53	15.10±0.758	8.39±0.312	8.63±0.592	57.47±2.853	54.78±2.505	1
	Raivavae (Australs)	5	16.60±0.430	9.54±0.527	9.10±0.523	55.96±5.596	58.39±5.839	15
	Rurutu (Australs)	36	17.84±0.786	10.30±0.339	9.58±0.449	57.77±1.403	53.72±2.019	15
	Papenoo	34	14.75±0.784	8.38±0.385	8.46±0.472	56.74±1.592	57.33±1.706	1
	Tuauru	12	14.95±0.510	8.60±0.277	8.72±0.472	57.58±1.754	58.08±2.465	1
	Pirai	37	15.25±0.612	8.90±0.411	8.99±0.403	58.07±1.953	58.82±1.946	1
	Fautaua	93	15.24±0.726	8.87±0.391	9.87±0.446	58.13±1.673	58.23±1.773	1
	Tipaerui	16	15.50±0.600	8.63±0.338	8.95±0.456	55.63±1.833	57.25±2.077	1
	Taapuna	12	15.45±0.911	8.63±0.519	8.95±0.659	59.17±1.700	57.83±1.700	1
	Maruapoo	9	15.12±0.447	8.83±0.422	8.82±0.365	58.06±2.793	57.83±2.571	1
	Atehi	11	15.32±0.772	8.86±0.545	9.00±0.714	58.05±2.996	58.95±1.050	1
	Aoua	19	14.67±0.495	8.44±0.344	8.42±0.343	57.39±1.334	57.45±1.700	1
	Otuna	18	15.09±0.754	8.58±0.354	8.98±0.421	56.94±1.771	59.17±1.347	1
	Tenaire	7	15.91±0.793	8.96±0.396	8.97±0.377	56.07±0.903	56.21±1.278	1
	Paraura	21	15.58±0.828	8.69±0.439	8.96±0.330	55.54±1.588	57.50±1.414	1
	Faarumai	10	14.96±0.554	8.44±0.311	8.64±0.250	56.20±1.487	57.80±0.900	1
	Tehoro	9	15.92±0.699	8.70±0.327	9.13±0.422	54.61±0.737	57.17±1.054	1
	Haavini	20	15.22±0.580	8.79±0.257	9.00±0.316	57.65±1.740	58.95±1.465	1
	Tautira	17	14.87±0.750	8.56±0.327	8.99±0.457	57.62±1.937	58.92±1.536	1
	Vavii	21	16.05±0.573	9.00±0.294	9.33±0.411	55.88±2.251	57.93±1.256	1
h 'marmorata'	Maara	3	15.85±0.616	8.57±0.189	8.33±0.189	53.83±0.943	55.83±0.943	1
	Tenaire	17	15.63±0.934	8.86±0.312	9.03±0.499	56.62±1.841	57.56±1.732	1

			N	Length	Width	Mouth H	W/L	Mouth H/L	
Partula	*h. 'marmorata'*	Vavii	19	15.46±0.717	8.91±0.324	9.15±0.364	58.18±1.416	59.24±0.965	1
		Titaviri	12	15.15±0.566	8.97±0.350	9.00±0.321	59.08±1.187	59.42±0.493	1
		Vaiaaia	15	15.63±0.640	9.14±0.294	9.18±0.449	58.43±1.612	58.70±1.511	1
		Tautira	10	15.17±0.306	8.98±0.299	9.06±0.250	58.90±1.744	59.80±1.005	1
		Haavini	30	15.07±0.731	8.77±0.333	8.83±0.367	58.23±1.982	58.83±1.325	1
		Paraura	19	15.42±0.541	8.74±0.408	9.94±0.392	56.66±1.724	57.97±1.758	1
	s. suturalis	Paia D	13	18.08±0.415	10.28±0.277	9.65±0.285	56.81±1.264	53.27±1.187	4
		Maatea D	530	17.72±0.786	10.01±0.378	9.12±0.421	56.86±1.980	51.51±1.733	4
		Haumi D	1685	17.56±0.784	9.93±0.405	9.07±0.416	56.47±2.113	51.60±1.800	4
		Hotutea D	761	17.64±0.988	9.95±0.465	9.08±0.540	56.27±2.385	51.42±1.975	4
		Atiraa D	301	17.62±0.852	10.01±0.429	9.27±0.423	56.85±1.972	52.19±1.861	4
	s. 'dendroica'	Vaihiaila S	39	20.62±0.802	11.43±0.453	10.74±0.450	55.29±2.338	51.99±1.657	4
		Matapoopoo S	16	20.81±0.601	11.29±0.254	10.96±0.387	54.11±1.902	52.50±2.000	4
		Tepatu S	16	20.86±0.912	11.31±0.572	10.70±0.374	54.25±3.437	51.25±2.194	4
		Pupuhi-Tevairoa S	30	20.03±0.761	11.17±0.360	10.64±0.366	55.80±1.763	53.17±1.616	4
	s. vexillum	Atimaha D	249	18.19±0.940	10.18±0.449	9.23±0.464	55.92±2.268	50.75±2.127	4
		Paahonu D	36	18.81±0.654	10.73±0.366	9.82±0.435	57.31±1.595	52.53±1.573	4
		Vaianai D	339	18.72±0.810	10.34±0.440	9.54±0.502	55.19±2.062	50.83±1.767	4
		Oio D	614	19.43±0.862	10.77±0.446	9.91±0.468	55.35±1.860	50.85±1.853	4
		Oio S	4	18.05±0.561	10.55±0.296	9.30±0.141	58.25±2.488	51.00±1.118	4
		Haapiti D	415	19.07±0.877	10.63±0.442	9.47±0.488	55.67±1.977	49.55±1.853	4
		Haapiti S	17	18.61±0.729	10.82±0.357	9.39±0.429	58.14±1.493	50.62±1.367	4
		Uufau D	158	18.75±0.796	10.56±0.441	9.59±0.426	56.19±2.207	50.93±1.922	4
		Uufau S	512	18.51±0.820	10.41±0.372	9.60±0.422	56.19±2.190	51.83±1.796	4
		Moruu D	1	18.35	10.10	9.10	54.50	50.50	4
		Moruu S	483	17.89±0.774	10.13±0.393	9.20±0.428	56.54±1.958	51.36±1.714	4
		Varari S	524	18.30±0.704	10.18±0.371	9.32±0.376	55.60±2.148	50.84±1.851	4
		Maraarii S	660	18.86±0.837	10.27±0.442	9.72±0.454	54.37±2.113	51.47±1.729	4
		Maatepua S	6	19.15±0.975	11.27±0.509	10.13±0.752	59.33±1.067	52.17±2.211	4
		Taapurau S	147	19.72±0.810	10.79±0.424	10.19±0.457	54.67±2.095	51.64±1.827	4
		Pafatu S	9	20.43±0.862	10.92±0.380	10.42±0.400	53.25±1.785	51.06±0.661	4
		Maramu D	163	20.57±0.805	11.12±0.450	10.35±0.495	54.15±2.115	50.27±1.889	4
		Mauaroa D	79	20.71±0.781	11.45±0.435	10.68±0.478	55.20±1.876	51.47±1.699	4
		Faataofe D	345	19.37±0.788	11.18±0.470	10.25±0.461	54.79±2.082	50.27±1.865	4
		Fanautaata D	165	20.35±0.760	11.04±0.405	9.97±0.420	54.21±1.809	48.93±1.706	4
		Roroie D	303	19.76±0.809	10.95±0.450	9.83±0.338	55.33±1.810	49.70±1.814	4
		Morioahu D	170	19.58±0.889	10.91±0.471	9.91±0.438	55.64±2.033	50.55±1.850	4
		Morioahu S	405	19.22±0.879	10.78±0.400	9.81±0.459	56.04±2.162	50.95±1.846	4
		Opunohu D	15	19.68±0.960	11.21±0.474	10.19±0.350	56.93±2.195	51.50±2.330	4
		Opunohu S	746	19.30±0.840	10.59±0.378	9.97±0.454	54.92±2.111	51.60±1.938	4
		Tupuna S	164	19.94±0.853	10.62±0.401	10.03±0.460	53.15±1.895	50.23±1.687	4
		Vaiana D	1	20.15	10.10	9.90	50.50	49.50	4
		Vaiana S	66	20.56±0.856	11.02±0.437	10.27±0.392	53.61±2.336	49.92±1.875	4
		Aareo S	12	20.35±0.732	10.72±0.398	10.09±0.327	52.80±1.826	49..51±0.817	4
		Urufara S	124	20.14±0.870	10.97±0.500	10.20±0.418	54.30±2.121	50.50±1.830	4
		Faatoai S	111	21.10±0.754	11.36±0.373	10.73±0.463	53.74±2.133	50.79±2.182	4
		Puutu D	121	19.57±0.810	10.73±0.383	10.08±0.434	54.76±2.075	51.42±1.974	4
		Puutu S	51	19.27±0.914	10.79±0.375	10.13±0.467	55.89±2.347	52.48±2.124	4
		Vaipohe D	157	20.34±0.767	11.15±0.516	10.48±0.444	54.70±2.239	51.47±1.824	4
		Vaipohe S	3	20.25±0.150	11.33±0.300	10.70±0.400	55.25±1.414	51.95±2.160	4
		Faamaariri D	448	19.62±0.843	10.78±0.435	10.08±0.467	54.86±1.768	51.52±1.726	4
		Faamaariri S	254	19.85±0.863	11.22±0.342	10.41±0.425	56.43±2.281	52.33±2.060	4
		Tepu S	13	19.04±0.720	10.04±0.285	10.73±0.387	56.88±1.643	52.58±2.055	4
		Maharepa S	57	19.52±0.924	10.01±0.417	10.73±0.453	55.17±2.089	51.30±1.475	4
		Vaianai D	94	18.93	10.62	9.64	56.40	51.14	4
		Uufau D	158	18.75	10.56	9.59	56.19	50.93	4
		Uufau S	512	18.51	10.41	9.60	56.19	51.83	4

		N	Length	Width	Mouth H	W/L	Mouth H/L	
s. vexillum	Maraarii S	660	18.86	10.27	9.72	54.37	51.47	4
	Faataofe S	345	20.36	11.18	10.26	54.79	50.27	4
	Roroie D	303	19.76	10.95	9.83	55.33	49.70	4
aurantia	Paraoro	462	18.14±0.712	9.73±0.367	9.27±0.382	53.54±1.972	50.89±1.585	4
	Paparoa	647	18.74±0.814	10.11±0.357	9.50±0.419	53.86±2.020	50.57±1.860	4
	Vaiare-Vaipohe	112	18.71±0.736	10.16±0.355	9.58±0.423	54.23±2.128	51.09±1.846	4
	Vaiare-Faamariri	159	18.08±0.801	9.73±0.358	9.23±0.411	53.53±2.203	50.95±1.796	4
mirabilis	Vaianai	73	16.62±0.734	9.40±0.366	8.70±0.409	56.50±1.828	52.24±2.233	4
	Faataofe	485	16.00±0.692	9.38±0.370	8.50±0.372	58.53±1.973	53.01±1.829	4
	Mouaroa	32	16.78±0.636	9.78±0.418	8.88±0.484	58.16±1.593	52.91±1.656	4
	Maramu	573	17.07±0.769	9.76±0.437	8.87±0.423	57.01±2.072	51.85±1.672	4
	Tefeo	455	16.77±0.759	9.67±0.401	8.69±0.417	57.61±2.245	51.67±1.672	4
	Tefeo east	769	16.62±0.664	9.64±0.367	8.63±0.379	57.96±1.933	51.86±1.673	4
	Fareaito	401	16.94±0.695	9.78±0.407	8.65±0.393	57.23±1.883	51.10±1.604	4
	Irua	494	16.47±0.764	9.59±0.275	8.60±0.446	58.12±2.110	52.05±1.668	4
	Ririofau	22	16.85±0.746	9.82±0.389	8.81±0.433	58.00±2.251	52.23±1.681	4
	Matapoopoo	884	16.50±0.690	9.63±0.373	8.68±0.363	58.29±2.109	52.51±1.696	4
mirabilis	S	7	16.17±0.747	9.73±0.328	8.67±0.361	50.00±2.711	49.38±1.852	4
taeniata elongata	Haapiti	172	15.48±0.839	9.63±0.481	9.27±0.430	55.99±2.208	53.77±1.934	4
	Oio	521	16.06±0.876	8.67±0.480	8.33±0.417	56.33±2.357	53.95±1.927	4
	Vaianai	471	16.37±0.842	9.22±0.404	9.07±0.443	56.20±2.142	55.28±1.954	4
	Paahonu	52	16.94±0.746	9.16±0.407	9.17±0.390	54.08±2.196	54.04±1.611	4
	Atimaha	523	16.02±0.852	8.91±0.417	9.16±0.476	55.59±2.082	54.87±1.809	4
	Maatea	100	16.42±0.947	9.16±0.499	9.21±0.515	55.68±2.253	55.78±1.942	4
	Haumi	154	15.80±0.843	8.77±0.513	8.68±0.452	53.43±2.382	54.77±1.753	4
	Maramu	447	16.09±0.767	8.52±0.402	8.81±0.458	52.89±1.994	54.74±1.756	4
	Tefeo	294	16.11±0.806	8.67±0.367	8.99±0.453	53.84±1.911	55.69±1.855	4
	Fareaito	1113	16.04±0.722	8.56±0.365	8.86±0.338	53.24±1.980	55.11±1.823	4
	Irua	384	16.51±0.762	8.95±0.412	9.19±0.410	53.75±2.163	55.78±1.850	4
	Ririofau	289	16.76±0.807	9.10±0.394	9.28±0.444	54.21±2.148	55.32±1.684	4
	Poua	63	17.51±0.876	9.45±0.460	9.44±0.502	53.67±1.915	55.00±1.580	4
	Mouaputa	466	16.07±0.908	8.69±0.484	9.04±0.438	54.06±2.114	56.11±1.928	4
	Matapoopoo	168	15.68±0.823	8.78±0.400	8.60±0.442	55.72±2.084	54.67±1.882	4
	Tepatu	6	16.30±0.585	9.00±0.252	9.10±0.231	55.00±1.118	55.83±1.598	4
t. simulans	Pafatu	434	17.63±0.896	10.43±0.415	9.49±0.467	59.15±2.542	53.75±2.056	4
	Taapurau	919	15.75±0.754	9.58±0.403	8.45±0.409	60.76±2.207	53.55±1.751	4
	Maatepua	98	16.73±0.760	10.03±0.416	9.11±0.366	59.91±2.346	54.46±1.818	4
	Maraarii	481	16.46±0.732	9.94±0.436	9.10±0.432	60.28±2.305	55.12±1.883	4
	Varari	488	15.87±0.764	9.57±0.375	8.60±0.381	60.26±2.210	54.03±1.773	4
	Maoruu	375	15.69±0.643	9.41±0.385	8.40±0.360	59.86±1.978	53.42±1.720	4
	Uufau	278	16.56±0.802	9.43±0.449	7.10±0.422	57.15±2.178	52.15±1.922	4
	Tiaorooro	40	17.36±0.944	9.30±0.446	9.25±0.520	53.60±1.729	53.95±2.133	4
	Mouaroa	185	15.76±0.631	9.08±0.348	8.51±0.367	57.51±2.004	53.70±1.739	4
	Faataofe	190	15.94±0.757	8.71±0.353	8.65±0.415	54.60±1.994	54.09±1.911	4
	Fanautaata	105	16.46±0.856	8.92±0.428	8.69±0.441	54.18±2.300	52.74±2.100	4
	Roroie	176	16.42±0.793	8.70±0.389	8.68±0.442	52.94±2.025	52.70±1.992	4
	Morioahu	389	16.79±0.818	9.18±0.419	8.84±0.424	54.61±1.997	52.51±1.720	4
	Opunohu	1529	15.34±0.739	8.48±0.370	8.03±0.376	55.67±2.221	52.61±1.855	4
	Tupuna	274	16.10±0.739	8.98±0.369	8.65±0.425	55.70±2.038	53.65±2.019	4
	Vaiana	154	16.78±0.772	9.30±0.367	8.97±0.298	55.38±2.073	53.34±1.612	4
	Hotutea	71	15.95±0.861	9.35±0.437	8.68±0.431	58.58±2.559	54.28±1.820	4
	Atiraa	558	16.59±0.816	9.87±0.440	8.92±0.443	59.36±2.241	53.59±1.908	4
	Vaipua	118	17.19±0.729	9.72±0.378	9.34±0.456	56.42±1.814	54.18±1.780	4
	Paopao	216	16.46±0.838	9.36±0.424	8.91±0.434	56.82±2.114	54.07±1.950	4
	Paparoa	946	16.67±0.837	9.29±0.409	9.12±0.478	55.67±2.454	54.53±1.987	4
	Vaiare-Vaipohe	228	15.75±0.776	8.69±0.377	8.72±0.412	55.14±2.169	55.17±1.999	4

			N	Length	Width	Mouth H	W/L	Mouth H/L	
Partula	*t. 'striolata'*	Vaiare-Puutu	537	15.38±0.707	9.59±0.385	8.14±0.377	62.18±2.269	53.36±1.736	4
		Vaiare-Vaipohe	155	15.87±0.746	9.06±0.430	8.07±0.412	57.61±2.151	54.75±1.867	4
		Vairare-Faamaariri	1792	15.08±0.663	8.99±0.357	8.08±0.360	59.54±2.336	53.39±1.815	4
		Tepu	438	15.28±0.663	9.58±0.386	8.17±0.389	62.67±2.304	53.38±1.848	4
		Temae	486	15.40±0.682	9.48±0.383	8.19±0.370	61.57±2.132	53.07±1.759	4
		Maharepa	1302	14.73±0.602	9.10±0.365	7.82±0.355	61.94±2.240	52.79±1.813	4
		Oroau	140	15.23±0.739	9.38±0.418	8.03±0.398	61.57±2.017	52.63±1.544	4
		Farehotu	423	15.54±0.644	9.55±0.344	8.16±0.359	61.41±2.120	52.42±1.692	4
		Paraoro	704	15.06±0.703	9.10±0.372	7.93±0.372	60.38±2.128	52.61±1.733	4
	t. 'nuclela'	Aareo	270	15.98±0.808	9.09±0.423	8.57±0.421	56.81±2.033	53.48±1.864	4
		Urufara	1939	15.95±0.794	9.14±0.402	8.54±0.461	57.19±2.231	53.38±1.889	4
		Poararua	344	16.28±0.809	9.50±0.417	8.78±0.445	58.32±2.320	53.77±1.907	4
		Vaitapi	151	16.40±0.788	9.67±0.402	8.92±0.381	58.98±2.357	54.34±2.168	4
		Faatoai	5424	15.09±0.606	9.05±0.322	8.09±0.326	59.99±2.211	53.51±1.745	4
	t. 'spadicea'	Uhene	364	15.81±0.807	10.00±0.431	8.67±0.427	63.21±2.378	43.22±1.989	4
		Vaihere	171	15.96±0.712	9.88±0.432	8.66±0.431	61.92±2.000	54.14±1.701	4
		Vaihiaiia	753	15.22±0.656	9.45±0.389	8.33±0.373	62.01±2.003	54.57±1.708	4
		Matapoopoo	553	15.85±0.787	9.20±0.411	8.67±0.445	58.01±2.184	54.58±1.974	4
		Maahumeru	32	15.93±0.703	9.55±0.422	8.37±0.408	59.91±2.120	52.38±1.692	4
		Punahara	464	16.24±0.777	9.65±0.421	8.69±0.418	59.32±2.059	53.41±1.728	4
		Marure	287	16.60±0.828	10.10±0.433	8.85±0.481	60.79±2.398	53.14±1.871	4
		Pupuhi-Teviroa	317	16.09±0.724	9.86±0.402	8.62±0.410	61.50±2.127	53.49±1.872	4
	t. 'propinqua'	Tepatu	189	17.23±0.762	9.85±0.425	9.08±0.458	57.07±1.810	52.56±1.957	4
		Matapoopoo	38	16.40±0.865	9.79±0.427	8.81±0.463	59.74±2.232	53.74±2.065	4
		Vaihiaiia	43	16.31±0.555	9.68±0.312	8.24±0.245	59.20±1.959	50.43±1.500	4
	incrassa exigua	Vaiare-Puutu	1	16.85	8.90	9.10	52.30	53.50	4
		Vaiare-Faamaariri	18	14..98±0.789	8.58±0.296	8.22±0.499	57.53±1.651	54.65±1.432	4
		Tepu	12	14.45±0.735	8.38±0.321	8.10±0.483	57.83±1.599	56.00±1.678	4
		Maharepa	26	15.00±0.767	8.53±0.401	8.36±0.344	55.78±1.645	55.66±1.598	4
		Paraoro	52	15.53±0.775	8.76±0.369	8.73±0.381	56.25±2.114	55.87±1.659	4
	i. incrassa	Tahiti (Aoua)	63	15.45±0.973	8.61±0.505	8.38±0.471	57.04±2.284	58.17±1.818	1
	mooreana	Haapiti	33	17.80±0.720	10.14±0.344	9.67±0.454	56.92±2.944	54.20±1.749	4
		Oio	217	18.00±0.717	10.12±0.390	9.75±0.394	56.12±2.403	54.14±1.733	4
		Vaianai	2036	17.45±0.669	9.88±0.357	9.50±0.392	56.52±2.173	54.32±1.774	4
		Paahonu	1	17.45	10.50	9.50	60.50	54.50	4
		Atimaha	1160	15.59±0.649	9.15±0.007	8.48±0.391	58.64±2.284	54.22±1.925	4
		Tefeo	88	18.32±0.697	19.20±0.371	9.71±0.431	55.55±2.005	52.92±1.629	4
		Maramu	1559	17.90±0.765	10.21±0.376	9.54±0.428	56.97±2.154	53.20±1.843	4
		Faataofe	384	18.26±0.668	10.09±0.356	9.80±0.373	55.16±2.073	53.56±1.557	4
		Fanautaata	6	18.05±0.490	10.10±0.383	9.83±0.377	55.83±2.211	54.33±1.462	4
		Roroie	51	18.21±0.630	10.19±0.292	9.98±0.396	55.97±1.405	54.66±1.526	4
	tohiveana olympia	Mouaputa	34	18.95±0.659	11.04±0.439	10.02±0.479	51.02±1.656	55.91±2.061	4
	t. tohiveana		553	21.74±0.989	12.16±0.495	11.11±0.473	51.12±1.659	56.06±2.061	4
	varia	Faahiti	342	17.45±0.801	10.50±0.456	9.50±0.431	54.44	60.17	14
		Atite	404	18.18±0.824	10.86±0.519	9.85	54.18	64.00	14
		Tareu	556	18.51±0.858	11.01±0.455	9.92±0.456	53.59	59.48	14
		Faie H	304	17.56±0.756	10.65±0.430	9.54±0.777	42.640±2.337	61.29±3.062	14
		Faanoo	101	17.09±0.745	10.57±0.385	9.42±0.408	55.12	61.85	14
		Vitou	264	17.56±0.722	10.70±0.416	9.30±0.421	52.96	60.93	14
		Haavaii Vairahi	19	16.85±0.583	10.50±0.385	9.10±0.408	54.00	62.31	14
		Havaii Mena Hui	92	17.11±0.789	10.50±0.481	9.30±0.433	54.35	61.37	14
		Haavai Hine	307	16.85±0.325	10.72±0.420	9.30±0.285	55.19	63.62	14
		Havaii centre	18	17.15±0.680	10.30±0.338	9.30±0.451	53.51±1.905	59.76±2.144	14
		Haavaii Tevaroa	103	17.45±0.746	10.50±0.479	9.10±0.448	52.15	60.15	14
		Hamene H	370	16.85±0.747	10.50±1.072	8.70±0.435	54.09±1.908	59.40±2.093	14
		Tepua H	726	17.14±0.829	10.27±0.518	9.40±0.723	40.60±3.855	59.41±3.486	14

		N	Length	Width	Mouth H	W/L	Mouth H/L	
varia	Tefari	4	17.65±0.881	10.50±1.049	7.18±0.463	40.68±1.976	59.15±2.118	15
	Mahuti	2	16.85	9.95	8.05	53.02	59.80	15
	Maioroa	355	18.49±0.725	11.06±0.415	9.81±0.388	53.06	59.82	14
	Parea	90	18.04±0.814	11.11±0.442	10.21±0.466	53.30±1.809	60.33±2.221	14
	Vaiatea	2	16.00	10.30	8.00±0.433	50.00	64.38	15
	Haapu D	118	17.64±0.705	10.9±0.387	9.90±0.516	56.12	61.79	14
rosea	Faahiti	11	21.81±0.966	12.57±0.502	8.85±0.537	42.21±1.987	58.09±1.589	15
	Tepepe	2	18.95	11.05	9.85±0.458	51.94	57.66	15
	Faie	5	22.00±1.838	13.04±0.378	11.50±0.306	41.20±2.425	59.094.518	15
	Fatotoro	10	22.97±0.955	14.07±0.751	9.80±0.430	42.66	61.25	15
	Vaaitou	3	22.03±0.808	13.00±0.721	9.07±0.706	44.92±2.937	60.52±2.579	15
	Hina, Havaii	7	22.56±0.927	13.64±0.594	10.13±0.888	38.47±2.918	59.54±2.675	15
	Mimihiu	10	22.99±0.717	13.69±0.749	8.86±0.832	38.75±3.687	59.52±2.786	15
	Vairahi	28	22.76±0.723	13.54±0.672	8.81±1.331	43.12±5.430	59.40±1.898	15
	Haavaii	11	23.20±1.163	13.77±0.637	10.00±0.870	42.74±3.233	59.02±2.387	15
	Haamene	30	20.45±1.101	12.9±0.654	9.25±0.808	40.57±2.913	57.22±3.070	15
	Tepare	13	21.80±1.761	12.46±1.030	8.83±0.976	44.51±4.958	59.31±2.099	15
	Mahuti	31	22.10±1.107	13.11±0.584	9.83±1.209	55.86±1.112	62.335.036	15
	Vaiarea	3	20.70±2.066	12.83±0.289	11.57±0.637	43.53±1.883	58.73±2.360	15
	Haapu	26	21.87±1.153	12.84±0.729	9.52±1.252	43.51±3.222	59.83±2.379	15
	Hopu	36	20.24±2.366	12.13±1.485	9.70	51.98	58.32	15
arguta		67	11.44±0.697	7.83±0.599	6.870±0.500	56.00±2.000	62.00±2.000	15
f. faba	Vaetahe	9	26.40±0.300	16.60±0.100	15.10±1.000	54.00±1.200	59.00±3.000	14
	Vaipao	367	24.90±0.600	15.50±0.800	14.10	56.62	62.20	14
	Vaitaporo	250	26.10±0.700	15.90±0.800	14.50±0.300	56.00±0.600	60.00±1.600	14
	Teupua	236	25.50±0.300	16.30±0.700	14.50±0.455	55.00±1.223	61.00±1.556	14
	Vairua	141	26.10±0.500	16.10±0.800	14.10±0.600	53.00±2.400	61.00±1.600	14
	Faaharoto	66	25.50±0.600	15.50±0.600	14.00±0.600	54.00±2.000	60.00±1.000	14
	Hamoa	327	25.50±0.900	15.40±0.700	13.90±0.600	54.00±1.800	58.00±2.000	14, 15
	Vairahi	210	25.80±0.800	15.50±0.500	13.50±0.400	53.00±2.200	58.50±2.000	14
	Averaiti	147	23.50±0.800	13.70±0.700	12.80±0.500	53.00±1.800	57.00±2.000	14
	Averarahi	118	23.40±0.700	13.5±0.900	12.50±0.500	57.00±3.000	61.00±2.000	14, 15
	Faaiti	44	22.80±1.120	13.97±1.091	12.50±0.600	53.00±1.600	58.00±2.000	14, 15
	Faaroa	247	24.60±0.900	14.9±0.900	13.50±0.400	54.00±2.200	61.00±1.500	14, 15
	Farepaiti	118	25.40±1.100	15.50±1.000	14.30±0.400	53.00±2.000	60.50±2.200	14
	Fareparahi	222	24.90±0.700	15.4±0.700	13.60±0.600	54.00±1.000	61.00±1.600	14
f. faba	Opoa	228	25.20±0.700	15.20±0.400	13.70±0.600	54.00±0.900	60.00±2.000	14, 15
	Hotopuu	148	25.50±0.500	15.50±0.400	13.50±0.600	55.00±1.200	63.00±3.000	14
	Faarahi	99	23.70±0.600	15.50±0.400	13.40±0.600	55.00±3.000	61.00±2.000	14
	Vaianae	25	25.00±0.400	15.30±0.900	13.90±0.400	55.00±0.600	59.00±1.400	14
	Faatemu	29	24.40±0.500	14.90±0.400	13.50±0.600	55.00±1.600	60.00±2.000	14
	Tuatau	117	24.80±0.400	15.30±0.600	13.90±0.500	55.00±0.800	61.00±1.400	14
	Vaetuuna	51	25.20±0.400	15.70±0.900	14.10±0.300	56.00±1.200	61.00±2.000	14
	Uturoto	92	25.20±0.500	15.50±0.400	14.30±0.600	55.00±1.600	61.00±2.000	14
	Varihuti	190	25.50±0.600	15.30±0.600	14.10±0.600	54.00±1.300	61.00±2.000	14
	Tirae	115	25.20±0.600	15.50±0.600	14.10±0.400	54.00±2.000	59.00±2.000	14
	Vaiaau	85	26.10±0.700	15.50±0.400	14.00±0.300	58.00±1.001	61.00±1.005	14
	Marahi	1	24.00	14.90	13.90	54.00	62.00	14
	Tehurui	1	24.30	15.10	13.50	54.00	60.00	14
	Utuara	23	24.90±0.700	15.10±0.400	13.70±1.300	54.00±1.010	61.00±1.232	14
	Ereeo	1	24.90	15.30	13.70	55.00±0.400	59.00±1.000	14
	Hanua	17	26.10±0.900	15.70±1.200	13.70±0.300	53.00±1.400	60.50±2.000	14
	Tooroa	274	24.60±0.600	15.20±0.700	13.50±0.355	53.23±1.765	62.95±1.332	14
	Tahara	44	24.60±1.200	15.10±0.600	13.30±0.400	55.00±2.000	63.00±2.400	14
	Huaru	262	24.00±0.900	15.50±0.600	13.30±0.400	53.00±1.800	62.00±2.000	14
	Pufau	117	23.70±0.600	14.00±0.400	12.70±0.600	54.00±1.800	62.50±2.200	14

			N	Length	Width	Mouth H	W/L	Mouth H/L	
Partula	*f. faba*	Faafau	377	23.90±0.800	15.10±0.600	13.00±0.500	54.00±2.000	62.00±1.800	14
		Apoiti	227	24.00±0.700	14.60±0.400	13.00±0.400	54.00±2.400	60.00±2.000	14
		S	1	25.8.	15.70	11.90	46.00	61.00	15
	f. subangulata	Patio	121	24.45±0.638	14.40±0.614	13.20±0.625	54.50±2.148	62.50±2.311	14
		Afatatea	57	24.65±1.253	15.10±0.587	13.80±0.667	54.50±1.938	60.50±2.278	14
		Pahure	88	24.65±1.079	14.90±0.618	13.30±0.565	54.50±1.873	60.50±2.184	14
		Rae	83	25.25±1.009	15.70±0.505	13.50±0.526	52.50±1.393	60.02±2.017	14
		Vaihuti	155	23.75±0.889	14.1±0.518	12.70±0.573	53.15±1.866	59.08±2.080	14
		Hamene A	121	23.91±1.064	14.13±0.608	12.71±0.625	54.68±2.277	60.77±2.402	14
		Hamene C	102	24.15±1.025	14.68±1.574	13.20±0.656	54.05±1.991	60.50±2.314	14
		Hamene D	62	24.65±1.104	15.10±0.572	13.50±0.678	52.50±1.987	60.50±2.390	14
		Motutiairi	133	23.87±1.119	14.45±0.623	12.93±0.611	53.50±1.943	60.50±2.360	14
		Faataoto	166	25.20±1.033	15.10±0.673	13.50±0.663	55.51±1.799	62.28±2.542	14
		Faahia	139	24.27±1.053	15.11±0.758	13.47±0.675	54.50±1.852	60.50±2.614	14
		Vaimai	82	24.65±1.004	15.10±0.784	13.50±0.545	54.95±1.875	61.67±2.262	14
		Apu Mao	82	24.08±0.956	14.88±0.634	13.24±0.630	55.45±1.984	60.93±2.230	14
		Apu Apu	110	23.26±1.013	14.19±0.639	12.91±0.577	55.86±1.851	62.18±2.267	14
		Apu Vaipiti	174	23.82±0.994	14.81±0.567	13.14±0.566	56.50±1.634	62.50±2.175	14
		Faafau	140	24.95±0.911	15.30±0.546	13.90±0.583	54.50±1.862	61.50±2.228	14
		Patii south	241	24.05±0.908	14.70±0.590	13.30±0.624	55.50±1.929	61.50±2.210	14
		Patii north	128	24.05±0.899	14.70±0.563	13.50±0.460	53.50±1.273	60.50±2.097	14
		Hurepiti C	13	25.45±0.800	15.30±0.351	13.90±0.585	52.50±1.851	58.50±2.286	14
		Hurepiti B	127	24.25±0.994	14.10±0.595	12.70±0.575	53.50±1.895	59.50±2.364	14
		Hurepiti A	227	23.15±1.007	13.70±0.578	12.50±0.582	53.50±1.700	59.50±2.039	14
		Hurepiti D	40	23.45±0.937	13.90±0.587	12.50±0.619	54.50±1.978	60.50±2.575	14
		Tapuamu south	66	22.45±0.893	14.10±0.558	12.90±0.618	54.50±1.980	60.50±2.075	14
		Tapuamu north	152	23.90±0.750	14.45±0.524	13.20±0.534	54.50±1.643	59.50±2.054	14
		Vaiore	40	24.65±1.111	14.50±0.512	13.50±0.642	55.50±1.892	61.50±2.070	14
		Tahure	99	24.65±1.120	14.90±0.709	13.10±0.526	53.50±1.787	61.50±2.430	14
		Atite	33	24.63±0.505	14.89±1.033	12.99±0.593	52.50±1.824	59.50±2.207	14
		Faaha	351	24.65±1.041	14.70±0.589	12.90±0.497	52.50±1.844	60.50±2.397	14
		Faaha B	40	24.65±0.966	14.90±0.505	13.30±0.564	52.94±1.816	59.08±2.215	14
		Maioroa	21	24.51±0.706	15.03±0.580	12.78±0.691	51.92±2.051	60.50±2.248	14
		Otune	165	25.18±0.760	14.10±0.695	12.70±0.629	53.80±1.982	62.62±2.341	14
	navigatoria 'protea'	Tepua	192	21.84±0.490	13.86±0.268	11.82±0.690	55.22±2.141	60.88±2.376	14
	n. radiata	Vairua	47	21.02±1.166	12.81±0.690	11.60±0.528	53.18±1.938	56.99±2.099	14
		Faararoto	2	19.85	12.00	11.00	52.86	56.39	14
		Haamoa	78	21.26±0.999	12.10±0.536	11.32±0.541	53.56±1.513	57.26±2.624	14
		Vairahi	22	22.68±1.110	13.08±0.536	12.14±0.865	52.44±1.886	58.57±2.826	14
	n.' vittata'	Averaiti	16	22.63±1.574	12.81±0.765	11.84±0.618	52.33±1.796	56.57±2.106	14
		Averarahi	63	22.51±1.138	12.75±0.668	11.80±0.269	52.67±1.724	57.00±1.708	14
	n. 'castanea'	Faaroa D	9	21.63±0.354	12.33±0.354	11.3±0.300	48.89±1.500	53.10±0.500	14, 15
		Faaroa S	1	20.70	110	10.10	54.10	58.20	15
		Farepaiti	30	22.30±0.150	13.00±0.999	12.08±0.560	52.64±1.884	57.03±1.178	14
	n. 'vittata'	Farepauti	11	22.25±1.187	12.63±0.645	11.83±0.546	51.78±1.513	56.00±1.808	14
	n . 'terrestris'	Opoa	7	21.14±0.712	12.13±0.623	11.04±0.479	52.50±1.791	59.00±1.874	14
		Hotopuu	2	20.30	11.90	10.70	52.91	56.43	14
	n. 'vittata'	Faarahi	32	21.54±0.761	12.16±0.410	11.30±0.208	53.70±1.992	55.67±1.429	14
		Faatemu	3	22.80±1.217	12.53±2.890	12.00±0.299	53.51±0.899	52.67±0.899	14
	n. 'variablis'	Tuatau	6	22.43±0.990	12.77±0.415	11.70±0.490	53.51±1.895	56.32±2.321	14
	n. 'approximata'	Vaetuuna	192	20.72±0.948	10.40±0.940	11.00±0.534	57.89±2.089	68.70±2.332	14
	n. 'vittata'	Uturoto	206	20.30±1.052	11.76±0.479	11.10	54.50	60.50	14
	n. levilineata	Marahi	2	18.30	11.10	9.80	54.10	59.05	5
	n. 'terrestris'	Tooroa	14	22.89±1.384	12.89±1.040	12.50±0.567	54.38±1.745	57.82±2.151	14, 15
	n. navigatoria	Uparu	37	22.14±0.861	12.8±0.534	12.06±0.443	55.82±2.034	59.09±2.201	14
	n. 'fusca'	Voaara	162	20.93±0.953	13.1±0.530	11.86±0.500	55.50±1.000	60.50±1.000	14

		N	Length	Width	Mouth H	W/L	Mouth H/L	
dentifera callifera		78	18.95±0.726	12.3±0.415	10.69±0.434	57.14±1.214	59.81±1.658	14, 15
d. candida	Farepaiti	45	19.64±0.701	11.79±0.501	11.24±0.405	57.57±1.741	59.61±1.812	14
	Fareparahi	27	20.04±0.554	11.97±0.389	11.54±0.445	58.50±1.446	61.50±1.717	14
	Maufenua	1	19.20	11.90	11.30	57.32	59.76	14
d. cedista	Hamoa	2	20.60	12.20	9.95	58.75	56.35	14, 15
d. citrina	Uparu	60	22.84±0.907	12.86±0.482	12.27±0.425	54.97±1.772	56.30±2.007	14
	Voaara	44	23.28±0.809	13.13±0.449	12.80±0.550	54.22±1.805	56.34±2.099	14
d. dentifera	Vairahi	42	19.61±0.773	11.46±0.445	11.01±0.493	39.80±1.667	57.98±1.191	14
	Averarahi	3	18.17±0.473	10.53±0.351	7.23±0.509	55.91±1.842	57.82±2.188	15
	Faaiti	11	19.96±1.119	10.67±0.509	11.05±0.618	54.72±2.529	55.83±3.432	14, 15
	Faaroa	14	20.11±0.607	11.31±0.433	10.72±0.438	40.31±2.124	57.17±2.470	14, 15
	Opoa	24	19.05±0.829	10.83±0.561	7.66±0.421	54.07±1.788	58.78±1.721	15
d. formosa	Faaroa	23	23.33±0.963	13.80±0.717	13.98±0.455	55.00±2.041	61.00±2.061	14
	Vaianae	4	21.80±0.779	13.50±0.428	11.95±0.523	56.59±2.010	57.54±2.080	14
	Faatemu	104	22.69±0.792	13.03±0.546	12.86±0.523	55.41±1.796	57.31±2.238	14
	Tuatau	21	20.69±0.919	11.91±0.438	11.48±0.560	54.94±1.734	58.31±1.982	14
	Vaihuti	116	20.35±0.951	11.89±0.506	11.20±0.513	55.56±1.747	57.33±1.740	14
	Tirae	18	20.52±0.758	11.77±0.382	11.42±0.543	56.14±1.893	57.70±2.019	14
	Vaiaau	368	21.32±0.931	12.32±0.521	11.98±0.765	55.93±1.942	57.84±2.078	14
d. imperforata	Vaiaau	7	20.11±0.955	11.55±0.611	11.19±0.534	56.13±1.505	59.50±1.849	14
	Marahi	62	20.44±1.069	12.03±0.476	11.44±0.503	56.52±1.719	58.67±2.178	14
	Utuara	466	20.72±0.874	12.16±0.476	11.73±0.482	55.93±1.808	60.14±2.269	14
	Ereeo	51	18.94±0.877	11.39±0.493	10.61±0.607	56.03±1.392	60.72±2.430	14
	Hanua	45	19.15±0.971	11.64±0.518	10.73±0.534	55.89±1.746	59.68±2.184	14
	Tooroa	106	19.61±0.996	11.69±0.530	10.98±0.390	56.39±1.909	58.59±2.342	14
	Tahurui	128	20.57±0.923	12.06±0.454	11.62±0.503	54.19±1.719	59.40±2.178	14
auriculata	Vaetahe	549	18.24±0.860	12.03±0.476	10.27±0.460	56.23±2.114	66.01±2.245	14
	Vaitaporo	162	18.20±0.867	11.93±0.461	10.04±0.457	55.32±1.088	65.48±2.093	14
	Tepua	7	19.68±0.420	13.07±0.225	10.70±0.262	58.80±1.414	66.34±1.552	14
	Vairua	23	20.09±0.656	13.43±0.472	11.34±0.300	56.24±1.325	66.84±2.238	14
	Hamoa	228	20.05±0.819	13.70±0.545	11.37±0.500	56.64±1.649	68.27±2.022	14
	Huaru	2	19.70	12.50	10.60	55.50	63.50	14
	Pufau	11	19.39±0.832	12.14±0.398	10.57±0.335	54.91±0.687	62.50±1.675	14
	Fanfau	6	19.10±0.512	11.93±0.398	10.33±0.440	53.67±1.768	62.33±1.931	14
	Vaipao	283	18.58±0.834	12.11±0.443	10.32±0.461	55.91±2.038	65.10±2.201	14
	Temehani	1	16.80	11.70	8.60	69.64	51.19	15
protracta		1	19.80	11.80	10.30	56.41	65.64	5
dolichostoma		1	19.50	12.80	11.00	41.10	59.40	5
magistri	Tooroa	1	22.40	13.30	9.20	53.99	58.83	15
l. lugubris	Utuara	516	17.57±0.869	10.35±0.456	9.51±0.468	54.62±1.936	59.50±1.894	14
l. ovalis	Vaiaau	34	19.90±0.857	11.84±0.541	10.87±0.507	54.12±2.142	58.98±2.282	14
leptochila		40	19.47±0.997	12.5±0.528	10.44±0.402	54.78±2.252	61.19±3.470	14
labrusca	Temehani	10	18.66±0.603	11.71±0.603	11.50±0.158	53.75±0.198	60.65±0.336	6, 15
dolorosa		5	17.33±0.487	10.26±0.542	9.38±0.160	54.10±1.356	58.90±1.743	14
g. garrettii	Tooroa	16	16.48±0.534	10.19±0.566	6.30±0.365	54.61±1.669	77.38±2.497	16
	Uparu	345	15.51±0.648	9.40±0.372	8.48±0.398	55.05±1.678	77.47±2.402	14
	Voaara	550	15.69±0.538	9.63±0.394	8.64±0.392	54.88±1.488	77.86±2.443	14
g. rustica	Ereeo	45	15.91±0.638	10.38±0.368	8.78±0.550	51.00±1.923	61.00±2.573	14
	Hanua	1	15.20	9.40	7.90	53.74	67.97	14
	Tooroa	86	16.1±0.871	10.95±0.519	8.66±0.510	54.23±2.019	66.89±2.980	14
g. thalia	Huaru	323	17.05±0.746	11.10±0.444	9.44±0.426	55.16±1.717	65.68±2.276	14
	Tahara	289	17.60±0.794	11.58±0.423	9.73±0.456	55.24±1.754	65.36±2.336	14
	Temehani	1	16.80	11.70	8.60	69.64	51.19	15
cuneata		1	17.30	10.20	8.70	57.23	59.04	5
levestriata		1	17.20	12.10	9.90	57.39	72.50	5
remota	Tuatau	9	16.98±0.568	12.28±0.475	9.74±0.160	57.30±0.748	72.30±0.748	14

			N	Length	Width	Mouth H	W/L	Mouth H/L	
Partula	*remota*	Vaatuna	5	16.37±0.407	11.9±0.179	9.38±0.115	56.50±0.471	70.50±1.732	14
		Uturoto	3	16.05±0.141	11.43±0.249	9.10±0.427	57.13±1.596	70.80±2.242	14
		Vaihuti	76	17.68±0.720	12.43±0.460	10.12±0.480	57.15±1.556	70.64±2.303	14
	tristis		5	16.62±0.041	10.82±0.073	9.30±0.464	55.66±2.227	65.89±1.931	6, 15
	atilis		15	18.17±0.900	11.99±0.541	10.11±0.332	57.46±1.660	71.60±2.166	14
	hebe	Vairahi	10	15.23±0.306	11.1±0.283	8.92±0.359	58.03±1.588	72.28±2.154	14
		Averarahi	81	14.85±0.525	10.77±0.382	8.65±0.332	58.30±1.416	73.35±2.172	14
		Faaiti	169	14.7±0.552	10.54±0.425	8.48±0.345	57.79±1.708	73.43±2.153	14
		Faaroa	14	15.33±0.561	11.24±0.578	8.87±0.360	56.19±1.580	71.63±2.193	14
		Farepaiti	289	15.64±0.623	11.21±0.460	8.81±0.422	56.81±1.453	72.22±2.226	14
		Fareparahi	757	15.81±0.690	11.42±0.465	9.00±0.349	55.93±1.477	70.15±2.179	14
		Opoa	667	14.65±0.554	10.28±0.377	8.21±0.380	56.20±1.539	69.94±2.042	14
		Hotopuu	616	14.55±0.695	10.19±0.489	8.19±0.422	58.50±1.683	73.50±2.655	14
		Vaianae	1	15.1	11.1	8.7	57.71	72.41	14
		Faatemu	33	14.44±0.642	10.43±0.384	8.37±0.359	56.73±1.619	70.13±2.224	14
		Vaiaau	95	14.69±0.560	10.44±0.418	8.46±0.289	58.70±1.470	72.90±1.800	14
		Maufena	670	14.7±0.521	10.79±0.413	8.58±0.486	56.85±1.812	71.50±2.528	14
	crassilabris	Tehurui	18	14.87±0.701	10.49±0.402	8.13±0.402	55.57±1.659	70.29±2.040	14
		Utuara	29	15.07±0.599	10.6±0.415	8.34±0.346	55.90±1.907	70.70±2.993	14
		Ereeo	10	14.48±0.703	10.28±0.449	8.10±0.425	55.25±1.689	70.27±2.240	14
	turgida	Tooroa	12	16.05±0.353	12.03±0.403	10.98±0.589	68.50±1.963	68.50±2.730	14
		Voaara	1	15.90	11.10	10.90	67.82	73.50	14
	clarkei		5	14.63±0.455	11.78±0.199	9.12±0.291	62.33±0.535	80.52±1.309	15
	meyeri		1	c16.5	10.50	7.24	49.10	69.23	15
	pearcekellyi		1	16.30	11.50	9.40	54.50	58.50	15
	planilabrum	Patio	130	21.65±0.791	12.7±0.315	11.70±0.162	53.50±1.905	58.50±3.142	14
		'virginea' Faatia	36	18.65±0.863	10.9±0.484	10.30±0.270	53.50±2.777	58.50±3.738	14
		Hamane	21	21.35±0.570	12.5±0.082	11.50±0.265	53.50±1.923	60.50±4.231	14
		'virginea' Faataoto	7	18.05±0.466	10.7±0.239	9.900±0.415	55.50±1.382	60.50±1.829	14
		'virginea' Apu Mao	97	18.05±0.729	10.9±0.418	10.10±0.400	55.50±1.621	58.50±1.821	14
		'virginea' Apu Apu	112	18.25±0.633	10.1±0.388	10.10±0.396	53.50±1.930	58.50±1.786	14
		Apu Vaipiti	13	21.05±0.845	12.7±0.262	11.30±0.348	52.50±1.412	58.50±2.003	14
		Patii	62	22.95±0.676	12.7±0.895	11.30±0.703	53.50±3.189	58.504.272	14
		'virginea' Hurepiti A	69	19.85±0.734	11.5±0.364	10.50±0.335	53.50±1.195	59.50±1.861	14
	planilabrum	*'virginea'* Hurepiti D	49	21.05±0.915	12.3±0.472	11.10±0.440	53.50±1.837	58.50±2.166	14
		'virginea' Otuone	122	19.85±0.784	11.5±0.504	10.50±0.522	52.50±1.595	59.50±2.223	14
		'virginea' Tapuamu south	159	20.45±1.073	11.9±0.580	10.70±0.247	52.50±0.536	58.50±0.268	14
		'virginea' Tapuamu north	187	20.45±0.856	12.1±0.402	10.80±0.465	52.50±1.555	58.50±1.862	14
		'virginea' Vaiore	36	21.35±0.791	12.5±0.390	11.30±0.462	55.50±1.462	57.50±2.145	14
		Puheru	64	22.15±1.046	12.7±0.388	11.85±0.335	53.51±2.257	60.01±4.167	14
		Pahure	132	21.35±0.905	12.9±0.274	11.50±0.353	53.50±2.222	58.50±4.370	14
	umbilicata	Patio	325	16.85	12.30	9.50	56.36	74.50	14
		Afatatea	47	17.45	12.50	9.50±0.381	57.70±1.700	75.50±2.315	14
		Haamene B	81	16.55	12.80	8.30	54.5	72.50	14
		Haamene A	506	16.55	12.10	8.10	56.5	73.50	14
		Faahia	6	17.45	12.70	9.70	55.5	73.50	14
		Apu Vaipiti	33	17.45	12.90	9.70	56.5	73.50	14
		Patii north	137	17.45	12.40	9.90±0.380	55.50±1.745	72.50±2.995	14
		Hurepiti B	59	17.15±0.711	12.5±0.432	9.30±0.400	55.50±1.606	73.50±2.445	14
		Hurepiti A	426	16.25±0.672	12.1±0.447	9.10±0.608	56.50±0.348	75.50±0.556	14
		Hurepiti D	46	16.85±0.699	12.7±0.739	9.50	54.50	72.50	14
		Otuona	99	16.55±0.648	12.5±0.380	9.50±0.392	55.50±1.693	74.50±2.594	14
		Tapuamu south	131	16.55±0.706	12.5±0.447	9.10±0.392	56.50±2.248	71.50±2.544	14
		Tapuamu north	268	16.55±0.493	11.7±0.252	9.30±0.178	55.50±2.000	71.50±2.000	14
		Vaiore	195	17.15	12.10	9.50	55.50	71.50	14
		Puheru south	35	17.45	12.30	9.70	55.50	73.50	14

		N	Length	Width	Mouth H	W/L	Mouth H/L	
umbilicata	Pahure	178	16.35	12.30	9.10±0.120	53.80±0.464	72.94±0.705	14
bilineata	Afatetia	183	18.65±0.828	12.5±0.505	10.50±0.467	55.50±1.511	66.50±2.217	14
	Faahaiti	102	17.75±0.743	11.7±0.493	9.40±0.495	55.50±1.589	66.50±2.263	14
eremita		1	16.90	11.00	9.10	48.24	56.24	7
sagitta		19	17.77±0.757	10.03±0.332	8.59±0.436	55.03±2.033	62.56±2.980	14
lutea	Faatadora	68	17.26±0.756	10.79±0.440	9.51±0.461	55.35±1.909	63.11±2.214	14
	Pahia	463	17.35±0.805	10.95±0.438	9.61±0.417	54.20±1.731	62.26±2.176	14
	Temanuu	247	17.83±0.728	11.10±0.408	9.68±0.454	54.96±1.938	62.79±2.222	14
makatea	Makatea	8*	20.98±1.042	12.54±0.734	10.36±0.115	50.74±1.858	55.96±0.635	15
cootei	Rimatara	2	20.70	11.60	10.50	49.23	51.67	15
	Rurutu	48	18.39±1.229	10.45±0.516	9.48±0.644	57.21±2.251	52.22±2.242	15
cootei × *hyalina*	Rurutu	61	17.54±1.042	10.21±0.446	9.33±0.393	58.31±2.681	53.26±1.689	15

* length n=8, width n=18

Appendix II – Radula variation

genus		species	half-row formula	total	N
Eua		*expansa*	10+123	133	1
		montana	(6-)10+(105-121)	115-131	6
		zebrina	(7-)8(-9)+(108-130)	115-139	6
Samoana		*abbrevita*	6+69	75	2
		attenuata	6+(78-87)	84-93	1
		annectens	5+86	91	1
		conica	7(-8)+106	114	3
		cramptoni	5+76	81	1
		fragilis	6+102	108	2
		ganymedes	6+70	76	1
		gonochila	6+(90-98)	96-104	2
		inflata	6+(87-90)	93-96	2
		margaritae	(5-7)+(100-108)	107-114	4
		medana	5+86	91	3
		minuta	6+96	102	1
		pilsbryi	6+(94-98)	100-104	2
		dryas	(5-) 6+(77-82)	83-88	4
		stevenoniana	6+(77-98)	83-104	3
		thurstoni	6+>60	>66	1
Palaopartula		*thetis*	(13-14)+>76	>89	2
Sphendone		*insolita*	10+80	90	2
Partula	Macronesia	*auraniana*	9+88	97	1
		flexuosa	7+(57-64)	64-73	2
		grisea	(7-10)+84	91	2
		micans	(5-)6(-9)+(53-82)	58-89	5
		pacifica	7(-8)+(71-96)	78-103	4
		pyramis	9+114	123	2
		similaris	8+75	83	1
		vanikorensis	8+73	82	1
	Micronesia	*emersoni*	?+86	>86	1
		*gibba**	(6-)7(-8)+73-99	88-107	21
		radiolata	(7-8)+(86-89)	94-96	2
		rufa	9+104	113	1
		salifana	7+86	93	1
		sp. (Rota)	9+90	99	1
	Fiji	*lanceolata*	(8-9)+?	?	3
	Futuna	*obesa*	(6-8)+88	94	3
	Cooks	*assimilis*	(8-)9+80	89	3
	Tahiti	*affins*	(7-)8(-10)+(92-95)	100-105	11
		clara	(6-)7(-9)+(68-98)	74-107	18
		compressa	(9-)10(-11)+102	111-113	4
		diminuta	(7-)8+(76-86)	83-94	5
		hyalina	(6-)7(-10)+(56-94)	63-101	36
		jackieburchi	9+95	104	1
		laevigata	9+91	100	1
		nodosa	(7-)8(-9)+(69-84	77-83	15
		otaheitana	(7-)8(-11)+(67-106)	74-114	37
		producta	(7-)8(-11)+(70-89)	77-99	28
	Moorea	*aurantia*	8(-10)+(74-100)	82-108	11
		mirabilis	(7-)8(-9)+(75-105)	83-113	46
		mooreana	(8-)9(-10)+(80-100)	88-109	18
		incrassa	(7-)8+(80-104)	79-112	9

	species	half-row formula	total	N
Partula				
Moorea	*suturalis*	(7-)8(-9)+(76-114)	79-121	124
	taeniata	(6-)7(-10)+(74-103)	83-112	104
	tohiveana	(7-)9(-12)+(89-123)	97-132	29
Huahine	*arguta*	7(-9)+81-98	88-106	4
	rosea	(10-)11(-12)+(85-104)	96-116	16
	varia	(6-)8(-9)+(67-105)	84-112	31
Raiatea	*auriculata*	(8-)9(-11)+(75-95)	84-104	11
	clarkei	9+(105-110)	114-119	3
	crassilabris	(7-)9+(77-95)	85-104	3
	dentifera	(8-)10(-11)+(72-107)	92-116	23
	dolorosa	8+100	108	1
	faba	(8-)10(-12)+(70-131)	78-141	69
	garrettii	(7-)9(-10)+(63-101)	72-111	29
	hebe	(8-)9(-12)+(74-100)	83-110	22
	labrusca	10+94	104	2
	leptochila	(7-)9(-11)+(83-112)	92-121	5
	lububris	(7-)9(-11)+(90-108)	101-119	9
	navigatoria	(8-)9(-12)+(66-115)	76-126	57
	remota	(9-)10+(81-83)	91-93	4
	turgida	7+94	101	1
Tahaa	*bilineata*	9+(86-102)	95-111	3
	planilabrum	(6-)9(-10)+(83-103)	100-113	12
	sagitta	8+>67	>75	1
	umbilicata	(9-)10(-11)+70-86	80-91	6
Bora Bora	*lutea*	(8-)9(-10)+78-93	86-102	12
Australs	*cootei*	(6-8)+?	?	2
hybrids	*aurantia* × *mirabilis*	8		1
	aurantia × *olympia*	7	86	1
	diminuta × *taeniata*	8	77	1
	mirabilis × *taeniata*	7-8		4
	mirabilis × *mooreana*	7-8	81-101	2
	mooreana × *olympia*	8	86	1
	mooreana x tohiveana	8-9	69-92	2
	otaheitana × *olympia*	7	69	1
	rosea × *varia*	9-12		9
	olympia × *tohiveana*	8	86	1

* anomalous data 12+102

Appendix III – Crampton's embryo data and comparison to Mendelian expectations

Embryo colour pattern frequency was compared to expectations derived from parental colours. Expectations were based on Mendelian assumptions, starting with the zonata-lyra-frenata allele system. Assuming random divison of genotypes within corresponding phenotypes, predicted frequncies were compared to embyronic data (χ^2 test). Non-signfiant results were taken to be in accordance with the propsed allele system. Signfiant values led to reanalysis with other allele combinations (e.g. allele loss).

Colour patterns: B – bisecta; C – cestata; D – dark; F – frenata; L – light; M – medium; P – purple; S – strigata; U – unbanded; W – white; X – apex; Z – zonata. Putative alleles in lower case, inequalities indicate direction of dominance. χ^2 d.f. for banding alleles = 15, shades = 3

		offspring patterns						alleles
		L	**S**	**M**	**D**	**B**	**Z**	χ^2
leptochila								
	C	6	0	0	0	0	0	
	D	32	0	0	0	0	0	
	S	40	0	22	0	0	0	
navigatoria								
Farepaiti	B	0	0	0	0	2	0	b>u 2.05
	D	3	0	0	2	2	0	l>d 1.36
	L	7	0	0	1	1	0	
imperforata	B	15	0	0	0	11	0	d>l 6.49
	D	11	0	0	0	1	0	
	L	311	0	0	0	4	0	
	M	41	0	0	0	1	1	
	Z	4	0	0	0	1	3	
umbilicata								
Hurepiti A	B	4	0	3	0	0	0	
	L	17	0	18	0	0	0	
	D	13	0	18	1	0	1	
	S	86	0	140	9	0	0	
	Z	4	0	1	0	0	70	
Tapuamu S	B	1	0	0	0	0	0	limited
	L	3	97	0	1	0	0	varition
	S	0	15	0	0	0	0	
Tapuamu N	B	0	0	12	1	0	1	
	D	0	0	14	1	0	1	
	L	1	0	123	2	0	1	
	S	0	0	73	5	0	1	
	Z	3	0	8	0	0	7	
Vaiore	B	0	0	10	0	0	0	a>f=z>u
	D	0	0	34	2	0	0	3.31
	L	7	0	18	0	0	0	
	S	5	0	166	2	0	0	
	Z	0	0	5	0	0	1	
Puheru	B	2	0	9	0	0	0	a>f=z>u
	D	0	0	2	2	0	0	33.18
	L	3	0	10	0	0	0	
	M	0	0	9	2	0	0	
Patio	B	1	0	0	1	0	0	a>f=z>u
	D	2	0	73	9	0	0	25.03
	L	5	0	311	7	0	0	
	Z	0	0	1	0	0	1	
Pahure	D	8	0	0	0	0	0	
	F	4	0	0	0	0	0	
	L	92	0	0	10	0	0	
	P	27	0	0	3	0	0	
	S	102	0	0	13	0	0	
Hurepiti	D	0	0	1	0	0	0	
	L	20	0	9	0	0	0	
	M	2	0	3	0	0	0	
	Z	0	0	2	0	0	0	
Hamene B	B	0	0	2	0	3	0	
	D	0	0	2	0	0	0	
	M	1	0	68	0	0	0	
	Z	0	0	2	3	0	0	
Hamene A	B	0	0	22	0	1	0	
	D	0	0	21	10	0	1	
	L	7	0	62	3	0	4	
	S	3	0	218	10	0	5	
	Z	0	0	14	2	0	14	
Fahaa	L	7	0	0	0	0	0	

		offspring pattern						alleles
		L	**M**	**D**	**B**	**F**	**Z**	χ^2
umbilicata								S, Z & F indistinct
Afatatia	D	0	0	1	3	0	0	d>l 3.90
	L	70	0	0	0	0	0	
Apu-Vaipiti	D	0	0	10	1	0	2	
	L	3	0	17	0	0	0	
	Z	0	0	1	1	0	2	
Patio N	D	3	0	11	17	1	0	
	L	3	0	0	0	0	0	
	S	40	0	58	2	0	2	
auriculata auriculata	B	5	0	1	0	0	0	l>d 7.61
	D	27	0	20	1	0	0	
	L	68	0	25	0	0	0	
robusta	B	19	0	0	4	0	0	
	D	70	0	129	2	0	0	
	L	210	0	174	8	0	0	
thalia	B	11	16	0	3	0	0	
	D	25	42	22	0	0	0	
	L	69	136	34	0	0	0	
	M	0	0	0	0	0	0	
planilabrum								
Hamene	D	0	0	16	0	0	0	l>d 0.97
	L	1	0	0	0	0	0	
faba subangulata								
Hurepiti D	D	4	3	0	0	0	0	a>f=z>u 6.08
	F	5	1	0	0	0	0	l>d 0.93
	L	13	2	0	0	1	1	
Tahure	B	0	11	0	11	2	3	a>f=z>u 34.11
	D	6	12	2	0	1	2	
	F	2	10	0	1	7	12	
	Z	4	22	1	4	4	28	
Faaha	B	1	2	1	4	0	2	a>f=z>u 15.80
	D	0	0	0	0	0	0	
	F	11	12	21	0	37	11	
	L	10	1	7	0	1	3	
	M	28	35	47	2	4	10	
	Z	1	4	11	0	3	9	
Faaha B	L	1	0	0	0	0	0	too few
	M	1	0	0	0	0	0	
	Z	0	1	0	0	0	0	
Hamene D	B	0	1	0	0	0	1	a>f=z>u 5.64
	D	4	27	16	0	2	1	m>l 4.52
	F	0	1	0	0	0	0	
	L	2	0	0	0	0	0	
	M	3	3	0	0	0	1	
hebe								
Averarahi	W	35	0	0	0	0	0	
	X	3	0	0	0	0	0	
Faaiti	W	65	0	0	0	0	0	
	X	4	0	0	0	0	0	
	M	2	0	0	0	0	0	
Faaroa	W	7	0	0	0	0	0	
	M	1	0	0	0	0	0	
Manfenua	W	321	0	0	0	0	0	
	M	8	0	0	0	0	0	
	W	1	0	0	0	0	0	
	R	1	0	0	0	0	0	
	B	13	0	0	0	0	0	

Appendix IV – Dissected material

Eua ***E. expansa*** NHMUK Reid 1995 (Savaii ×1). ***E. globosa*** BPBM 115426 ('Eua ×1). ***E. montana*** BPBM 1967-144.2 (x1); FMNH 152750 ('Upolu ×3); FMNH 152833 ('Upolu ×3). ***E. sp.*** BPBM 108175 (Savai'i 'expansa' ×2). ***E. zebrina*** ANSP A988 (Tutuila ×6); BPBM 84467 (Tutuila ×1).

Samoana ***S. abbreviata*** ANSP A1761f (Tutuila ×1). ***S. alabastrina*** ANSP A1762f (Moala ×1); BPBM 180299 (Moala ×1). ***S. annectens*** BPBM 136060 (Huahine ×1). ***S. attenuata*** BPBM 136074 (Raiatea ×1); BPBM 136159 (Tahaa ×1).; ZSL 750 (unknown ×1); ZSL 662 (unknown ×1); ZSL 526 (unknown ×1); ZSL 593 (Mt. Marau ×1). ***S. bellula***BPBM 1970-36 (Ua Pou ×1); ZSL 661 (Ua Pou ×1). ***S. burchi*** BPBM 145006 (Tahiti ×1). ***S. conica*** ANSP A1382 (Tutuila ×1); ANSP A1762e (canalis Savai'i ×1); ANSP A1763a (Tutuila ×1). ***S. cramptoni*** ANSP A1763b ('Eua ×1); BPBM 118453 ('Eua ×1). ***S. decussatula*** BPBM 96082 (Nuku Hiva ×1); BPBM 96128 (Nuku Hiva ×1); BPBM 96130 (Nuku Hiva ×1); BPBM 96132 (Nuku Hiva ×1); BPBM 99834 (Tahauta ×1); BPBM 99878 (Tahauta ×1); BPBM 136042 (Tahauta ×1); BPBM 1969-48 (Tahauta ×2); BPBM 1969-47.5 (Tahauta ×2);ZMH 9798 (unknown ×1). ***S. fragilis*** ANSP A1434d (Guam ×1); FMNH 24965 (Guam ×1); UF 481329 (Rota ×1). ***S. dryas*** BPBM 135957 (Raivavae dryas ×2); BPBM 135981 (Raivavae oreas ×5); BPBM 144843 (Raivavae hamadryas ×1); MNHN (Rv50.1) (Raivavae dryas ×1); MNHN (Rv62) (Raivavae hamadryas ×3); MNHN (Rv63) (Raivavae hamadyras ×2); MNHN (Rv83) (Raivavae ×1). ***S. ganymedes*** BPBM 98547 (Hiva Oa ×1); BPBM 98599 (Hiva Oa ×1); BPBM 98598 (Hiva Oa ×1); BPBM 1969-46 (Hiva Oa ×1); ZMH 9805 (Hiva Oa ×1); ZSL 694 (Hiva Oa ×2); ZSL 833 (Hiva Oa ×5); ***S. inflata*** BPBM 99735S (Tahauta ×1); ZMH 9808 (Tahauta ×1). ***S. minuta*** BPBM 100103 (Fatu Hiva magdalinae ×1). ***S. margaritae*** MNHN (Rp04) (Rapa ×1); MNHN (Rp10) (Rapa ×1); MNHN (Rp32) (Rapa ×2); MNHN (Rr32a) ('Rurutu' ×1). ***S. medana*** BPBM 11875 (Fatu Hiva lebrounneci ×3); BPBM 11876 (Hiva Oa ×2); BPBM 98585 (Hiva Oa medana/lebrounneci ×1); BPBM 99794 (Tahauta medana var. ×1); BPBM 99827 (Hiva Oa ×1); BPBM 99875 (Tahauta medana cincta ×1); BPBM 99823 (Tahauta diaphana ×1); BPBM 100053 (Fatu Hiva ×1); BPBM 100068 (Fatu Hiva lebrounneci ×1); BPBM 100069 (Fatu Hiva medana/lebrounneci ×1); BPBM 1964-4 (Fatu Hiva ×1); BPBM 1969-41 (Tapeata ×1). ***S. pilsbryi*** BPBM 11874 (Fatu Hiva ×2). ***S. stevensoniana*** FMNH 152740 (Savaii ×2); FMNH 152990 (Savaii ×1). ***S. gonochila*** BPBM 10363 (Ua Huka ×2); BPBM 100057 (Fatu Hiva ×1); BPBM 105026 (Ua Huka ×2); BPBM 105030 (Ua Huka ×1); BPBM 105040 (Ua Huka ×1); BPBM 108040 (Ua Huka ×1); BPBM 108048 (Ua Huka ×1); ZSL '497' (Ua Huka ×1). ***S. thurstoni*** FMNH 181109 (Ofu ×1).

Palaopartula ***P. calypso*** BPBM 212199 (Babelthaup ×1); ***P. leucothoe*** BPBM 159776 (Peleiu ×1). ***P. thetis*** BPBM 160543 (Babelthaup ×1); BPBM 211296 (Koror ×1); FMNH 271887 (Ngeruktabel ×1); ZMH 9828 ('Tahiti' ×1).

Partula ***P. affinis*** ANSP 318706 (Vaii to Oofu ×4); BPBM 89584 (Papenoo'solidula' ×2); NHMUK (Cheeseman 1925) ('Raiatea' ×3); ZSL 612 (Fareara ×1); ZSL 632 (Faaroa ×1); ZSL 1410 (x2); ZSL 'T6' (Faaroa ?, Tahiti ×3). ***P. arguta*** ANSP A1425 (Tefari ×4); ZSL 159 (Fareaito ×3); ZSL 359 (Fitii ×1). ***P. assimilis*** FMNH 144697 (Rarotonga ×1); FMNH 153787 (Rarotonga ×2); FMNH 296430 (Rarotonga ×1). ***P. atilis*** BPBM 215302 (x1). ***P. auraniana*** FMNH 109348 (Vanua Lava ×1). ***P. aurantia*** BPBM 1970-268 (Paparoa ×1); NHMUK Murray & Clarke (267 ×4); NHMUK Murray & Clarke (269 ×2); NHMUK Murray & Clarke (270 ×4); NHMUK Murray & Clarke (342 ×2). ***P. auriculata*** ANSP A959 (Tepua ×3); ANSP A1415 (Vaitaporo robusta ×7); ZSL 185 (Temehani ×1). ***P. bilineata*** ANSP A797 (Fahaa ×1); ANSP A798 (Afatatea ×3). ***P. clara*** ANSP A890 (Tereehia angusta ×8); ANSP A938C (Vaipuarii minor ×3); ANSP A938D (Tiamao parva ×2); ANSP A938E (Papeiti prima ×3); BPBM 1970-178 (Tereehia angusta ×1); ZSL 495 (Fareara ×1); ZSL 737 (Papenoo ×1); ZSL 1165 (Papenoo ×1). ***P. clarkei*** ZSL 177 (Tevaitoa ×2); ZSL 250 (Tevaitoa ×1); ZSL 316 (Tevaitoa ×1). ***P. compressa*** BPBM 142356 (Orofena ×2); BPBM 145548 (Aorai ×2); ZSL 675 (Viriviriterai ×4); ZSL 676 (Viriviriterai ×2). ***P. cootei*** BPBM 136800 (Rurutu ×2); BPBM 136850 (Rurutu ×1); BPBM 136861 (Rurutu ×3); BPBM 136865 (Rurutu ×4). ***P. cootei × hyalina*** BPBM 136795 (Rurutu ×1); BPBM 136800 (Rurutu ×1); BPBM 136850 (Rurutu ×2); BPBM 136856 (Rurutu ×1); BPBM 136858 (Rurutu ×1); BPBM 136863 (Rurutu ×2); BPBM 136871 (Rurutu ×3). ***P. cramptoni*** BPBM 215286 (x1). ***P. crassilabris*** ANSP A789 (Tehuri ×2). ***P. cytherea*** BPBM 87053 (x3). ***P. denitfera citrina*** ANSP A946 (Uparu ×1); ANSP A951 (Voaara ×4). ***P. d. dentifera*** ANSP A940 (Averarahi ×6). ***P. dentifera formosa*** ANSP 294480 (Tutira ×4). ***P. dentifera imperforata*** ANSP A960 (Tooroa raiatensis ×3); ANSP A1391 (Utuara ×6). ***P. diminuta*** ANSP A880 (Pirai ×5). ***P. dolorosa*** BPBM 139341 (x1); ZSL 245 (x1). ***P. emersoni*** BPBM 154221 (x1). ***P. f. faba*** ANSP A762 (Hotopu ×5); ANSP A795 (Vaenae ×2); ANSP A820 (Averarahi ×6); ANSP A826 (Vaitaporo ×2); ANSP A830 (Tuatau ×4); ANSP A832 (Tepua ×6); ANSP A834 (Faaharoto ×1); ANSP A1380 (Averaiti ×6); ANSP A1383 (Farepaiti ×1); NHMUK 2231 (northern hills ×3); ZSL 221 (Hotopuu ×1); ZSL 262 (Hamoa ×2); ZSL 276 (Temehanirahi ×2); ZSL 279 (nr Pufau ×2). ***P. faba subangulata*** ANSP A802 (Hurepiti D ×2); ANSP A837 (Hurepiti A ×4); ANSP A839 (Tapuamu N ×2); ANSP A993 (Vaimai ×6); ANSP A1378 (Hamene B ×1); ANSP A1384 (Farepiti ×1); ANSP A1386 (Hamene C ×5); ANSP A1394 (Patio ×5); ANSP A1396 (Puheru ×1); ANSP A1401 (Tehuare ×1); ANSP A1421 (Haamene A ×3); ANSP A1419 (Tahaa ×1). ***P. flexuosa*** BPBM 194055 (x1); UMMZ 304355 (Guadalcanal regularis ×3). ***P. g. garrettii*** ANSP A746 (Uparu ×10). ***P. garrettii rustica*** ANSP A791 (Ereeo ×6); ZSL 597 (Tevaitoa ×1); ZSL 810 (Tevaitoa ×3). ***P. garrettii thalia*** ANSP A1431A (Tahara ×2); ANSP A1431B (Hauru ×6); ZSL 185 (Temehani ×1). ***P. gibba*** ANSP A852 (Guam ×5); ANSP A987 (Guam ×7); ANSP A1426 (Saipan ×1); BPBM 188957 (Rota elongate form ×1); BPBM 188959 (Rota ×1); FLMNH 45844 (Guam ×1); ZSL 683 (Saipan ×1); ZSL 1348 (Saipan ×1); ZSL (Saipan ×10). ***P. grisea*** NHMUK 1884.6.12.224 (hartmani ×1); FLMNH 366509 (New Britain carteriensis ×2). ***P. guamensis*** BPBM 156441 (Ponape ×1). ***P. hebe*** ANSP A764 (Faaroa ×4); ANSP A765 (Farepiti ×5); ANSP A793 (Faatemu ×6); ANSP 272359 (Farepaiti ×4); ZSL 442 (Hotopuu ×1); ZSL 554 (Hotopuu ×2). ***P. hyalina*** ANSP A730 (Punaruu ×4); ANSP A889 (Pirai ×4); ANSP A937 (Faaruami ×4); ANSP A937A (Papenoo ×4); ANSP A937B (Paraura ×6); ANSP 318706 (Vavii to Oofu ×4); BPBM 126887 (Rurutu ×1); BPBM 136767 (x1); BPBM 136858 (Rurutu ×3); BPBM 136867 (Rurutu ×1); BPBM 215795 (Cooks islands ×3); UMMZ PHRM312 (Rimatara ×1); UMMZ PHRR (Rurutu ×1); UMMZ PHRV (Raivavae ×1); UMMZ PHTB (Tubuai ×1); MNHN (Rv39) (Raivavae ×1); MNHN (Rv621) (Raivavae ×1); MNHN (Rv632) (Raivavae ×1); MNHN (Th40) (Tahiti ×1); NHMUK (Cheeseman 1925) ('Raiatea' ×2); ZSL 539 (South of Baie Pirae ×1); ZSL 730 (unknown ×2); ZSL 751 (South of Baie Pirae ×2); ZSL 1124 (unknown ×4). ***P. incrassa exigua*** NHMUK Murray & Clarke (267 ×3); NHMUK Murray & Clarke (269 ×3); NHMUK Murray & Clarke (270 ×1); NHMUK Murray & Clarke (339 ×1); NHMUK Murray & Clarke (366 ×1). ***P. i. incrassa*** ANSP A938A (Atehi ×1). ***P. jackieburchi*** BPBM 86697 (Maire [in Aivaro] ×2); BPBM 86749 (Aiavaro & Tiirahi otaheitana ×1); BPBM 86880 (Tautira ×1); BPBM 86882 (Tautira ×3); BPBM 1970-189 (1970-189 rubescens ×1); BPBM 1970-190 (1970-190 ×1). ***P. labrusca*** BPBM 139357 (Temehani ×1); ZSL 388 (Temehanirahi ×2). ***P. laevigata*** ZSL 695 (Teihomonono ×1); ZSL 986 (Teihomonono ×1). ***P. lanceolata*** ANSP A1763e (Mango ×1); BPBM 78940 (Mango sarissa ×1); BPBM 78942 (Mango ×4); BPBM 78943 (Mango cf. sarissa ×1); FMNH 168256 (NW Thithia cf. sarissa ×4). ***P. langfordi*** BPBM 213183 (Rota ×1). ***P. leefei*** BPBM 175135 (Rotuma ×1). ***P. leptochila*** ANSP A978 (Upper Huaru ×4); ZSL 184 (Temehanirahi ×2); ZSL 245 (Temehanirahi ×3). ***P. lirata*** BPBM 18068 (Mango ×2); BPBM 78936 (Mango ×1); BPBM 78924 (Thithia ×1); BPBM 180142 (Oneata ×1); BPBM 180218 (Naiau ×2); FMNH 168151 (NW Vanua Mbalavu ×4); ZSL 839 (Cikambia-i-lau ×3); ZSL 901

(Cikambia-i-lau ×3). ***P. l. lugubris*** BPBM 87366 ('highest mountain' ×2); BPBM 215312 (Vaiau ×1). ***P. lugubris ovalis*** ANSP A970 (Vaiaau ×7); BPBM 215312 (Utuara lugubris ×1); BPBM 215313 (Utuara ×1). ***P. lutea*** NHMUK (Cheeseman 1925) (Bora Bora ×3); ANSP A1414 (Pahia ×5); ANSP 290761 (Temauu ×4). ***P. micans*** NHMUK 2184 (Sta. Isabel coxi ×1); NHMUK 1957.11.8.63 (Ontong Java coxi ×1); NHMUK 1957.11.8.86-90 (Ontong Java coxi ×1); UMMZ (NilaP1) (Shortlands ×2). P. ***mirabilis*** NHMUK Murray & Clarke (151 ×6); NHMUK Murray & Clarke (159 ×1); NHMUK Murray & Clarke (371 ×14); NHMUK Murray & Clarke (373 ×2); NHMUK Murray & Clarke (374 ×1); NHMUK Murray & Clarke (394 ×1); NHMUK Murray & Clarke (399 ×6); NHMUK Murray & Clarke (510 ×5); NHMUK Murray & Clarke (521 ×1); NHMUK Murray & Clarke (161-2 ×6); ZSL 446 (Far. X Mar. ×2); ZSL 939 (mixed ×1). ***P. mirabilis propinqua*** ANSP A691 (Matapoopoo ×1). ***P. mooreana*** NHMUK Murray & Clarke (151 ×6); NHMUK Murray & Clarke (345 ×4); NHMUK Murray & Clarke (431 ×2); NHMUK Murray & Clarke (434 ×2); NHMUK Murray & Clarke (510 ×6); ZSL 202 (Maatea ×2); ZSL 300 (Atimaha ×2). ***P. navigatoria*** ANSP A736 (Faarai castanea ×2); ANSP A738 (Farapaiti vittata ×8); ANSP A1371 (Vaetuna vittata ×5); ANSP A1413 (Uturoto terrestris ×5); ANSP A1432 (Tepua protea ×6); ANSP 294465 (Tuatua vittata ×4). ***P. navigatoria levilineata*** BPBM 215244 (levilineata ×1). ***P. n. navigatoria*** ANSP A944 (Uparu ×8). ***P. navigatoria radiata*** ANSP A952 (Vairahi ×7); ANSP A964 (Vairua ×4); ANSP A2747 (Faaharota ×1); ZSL 420 (Hamoa ×1); ZSL 1351 (Hamoa ×2). ***P. nodosa*** ANSP A757 (Taapuna nodosa composita ×6); ANSP A770 (Maruapoo nodosa laeva ×4); ANSP A772 (Aoua nodosa intermedia ×3); ZSL 1365 (Papehue ×2). ***P. obesa*** BPBM 115471 (Futuna subgonochila ×2); BPBM 115481 (Alofi orthochila ×4); BPBM 115488 (Alofi subgonochila ×2); BPBM 115487 (Alofi orthochila ×2); BPBM 115502 (Alofi subgonochila ×7); BPBM 115511 (Alofi orthochila ×2); BPBM 1974-20 (Futuna subgonochila ×2); BPBM 1974-21 (Futuna subgonochila ×5); BPBM 1974-25 (Futuna subgonochila ×3); BPBM 1974-29 (Alofi subgonochila ×2). ***P. otaheitana*** NHMUK (Cheeseman 1925) ('Raiatea' ×2); ANSP A730 (Punaruu crassa ×1); ANSP A782 (Ahonu ×6); ANSP A824 (Vaipoe ×4); ANSP A835 (Vairaharaha ×6); ZSL 360 (unknown ×1); ZSL 467 (Mt. Marau ×1); ZSL 472 (Aanoha ×3); ZSL 519 (Mt. Marau ×4); ZSL 588 (Mt. Marau ×1); ZSL 633 (south of Baie Pirae ×1); ZSL 665 (Mt. Marau ×2); ZSL 680 (Teihomonono ×1); ZSL 775 (Mt. Marau ×1); ZSL 803 (Papehue ×1); ZSL 1115 (Mt. Marau ×1); ZSL 1116 (Mt. Marau ×2); ZSL 1247 (Mt. Marau ×1); MNHN (Th118) (x2); BPBM 86721 (Aiavaro ×1); BPBM 86758 (Aiavaro & Tiirahi ×1); BPBM 1970-107.3 (Apiririmarue ×1); ZMH 9832 (Taharuu ×1). ***P. otaheitana rubescens*** ANSP A781 (Papenoo ×3). ***P. otaheitana sinistrorsa*** ANSP 822 (Vavii ×2); ANSP 318706 (Vavii to Oofu ×5); ZSL 676 (Viririviriterai ×1). ***P. pacifica*** ANSP A1762c (Tanna turneri ×1); BPBM 81632 (Tanna turneri ×1); BPBM 1967-23 (Aore mcgillivrayi ×1); FMNH 175201 (Efate caleconica ×2); FMNH 175349 (Tanna turneri ×2). ***P. pearcekellyi*** BPBM 136082 ('?hebe' ×1). ***P. planilabrum*** ANSP A806 (Tapuaamu S virginea ×6); ANSP A808 (Vaiare ×6). ***P. producta*** ANSP A745 (Teoha otaheitana ×5); ANSP A754 (Vaihiria otaheitana ×4); ANSP A821 (Moaroa otaheitana ×3); ANSP A823 (Apirimaue ×5); ANSP A864 (Faarahi ×3); BPBM 1970-107.2 (Apiririmrue ×1); ZMH 9832 (Taharuu otaheitana ×7). ***P. pyramis*** FMNH 175215 (Malekula ×3). ***P. radiolata*** BPBM 152805 (Guam ×1); ZSL 700 (Navy Hill, Guam ×1); ZSL 742 (Navy Hill, Guam ×1). ***P. remota*** ANSP A790 (Vaihuti ×4). ***P. rosea*** ANSP A1388 (Haapu ×4); ANSP A1410 (Tepare ×4); ANSP A1425 (Tefari ×4); ZSL A1027 (Mauhuti ×3); ZSL 156 (Fareaito ×2); ZSL 448 (Mauhuti ×1). ***P. sp. (Rota)*** UF458524 (Rota ×3). ***P. rufa*** BPBM 156204 (x1); ***P. sagitta*** BPBM 136150 (Huahine ×3). ***P. salifana*** ANSP A1434a (Mt. Salifan, Guam ×1); BPBM 152860 (Mt. Saifan, Guam ×2). ***P. similaris*** BPBM 1966-429 (Woodlark ×1); ZMH 9819 (New Britain kubaryi ×1). ***P. suturalis*** NHMUK Murray & Clarke (82 ×1); NHMUK Murray & Clarke (150 ×2); NHMUK Murray & Clarke (151 ×7); NHMUK Murray & Clarke (159 ×3); NHMUK Murray & Clarke (161-2 ×3); NHMUK Murray & Clarke (170 ×1); NHMUK Murray & Clarke (267 ×4); NHMUK Murray & Clarke (269 ×1); NHMUK Murray & Clarke (326 ×1); NHMUK Murray & Clarke (339 ×13); NHMUK Murray & Clarke (342 ×5); NHMUK Murray & Clarke (345 ×2); NHMUK Murray & Clarke (348 ×4); NHMUK Murray & Clarke (354 ×3); NHMUK Murray & Clarke (361 ×3); NHMUK Murray & Clarke (366 ×2); NHMUK Murray & Clarke (371 ×14); NHMUK Murray & Clarke (373 ×2); NHMUK Murray & Clarke (374 ×2); NHMUK Murray & Clarke (379 ×5); NHMUK Murray & Clarke (382 ×2); NHMUK Murray & Clarke (394 ×2); NHMUK Murray & Clarke (399 ×6); NHMUK Murray & Clarke (431 ×11); NHMUK Murray & Clarke (434 ×3); NHMUK Murray & Clarke (445 ×10); NHMUK Murray & Clarke (501 ×10); NHMUK Murray & Clarke (521 ×2); NHMUK Murray & Clarke (525 ×7); NHMUK Murray & Clarke (551 ×1). ***P. s. suturalis*** ZSL 717 (Haumi ×2). ***P. suturalis vexillum*** ZSL 686 (Fareaito ×2). ***P. taeniata*** ANSP A939 (x1); NHMUK Murray & Clarke (63 nucleola ×2); NHMUK Murray & Clarke (73 ×1); NHMUK Murray & Clarke (151 ×6); NHMUK Murray & Clarke (159 ×1); NHMUK Murray & Clarke (267 ×4); NHMUK Murray & Clarke (269 ×2);NHMUK Murray & Clarke (339 ×9); NHMUK Murray & Clarke (342 ×2); NHMUK Murray & Clarke (345 ×4); NHMUK Murray & Clarke (348 ×2); NHMUK Murray & Clarke (354 ×2); NHMUK Murray & Clarke (361 ×2); NHMUK Murray & Clarke (366 ×3); NHMUK Murray & Clarke (371 ×9); NHMUK Murray & Clarke (373 ×3); NHMUK Murray & Clarke (374 ×2); NHMUK Murray & Clarke (379 ×9); NHMUK Murray & Clarke (380 ×2); NHMUK Murray & Clarke (382 ×1); NHMUK Murray & Clarke (394 ×2); NHMUK Murray & Clarke (399 ×6); NHMUK Murray & Clarke (431 ×3); NHMUK Murray & Clarke (445 ×6); NHMUK Murray & Clarke (501 ×3); NHMUK Murray & Clarke (510 ×6); NHMUK Murray & Clarke (521 ×2); NHMUK Murray & Clarke (525 ×2); NHMUK Murray & Clarke (551 ×1); ZSL 324 (Fareaito elongata ×2); ZSL 1209 (Haapiti simulans ×2); ZSL 1251 (Faatoai nucleola ×1). ***P. taeniata propinqua*** ANSP A756 (Tepatu ×4). ***P. tohiveana olympia*** NHMUK Murray & Clarke (192 ×7); NHMUK Murray & Clarke (339 ×5); NHMUK Murray & Clarke (382 ×2); NHMUK Murray & Clarke (383 ×1); NHMUK Murray & Clarke (399 ×3). ***P. t. tohiveana*** NHMUK Murray & Clarke (373 ×3); NHMUK Murray & Clarke (374 ×2); NHMUK Murray & Clarke (399 ×2); NHMUK Murray & Clarke (551 ×2); ZSL 560 (Fareaito ×2). ***P. tristis*** BPBM 139607 (Tooroa ×2); BPBM 139608 (Tooroa ×1). ***P. turgia*** BPBM 136088 (Raiatea ×1). ***P. umbilicata*** ANSP A814 (Haamene A ×6). ***P. vanikorensis*** BPBM 189717 (Santa Cruz ×1); BPBM 1978-22 (Santa Cruz ×1). ***P. varia*** ANSP A719 (Tepua ×6); ANSP A732 (Faahiti ×3); ANSP A734 (Tarea ×7); ANSP A742 (Haavau ×4); ANSP A1423 (Atita ×3) ANSP A1425 (Tefari ×4); ZSL 1169 (mixed ×4)

Hybrids ***P. aurantia × mirabilis*** ANSP (MJ S63B) (laboratory hybrid ×1). ***P. aurantia × olympia*** ANSP (MJ S22) (laboratory hybrid ×1). ***P. diminuta × taeniata*** ANSP (MJ S107) (laboratory hybrid ×1). ***P. hyalina × taeniata*** ANSP (JM S103) (laboratory hybrid ×1). ***P. mirabilis × mooreana*** ANSP (JM S65) (laboratory hybrid ×1). ***P. mirabilis × taeniata*** ZSL 102 (laboratory hybrid ×2); ZSL 741 (laboratory hybrid ×1); ZSL CSJ001 MT (laboratory hybrid ×3); ZSL CSJ002 MT (laboratory hybrid ×1). ***P. mooreana × tohiveana*** ANSP (JM S74) (laboratory hybrid ×1). ***P. n. navigatoria × radiata*** ZSL 974 (zoo hybrid ×2). ***P. olympia × mooreana*** ANSP (JM S12) (laboratory hybrid ×1). ***P. olympia × otaheitana*** ANSP (MJ S99) (laboratory hybrid ×1). ***P. olympia × suturalis*** ANSP (MJ S36) (laboratory hybrid ×1). ***P. olympia × tohiveana*** ANSP (JM S5) (laboratory hybrid ×1). ***P. rosea × varia*** ZSL 1461 (zoo hybrid ×6). ***P. suturalis × tohiveana*** ANSP (JM S28B) (laboratory hybrid ×1)

References

Dates of publication for Proceedings of the Zoological Society of London are from Duncan (1937) and Dickinson (2005).

Adams. C.B. 1851. Descriptions of New Species of *Partula* and *Achatinella. Contrib. Conchol.* **8**: 125-128 [reprinted in *Ann. Lyceum Nat. Hist. NY* **5**: 41-44]

Albers, J.C. 1859. *Dei Heliceen, nach naturlicher Verwandtschaft systematisch geordnet*. T.C.F. Enslin, Berlin

Albers, J.C. 1857. Diagnosen neuer Heliceen mit gelegentlicher Berichtigung einiger alteren Arten. *Malakozool. Blatt.* **4**: 89-100

Amin, R. 1998. Group based demographic management. In: Mace, G., H. Hall & A. Vedmar (compilers) *Proceedings of the Population Group Management Workshop 5th-7th October 1998.* Zoological Society of London

Amin, R. 1997. *Captive Bred* Partula *Population Dynamic Analysis*. BSc thesis, University of London

Ancey, C.F. 1884 Etude sur quelques mollusques terrestres inèdits ou mal connus. *Natur. Sicil.* **3**: 344-346

Anton, E. 1893. *Verzeichniβ der Conchylien, welche sich in der Sammlung von H. E. Anton befinden.* Halle

Bailey, D.W. 1956. Re-examination of the diversity in *Partula taeniata. Evolution* **10**: 360-366

Bakeo, R. & F. Qarani. 2005 Country report on the forestry invasive species situation in Vanuatu. In: McKenzie, P., C. Brown, S. Jianghua & W. Jian (eds). The unwelcome guests. *Proceedings of the Asia-Pacific Forest Invasive Species Conference, Kunming, Yunnan Province, China, 17-23 August 2003.* FAO

Baker, H.B. 1963. Type land snails in the Academy of Natural Sciences of Philadelphia. Part II. Land Pulmonata, exclusive of North America north of Mexico. *Proc. Acad. Nat. Sci. Phil.* **115**(8): 191-259

Baum, D.A. & K.L. Shaw. 1995. Genealogical perspectives on the species problem. In: Hoch, P.C. & A.G. Stephenson (eds.). *Experimental and molecular approaches to plant biosystematics*. Missouri Botanical Garden, St. Louis

Bauman, S. 1996. Diversity and decline of land snails on Rota, Mariana islands. *Amer. Malacol. Bull.* **12**(1-2): 13-27

Bauman, S. & A.M. Kerr. 2013. *Partula desolata* sp. nov. (Pulmonata: Partulidae), an extinct land snail from Rota, Mariana Islands, Micronesia. *Micronesica* **08/2013**: 1-12

Beck, H. 1837. *Index Molluscorum*. Mus. Ch. Fred, (i)

Berovides Alvarez, V. & G. Valdes Zamora. 1986. Polimorphismo genetico de *Polymita picta roseolimbata* Torre, 1950 en la region de Maisi, Cuba. *Carib. J. Sci.* **22**: 179-189

Bertsch, H. 1976. Intraspecific and ontogenetic radular variation in opisthobranch systematics (Mollusca: Gastropoda). *Syst. Zool.* **25**: 117-122

Binney, W.G. 1876 (for 1875). On the lingual dentition, jaw, and genitalia of *Carelia, Onchidella*, and other Pulmonata. *Proc. Acad. Nat. Sci. Phil.* **28**: 183-192

Bloxam, Q.M.C., S.J. Tonge & R. Horton. 1984. A preliminary account of the history and management of populations of endangered Moorean tree snail *Partula* spp. at Jersey Wildlife Preservation Trust. *Dodo* **20**:73-79

Bloxham,Q.M.C. & S.J. Tonge. 1986. Breeding programs for reptiles and snails at Jersey zoo: An appraisal. *Intl. Zoo Yearb.* **24/25**: 49-56

Boettger, C. 1916. Die Molluskenausbeute der Hanseatischen Südsee-Expedition 1909. *Abh. Senckenberg. naturforsch. Gesell.* **36**(3): 287-308

Boettger, C.R. 1921. *Carabus morbillosus* Fabr. und *Otala tigri* Gerv., eine Studie gegenseitiger Anpassung. *Abh. Senckenb. Naturforsch. Ges.* **37**: 321- 325

Boettger, C.R. 1935. Die Abwehrder Landschnecken gegen Kafer. *Sber. Ges. Naturf. Freunde Berl.* (**1935**): 93-102

Bolton, P.E., L.E. Rollins & S.C. Griffin. 2015. The danger within: the role of genetic, behavioural, and ecological factors in population persistence of colour polymorphic species. *Mol. Ecol.***24**(12): 2907-2915

Boycott, A.E. & C. Diver. 1923. On the inheritance of sinistrality in *Limnaea peregra. Proc. R. Soc. Lond.* B **95**: 207-213

Brazier, J. 1871. Notes on the localities of *Dolium melanostoma* and other shells found in Australia and the adjacent islands in the Australian Seas. *Proc. zool. Soc. Lond.* (**1871**): 585-587

Brooks DR, McLennan DA. 2002. *The nature of diversity: an evolutionary voyage of discovery*. University of Chicago Press, Chicago

Boss, K.J. 1982. Mollusca. In: Parker, S.P. (ed.) *Synopsis and Classification of Living Organisms*. Vol. **2**. McGraw-Hill, New York

Bouchet P. & J.-P. Rocroi. 2005. Classification and nomenclator of gastropod families. *Malacologia* **47**(1-2): 1-397

Bouchet, P., H. Le Guyader & O. Pascal (eds) 2007. *The Natural History of Santo*. MNHN, Paris

Bowden, J. 2000. Snails in Vanuatu. *Tentacle* **9**: 11

Broderip, W.J. 1832. New species of shells collected by Mr. Cuming on the western coast of South America and in the islands of the South Pacific Ocean. *Proc. Comm. Sci. Corr. Zool. Soc. Lond.* **2**: 124-126

Brodie, G., F. Stevens & G. Barker. 2012. *Report on a preliminary survey of the land snail fauna of Rotuma*. Unpublished report, University of the South Pacific

Brodie, G., G.M. Barker, F. Stevens & M. Fiu. 2014. Preliminary re-survey of the land snail fauna of Rotuma: conservation and biosecurity implications. *Pacific Cons. Biol.* **20**: 94-107

Brook, F.J. 2010. Coastal landsnail fauna of Rarotonga, Cook Islands: systematics, diversity, biogeography, faunal history and environmental influences. *Tuhinga* **21**: 161-252

Brook, F. 2012. Report on the landsnail fauna of Upland Savai'i. In: Atherton, J. & B. Jefferies (eds.) *Rapid biodiversity assessment of upland Savai'i, Samoa*. SPREP, Samoa

Bruguière, J.G. [ed.] & C.H. Hwass, 1792. *Encyclopédie méthodique. Histoire naturelle des vers. Tome première partie 2.* Panckoucke, Paris

Burch, J.B. 1965. Chromosome numbers and systematic in euthyneran snails. *Proc. First Europ. Malacol. Congr.* (**1962**): 215-241

Burch, J.B. 2007. A new species of land snail (Stylommatophora: Partulidae) from Raiatea, French Polynesia, Oceania. *Occ. Pap. Mus. Zoo. Univ. Michigan* **740**: 1-8

Burlingham-Johnson, A., D. Clarke & P. Pearce-Kelly. 1994. CERCI©: a computer system for the demographic and genetic analysis of captive invertebrates, fish and other populations of colony animals. *Internl. Zoo Yrb.* **33**(1): 278-283

Cain, A.J. 1977. The uniqueness of the polymorphism of *Cepaea* (Pulmonata: Helicidae) in Western Europe. *J. Conchol.* **29**: 129-136

Cain, A.J. & P.M. Sheppard. 1950. Selection in the polymorphic land snail *Cepaea nemoralis. Heredity* **4**: 275-294

Cain, A.J. & J.D. Currey. 1963. Area effects in *Cepaea. Phil. Trans. R. Soc. Lond. B* **246**: 1-81

Cameron, R.A.D. 2013. The poor relation? Polymorphism in *Cepaea hortensis* (O. F. Müller) and the Evolution Megalab. *J. Mollus. Stud.* **79**(2): 112-117

Cameron, R.A.D., M.A. Carter, M.A. Palles-Clark. 1980. *Cepaea* on Salisbury Plain: patterns of variation, landscape history and habitat stability. *Biol. J. Linn. Soc.* **14**: 335-358

Cameron, R.A.D. & L.M. Cook. 2013. Temporal morph frequency changes in sand-dune populations of *Cepaea nemoralis* (L.) *Biol. J. Linn. Soc.* **108**: 315-322

Cameron, R.A.D., L.M. Cook & J.J.D. Greenwood. 2013. Change and stability in a steep morph-frequency cline in the snail *Cepaea nemoralis* (L.) over 43 years. *Biol. Journ. Linn. Soc.* **108**: 473-483

Cameron, R.A.D., L.M. Cook & J.D. Hallows. 1996. Land Snails on Porto Santo: Adaptive and Non-Adaptive Radiation. *Phil. Trans. Biol. Sci.* **351**(1337): 309-327

Carpenter, P.P. 1865 (for 1864). In Pease, W.H. 1865

Catlow, A. & L. Reeve. 1845. *The Conchologist's Nomenclator: A Catalogue of All the Recent Species of Shells... with Their Authorities...* Reeve Brothers, London

Centre for Biological Diversity. (undated). Candidate Petition Project – Mollusks. Petitions to list as a federally endangered species. http://www.biologicaldiversity.org/campaigns/candidate_ project/pdfs/mollusk-petition.pdf

Chemnitz, J.H. 1786. *Neues systematisches Conchylien-Cabinet / geordnet und beschrieben von Friedrich Heinrich Wilhelm Martini und unter dessen Aufsicht nach der Natur gezeichnet und mit lebendigen Farben erleuchtet.* **9**

Chenu, J.C. 1845. *Bibliothèque conchyliologique*. A. Franck, Paris

Chiba, S. 1997. Novel colour polymorphisms in a hybrid zone of *Mandarina. Biol. J. Linn. Soc.* **61**: 369-384

Chiba, S. 1998. Genetic variation derived from natural gene flow between sympatric species in land snails (*Mandarina*). *Heredity* **90**: 617-623

Chiba, S. 1999. Accelerated evolution of land snails *Mandarina* in the oceanic Bonin Islands: evidence from mitochondrial DNA sequences. *Evolution* **53**: 460-471

Clarke, B. 1962. Balanced polymorphism and the diversity of sympatric species. In: Nichols, D. (ed.) *Taxonomy and Geography*. Systematics Association, Oxford.

Clarke, B. 1966. The evolution of morph-ratio clines. *Am. Nat.* **100**: 389-402

Clarke, B. 1968. Balanced polymorphism and regional variation in land snails. In: Drake, E.T. (ed) *Evolution and environment*. Yale University Press, New Haven

Clarke, B.C. & J. Murray. 1962. Changes of gene-frequency in *Cepaea nemoralis (L). Heredity* **17**: 445-465

Clarke, B. & J. Murray. 1969. Ecological genetics and speciation in land snails of the genus *Partula. Biol. J. Linn. Soc.* **1**: 31-42

Clarke, B. & J. Murray. 1971. Polymorphism in a Polynesian Land Snail *Partula suturalis vexillum*. In: Creed, R. (ed.). *Ecological Genetics and Evolution*. Blackwell, Oxford

Clarke, B.C., J. Murray & M.S. Johnson 1984. The extinction of endemic species by a program of biological control. *Pacific*

Science **38**(2): 97-104

Clarke, B.C., A.W. Horsley, D.T. Parkin. 1978. Genetic variation and natural selection in pulmonate snails. In: Fretter, V. & J. Peake (eds) *Pulmonates*. Academic Press, New York

Clarke, B. & S.M. Wells. 1992. Extinction of *Partula* snails in French Polynesia – is captive breeding the only solution? In: Gittenberger, E. & J. Gould, J. *Proceedings of the Ninth International Malacological Congress 1986*. Unitas Malacologica, Leiden

Clarke, D., A. Burlingham-Johnson & P. Pearce-Kelly. 1998. A colony management and analysis database relevant for mollusc conservation. *Journ. Conch.* **36**(2): 97-100

Clench, W.J. 1933. Two new land snails from the Bismark Archipelago. *Nautilus* **47** (1): 23-24

Clench, W.J. 1941. The land Mollusca of the Solomon islands (Succineidae, Bulimulidae and Partulidae). *Amer. Mus. Novit.* **1129**: 1-22

Cockerell, T.D.A. 1925. *Partula* in Guam. *Nautilus* **39**: 19-21

Cook, L. M. 2008. Species richness in Madeiran land snails, and its causes. *Journ. Biogeogr*. **35**: 647-653

Cook, L.M. & M. Oźgo. 2015. Colour and pattern disequilibrium in *Cepaea nemoralis* on a northern European transect. *Biol. J. Linn. Soc.* **116**(2): 260-266

Cooke, C.M. Jr. & Crampton, H.E. 1930. New species of *Partula*. *Occ. Pap. B.P. Bishop Mus.* **9**(11): 3-9

Cooke, C.M. & Y. Kondo 1960. Revision of Tornatellinidae and Achatinellidae (Gastropoda, Pulmonata). *B.P. Bishop Mus. Bull.* **221**

Cooper, J.E. & C. Knowler. 1992. Investigations into causes of death of endangered molluscs (*Partula* species). *Vet. Rec.* **131**(15): 342-344

Coote, T. 1999. *The Genetics and Conservation of Polynesian Tree Snails* (*Family Partulidae*). PhD Thesis, University of London

Coote, T. 2007. Partulids on Tahiti: Differential persistence of a minority of endemic taxa among relict populations. *Amer. Mal. Bull.* **22**(1): 83-87

Coote, T. 2009. *Samoana decussatula*. The IUCN Red List of Threatened Species 2009: e.T168196A6464638

Coote, T. 2011. *Polynesian Diaries: Expedition to French Polynesia, 1995*. Lulu.com

Coote, T. 2013. *Polynesian Diaries: Around the Islands, 2000-2005*. TeeCee Book

Coote, T. 2014. *Polynesian Diaries: Around the Islands, 2006-2011*. TeeCee Books

Coote, T., D. Clarke, C.J.S. Hickman, J. Murray & P. Pearce-Kelly. 2004. Experimental release of endemic *Partula* species, extinct in the wild, into a protected area of natural habitat on Moorea. *Pac. Sci.* **58**: 429-434

Coote, T. & E. Loève. 2003. From 61 species to five: endemic tree snails of the Society Islands fall prey to an ill-judged biological control programme. *Oryx* **37**(1): 91-96

Coote, T., E. Loève, J.-Y. Meyer & D. Clarke. 1999. Extant populations of endemic partulids on Tahiti, French Polynesia. *Oryx* **33**(3): 215-222

Coote, T., B. Merz & P. Pearce-Kelly. 2015. *Action Plan for the conservation of French Polynesia's endemic tree snails of the genera* Partula, Samoana, & Trochomorpha. Unpublished

Cowie, R.H. 1992. Evolution and extinction of Partulidae, endemic Pacific island land snails. *Phil. Trans. R. Soc. Lond. B* **335**: 167-191

Cowie, R.H. 1998. Catalog of the nonmarine snails and slugs of the Samoan Islands. *Bishop Mus. Bull. Zool.* **3**: 1-122

Cowie, R.H. & R.P. Cook. 1999. The distribution and abundance of land snails in the National Park of American Samoa, with particular focus on Partulidae. *Cooperative National Park Resources Study Unit, University of Hawai'i at Manoa. Technical Report* **125**

Cowie, R.M. & R.P. Cook. 2001. Extinction or survival: partulid tree snails in American Samoa. *Biodiv. Conserv.* **10**: 143-159

Cowie, R.H. & A.C. Robinson. 2003. The decline of native Pacific island faunas: changes in the land snails of Samoa through the 20th century. *Biol. Conserv.* **110**: 55-65

Cowie, R.H., R.J. Rundell, F. Mika & P. Setu. 2002. The endangered partulid tree snail *Samoana thurstoni* (Cooke and Crampton, 1930) on Olosega and the land snail diversity of the Manu'a Islands, American Samoa. *Am. Malacol. Bull.* **17**(1/2): 37-43

Cox, J.C. 1868. *Dr. James C. Cox's Exchange List of Land and Marine Shells Australia and the Adjacent Islands*. Privately published

Coyne, J.A. & H.A. Orr. 2004. *Speciation*. Sinauer, Sunderland

Crampton, H.E. 1908. Physiological problems of the geographical distribution of *Partula* in Polynesia, with demonstration of specimens. *Proc. Soc. Exp. Biol. Med.* **5**: 57-58

Crampton, H.E. 1916. *Studies on the variation, distribution and evolution of the genus* Partula. *The species inhabiting Tahiti.* Carnegie Institution of Washington **228**: 1-311

Crampton, H.E. 1921. A journey to the Mariana Islands Guam and Saipan. *Natural History* **21**(2): 126-145

Crampton, H.E. 1924a. New and significant species of *Partula* from Moorea, Society Islands. *Nautilus* **37**: 109-118.

Crampton, H.E. 1924b. The coincident production of dextral and sinistral young in the land-gasteropod *Partula. Science* **59**: 558-559

Crampton H. E. 1925. *Studies on the variation, distribution and evolution of the genus* Partula. *The species of the Mariana Islands, Guam and Saipan.* Carnegie Institution of Washington, **228a**: 1-116

Crampton, H.E. 1928. The differentiation of species. *Science* **67**: 615-619

Crampton, H.E. 1929 Continuation of studies concerning variation of the genus *Partula* with relation to geographical distribution. *Carnegie Inst. Wash. Yearbook* **28**: 336

Crampton, H.E. 1932. *Studies on the variation, distribution, and evolution of the genus* Partula. *The species inhabiting Moorea.* Carnegie Institute of Washington, Washington, D.C., Publ. No. **410**: 1-335

Crampton, H.E. 1956. New species of land snails of the genus *Partula* from Raiatea. *Am. Mus. Novit.* **1761**: 1-17

Crampton, H.E. & Cooke, C.M. J. 1953. New Species of *Partula* from Southeastern Polynesia. *Occ. Pap. B.P. Bishop Mus.* **21**(8): 135-159

Crosse, H., 1865. Addition à la note de M. Mousson sur la faune malacologique et fluviatile des archipels Viti et Samoa. *Journ. Conchyl.* **13**: 430-432

Crosse, H. 1871. Des espèces terrestres et fluviatiles que l'on a consideérées, á tort, comme appurtenant à la Faune malacologique de la Nouvelle-Calédonie. *Journ. Conchyl.* **19**: 170-187

Crosse, H. 1894. Faune malacologique terrestre et fluviatile de la Nouvelle-Caledonie et ses dependences. *Journ. Conchyl.* **42**: 161-473

Cunningham, A.A. & P. Daszak. 1998. Extinction of a species of land snail due to infection with a microsporidian parasite. *Conserv. Biol.* **12**: 1139-1141

Cunningham, A.A., P. Daszak, S.K. Macgregor, I. Foster, D. Clarke & P. Pearce-Kelly. 1996. Mortality of endangered snails of the genus *Partula*: preliminary results of pathologic investigations. *Journ. Zoo Wildl. Med.* **27**: 19-27

Cwiertnia, P., A. Jozefowicz & M. Urbanska. 2010. Condition of populations of the snails of the genus *Partula* and problems of their ex situ protection. *Folia Malacol.* **18**(3)

Dall, W.H. 1882. American work on recent Mollusca in 1881. *Am. Nat.* **16**(12): 953-968

Dall, W. H. 1905. Thomas Martyn and the Universal Conchologist. *Proc. U.S. nat. Mus.* **19**: 415-432

Dall, W.H. 1910. Description of two new pulmonate molusks with a list of other species from the Solomon Islands, collected by Dr. George A. Dorsey. *Field Mus. Nat. Hist. Pubs.* **139** (Zool. 7): 215-221

Dance, P. 1986. *A history of shell collecting.* E.J. Brill – Dr. W. Backhuys, Leiden

Dance, S.P. 1980. Hugh Cuming (1791-1865) Prince of collectors. *J. Soc. Biblphy. Nat. Hist.* **9**(4): 477-501

Darwin, F. 1887. *The life and letters of Charles Darwin, including an autobiographical chapter.* **2**. John Murray, London

Davison, A., N. Constant, H. Tanna, J. Murray & B. Clarke. 2009. Coil and shape in *Partula suturalis*: the rules of form revisited. *Heredity* **103**(3):268-78

Davison, A., N.H. Barton & B. Clarke. 2009. The effect of coil phenotypes and genotypes on the fecundity and viability of *Partula suturalis* and *Lymnaea stagnalis*: implications for the evolution of sinistral snails. *Journ. Evol. Biol.* **22**(8): 1624-1635

Davison, A., G.S. McDowell, J.M. Holden, H.F. Johnson, G.D. Koutsovoulos, M.M. Liu, P.Hulpiau, F. Van Roy, C.M. Wade, R. Banerjee, F. Yang, S. Chiba, J.W. Davey, D.J. Jackson, M. Levin, M.L. Blaxter. 2016. Formin is associated with left-right asymmetry in the pond snail and the frog. *Current Biology* **25**: 654-660

Dennis, R.L.H., L. Dapporto, S. Fattorini & L.M. Cook. 2011. The generalism-specialism debate: the role of generalists in the life and death of species. *Biol. J. Linn. Soc.* **104**(4): 725-735

Deshayes, G.P. & H. Milne-Edwards in Lamarck, J.B.P.A.M. de, G.P. 1839. *Histoire naturelle des animaux sans vertèbres.* Meline, Cans & Co., Brussels. 3rd edition. **3**

Dickinson, E.C. 2005. The Proceedings of the Zoological Society of London, 1859-1900: an exploration of breaks between calendar years of publication. *Journ. Zool.* **266**(4): 427-430

Dillwyn, L.W. 1817. *A descriptive catalogue of recent shells, arranged according to the Linnæan method; with particular attention to the synonymy.* J.&A. Arch, Cornhill. **1**

Dingwall, P.R. 2013. *Report on the reactive monitoring mission to East Rennell, Solomon Islands. 21-29 October 2012.* Report **854** to World Heritage Committee, UNESCO. Unpublished

Diver, C. 1939. Aspects of the study of variation of snails. *Journ. Conch.* **21**: 91-141

Dobzhansky, T. 1937. *Genetics and the origin of species*. Columbia University Press, New York

Douris, V., S. Giokas, D. Thomaz, R. Lecanidou & G.C. Rodakis. 2007. Inference of evolutionary patterns of the land snail *Albinaria* in the Aegean archipelago: is vicariance enough? *Mol. Phyl. Evol.* **44**(3): 1224-1236

Drake, D.R., W.A. Whistler, T.J. Motley & C.T. Imada. 1996. Rain forest vegetation of 'Eua Island, Kingdom of Tonga. *N.Z. Journ. Bot.* **34**: 65-77

Duncan, F.M. 1937. On the dates of publication of the Society's 'Proceedings,' 1859-1926. *Proc. Zool. Soc. Lond.* **A107**(1): 71-84

Duncan, R.A. & I. McDougall. 1976. Linear volcanism in French Polynesia. *J. Volcanol. Geotherm. Res.* **1**: 197-227

Economo, E.P. & E.M. Sarnat. 2012. Revisiting the Ants of Melanesia and the Taxon Cycle: Historical and Human-Mediated Invasions of a Tropical Archipelago. *Am. Nat.* **180**(1): E1-E16

Eldredge, J.G. 1994. Introductions and transfers of the triclad flatworm *Platyemus manokwari*. *Tentacle* **3**: 8

Emberton, K.C. 1982. Environment and shell shape in the Tahitian land snail *Partula otaheitana*. *Malacologia* **23**(1): 23-35

Ermentrout, B. & J. Campbell. 1986. A model for shell patterns based on neural activity. *Veliger* **28**(4): 369-388

Ereshefsky, M. 1992. *The Units of Evolution: Essays on the Nature of Species*. MIT Press, Cambridge, Massachusetts

Ereshefsky, M. 2001. The Poverty of the Linnaean Hierarchy: a Philosophical Study of Biological Taxonomy. Cambridge University Press, Cambridge

FAO-SAPA (Food and Agriculture Organization Sub-regional officer for the Pacific Islands). 2002. Bio-control: flatworms and nemertean worms collected and identified from Cook Islands, Niue, Tonga and Vanuatu. *SAPA Newsletter* **6**: 3

Férussac, A.E.J.P.J.F. d'A. 1821-22. *Tableaux systématiques des animaux mollusques classés en familles naturelles, dans lesquels on a établi la concordance de tous les systèmes; suivis d'un prodrome général pour tous les mollusques terrestres ou fluviatiles, vivants ou fossiles.* Bertrand, Paris

Férussac, A.E.J.P.J.F. d'A. & G.P. Deshayes 1819-51. *Histoire naturelle générale et particulière des mollusques terrestres et fluviatiles tant des espèces que l'on trouve aujourd'hui vivantes, que des dépouilles fossiles de celles qui n'existent plus ; classés d'après les caractères essentiels que présentent ces animaux et leurs coquilles.* J.-B. Balliere, Paris

Fischer, P. 1885. *Manuel de Conchyliologie et de Paléontologie Conchyliologique*. Savy, Paris

Fisher, K. 2006. Rank-free monograph: a practical example from the moss clade *Leucophanella* (Calymperaceae). *Syst. Biol.* **31**: 13-30

Fontaine, B.F. & O. Gargominy. 2002. *Inventaire des mollusues continentaux de l'archipel des Australe* (*Raivavae et Rapa*). MNHN, Paris. Unpublished

Fontaine, B., O. Gargominy & V. Prié. 2011. Indigenous land snails. In: Bouchet, P., H. Le Guyader & O. Pascal (eds) *The Natural History of Santo*. MNHN, Paris

Ford, E.B. 1964. *Ecological genetics*. Methuen & Co., London

Fulton, H.C. 1917. Molluscan notes. III. No. 11. On the type-specimen of *Partula newcombianum*, Hartman. *Proc. Malacol. Soc. Lond.* **12**: 238

Gargominy, O. & B. Fontaine. 2003. *Inventaire des mollusques continentaux de l'archipel des Australe* (*Tubai, Rurutu*). MNHN, Paris. Unpublished

Gargominy, O., B. Fontaine & V. Prié. 2011. Focus on alien land snails. In: Bouchet, P., H. Le Guyader & O. Pascal (eds) *The Natural History of Santo*. MNHN, Paris

Gargominy, O. & B. Fontaine. 2014. A global overview of the terrestrial and freshwater molluscs. In: Meyer, J.-Y. & E.M. Claridge (eds) *Terrestrial Biodiversity of the Austral Islands*. MNHN, Paris

Garrett, A. 1879. List of land shells inhabiting Rurutu, one of the Austral islands, with remarks on their synonymy, geographical range, and description of a new species. *Proc. Acad. Nat. Sci., Phil.* **31**(1): 17-30

Garrett, A. 1884. The terrestrial Mollusca inhabiting the Society Islands. *Journ. Acad. Nat. Sci., Phil.* (*ser. II*) **9**: 17-114

Garrett, A. 1887a. The terrestrial Mollusca inhabiting the Samoa or Navigator Islands. *Proc. Acad. Nat. Sci. Phil.* **39**(2): 124-153

Garrett, A. 1887b. Mollusques terrestres des Iles Marquises (Polynesie). *Bull. Soc. Malacol. Fr.* **4**: 25

Gerlach, J. 1994. *The ecology of the carnivorous snail* Euglandina rosea. DPhil thesis, Oxford University

Gerlach, J. 1999. Snails of the genus *Pachnodus* (Mollusca; Gastropoda; Enidae): their origins and evolution. *Journ. Biogeogr.* **26**(2): 251-255

Gerlach, J. 2001. Predator, prey and pathogen interactions in introduced snail populations. *Anim. Conserv.* **4**: 203-209

Gerlach, J. 2014. *Snailing Round the South Seas – the* Partula *story*. Phelsuma Press, Cambridge

Ghiselin, M.T. 1974. A radical solution to the species problem. *Syst. Zool.* **23**: 536-544

Gittenberger, E. 1996. Adaptations of the aperture in terrestrial gastropod-pulmonate shells. *Neth. J. Zool.* 46: 191-205

Gmelin, J.F. 1791. *Caroli a Linnaei Systema Naturae per Regna Tria Naturae, Editio Decima Tertia, Aucta Reformata.* **1**(6

Vermes). G.E. Beer, Leipzig

Gloyne, C.P. 1878. Notes on the genus *Partula. Quart. Journ. Conchol.* **1**(15): 337-338

Golikov, A.N. & Ya.I. Starobogatov. 1989. Voprosy filogenii i sistemy perednezhabernykh brjukhonogikh molljuskov. [Problems of phylogeny and system of the prosobranchiate gastropods]. *Trudy Zoologicheskogo Instituta* **187**: 4-77

Goodacre, S.L. 2001. Genetic variation in a Pacific Island land snail: population history versus current drift and selection. *Proc. R. Soc. Lond. B* **268**(1463): 121-126

Goodacre, S.L. 2002. Population structure, history and gene flow in a group of closely related land snails: genetic variation in *Partula* from the Society Islands of the Pacific. *Mol. Ecol.* **11**: 55-68

Goodacre, S.L. & C.M. Wade. 2001a. Molecular evolutionary relationships between partulid land snails of the Pacific. *Proc. R. Soc. Lond.* B **268**: 1-7

Goodacre, S.L. & C.M. Wade. 2001b. Patterns of genetic variation in Pacific island land snails: the distribution of cytochrome b lineages among Society Island *Partula. Bio. J. Linn. Soc.* **73**:131-138

Goodhart, C.B. 1963. "Area effects" and non-adaptive variation between populations of *Cepaea* (Mollusca). *Heredity* **18**: 459-465

Goodhart CB. 1987. Why are some snails visibly polymorphic, and others not? *Biol. J. Linn. Soc.* **31**: 35-58

Goodin, J.T. 1990. *Patterns of morphological covariation in* Partula. MSc thesis, University of Western Australia

Goodin, J.T. & M.S. Johnson. 1992. Patterns of morphological covariation in *Partula. Syst. Biol.* 41(3): 292-304

Goodfriend, G.A. 1986. Variation in land-snail form and size and its causes: a review. *Syst. Zool.* **35**(2): 204-223

Gould, A.A. 1848 (for 1847). [Untitled]. *Proc. Boston Soc. Nat. Hist.* **2**: 196-198

Gould, A.A. 1852. Mollusca & shells. With an atlas of plates. *United States exploring expedition. During the years 1838, 1839, 1840, 1841, 1842. Under the command of Charles Wilkes, U.S.N.* **12**. Sherman, Philadelphia

Gould, A.A. 1862 Expedition shells described for the work of the United States Exploring Expedition during the years 1838-1842. [reprinted from *Proc. Boston Soc. Nat. Hist.*]. In: *Otia conchologica: descriptions of shells and molluscs, from 1839 to 1862.* Gould & Lincoln, Boston

Gould, S.J. & D.S. Woodruff. 1990. History as a cause of area effects – an illustration from *Cerion* on Great Inagua, Bahamas. *Biol. J. Linn. Soc.* **40**: 67-98

Gouviea. A. 2012. *Investigation of the Factors Affecting the Populations Dynamics of Captive Partula Snails.* PhD thesis, Imperial College, London

Gouveia, A.R., P. Pearce-Kelly, D.L.J. Quicke & S.R. Leather. 2011. Effects of different calcium concentrations supplemented on the diet of *Partula gibba* on the morphometric growth parameters. Weight and reproduction success. *Malacologica* **54**(1): 139-146 [also in *Proc. 8th Int. Congr. Behav., Physiol. Gen. Wildl.*]

Gouveia, A., R. Pizzi & P. Pearce-Kelly, D. MacFarlane, D. Clarke, S. MacGregor, B. Clark & W. Bordman. 2006. Investigating the role of environmental factors in captive *Partula clara* and *Partua tohiveana*. In: Dow, S. & F. Clark (eds) *Proceedings 9th Annual Symposium of Zoo Research.* Zoological Gardens of Great Britain and Ireland, London

Gouveia, A., R. Pizzi & P. Pearce-Kelly. 2007. Environmental Influences in Captive Population Declines of two Endangered Polynesian Tree Snails (*Partula* Species). In: *Proceedings 9th Annual Symposium of Zoo Research*

Grant, V. 1981. *Plant Speciation.* 2nd ed. Columbia University Press, New York

Grateloup, J.P.S. de 1840a (for 1839). Mémoire sur plusiers expèces de coquilles nouvelles ou peu connues de Mollusques exotiques vivants, terrestres, fluviatiles et marins. *Actes Soc. Linn. Bordeaux* **11**: 394-455

Grateloup, J.P.S. de 1840b (for 1839) Note sur un memoire relatif à des mollusques exotiques nouveaux ou peu connus. *Actes Soc. Linn. Bordeaux* **11**: 161-170

Gray, J.E. 1839. Molluscous animals, and their shells. *In*: Richardson, J., N.A. Vigors, G.T. Lay, E.T. Bennett, R. Owen, J.E. Gray, W. Buckland & G.B. Sowerby (eds). *The zoology of Captain Beechey's voyage; compiled from the collections and notes made by Captain Beechey, the officers and naturalist of the expedition, during a voyage to the Pacific and Behring's Straits performed in His Majesty's ship Blossom, under the command of Captain F.W. Beechey, R.N., F.R.S., &c. &c. in the years 1825, 26, 27, and 28.* Henry G. Bohn, London

Gulick, J.T. 1872. On the variation of species as related to their geographical distributions, illustrated by the Achatinellinae. *Nature* **6**: 222-224

Haas, F. 1939. Malacological notes. *Field Mus. nat. Hist., Zool.* **24**: 93-103

Hadfield, M.G. 2015. The occurrence of the endangered tree snail *Partula gibba* in the Mariana Islands, with a focus on Pagan Island. *Bishop Mus. Bull. Zool.* **9**: 147-167

Hadfield, M.G. & S.E. Miller. 1989. Demographic studies on Hawai"s endangered tree snails: *Partula proxima. Pac. Sci.* **43**:1-16

Hafiz Ur, R., N. Hideo & K. Kei. 2013. Geological Origin of the Volcanic Islands of the Caroline Group in the Federated

States of Micronesia, Western Pacific. *South Pacific Stud.* **33**(2): 101-118
Hall, K.T. & M.G. Hadfield. 2009. Application of harmonic radar technology to monitor tree snail dispersal. *Invertebr. Biol.* **128**(1): 9-15
Hall, K.T., M.B. Baker & M.G. Hadfield. 2010. Using dispersal rates to guides translocation across impermeable wildlife reserve boundaries: Hawaiian tree snails as a practical example. *Malacologia* **52**(1): 67-80
Hart, S. &. M.G. Jackson. 2014. Ta'u and Ofu/Olosega volcanoes: The "Twin Sisters" of Samoa, their P, T, X melting regime, and global implications. *Geochem. Geophys. Geosyst.* **15**(6): 2301-2318
Hartman, W. D. 1880. Description of a *Partula* supposed to be new, from the island of Moorea. *Proc. Acad. Nat. Sci. Phil.* **32**: 229
Hartman, W.D.1881a. Observations on the species of the genus *Partula*. *Bull. Mus. Comp. Zool., Harv.* **9**: 171-190
Hartman, W.D. 1881b. *A catalogue of the Genus* Partula *Fér*. F.S. Hickman, West Cheser.
Hartman, W.D. 1885. Descriptions of New Species of *Partula* and a Synonymic Catalogue of the Genus. *Proc. Acad. Nat. Sci. Phil.* **37**: 203-223
Hartman, W.D. 1886. New species of *Partula* from the New Hebrides and Solomon islands. *Proc. A.N.S.P.* **38**: 30-35
Hartman, W.D. 1888. New species of shells from the New Hebrides and Sandwich Islands. *Proc. A.N.S.P.* **40**: 250-391
Hartman, W.D. 1893 (for 1892). Catalogue of the genus *Partula*. *Nautilus* **6**(7): 73-76
Hartman, W.D. 1897. *Partula*: Notes and Corrections. *Nautilus* **10**: 44
Hayashi, M. & S. Chiba. 2000. Intraspecific diversity of mitochondrial DNA in the land snail *Euhadra peliomphala* (Bradybaenidae). *Biol. J. Linn. Soc.* **70**: 391-401
Heath, C.J. 1975. Colour, sunlight and internal temperatures in the land-snail *Cepaea nemoralis* (L.). *Oecologica* **19**: 29-38
Hedley, C. 1892 (for 1891). The land molluscan fauna of British New Guinea. *Proc. Linn. Soc. N. S. Wales* **6**: 67-116
Hendry, A., S. Vamosi, S. Latham, J. Heilbuth & T. Day. 2000. Questioning species realities. *Conserv. Genet.* **1**: 67-76
Herbert, D.G. & A. Mitchell. 2009. Phylogenetic relationships of the enigmatic land snail genus *Prestonella*: the missing African element in the Gondwanan superfamily Orthalicoidea (Mollusca: Stylommatophora). *Bio. J. Linn. Soc.* **96**: 203-221
Herrmannsen, A.N. 1852. *Indicis generum Malacozoorum primordia*. Casell (Theodor Fischer). **2**
Heynemann, F.D. 1867. Die Zungen von *Partula* und *Achatinella.Malak. Blatt.* 14: 146-150
Holland, B.S. & M.G. Hadfield. 2002. Islands within an island: phylogeography and conservation genetics of the endangered Hawaiian tree snail *Achatinella mustelina*. *Mol. Ecol.* **11**: 365-375
Holland, B.S. & M.G. Hadfield. 2004. Origin and diversification of the endemic Hawaiian tree snails (Achatinellidae: Achatinellinae) based on molecular evidence. *Mol. Phyl. Evol.* **32**: 588-6
Hopper, D.R. & B.D. Smith. 1992. The status of tree snails (Gastropoda: Partulidae) on Guam, with a resurvey of sites studied by H. E. Crampton in 1920. *Pacific Sci.* **46**: 77-85
Hugall, A., C. Moritz, A. Moussalli & J. Stanisic. 2002. Reconciling paleodistribution models and comparative phylogeography in the Wet Tropics rainforest land snail *Gnarosophia bellendenkerensis* (Brazier 1875). *Proc. Natl. Acad. Sci. USA* **99**: 6112-6117
Huxley, J. 1940. *The new systematics*. Oxford University Press, Oxford
Huxley, J. 1942. *Evolution. The modern synthesis.* George Allen & Unwin, London
International Commission on Zoological Nomenclature. 1957. Opinion 456: Rejection of the works by Thomas Martyn published in 1784 with the title "the universal conchologist" as a work, which does not comply with the requirements of article 25 of the "règles" and which therefore possesses no status in zoological nomenclature and rejection also of a proposal that the foregoing work should be validated under the plenary powers. *Opinions and Declarations* **15**: 393-418
International Commission on Zoological Nomenclature. 1999. *International Code of Zoological Nomenclature.* 4th edition. Natural History Museum, London
Jay, J.C. 1850. *A Catalogue of the Shells, Arranged According to the Lamarckian System*. London
Johnson, R.I. 1994. Types of shelled Indo- Pacific mollusks described by W. H. Pease. *Harv. Univ. Bull. Mus. Comp. Zool.* **154**(1): 1-61
Johnson, M.S. 1982. Polymorphism for direction of coil in *Partula suturalis* – behavioral isolation and positive frequency-dependent selection. *Heredity* **49**: 145-151
Johnson, M.S. 1987. Adaptation and rules of form: chirality and shape in *Partula suturalis*. *Evolution* **41**(3): 672-675
Johnson, M.S. 2011. Thirty-four years of climatic selection in the land snail *Theba pisana*. *Heredity* **106**(5): 741-748
Johnson, M.S. 2012. Epistasis, phenotypic disequilibrium and contrasting associations with climate in the land snail *Theba pisana*. *Heredity* **108**(3): 229-235

Johnson, M.S., B. Clarke & J. Murray. 1977. Genetic variation and reproductive isolation in *Partula. Evolution* 3l: 116-126
Johnson, M.S., B. Clarke & J. Murray. 1988. Discrepancies in the estimation of gene flow in *Partula. Genetics* **120**(1): 233-8
Johnson, M.S., B. Clarke & J. Murray. 1990. The coil polymorphism in *Partula suturalis* does not favor sympatric speciation. *Evolution* **44**: 459-464
Johnson, M.S., J. Murray & B. Clarke. 1986a. Allozymic similarities among species of *Partula* on Moorea. *Heredity* **56**(3): 319-327
Johnson, M.S., J. Murray & B. Clarke. 1986b. An electrophoretic analysis of phylogeny and evolutionary rates in the genus *Partula* from the Society Islands. *Proc. R. Soc. Lond. B. Biol. Sci.* **227**: 161-177
Johnson, M.S., J. Murray & B. Clarke. 1987. Independence of genetic subdivision and variation for coil in *Partula suturalis. Heredity* **58**: 307-313
Johnson, M.S., J. Murray & B. Clarke. 1993a. Evolutionary relationships and extreme genital variation in a closely related group of *Partula. Malacologia* **35:** 43-61
Johnson, M.S., J. Murray & B. Clarke. 1993b. The ecological genetics and adaptive radiation of *Partula* on Moorea. *Oxf. Surv. Evol. Biol.* **9:** 167-238
Johnson, M.S., J. Murray & B. Clarke. 2000. Parallel evolution in Marquesan partulid land snails. *Biol. J. Linn. Soc.* **68**: 577-598
Jones, J.S. 1973. Ecological genetics and natural selection in molluscs. *Science* **1182**: 546-552
Jønsson, K.A., P.-H. Fabre, S.A. Fritz, R.S. Etienne, R.E. Ricklefs, T.B. Jørgensen, J. Fjeldså, C. Rahbek, P.G.P. Ericson, F. Woog, E. Pasquet & M. Irestedt. 2012. Ecological and evolutionary determinants for the adaptive radiation of the Madagascan vangas. *Proc. Natl. Acad. Sci. USA* **109**(17): 6620-6625
Kerr, A.M. 2013. The partulid tree snails (Partulidae: Stylommatophora) of the Mariana Islands, Micronesia. *Univ. Guam Marine Lab. Tech. Rep.* **152**
Kerr, A.M. & G.C. Fielder. 2015. Sinistral coiling in the arboreal snail *Partula gibba* Férussac, 1821(Stylommatophora: Partulidae) from Guam, Mariana Islands. *Amer. Malac. Bull.* **33**(2): 1-5
Kilias, R. 1969. Die Typen und Typoide der Mollusken-Sammlung des Zoologischen Museums in Berlin (5). *Mitt. Zool. Mus. Berl.* **45**(2): 245-258
Kobayashi, S.R. & M.G. Hadfield. 1996. An experimental study of growth and reproduction in the Hawaiian tree snails *Achatinella mustelina* and *Partulina redfiledii* (Achatinellinae). *Pac. Sci.* **50**(4): 339-354
Kobelt, W. 1890-2. *Systematisches Concylien Cabinet* **13a**. Bulimus *continuation,* Placostylus.
Kohn, A.J. 2009. Type specimens of *Conus* (Mollusca: Gastropoda) in the Zoological Museum of the University of Copenhagen: A Historical Account. *Steenstrupia* **30**(2): 97-113
Kondo, Y. 1948. Anatomy of *Diplomorpha delatouri* (Hartman) and four species of *Placostylus* (Pulmonata, Bulimulidae). *Nautilus* **61**(4): 119-126
Kondo, Y. 1955. *A revision of the family Partulidae* (*Gastropoda, Pulmonata*). PhD thesis, Harvard University, Cambridge, Massachusetts
Kondo, Y. 1968. Partulidae: Preview of anatomical revision. *Nautilus* **81**(3): 73-77
Kondo, Y. 1970. Some aspects of Mariana islands Partulidae (Mollusca, Pulmonata). *Occ. Pap. B.P. Bishop. Mus*. **24**(5): 73-90
Kondo, Y. 1973. *Samoana* of the Society Islands (Pulmonata: Partulidae). *Malacol. Rev.* **6**: 19-33
Kondo, Y. 1981 (for 1980). *Samoana jackieburchi*, new species (Gastropoda: Pulmonata: Partulidae). *Malacol. Rev.* **13**(1-2): 25-32
Kondo, Y. & J.B. Burch 1979. Extrusive genital anatomies and their internal postures in *Partula affinis* of Tahiti. *Malacol. Rev.* **12**: 79-84
Kondo, Y. & J.B. Burch. 1983. Two amendments to Crampton's monograph on Tahitian Partulidae. *Malacol. Rev.* **16**: 101-106
Kondo Y. & J.B. Burch. 1989. *Samoana medana* new species (Gastropoda: Pulmonata: Partulidae). *Bishop Mus. Occ. Pap.* **29**: 131-137
Konter, J.G. & M.G. Jackson. 2012. Large volumes of rejuvenated volcanism in Samoa: evidence supporting a tectonic influence on late-stage volcanism. *Geochem. Geophy. Geosyst.* **13**(6): unpaginated
Kornilios, P., E. Stamataki & S. Giokas. 2015. Multiple reversals of chirality in the land snail genus *Albinaria*(Gastropoda, Clausiliidae). *Zoologica Scripta* **44**(6): 603-611
Lamarck, J.B.P.A.M. de, G.P. 1839. *Histoire naturelle des animaux sans vertèbres*. Meline, Cans & Co., Brussels. [3rd ed.] **3**
Lee, T., J. B Burch, T. Coote, B. Fontaine, O. Gargominy, P. Pearce-Kelly, D. Ó Foighil. 2007a. Prehistoric inter-archipelago

trading of Polynesian tree snails leaves a conservation legacy. *Proc. R. Soc. B: Biol. Sci.* **274**: 2907-2914

Lee, T., J.B. Burch, Y. Jung, T. Coote, P. Pearce-Kelly & D. Ó Foighil. 2007b. Tahitian tree snail mitochondrial clades survived recent mass extirpation. *Current Biol.* **17**(13): R502-3

Lee, T., J.-Y. Meyer, J.B. Burch, P. Pearce-Kelly & D. Ó Foighil. 2008. Not completely lost: two partulid tree snail species persist on the highest peak of Raiatea, French Polynesia. *Oryx* **42**: 615

Lee, T., J.B. Burch, T. Coote, P. Pearce-Kelly, C. Hickman, J.-Y. Meyer & D. Ó Foighil. 2009. Moorean tree snail survival revisited: a multi-island genealogical perspective. *BMC Evol. Biol.* **9**(204): unpaginated

Lee, T., J. Li, C.K.C. Churchill & D. Ó Foighil. 2014. Evolutionary history of a vanishing radiation: isolation-dependent persistence and diversification in Pacific Island partulid tree snails. *BMC Evo. Biol.* **14**: unpaginated

Leschke M. 1912 (for 1911). Mollusken der Hamburger Südsee Expedition 1908/09. (Admiralitatsinsel, Bismarch-Archipel, Deutsch-Neugiunea). *Mitt. Naturhist. Mus. Hamb.* **29**: 89-172

Lesson, R.P. 1831. In: L.I. Duperrey.*Voyage autour du monde, execute par ordre du Roi, sur la corvette de La Majeste La Coquille, pendant les annees 1822, 1823, 1824 et 1825*. Vol. 2.

Lightfoot, J. 1786. *A catalogue of the Portland Museum: lately the property of the Duchess Dowager of Portland deceased Which will be Sold by Auction,by Mr. Skinner and Co. on Monday the 24th of April, 1786, and the thirty-seven following days, At Twelve O'Clock, Sundays, and the 5th of June, (the Day his Majesty's Birth-Day is kept) excepted. At the late Dwelling-House, In Privy-Garden, Whitehall; By Order of the Acting Executrix.* Skinner, London

Lipton, C.S. & J. Murray. 1979. Courtship of the land snails of the genus *Partula. Malacologica* **19**(1): 129- 146

Lovenburg, V. 2009. *Terrestrial Gastropod Distributional Factors: Native and Nonnative Forests, Elevation and Predation on Mo'orea, French Polynesia.* Undergraduate project, UCB Moorea Class: Biology and Geomorphology of Tropical Islands, unpublished

Lundman, H. 1947. Maps of the racial geography of some Partulae of the Society Islands based upon the material published by H. E. Crampton. *Zool. Bid. Uppsala* **25**: 517-533

Lydeard, C., R.H. Cowie, W.F. Ponder, A.E. Bogan, P. Bouchet, S.A. Clark, K.S. Cummings, T.J. Frest, O. Gargominy, D.G. Herbert, R. Hershler, K.E. Perez, B. Roth, M. Seddon, E.E. Strong & F.G. Thompson. 2004. The Global Decline of Nonmarine Mollusks. *BioScience* **54**(4): 321-330

MacArthur, R.H. & E.O. Wilson. 1963. An equilibrium theory of insular zoogeography. *Evolution* **17**(4): 373-387

Mace, G.M., P. Pearce-Kelly & D. Clarke. 1998. An integrated conservation programme for the tree snails (Partulidae) of Polynesia: A review of captive and wild elements. *Journ. Conch.* (Special issue **1**): 89-96

Martens, E. von. 1860. In: Albers, J.C. *Die Heliceen nach natürlicher Verwandtschaft systematisch geordnet von Joh. Christian Albers. Zweite Ausgabe nach dem hinterlassenen Manuskript besorgt von Eduard von Martens.* Engelmann, Leipzig

Martens, E.C. von. 1877. Übersicht der während der Reise um die Erde in den Jahren 1874-1876 auf. S.M. Schiff *Gazelle* gesammelten Land und süsswasser Mollusken. *Monatsb. K. Preuss. Akad. Wissensch. Berlin* (**1877**): 261-291

Martens, E. von. 1881. *Conchologische Mittheilungen als forsetzung der Novitates conchologicae.* **1**. T. Fischer, Kassel.

Martens, E. von & B. Langkavel. 1871. Eine Sammlung von Südsee-Conchylien. *Donum Bismarkianum.* F. Berggold, Berlin

Martyn, T. 1784. *Universal Conchologist, exhibiting the figure of every known shell...drawn...and painted...with a new systematic arrangement.* London

Martyn, T. 1789. *Figures of non descript shells, collected in the different voyages to the South Seas since the year 1764. Published by Thomas Martyn, and sold at his house, No. 16, Great Marlborough Street, London.* The Universal Conchologist [2nd ed.]

Maury, R.C., G. Ille, H. Guillou, C. Chauvel, P. Rossi, C. Pallares & C. Legendre. 2013. Temporal evolution of a Polynesian hotspot: New evidence from Raivavae (Austral islands, South Pacific ocean). *Bull. Soc. Géol. Fr.* **184**: 557-567

Mayden, R.L. 1997. A hierarchy of species concepts: the denouement in the saga of the species problem. In: Claridge, M.F., H.A. Dawah & M.R. Wilson (eds.) *Species: the units of biodiversity.* Chapman and Hall, London

Mayden, R. 1999. Consilience and a hierarchy of species concepts: advances toward closure on the species puzzle. *Journ. Nematol.* **31**(2): 95-116

Mayden, R.L. & R.W. Wood. 1995. Systematics, species concepts, and the evolutionarily significant unit in biodiversity and conservation biology. In: Nielson, J.L. (ed) *Evolution and the aquatic ecosystem: defining unique units in population conservation.* Symposium 17. American Fisheries Society, Bethesda, Maryland

Mayr, E. 1942. *Systematics and the origin of species.* Columbia University Press, New York.

Mayr E. 1947. Ecological factors in speciation. *Evolution* **1**: 263-288

Mayr, E., E. Gorton Linsley & R.L. Usinger. 1953. *Methods and principles of systematic zoology.* McGraw-Hill, New York
McKay, B.D. & R.M. Zink. 2014. Sisyphean evolution in Darwin's finches. *Biol. Rev.* **90**(3): 689-698
Menke, C.T. 1828. *Synopsis methodical molluscorum generum omnium et specierum earum quae in museo Menkeano adservantur.* Ed. 2. Uslar, Pyrmonti
Meyer, J.-Y. 2003. *Rapport de mission botanique à Ua Pou (Marquises nord) du 16 au 23 Aout 2003.* Unpublished report. http://www.li-an.fr/jyves/Meyer_2003_Rapport_mission_Ua_Pou.pdf
Meyer, J.-Y. 2004. *Végétation et flore des hauts-plateaux humides de Terepo (Papara) et Viriviriterai (Hitiaa). Contribution à la Biodiversité de Polynésie française N°10 : Sites Naturels d'Intérêt Ecologique IV.* Délégation à la Recherche/ Institut Louis Malardé, Papeete.
Meyer, J.-Y. 2006. Rapport de mission botanique à Ua Huka (Marquises, groupe nord) du 24 au 31 Juillet 2005 et description de la végétation et de la flore de la reserve naturalle de Vaikivi. Unpublished report. http://www.li-an.fr/jyves/Meyer_2005_Rapport_mission_Ua_Huka.pdf
Meyer, J.-Y., H. Jourdan & A. Malau. 2008. Recent incursion of invasive animal and plant species in the small and remote archipelagos of Wallis and Futuna (South Pacific). *Aliens* 8-9
Michener, C. D. 1970. Diverse approaches to systematics. *Evol. Biol.* **4**:1-38
Miller, S. 1993. *Final report on surveys of the arboreal and terrestrial snail fauna of American Samoa.* Unpublished report to USFWS
Mishler, B., 2003. The advantages of rank-free classification for teaching and research. *Cladistics* **19**: 157
Moellendorff, O.F. in J.F. Quadras & O.F. Moellendorff. 1894. Diagnoses specierum novarum a J.F. Quadras in insulis mariannis collectarum scripserunt. *Nach. Deutsch. Malakozool. Ges.* **26**(1-2): 13-22
Moellendorff, O.F. 1900. The land-shells of the Caroline Islands. *Journal of Conchology* 7(5): 101-126
Mörch, O.A.L. 1850. *Catalogus conchyliorum quae reliquit C.P. Kierulf.* Trieri, Hafniae
Morelet, A. 1853. Description de coquilles nouvelles de l'Australie. *Journ. Conchyl.* **4**: 369-374
Mousson, A. 1865. Coquilles terrestres et fluviatiles de quelques îles de l'océan Pacifique, recueilles par M. le Dr E. Graeffe. *Journ. Conchyl.* **13**(2): 172
Mousson, A. 1869. Faune malacologique terrestre et fluviatile des îles Samoa, publiée d'après les envois de M. le Dr E. Graeffe. *Journ. Conchyl.* **17**(4): 338
Mousson, W.T. 1871. Faune malacologique terrestre et fluviatile ses îles Tonga, d'après les envois de M. le docteur Ed. Graeffe. *Journ. Conchyl.* **19**: 5-33
Mueller-Dombois, D. & F.R. Fosberg. 1998. *Vegetation of the Tropical Pacific Islands.* Springer-Verlag, New York
Murray, E. 1993. The sinister snail. *Endeavour* **17**(2): 78-83
Murray, J. 1972. *Genetic diversity and natural seletion.* Oliver & Boyd, Edinburgh
Murray, J. 1975. The genetics of the Mollusca. In: King, R.C. (ed.) *Handbook of Genetics.* **3**. *Invertebrates of genetic interest.* Plenum, New York
Murray, J. & B. Clarke. 1966. The inheritance of polymorphic shell characters in *Partula* (Gastropoda). *Genetics* **54**: 1261-1277
Murray, J. & B. Clarke. 1976a. Supergenes in polymorphic land snails. I. *Partula taeniata. Heredity* **37**: 253-269
Murray, J. & B. Clarke. 1976b. Supergenes in polymorphic land snails. II. *Partula suturalis. Heredity* **37**: 271-282
Murray, J. & B. Clarke. 1980. The genus *Partula* on Moorea: speciation in progress. *Proc. R. Soc. Lond.* B **211**: 83-117
Murray, J. & B. Clarke. 1984. Movement and gene flow in *Partula taeniata. Malacologia* **25**: 343-348
Murray, J., B. Clark & M.S. Johnson. 1993. Adaptive radiation and community structure of *Partula* on Moorea. *Proc. R. Soc. B: Biol. Sci.* **254**: 205-211
Murray, J., M.S. Johnson & B. Clarke. 1982. Microhabit differences among genetically similar species of *Partula. Evolution* **36**(2): 316-32
Murray, J., E. Murray, M.S. Johnson & B. Clarke. 1988. The extinction of *Partula* on Moorea. *Pacific. Sci.* **42**(3-4): 150-153
Murray, J., O.C. Stine & M.S. Johnson. 1991. The evolution of mitochondrial DNA in *Partula. Heredity* **66**: 93-104
Myers, N., R.A. Mittermeier, C.G. Mittermeier, G.A.B. da Fonseca & J. Kent. 2000. Biodiversity hotspots for conservation priorities. *Nature* **403**: 853-858
Naomi, S.-I. 2011. On the integrated frameworks of species concepts: Mayden's hierarchy of species concepts and de Queiroz's unified concept of species. *J. Zool. Syst. Evol. Res.* **49**(3): 177-184
Neall, V.E. & S.A. Trewick. 2008. The age and origin of the Pacific islands: a geological overview. *Phil. Trans. R. Soc. Lond. B. Biol. Sci.* **363**(1508): 3293-3308
Neiber, M.T. & B. Hausdorf. 2015. Molecular phylogeny reveals the polyphyly of the snail genus *Cepaea* (Gastropoda:

Helicidae). *Mol. Phyl. Evol.* **93**: 143-149
Nelson, G. & N. I. Platnick. 1981. *Systematics and biogeography*. Columbia University Press, New York
Noor, M.A.F. 2002. Is the biological species concept showing its age? *Trends Ecol. Evol.* **17**: 153-154
Nordsieck, H. 1986. The system of the Stylommatophora (Gastropoda), with special regard to the systematic position of the Clausiliidae, II. Importance of the shell and distribution. *Arch. Mollusk.* **117**(1-3): 93-116
Nunn, P.D. 1990. Coastal Processes and Landforms of Fiji: Their Bearing on Holocene Sea-Level Changes in the South and West Pacific. *Journ. Coast. Res.* **6**(2): 279-310
Ó Foighil, D. 2012a. *Eua globosa*. The IUCN Red List of Threatened Species. Version 2014.3. <www.iucnredlist.org>. Downloaded on 29 January 2015
Ó Foighil, D. 2012b. *Partula subgonochila*. In: The IUCN Red List of Threatened Species. Version 2014.3. <www.iucnredlist.org>. Downloaded on 29 January 2015
Ó Foighil, D. 2012c. *Partula auraniana*. The IUCN Red List of Threatened Species. Version 2014.3. <www.iucnredlist.org>. Downloaded on 29 January 2015
Ó Foighil, D. & R.J. Rundell. 2012a. *Partula calypso*. In: The IUCN Red List of Threatened Species. Version 2014.3. <www.iucnredlist.org>. Downloaded on 29 January 2015
Ó Foighil, D. & R.J. Rundell. 2012b. *Partula thetis*. In: The IUCN Red List of Threatened Species. Version 2014.3. <www.iucnredlist.org>. Downloaded on 29 January 2015
Ó Foighil, D. & R.J. Rundell. 2012c. *Partula leucothoe*. In: The IUCN Red List of Threatened Species. Version 2014.3. <www.iucnredlist.org>. Downloaded on 29 January 2015
Ó Foighil, D., T. Lee & J. Slapcinsky. 2011. Prehistoric anthropogenic introduction of partulid tree snails in Papua New Guinean archipelagos. *Journ. Biogeogr.* **38**(8): 1625-1632
Ochman, H., J.S. Jones & R.K. Selander. 1983. Molecular area effects in *Cepaea*. *Proc. Natl. Acad. Sci. USA* **80**: 4189-4193
Ohbayashi, T. & K. Taeuchi. 2007. Fluctuation of the Distribution and Population of the Giant African Snail, *Achatina fulica* Bowdich (Pulmonata: Achatinidae) in Chichijima and Hahajima Islands, of the Ogasawara (Bonin) Islands, from the Years 1995 to 2001. *Jpn. J. Appl. Entomol. Zool.* **51**: 221-230
Ono, R., S. Soegondho & M. Yoneda. 2009. Changing Marine Exploitation During Late Pleistocene in Northern Wallacea: Shell Remains from Leang Sarru Rockshelter in Talaud Islands. *Asian Persp.* **48**(2): 318-341
Orr, H.A. 2001. Some doubts about (yet another) view of species. *J. Evol. Biol.* **14**: 870-871
Ożgo, M. 2011. Rapid evolution in unstable habitats: a success story of the polymorphic land snail *Cepaea nemoralis* (Gastropoda: Pulmonata) *Biol. J. Linn. Soc.* **102**: 251-262
Ożgo, M. 2014. Rapid evolution and the potential for evolutionary rescue in land snails. *Journ. Moll. Stud.* **80**(3): 286-290
Ożgo, M. & M. Schilthuizen. 2012. Evolutionary change in *Cepaea nemoralis* shell colour over 43 years. *Global Change Biol.* **18**(1): 74-81
Paetel, F. 1873. *Catalog der Conchylien-Sammlung: nebst Uebersicht des angewandten Systems*. Gebruder Paetel, Berlin
Paetel, F. 1883. *Catalog der Conchylien-Sammlung*. Gebruder Paetel, Berlin
Palmer, D.H. & M.R. Kronforst. 2015. Divergence and gene flow among Darwin's finches: A genome-wide view of adaptive radiation driven by interspecies allele sharing. *Bioessays* **37**: 968-974
Parent, C.E. & B.J. Crespi. 2006. Sequential colonisation and diversification of Galápagos endemic land snail genus *Bulimulus* (Gastropoda, Styommatophora). *Evolution* **60**(11): 2311-2328
Parent, C.E. & B.J. Crespi. 2009. Ecological opportunity in adaptive radiations of Galápagos endemic island snails. *Amer. Nat.* **174**(6): 989-905
Paterson, H. E. H. 1985. The recognition concept of species. In: Vrba, E.S. (ed.). *Species and speciation*. Transvaal Museum, Pretoria
Partula Propagation Group. 1991. *Operation Partula: an expedition to the Island of Moorea to assess the current status of the indigenous and alien snail species. Organised by the Zoological Society of London and Nottingham University on behalf of the Partula Propagation Group.* Unpublished
Pearce-Kelly, P. 1994. Invertebrate propagation and re-establishment programmes: the conservation and education potential for zoos and related institutions. In: Olney, P.J.S., G.M. Mace & A.T. Feistner (eds.) *Creative Conservation*. Springer, Leiden
Pearce-Kelly, P. & D. Clarke. 1992. *Partula '92. Proceedings of the annual meeting of the Partula Propagation Group. London Zoo 28th August 1992*. Unpublished
Pearce-Kelly, P., D. Clarke & G.M. Mace. 1994. *Partula '94: An action plan for the conservation of the family Partulidae.* Zoological Society of London
Pearce-Kelly, P., G.M. Mace & D. Clarke. 1995. The release of captive-bred snails (*Partula taeniata*) into a semi-natural

environment. *Biodiv. Conserv.* **4**: 645-663
Pease, W.H. 1865 (for 1864). Descriptions of New Species of Land Shells from the Islands of the Central Pacific. *Proc. Zool. Soc. Lond.* **32**: 668-676
Pease, W.H. 1866. Descriptions of new species of *Partula*, inhabiting the Tahitian Archipelago. *Am. Journ. Conch.* **2**(3): 193-203
Pease, W.H. 1867. Illustrations of new species of *Partula*. *Amer. Journ. Conch.* **3**(1): 81
Pease, W.H. 1868a. Descriptions of New Species of Land Shells Inhabiting Polynesia. *Am. Journ. Conch.* **3**: 223-230
Pease, W.H. 1868b. Description of a New Genus and Eleven Species of Land Shells, Inhabiting Polynesia. *Am. Journ. Conch.* **4**: 153-160
Pease, W.H. 1870. Remarques sur certaines especes de coquilles terrestres, habitant la Polynesie, et description d'especes nouvelles. *Journ. Conchyl.* **18**: 393-403
Pease, W.H. 1871a. Descriptions of new species of land shells. *Amer. Journ. Conch.* 7(1): 26
Pease, W.H. 1871b. Catalogue of the land-shells inhabiting Polynesia, with remarks on their synonymy, distribution, and variation, and descriptions of new genera and species. *Proc. Zool. Soc. Lond.* (**1871**): 473
Pease, W.H. 1872. Descriptions of four species of land shells inhabiting the Papuan Islands. *Am. Journ. Conch.* **7**: 196-197
Peijel, F. & G. Rouse., 2000. Least-inclusive taxonomic unit: a new taxonomic concept for biology. *Proc. R. Soc. Lond. B Biol. Sci.* **267**: 627-630
Pelep, P.O. & M.G. Hadfield. 2011. The status of the endemic snails of the genus *Partula* (Gastropoda: Partulidae) on Pohnpei, Federated States of Micronesia. *Micronesica* **41**(2): 253-262
Petit de la Saussaye, S. 1850. Description de coquilles nouvelles. *Journ. Conchyl.* **1**: 169-171
Petit, J. & G. Prudent. (eds). 2010. *Climate Change and Biodiversity in the European Union Overseas Entities*. IUCM, Gland and Brussels
Pfeiffer, L. 1841. *Symbolae ad Historiam Heliceorum*. Th. Fisher **1**
Pfeiffer, L. 1842. *Symbolae ad Historiam Heliceorum*. Th. Fisher **2**
Pfeiffer, L. 1846. Descriptions of twenty new species of Helicea, in the collection of H. Cuming, Esq. *Proc. Zool. Soc. Lond.* **13**: 37-43
Pfeiffer, L. in Menke, K.T. & L. Pfeiffer. 1847. Diagnosen neuer Heliceen. *Zeitsch. Malakozool.* **4**: 81-84
Pfeiffer, L. 1848. *Monographia heliceorum viventium, sistens descriptions systematicas et criticas omnium huis f*amiliae generum et specierum hodie cognitarum. F.A. Brockhaus, Leipzig. Vol. **2**
Pfeiffer, L. 1849 (for 1848). Descriptions of twenty-four new species of Helicea, from the collection of H. Cuming, Esq. *Proc. Zool. Soc. Lond.* **16**: 126-131
Pfeiffer, L. 1850. Beschreibungen neuer Landschnecken. *Zeitsch. Malakozool.* 7(6): 65-80
Pfeiffer, L. 1853a (for 1851). Descriptions of fifty-four species of land shells from the collection of H. Cuming, Esq. *Proc. Zool. Soc. Lond.* **19**: 252-263
Pfeiffer, L. 1853b. *Monographia heliceorum viventium, sistens descriptions systematicas et criticas omnium huis f*amiliae generum et specierum hodie cognitarum. **3.** F.A. Brockhaus, Leipzig
Pfeiffer, L. 1854a (for 1852). Descriptions of sixty-six new land shells, from the collection of Hugh Cuming Esq. *Proc. Zool. Soc. Lond.* **20**: 56-70
Pfeiffer, L. 1854b (for 1852). Descriptions of eighteen new species of land shells from the collection of H. Cuming, Esq. *Proc. Zool. Soc. Lond.* **20**: 83-87
Pfeiffer, L. 1854c (for 1852). Descriptions of fourteen new species of land shells, from the collection of Hugh Cuming, Esq. *Proc. Zool. Soc. Lond.* **20**: 135-138
Pfeiffer, L. 1854d (for 1852). Descriptions of eighteen new species of land shells from the collection of H. Cuming, Esq. *Proc. Zool. Soc. Lond.* **20**: 381-387
Pfeiffer, L. 1854e (for 1853). Descriptions of sixteen new species of Helicea from the collection of H. Cuming, Esq. *Proc. Zool. Soc. Lond.* **21**: 122-126
Pfeiffer, L. 1855a. Descriptions of Forty-seven New Species of Helicea, from the Collection of H. Cuming, Esq. *Proc. Zool. Soc. Lond.* **23**: 91-101
Pfeiffer, L. in Küster, H.C. & L. Pfeiffer. 1855b. Die Gattungen Bulimus, Partula, Achatinella, Achatina und Azeca: in Abbildungen nach der Natur mit Beschreibungen. *Volume 1, Part 13 of Sytematisches Conchylien-Cabinet von Martini und Chemnitz*. Verlag von Bauer und Raspe (Julius Merz)
Pfeiffer, L. 1856. Versuch einer Anordnung der Heliceen nach naturlichen Gruppen. *Malakozool. Blatt.* **2**: 145-185
Pfeiffer, L. 1857a. Zur Fauna der Admiralitats-Inseln. *Malakozool. Blatt.* **3**: 237-248
Pfeiffer, L. 1857b (for 1856). Descriptions of fifty-eight new species of Helicea from the collection of H. Cuming, Esq. *Proc.*

Zool. Soc. Lond. **24**: 324-336
Pfeiffer, L. 1857c (for 1856). Descriptions of eighteen new species of land-shells collected on the Admiralty Islands, from the collection of Mr. H. Cuming. *Proc. Zool. Soc. Lond.* **24**: 381-385
Pfeiffer, L. 1857d (for 1856). Descriptions of thirty-three new species of landshells from the collection of H. Cuming, Esq. *Proc. Zool. Soc. Lond.* **24**: 385-392
Pfeiffer, L. 1857e. Descriptions of Thirty-one New Species of Land-shells, from Mr. Cuming's Collection. *Proc. Zool. Soc. Lond.* **25**: 107-113
Pfeiffer, L. 1857f. *Systematisches Conchylien-Cabinet von Martini und Chemnitz*. J.H. Casselis. 2nd ed **1**(13)
Pfeiffer, L. 1859a. *Monographia heliceorum viventium, sistens descriptions systematicas et criticas omnium huis* familiae generum et specierum hodie cognitarum. F.A. Brockhaus, Leipzig. **4**
Pfeiffer, L. 1859b. *Novitates conchologicae. Series prima. Mollusca extramarina. Beschreiber und Abbildung, neuer oder kritischer Land- und Susswasser-Mollusken.* **1**(10)
Pfeiffer, L. 1860. Descriptions of Thirty-six New Species of Land-shells from Mr. H. Cuming's Collection. *Proc. Zool. Soc. Lond.* **28**: 133-141
Pfeiffer, L. 1861. Diagnosen neuer Heliceen. *Malakozool. Blatt.* **8**: 11-16
Pfeiffer, 1862 (for 1861). Description of sixteen new species of land-shell from the collection of H. Cuming, Esq. *Proc. Zool. Soc Lond.* **29**: 386-389
Pfeiffer, L. 1868. *Monographia heliceorum viventium, sistens descriptions systematicas et criticas omnium huis familiae generum et specierum hodie cognitarum.* F.A. Brockhaus, Leipzig. **6**
Pfeiffer, L. 1877. *Monographia heliceorum viventium, sistens descriptions systematicas et criticas omnium huis familiae generum et specierum hodie cognitarum.* F.A. Brockhaus, Leipzig. **8**
Pfeiffer, L. 1881. *Nomenclator heliceorum viventium quo continentur nomina omnium hujus familiae generum et specierum hodie cognitarum disposita ex affinitate naturali.* T. Fischer, Cassel.
Pfenninger, M. & F. Magnin. 2001. Phenotypic evolution and hidden speciation in *Candidula unifasciata* (Helicellinae, Gastropoda) inferred by 16S variation and quantitative shell traits. *Mol Ecol* **10**: 2541- 2554
Pilsbry, H.A. 1900. On the zoological position of *Partula* and *Achatinella. Proc. Acad. Nat. Sci. Phil.* **3**: 561-567
Pilsbry, H.A. 1909-1910. *Manual of conchology, structural and systematic. Vol. 20, Caecilioides, Glessula and Partulidae.* Conchological Department, Academy of Natural Sciences of Philadelphia
Pilsbry, H.A. 1913. New species of *Clausilla* and *Partula* from the collection of Mr. J.S. Emerson. *Nautilus* **27**: 65-67
Pilsbry, H.A. 1924. Land and fresh water mollusks from Mindanao. *Proc. ANSP*. **76**: 15-17
Pilsbry, H.A. & Cooke, C.M. Jr. 1934. Partulidae of Tonga and related forms. *Occ. Pap. B.P. Bishop Mus.* **10**(14): 1-22
Pippard, H. 2012. *The current status and distribution of freshwater fishes, land snails and reptiles in the Pacific Islands of Oceania.* IUCN, Gland, Switzerland
Price, M.R. & M.G. Hadfield. 2014. Population genetics and the effects of a severe bottleneck in an *ex situ* population of Critically Endangered Hawaiian tree snails. *PLoS ONE* **9**(12): e114377
Price, A.A., M.G. Jackson, J. Blichert-Toft, P.S. Hall, J.M. Sinton, M.D. Kurz & J. Blusztajn. 2014. Evidence for a broadly distributed Samoan-plume signature in the northern Lau and North Fiji Basins. *Geochem. Geophys. Geosyst.* **15**: 986-1008
Purea, M., S. Matalavea, T. Bourke & D. Hunter. 1998. *Platydemus manokwari* de Beauchamp, a flatworm predator of the giant African snail (*Achatina fulica* Bowdich) recorded in Samoa. *Journ. South Pac. Agric.* **5**(2): 71-72
de Quieroz, K. 2005. Different species problems and their resolution. *BioEssays* **27**: 1263-1269
de Queiroz, K. 2007. Species concepts and species delimination. *Sys. Biol.* **56**(6): 879-896
Quoy, J.E. & J.P. Gaimard. 1824. Zoologie. In Freycinet, M.L. de: *Voyage autour du monde: enterpris par ordre du roi sous le ministère et conformément aux instructin de S. Exc. M. Le Vicomte de Nouchage Secrétaire d'État au Départment de La Marine execute sur les corvettes de S.M. l'Uranie et la Physiciene, pendant les années 1817, 1819 et 1820.* Pillet Ainé, Paris
Quoy J.E. & J.P. Gaimard. 1832. *Voyage de découvertes de l' "Astrolabe" exécuté par ordre du Roi, pendant les années 1826-1829, sous le commandement de M. J. Dumont d'Urville. Zoologie* Paris **2**(1)
Rabosky, D.L. & U. Lovett. 2008. Density-dependent diversification in North American wood warblers. *Proc. Biol. Sci.* **275**: 2363-2371
Rauschenberg, R.A. 1968. Daniel Carl Solander: Naturalist on the "Endeavour". *Trans. Am. Phil. Soc.* N.S. **58**(8): 1-66
Raymundo, M.L. & R.H. Miller. 2012. Little Fire Ant, *Wasmannia auropunctata* (Roger) (Hymenoptera: Formicidae), Established at Several Locations on Guam. *Proc. Hawaii. Entomol. Soc.* **44**: 85-87
Reeve, L. 1842. *Conchologia systematica, or Complete system of conchology: in which the Lepades and conchiferous*

Mollusca are described and classified according to their natural organization and habits. Longman, Brown, Green and Longmans, London. **2**

Reeve, L.A. 1843. Descriptions of new species of shells figured in the 'Conchologica Systematica'. *Proc. Zool. Soc. Lond.* **10**: 197-210

Reeve, L.A. 1849-51. Monograph of the genus *Partula. Conchologia Iconica, or, Illustrations of the shells of molluscous animals.* **6**. Savill, Edwards & Co., London

Régnier, C. G. Achaz, A. Lambert, R.H. Cowie, P. Bouchet & B. Fontaine. 2015a Mass extinction in poorly known taxa. *Proc. Natl. Acad. Si. USA* **112**(25): 7761-7766

Régnier, C., P. Bouchet, K.A. Hayes, N.W. Yeung, C.C. Christensen, D.J.D. Chung, B. Fontaine & R.H. Cowie. 2015b. Extinction in a hyperdiverse endemic Hawaiian land snail family and implications for the underestimation of invertebrate extinction. *Conserv. Biol.* **29**: 1715-1723

Rensch, I. 1937. Systematische und tiergeographische untersuchungen über die landschneckenfauna des Bismarck-Archipels II. *Arch. Naturges.* **6**: 526-644.

Rensch, I. & B. Rensch 1935. Systematische und tiergeographische Untersuchungen über die Landschnecken der Salomonen auf Grund der Sammlungen von Dr. E. Paravicini und Dr. H. Hediger. Teil I. *Rev. Suisse Zool.* **42**: 51-86

Rex, M.A. 1972. The relationship of island area and isolation to color polymorphism in *Liguus fasciatus* (Pulmonata, Bulimulidae). *Breviora* **11**(391): 1-15

Richardson, C.L. 1990. Partulidae: catalog of species. *Tryonia* **19**(i): 1-96

Ricklefs, R.E. & G.C. Cox. 1972. Taxon cycles in the West Indian avifauna. *Amer. Natur.* **106**: 195-219

Rosin, Z.M., J. Kobak, A. Lesicki & P. Tryjanowski. 2013. Differential shell strength of *Cepaea nemoralis* color morphs – implications for their anti-predator defence. *Naturwiss.* **100**: 843-851

Rosindell, J. & A.B. Phillimore 2011. A unified model of island biogeography sheds light on the zone of radiation. *Ecol. Lett.* **14**(6): 552-560

Rundell, R.J. 2005. *The Land Snails of Belau: Survey of the 16 States*. Unpublished report.

Rundell, R.J. 2010. Diversity and conservation of the land snail fauna of the western Pacific islands of Belau (Republic of Palau, Oceania). *Amer. Malac. Bull.* **28**: 81-90

Rundell, R.J. & J.E. Czekanski-Moir. 2015. A survey of the land snails of Kosrae (Caroline Islands, Micronesia) including the rediscovery of endemic *Delos oualanensis* (Pease, 1866) (Mollusca: Pulmonata: Rhytididae). *Malacologia* **59**(1): 13-20

Rundell, R.J. & T.D. Price. 2009. Adaptive radiation, nonadaptive radiation, ecological speciation and non-ecological speciation. *Trends Ecol. Evol.* **24**(7): 394-340

Sartori, A.F., O. Gargominy & B. Fointaine. 2014. Radiation and decline of endodontid land snails in Makatea, French Polynesia. *Zootaxa* **3772**(1): 1-68

Sauter, R., P. ten Brink, S. Withana, L. Mazza, F. Pondichie, A. Lopes, J. Clinton & K. Bego. 2013. Five case studies on the impacts of climate change on European islands, Annex II to the final report '*Impacts of climate change on all European islands*', Institute for European Environmental Policy (IEEP) for the Greens/EFA of the European Parliament. Draft Final Report. Brussels

Schaufuss, L.W. 1869. *System und Aufzählung sämmtlicher Conchyliden der Sammlung von Fr. Paetel. Zur Belebung des Interesses für Malakozoologie.* Commission bei O. Weiske, Dresden

Scheil, A.E., S. Hilsman, R. Triebskorn & H.-R. Köler. 2013. Shell colour polymorphism, injuries and immune defence in three helicid snail species, *Cepaea hortensis, Theba pisana* and *Cornu aspersum maximum*. *Res. Immunol.* **3**: 73-78

Schileyko, A.A. 1979. Sistema Otryada Geophila (=Helicida) (Gastropoda Pulmonata) Morfologiya, Sistematika i Filogeniya Mollyuskov. *Trudy Zoologicheskogo Instituta, Akademiya Nauk SSSR* **80**: 1-69 [edited translation: Boss, K.J. & N.E. Jacobson. 1985. The system of the order Geophila (=Helicida) (Gastropoda Pulmonata). *Morphology, systematics and phylogeny of Mollusks*. Special Occasional Publication of the Department of Mollusks, Harvard University]

Schileyko, A.A. 1999. Treatise on Recent terrestrial pulmonate molluscs, 3. Partulidae, Aillyidae, Bulimulidae, Orthalicidae, Megaspiridae, Urocoptidae. *Ruthenica, Supplement* **2**: 263-436

Schilthuizen, M., R.F. Hoekstra & E. Gittenburger. 1999. Selective increase of a rare haplotype in a land snail hybrid zone. *Proc. R. Soc. Lond. B* **268**: 2181-2185

Schilthuizen, M. 2012. Scrutinising snail shells. *Heredity* **108**(4): 364-365

Schmeltz, J.D.E. 1865. *Catalog II der zum Verkauf stehenden Doubletten aus dem naturhistorischen Expeditionen der Herren Joh. Ces. Godeffroy & Sohn in Hamburg* [no publisher given], Hamburg **25**

Schmeltz, J.D.E. 1869. *Museum Godeffroy. Catalog IV, nebst einer Beilage, enthaltend: topographische Notizen; Beschreibung neuer Bryozoen von Senator Dr. Kirchenpauer zu Hamburg und einer neuen Asteriden-Gattung von Dr. Chr. Lütken zu Kopenhagen.* Wilhelm Mauke Söhne, vormals Perthes-Besser & Mauke, Hamburg **71**

Schmeltz, J.D.E. 1874. *Museum Godeffroy Catalog V. Nebst einer Beilage enthaltend topographische und zoologische Notizen.* L. Friederichsen & Co., Hamburg

Schreiner, I. 1989. *Biological Control Introductions in the Caroline and Marshall Islands.* College of Agriculture and Life Sciences, University of Guam, Guam

Schwable, G. & J. Murray. 1970. Electrophoresis of proteins in natural populations of *Partula* (Gastropoda). *Evolution* **24**: 424-430

Scvortzoff, E. 1966. *Chromosome numbers in land snails of the genus* Partula *that inhabit the island of Moorea.* M.A. Thesis, University of Virginia

Selman, B.J., A.A. Jones & G.M. Barker. 2004. Microsporidia (Microspora) parasitic in terrestrial gastropods. In: Barker, G.M. (ed) *Natural Enemies of Terrestrial Molluscs.* CABI, Wallingford

Semper, O. 1865. Description de trois espèces nouvelles du genre *Partula. Journ. Conchyl.* **13**: 417-422

Shimizu, Y. & R. Ueshima. 2000. Historical biogeography and interspecific mtDNA introgression in *Euhadra peliomphala* (the Japanese land snail). *Heredity* **85**: 84-96

Silvertown, J., L. Cook, R. Cameron, M. Dodd, K. McConway, J. Worthington, *et al*. 2011. Citizen science reveals unexpected continental-scale evolutionary change in a model organism. *PLoS ONE* **6**: e18927

Simpson, G.G. 1951. The species concept. *Evolution* **5**: 285-298

Simpson, G.G. 1953. *The Major Features of Evolution.* Columbia University Press, New York

Sischo, D.R., M.R. Price, M.-A. Pascua & M.G. Hadfiled. 2016. Genetic and demographic insights into the decline of a captive population of the endangered Hawaiian tree snail *Achatinella fuscobasis* (Achatinellinae). *Pac. Sci.* **70**: 133-141

Slapcinsky, J. & F. Kraus. 2016. Revision of Partulidae (Gastropoda, Stylommatophora) of Palau, with description of a new genus for an unusual ground-dwelling species. *Zookeys* **614**: 27-49

Smith, E.A. 1884. An account of the land and freshwater Mollusca collected during the voyage of the "Challenger" from December 1 872 to May 1876. *Proc. Zool. Soc. Lond.* (**1884**): 258-281

Smith, E.A. 1897. On a collection of land and freshwater shells from Rotuma Island. *Ann. Mag. Nat. Hist.* (6) **20**: 519-523

Smith H.H. 1902. XVII. An annotated catalogue of shells of the genus *Partula* in the Hartman Collection belonging to the Carnegie Museum. *Ann Carnegie Mus.* **1**(3-17): 422-454

Smith, B.D. 1995. Tree snails, tropical storms, and drought in the Mariana Islands. *Programs and abstracts, American Malacological Union, 61st Annual Meeting, Hilo, Hawaii*

Smith, B.D. 2008. Preliminary assessment of endemic arboreal snails in three forest types in Sarigan, with notes on ground-dwelling species. *In*: Martin, G. (ed), Wildlife and vegetation surveys of Sarigan Island, April 13-25, 2006. *Commonwealth of the Northern Mariana Islands, Division of Fish and Wildlife, Technical Report* **14**: 8-23

Smith, B. 2013. Taxonomic inventories and assessments of terrestrial snails on the islands of Tinian and Aguiguan in the Commonwealth of the Northern Mariana islands. *Univ. Guam Marine Lab. Tech. Rep.* **154**

Smith, B.D. & D.R. Hopper. 1994. The Partulidae of the Mariana Islands: Continued threats and declines. *Hawaiian Shell News* **42**(6): 10-11

Smith, B.D., R. Cooper-Nurse & A.M. Gawle. (undated). *Survey of endangered tree snails on navy-owned land in Guam.* Unpublished

Smith, H.H. 1903. Annotated Catalogue of shells of the genus *Partula* in the Hartman collection, belonging to the Carnegie Museum. *Ann. Carnegie Mus.* **1**: 422-485

Solem, A. 1955. Mexican mollusks collected for Dr. Bryant Walker in 1926, XI: *Drymaeus. Occ. Pap. Mus. Zool., Univ. Mich.* **566**: 1-20

Solem, A. 1959. Systematics of the land and freshwater Mollusca of the New Hebrides. *Fieldiana Zool.* **43**: 1-359

Solem, A. 1972. Microarmature and barriers in the apertures of land snails. *Veliger* **15**: 81-87

Solem, A. 1974. Apertural barriers in Pacific island land snails of the families Endodontidae and Charopidae. *Veliger* **15**(4): 300-306

Solem, A. 1982. *Endodontoid land snails from Pacific islands (Mollsuca: Pulmonata: Sigmurethra). Part 1.* Field Museum of Natural History, Chicago.

Soós, L. 1911. On a collection of land shells from New Guinea and adjacent islands. *Ann. Mus. Natn. Hungar.* **9**: 345-356

Sowerby, G.B. 1841. Conchological illustrations, or coloured figures of all the hitherto unfigured recent shells, *Bulinus.* Sowerby, London

Stallard, C. & T.C. Gilbert. 2003. An investigation into the nutrient intake, output and approximate digestibilities of the Polynesian tree snail (*Partula clara*). *Proceedings of the 5th Annual Symposium on Zoo Research, Marwell Zoological Park, Winchester, UK, 7-8th July 2003.* Federation of Zoological Gardens of Great Britain and Ireland

Stattersfield, A.J., M.J. Crosby, A.J. Long & D.C. Wege. 1998 *Endemic Bird Areas of the World. Priorities for biodiversity conservation.* BirdLife Conservation Series **7**. BirdLife International, Cambridge

Sterki, V. 1894. Growth changes of the radula in land-mollusks. *Proc. Acad. Nat. Sci. USA* **45**: 388-400

Sugiura, S. 2009. Seasonal fluctuation of invasive flatworm predation pressure on land snails: implications for the range expansion and impacts of invasive species. *Biol. Conserv.* **142**: 3013-3019

Sugiura, S. & Y. Yamaura. 2008. Potential impacts of the invasive flatworm *Platydemus manokwari* on arboreal snails. *Biol. Invasions.* **11**(3): 737-742

Sykes, E.R. 1903. On a collection of non-marine shells, formed by Mr. J. J. Walker, in the New Hebrides. *Proc. Malac. Soc. London* **5**: 196-200

Taputuarai, R. & J.-Y. Meyer. 2007. *Rapport de mission a Fatu Hiva et Hiva Oa (Marquises sud) du 13 au 24 Aout 2007.* Unpublished report

Thacker, R.W. & M.G. Hadfield. 2000. Mitochondrial phylogeny of extant Hawaiian tree snails (Achatinellinae). *Mol. Phylogenet. Evol.* **16**: 263-270

Thiele, J. 1931. *Handbuch der Systematishcen Weichtierkunde* **II**. Fischer, Jena

Tillier, S. 1989. Comparative morphology, phylogeny and classification of land snails and slugs (Gastropoda: Pulmonata: Stylommatophora). *Malacologia* **30**(1-2): 1-303

Tillier, S. & B.C. Clarke. 1980. Lutte biologique et destruction du patrimoine génétique: le cas des Mollusques gastéropodes pulmones dans les territoires français du Pacifique. *Genet. Sel. Evol.* **15**: 559-566

Tillier, S. & P. Mordan. 1983. The conchological collections of Bruguière and Olivier from the Ottoman Empire (1792-1798). *Journ. Conch.* **31**: 153-160

Tillier, S. & P.B. Mordan. 1995. The anatomy and systematics of the New Caledonian land snail genus *Draparnaudia* Montrouzier, 1859 (Pulmonata: Orthurethra). *Zool. Journ. Linn. Soc.* **113**: 47-91

Tonge, S. 1991. *Partula '91: proceedings of the Partula Propagation Group meeting held at London Zoo 16th May 1991.* Unpublished

Tonge, S., & Q. Bloxham. 1991. A review of the captive-breeding programme for Polynesian tree snails Partula spp. *Internl. Zoo Yrb.* **30**(1), 51-59

Tuskes, P.M. 1981. Population structure and biology of *Liguus* tree snails on Lignumvitae Key. *Nautilus* **95**: 162- 169

Vaught, K.C. 1989. *A classification of the living Mollusca.* American Malacologists Inc., Melbourne, Florida

Wada, S. & S. Chiba. 2013. The dual protection of a micro land snail against a micro predatory snail. *PLoS ONE* **8**(1): e54123

Wade, C.M. & P.B. Mordan. 2000. Evolution within the gastropod molluscs; using the ribosomal RNA gene-cluster as an indicator of phylogenetic relationships. *J. Moll. Stud.* **66**: 565-570

Wade, C.M., P.B. Mordan & B.C. Clarke. 2001. A phylogeny of the land snails (Gastropoda: Pulmonata). *Proc. R. Soc. Lond. B* **268**: 413-422

Wall, S., M.A. Carter & B. Clarke. 1980. Temporal changes of gene frequencies in *Cepaea hortensis*. *Bio. J. Linn. Soc.* **14**: 303-317

Watanabe, Y. & S, Chiba. 2001. High within-population mitochondrial DNA variation due to microvicariance, population mixing in the land snail *Euhadra quaesita* (Pulmonata: Bradybaenidae). *Mol. Ecol.* **10**: 2635-2645

Wells, S.M. 1995. Endangered Mollusca and methods of conservation. *The Conservation Biology of Molluscs: Proceedings of a Symposium Held at the 9th International Malacological Congress, Edinburgh, Scotland, 1986.* No. 9. IUCN

Wetterer, J.K. & S.D. Porter. 2003. The little fire ant, *Wasmannia auropunctata*: distribution, impact and control. *Sociobiol.* **41**: 1-41

Wilson, E.O. 1959 Adaptive shift and dispersal in a tropical ant fauna. *Evolution* **13**: 122-144

Wood, W. 1818 *Index testaceologicus, or, A catalogue of shells, British and foreign: arranged according to the Linnean system: with the Latin and English names, references to authors, and places where found : illustrated with 2,300 figures.* W. Wood, London (2nd ed. 1828)

Woodruff, D.S. 1989. Genetic anomalies associated with *Cerion* hybrid zones: the origin and maintenance of new electrophoretic variants called hybrizymes. *Biol. J. Linn. Soc.* **36**: 281-294

Wright, S. 1940. The statistical consequences of Mendelian heredity in relation to speciation. In: Huxley, J. (ed.) *The New Systematics*. Oxford University Press, London

Wright, S. 1955. Classification of the factors of evolution. *Cold Spring Harbor Symposia on Quantitative Biology* **20**: 16-24

Wright, S. 1965. Genetics and twentieth century Darwinism. *Amer. J. Hum. Genet.* **12**(2): 365-372

Zilch, A. 1962. Die Typen und Typoide des Natur-Museums Senckenberg, 26: Mollusca, Achatinellacea. *Archiv Mollusk.* **91**(1/3): 77-94

Index and synonymy of Partulidae

Valid names shown in bold

A

B

C

G

H

I

M

N

Q

R

S

T

U

V

W

Z

Topical, geographical and non-partulid index

A

B

C

D

E

F

G

H

I

J

K

L

M

N

O

P

R

S

T

U

V

W

Y

Z

www.ingramcontent.com/pod-product-compliance
Lightning Source LLC
LaVergne TN
LVHW071243210926
848486LV00007B/331
* 9 7 8 0 9 9 3 2 2 0 3 3 3 *